Asset Analytics

Performance and Safety Management

Series Editors

The main aim of this book series is to provide a floor for researchers, industries, asset managers, government policy makers and infrastructure operators to cooperate and collaborate among themselves to improve the performance and safety of the assets with maximum return on assets and improved utilization for the benefit of society and the environment.

Assets can be defined as any resource that will create value to the business. Assets include physical (railway, road, buildings, industrial etc.), human, and intangible assets (software, data etc.). The scope of the book series will be but not limited to:

- Optimization, modelling and analysis of assets
- Application of RAMS to the system of systems
- Interdisciplinary and multidisciplinary research to deal with sustainability issues
- Application of advanced analytics for improvement of systems
- Application of computational intelligence, IT and software systems for decisions
- Interdisciplinary approach to performance management
- Integrated approach to system efficiency and effectiveness
- Life cycle management of the assets
- Integrated risk, hazard, vulnerability analysis and assurance management
- Adaptability of the systems to the usage and environment
- Integration of data-information-knowledge for decision support
- Production rate enhancement with best practices
- Optimization of renewable and non-renewable energy resources

More information about this series at http://www.springer.com/series/15776

P. K. Kapur · Gurinder Singh ·
Yury S. Klochkov · Uday Kumar
Editors

Decision Analytics Applications in Industry

Springer

Editors
P. K. Kapur
Centre for Interdisciplinary Research
Amity University
Noida, India

Gurinder Singh
Group VC
Amity University
Noida, Uttar Pradesh, India

Yury S. Klochkov
Department of Economics and Management in Machine Building
Peter the Great St. Petersburg Polytechnic
Saint Petersburg, Russia

Uday Kumar
Operation and Maintenance Engineering
Luleå University of Technology
Luleå, Sweden

ISSN 2522-5162 ISSN 2522-5170 (electronic)
Asset Analytics
ISBN 978-981-15-3642-7 ISBN 978-981-15-3643-4 (eBook)
https://doi.org/10.1007/978-981-15-3643-4

This Springer imprint is published by the registered company Springer Nature Singapore Pte Ltd.
The registered company address is: 152 Beach Road, #21-01/04 Gateway East, Singapore 189721, Singapore

Preface

Decision-making is a process of selecting the best among the different alternatives. Decision analysis (DA) is the discipline comprising the philosophy, methodology, and professional practice necessary to address important decisions in a formal manner. Decision analytics refers to identification and assessment of important aspects of a decision to prescribe a recommended course of action translating it further in a formal manner to give the recommendation and insights to the decision-maker and other stakeholders. This book presents decision-based frameworks for solving complex industrial and scientific problems. It addresses research gaps in the field of system sciences, operations, and management. This book volume contains contributed work from researchers across the globe. It offers guidance for the implementation of decision-based methods in business and asset management. This volume provides varied qualitative as well as quantitative analysis in areas such as cybersecurity, sustainability, multivariate analysis, customer satisfaction, parametric programming, software reliability growth modeling, and blockchain technology to name a few. This book also renders integrated approaches and practices in the field of machine learning and genetic algorithm. It also offers the application in the fields of supply chain and logistics, cloud computing, six sigma, production management, big data analysis, satellite imaging, game theory, biometric systems, quality, and system performance.

The volume editors are grateful to various authors who have made significant contributions to this volume through their research work. The editors are also grateful to several reviewers for their comments and suggestions, which helped in improving the quality of the papers. We hope that this volume makes significant contributions in the field of applications of analytics in decision-making.

Noida, India — P. K. Kapur
Noida, India — Gurinder Singh
Saint Petersburg, Russia — Yury S. Klochkov
Luleå, Sweden — Uday Kumar

Acknowledgements

This book is a result of sincere efforts of many people who have helped in the successful completion of this book. We wish to express our deepest gratitude to all the researchers who have contributed their valuable research work to write this book. We would like to take this opportunity to thank all the notable reviewers who have taken out their valuable time in reviewing the research work and providing their insightful comments and recommendations. Their thoughtful suggestions have assisted in improving the content of the book. A special thanks to our research scholars, especially, Saurabh Panwar and Vivek Kumar who have provided immeasurable assistance in expediting the publication process. We are also indebted to all the professionals who helped in the successful publication of the book. We humbly extend our heartfelt gratitude to our parents, spouses, children, and other family members, for their unconditional support and continuous motivation.

Lastly, we apologize for any omissions.

P. K. Kapur
Gurinder Singh
Yury S. Klochkov
Uday Kumar

Contents

Editors and Contributors

About the Editors

P. K. Kapur is Director of the Amity Centre for Interdisciplinary Research, Amity University, Noida, and former Dean of the Faculty of Mathematical Sciences, University of Delhi, India. He has been the President of the SREQOM (Regd.) since 2000 and is a former President of ORSI. He is the Editor-in-Chief of International Journal of System Assurance Engineering and Management, Springer (India), has published over 350 papers in international journals and proceedings, and has co-authored two books on Software Reliability with OR Applications.

Gurinder Singh is Group Vice Chancellor of Amity Universities, Director General of the Amity Group of Institutions, India, and Vice Chairman of the Global Foundation for Learning Excellence. He holds doctorate and postgraduate degrees from the Indian Institute of Foreign Trade, Delhi and has more than 20 years of experience in institutional building, teaching, research, and industry. He has given keynote lectures at various international forums which include the prestigious Million Dollar Round Table Conference at Harvard Business School, NYU, University of Leeds, University of Berkeley, National University of Singapore, and many more.

Yury S. Klochkov is Director of the Monitoring Center for Education and Research and a Professor at the Department of Economics and Management in Mechanical Engineering, St. Petersburg Polytechnic University, Russia. He has completed his Ph.D. and postdoctoral degree in Standardization and Quality Management in 2006 and 2012, respectively. In the past, he has held several job positions at Samara State Aerospace University, including Director of Quality Management and Vice-President of the Thesis Committee. He has authored over 100 publications in various areas of quality management.

Uday Kumar is the Chair Professor of Operation and Maintenance Engineering, and Director of Research and Innovation (Sustainable Transport) and of Luleå Railway Research Center at Luleå University of Technology, Sweden. He has published more than 350 papers in international journals and conference proceedings, co-authored 4 books on maintenance engineering, and contributed to the World Encyclopedia on Risk Management. He is one of the Editors-in-Chief of the International Journal of System Assurance and Management (Springer) and an Area Editor (Europe) for the Journal of Quality in Maintenance Engineering (Emerald).

Contributors

Tanya Agarwal Netaji Subhas University of Technology, Delhi, India

P. Agarwal Department of Applied Sciences, SRMIST, NCR Campus, Modinagar, India

Anu Gupta Aggarwal Department of Operational Research, University of Delhi, Delhi, India

Sanchita Aggarwal University of Delhi, Delhi, India

K. M. Ahamed Sheriff Sethu Institute of Technology, Virudhunagar District, Virudhunagar, Tamil Nadu, India

Anjali Ahuja Manipal University Jaipur, Jaipur, Rajasthan, India

Jyoti Ahuja Government Post Graduate College for Women, Rohtak, India

Sameer Anand Department of Computer Science, Shaheed Sukhdev College of Business Studies, University of Delhi, Delhi, India

Anirudh ASET, Amity University, Noida, Uttar Pradesh, India

Ekta Raphael Anthony Integral University, Lucknow, Uttar Pradesh, India

Nitin Arora Amity International Business School, Amity University Campus, Noida, India

S. Bansal Department of Statistics, Ram Lal Anand College, University of Delhi, New Delhi, India

Vikas Singh Bhadoria Department of Electrical and Electronics Engineering, ABES Engineering College, Ghaziabad, UP, India

Ashish Bhargava ABV-Indian Institute of Information Technology and Management, Gwalior, MP, India

Vishnu Bhaskar Netaji Subhas University of Technology, Delhi, India

Rekha Bhatia Department of Computer Science, Punjabi University Regional Centre for IT and Management, Mohali, India

J. Chattopadhyay Reactor Safety Division, Bhabha Atomic Research Centre, Mumbai, India

Pradnya Chitrao Symbiosis Institute of Management Studies (SIMS), Pune, India

Umesh Kumar Chopra Amity University Uttar Pradesh, Noida, India

S. C. Dabas Department of Hindi, Ram Lal Anand College, University of Delhi, New Delhi, India

D. Das Department of Statistics, Ram Lal Anand College, University of Delhi, New Delhi, India

Swapnil Dhanwal Netaji Subhas University of Technology, Delhi, India

Joydip Dhar ABV-Indian Institute of Information Technology and Management, Gwalior, MP, India

J. Dhingra Department of Statistics, Ram Lal Anand College, University of Delhi, New Delhi, India

M. Elango Thiagarajar College of Engineering, Madurai, Tamil Nadu, India

B. G. Fernandes Department of Electrical Engineering, Indian Institute of Technology, Mumbai, India

Aaquib Firdous Electrical & Renewable Energy Engineering, BGSBU, Rajouri, India

A. K. Ghosh Health Safety & Environment Group, Bhabha Atomic Research Centre, Mumbai, India

V. Gopika Reactor Safety Division, Bhabha Atomic Research Centre, Mumbai, India

S. Gulati Department of Statistics, Ram Lal Anand College, University of Delhi, New Delhi, India

A. Gupta Department of Statistics, Ram Lal Anand College, University of Delhi, New Delhi, India

Amarnath Gupta Research Scholar, Symbiosis International University, Pune, India

Rahul Gupta Department of IT, ASET, Amity University, Noida, UP, India

Anjali Harit DN College, Meerut, India

Anamika Jain Manipal University Jaipur, Jaipur, Rajasthan, India

M. Jain Department of Mathematics, Indian Institute of Technology Roorkee, Roorkee, India

R. Jain Department of Statistics, Ram Lal Anand College, University of Delhi, New Delhi, India

Ajay Jaiswal Shaheed Sukhdev College of Business Studies, Delhi, India

Shiva Pujan Jaiswal Depatrment of Electrical and Electronics Engineering SET, Sharda University, Noida, India

Singh Jaspreet Terminal Ballistics Research Laboratory, Chandigarh, India

Kaur Jatinder Terminal Ballistics Research Laboratory, Chandigarh, India

Sheetal Joshi ASET, Amity University, Noida, Uttar Pradesh, India

Mahapara Khurshid Central University of Jammu, J&K, Samba, India

Nellie V. Klassen Peter the Great St. Petersburg Polytechnic University, St. Petersburg, Russia

Evgenii A. Konnikov Peter the Great St. Petersburg Polytechnic University (SPbPU), St. Petersburg, Russia

Olga A. Konnikova St. Petersburg State University of Economics, St. Petersburg, Russia

Ajay Kumar ABV-Indian Institute of Information Technology and Management, Gwalior, MP, India

Sachin Kumar Cluster Innovation Centre, University of Delhi, Delhi, India

Uday Kumar Luleå University of Technology, Luleå, Sweden

Madhu Kumari Delhi College of Arts & Commerce, University of Delhi, Delhi, India

Naveen Kumari Department of Computer Science, Punjabi University Regional Centre for IT and Management, Mohali, India

Kavita Laghate Jamnalal Bajaj Institute of Management Studies, Mumbai University, Mumbai, India

K. M. Mahaboob Sheriff Nawab Shah Alam Khan College of Engineering and Technology, Hyderabad, TS, India

S. Malik Department of Statistics, Ram Lal Anand College, University of Delhi, New Delhi, India

Umar Maqbool Electrical & Renewable Energy Engineering, BGSBU, Rajouri, India

Ashok G. Matani Mechanical Engineering Department, Government College of Engineering, Amravati, India

Harsh Maurya Department of Electrical Engineering, Gautam Buddha University, Greater Noida, UP, India

Bhawna Mehta Delhi Technological University, Delhi, India

Kunal Mehta Sapient Corporation, New Delhi, India

Bishal Mishra Department of Electronics and Communication, Amity University, Noida, India

Pragya Mishra AIAS, Amity University, Noida, India

B. Mitra Department of Statistics, Ram Lal Anand College, University of Delhi, New Delhi, India

Sheikh Suhail Mohammad Electrical Engineering Department, NIT, Srinagar, India

A. Mohan Department of Statistics, Ram Lal Anand College, University of Delhi, New Delhi, India

Shafqat Nabi Mughal Electrical & Renewable Energy Engineering, BGSBU, Rajouri, India

K. Muralidharan Department of Statistics, Faculty of Science, The MSU, Vadodara, India

Naina School of Computational and Integrative Sciences, Jawaharlal Nehru University, New Delhi, India

Nidhi Shridhar Natrajan Symbiosis Centre of Management Studies, Noida, India

Tahleela Navid Electrical & Renewable Energy Engineering, BGSBU, Rajouri, India

Srivastava Niraj Terminal Ballistics Research Laboratory, Chandigarh, India

Vinayak Oak Jamnalal Bajaj Institute of Management Studies, Mumbai University, Mumbai, India

Zahid Nazir Padder Electrical & Renewable Energy Engineering, BGSBU, Rajouri, India

Nidhi Singh Pal Department of Electrical Engineering, Gautam Buddha University, Greater Noida, UP, India

Rajiv Pandey Amity University Uttar Pradesh, Noida, India

Vishal Pradhan ABV-Indian Institute of Information Technology and Management, Gwalior, MP, India

Jagdish Prasad Department of Statistics, University of Rajasthan, Jaipur, India; Presently Amity School of Applied Sciences, Amity University Rajasthan, Jaipur, India

Z. Rahman Department of Statistics, Ram Lal Anand College, University of Delhi, New Delhi, India

Rajesh Department of Statistics, University of Rajasthan, Jaipur, India

Arpit Rajput ASET, Amity University, Noida, Uttar Pradesh, India

Anil Rana Bhartiya Skill Development University, Jaipur, India

Seema Rathee Guru Jambheshwar University of Science & Technology, Hisar, India

Ayush Kumar Rathore Amity University Uttar Pradesh, Noida, India

Neha Raval Department of Statistics, Faculty of Science, The MSU, Vadodara, India

Sharma Ritu Terminal Ballistics Research Laboratory, Chandigarh, India

Rinku Sanjeev Symbiosis Centre of Management Studies, Noida, India

T. V. Santhosh Reactor Safety Division, Bhabha Atomic Research Centre, Mumbai, India

Arvind Selwal Central University of Jammu, J&K, Samba, India

Anurag Sharma DN College, Meerut, India

Meera Sharma Swami Shraddhanand College, University of Delhi, Delhi, India

Shivam Sharma Indraprastha Institute of Information Technology, New Delhi, India

Savita Kumari Sheoran Department of Computer Science & Engineering, Indira Gandhi University Meerpur, Rewari, India

Avinash K. Shrivastava International Management Institute, Kolkata, West Bengal, India

Vivek Shrivastava Department of Electrical Engineering, National Institute of Technology, New Delhi, India

Rehan Ahmed Siddiqui Department of Electronics and Communication, Amity University, Noida, India

Aditya Kumar Singh Ramgarh Engineering College, Ramgarh, Jharkhand, India

Archana Singh Department of IT, ASET, Amity University, Noida, UP, India

Archana Singh Delhi Technological University, Delhi, India

Gajendra Pratap Singh School of Computational and Integrative Sciences, Jawaharlal Nehru University, New Delhi, India

Mamtesh Singh Department of Zoology, Gargi College, University of Delhi, Delhi, India

Nandita Singh Amity International Business School, Amity University Campus, Noida, India

Shivraj Singh CCS University, Meerut, India

V. B. Singh Delhi College of Arts & Commerce, University of Delhi, Delhi, India

Ekta Singhal Fortune Institute of International Business, New Delhi, India

Ashok Kumar Sinha Defence Terrain Research Laboratory, Defence Research and Development Organisation, New Delhi, India

R. Solanki Department of Applied Sciences, SRMIST, NCR Campus, Modinagar, India

A. Suneja Department of Statistics, Ram Lal Anand College, University of Delhi, New Delhi, India

Gulshan Taneja Department of Mathematics, Maharshi Dayanand University, Rohtak, India

Adithya Thaduri Luleå University of Technology, Luleå, Sweden

Malay Ranjan Tripathy Department of Electronics and Communication, Amity University, Noida, India

Ajit Kumar Verma Western Norway University of Applied Sciences, Haugesund, Norway

Vandani Verma AIAS, Amity University, Noida, India

Vibha Verma Department of Operational Research, University of Delhi, Delhi, India

Romika Yadav Department of Computer Science & Engineering, Indira Gandhi University Meerpur, Rewari, India

Chapter 1
Application of Demand-Side Management Techniques for Sustainable Energy

Sheikh Suhail Mohammad, Umar Maqbool, Aaquib Firdous, Tahleela Navid, Zahid Nazir Padder, and Shafqat Nabi Mughal

1.1 Introduction

At present times commercial, residential, and industrial sectors are making a transition from conventional energy resources to green energy resources for energy needs. The disadvantages of conventional energy resources are, they are exhaustible energy resources and produce a lot of emissions which go against environmental policies. The disadvantage of green energy resources is that they are intermittent in nature, hence not a reliable source of power. To increase the reliability of green energy resources, these are integrated with grid or energy storage devices [1]. All these green energy resources are interconnected with the grid at different voltage levels like 33 kV, 11 kV, and 400 V [2]. However, before integrating many energy resources storage devices with grid, there is a need to find an optimal combination and for obtaining that many tools like Homer, etc. are used [3–5]. Lots of case studies are present where Homer simulating tool was used to find the best ideal combination among the available energy resources for electrifying the residential areas and for campus electrification [6]. Many international agencies like the International Renewable Energy Agency (IRENA) provide financial support in switching to renewable energy resources for energy needs [7].

The aim of this study is to find the optimal combination of solar PV with the electric utility grid for the electrification of Baba Ghulam Shah Badshah University campus. The optimal load demand was calculated after the implementation of demand-side management techniques. Based on the optimal load demand, pre-homer analysis, and post-homer analysis, an optimal combination that will provide cost-effective and reliable electricity to the campus was found.

S. S. Mohammad (✉)
Electrical Engineering Department, NIT, Srinagar, India
e-mail: ssuhail73@gmail.com

U. Maqbool · A. Firdous · T. Navid · Z. N. Padder · S. N. Mughal
Electrical & Renewable Energy Engineering, BGSBU, Rajouri, India

P. K. Kapur et al. (eds.), *Decision Analytics Applications in Industry*, Asset Analytics,
https://doi.org/10.1007/978-981-15-3643-4_1

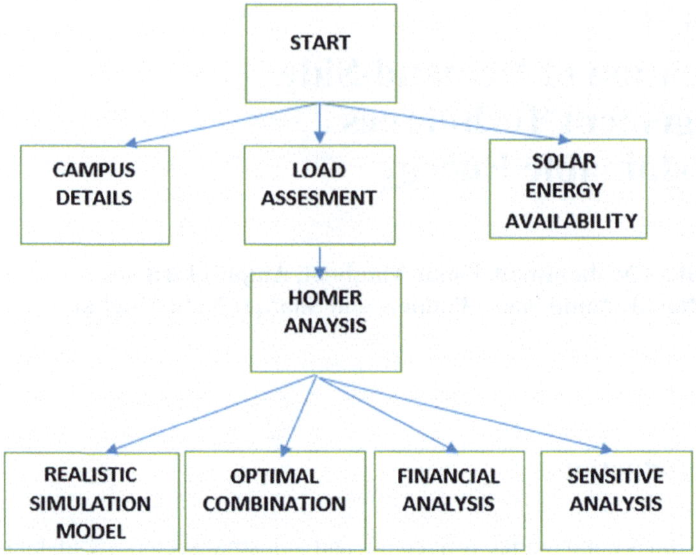

Fig. 1.1 Methodology flow chart

1.2 Methodology

A flowchart showing the methodology adopted for obtaining the optimal combination of solar PV and grid is given above in Fig. 1.1 [8].

1.3 Campus Details and Solar Energy Availability

The campus details are given in Table 1.1. Figure 1.2 shows the yearly solar profile of the location under consideration [9]. The campus has one 100 kW solar power plant already installed. In the simulation model, 200 kW solar photovoltaic power

Table 1.1 Campus details

Particulars	Details
Campus name	Baba Ghulam Shah Badshah University
District	Rajouri
State	J & k
Country	India
Latitude	74.3471°E
Longitude	33.3969°N
Grid electricity	Available

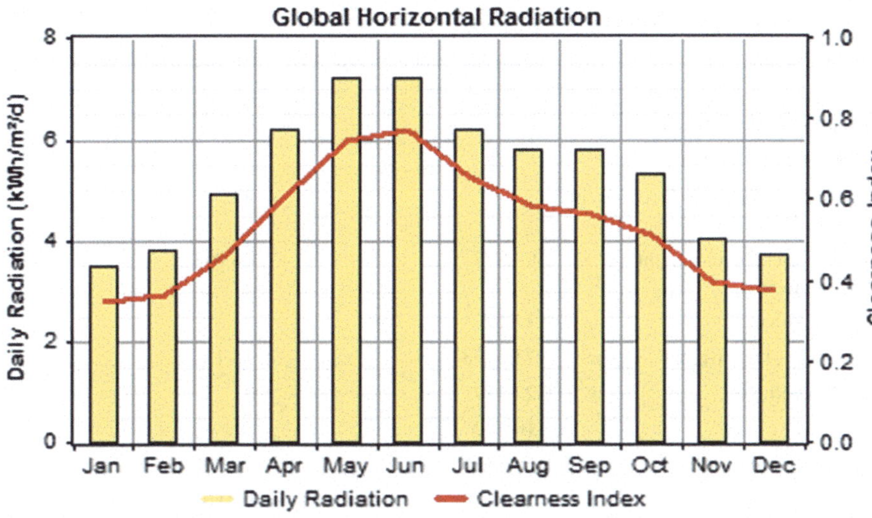

Fig. 1.2 Solar irradiance profile

plant was grid interconnected. The temperature remains between 25 and 32 °C for the region where the campus is located, this temperature range is considered ideal for the solar photovoltaic power generation.

1.4 Campus Load Estimation

In the case study, the load of the university has been calculated by using the energy audit methodology. An energy audit is a systematic approach to decision-making in the area of energy management. The energy audit is also defined as the "verification, monitoring and analysis of the use of energy including energy efficiency with cost-benefit analysis and an action plan to reduce energy consumption" as per energy act 2001. In this study, two load profiles were under consideration, one load profile was based on the existing load demand and the other was determined based on the demand-side management techniques like replacement of inefficient technology with energy efficient technology. Table 1.2 summarizes the load demand of the campus before and after the demand-side management techniques were applied, respectively. The shifted load profile as given in Fig. 1.3, was determined by replacing inefficient heating, lighting, and other electronic gadgets by efficient ones. The daily load profile that was used for the simulation purposes is given in Fig. 1.4, for the simulation purpose load of academic blocks, was considered, excluding the hostel blocks (Figs. 1.5 and 1.6).

Table 1.2 Calculated load profile of the campus

Building name	Existing load demand (kW)	Shifted load demand (kW)
School of Engineering and Technology (Main-Block)	23.872	6.202
Electrical-Lab (Block)	6.378	3
Electrical Engineering Department (Main-Building)	7	3
Research and Development (R&D)-Block	98.887	46.873
Nursing college	7.717	3
Biosciences-Department	73.6758	15.164
Diploma college	22.319	7
Workshop	10.162	5
Administration block	68.3826	24.776
Canteen	18.366	6.886
Police quarters	32.932	20.181
TAJ hostel	105.667	26.324
Sabrang Residential quarters	104.782	27.357
PG hostel	133.349	24.742
Girls hostel	101.980	17.245
University hospital	24.970	4.489

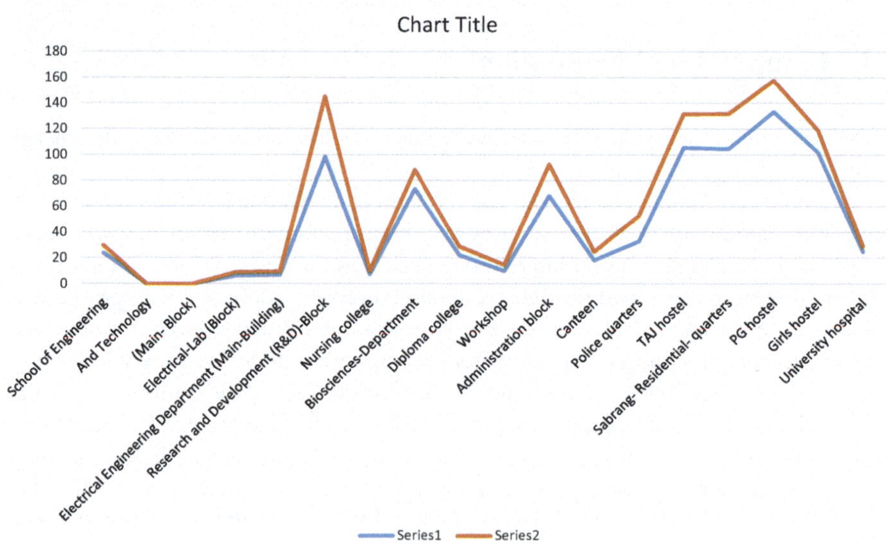

Fig. 1.3 Shifted load demand

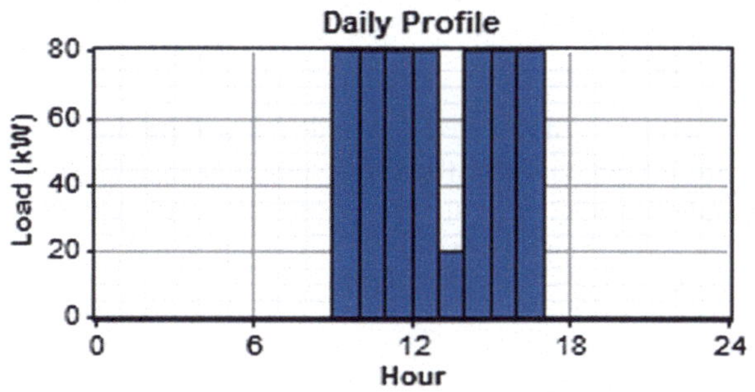

Fig. 1.4 Daily load profile

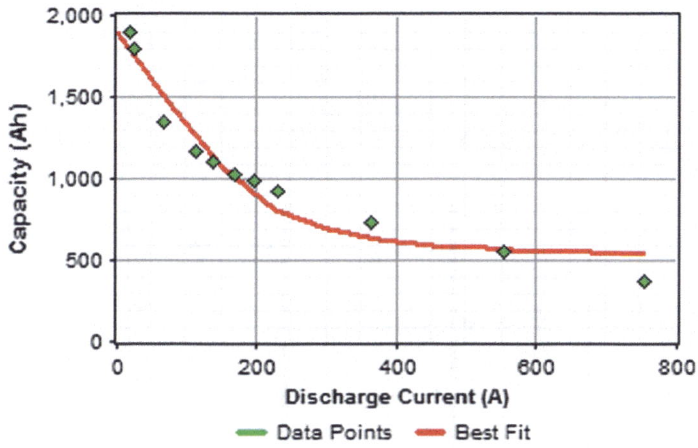

Fig. 1.5 Capacity curve

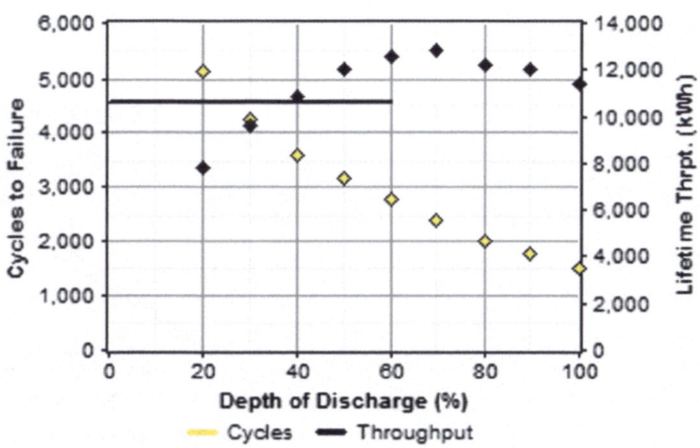

Fig. 1.6 Lifetime curve

1.5 Simulation Model

The simulation model is given in Fig. 1.7 was developed on the basis of information collected during pre-homer analysis, the PV system is considered to be grid interconnected. The simulation model consists of a photovoltaic system (PV), battery bank, converters, and load. The component details are given below.

1.5.1 Photovoltaic System (PV)

PV is one of the technologies that translate solar energy into DC electricity. The size of the PV system in the simulation model was taken to be 200 kW. The capital cost for each 1 kW PV panel was taken to be 563$ and the replacement cost for the same size was taken as $480, Fig. 1.8 shows the cost curve for the 200 kW PV plant. The lifetime for the PV plant was considered to be 20 years.

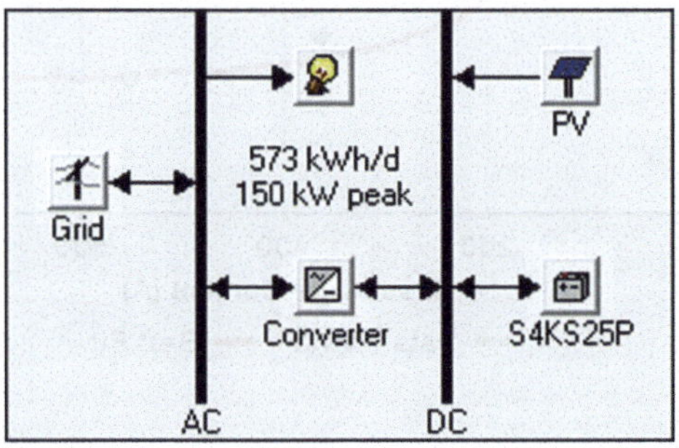

Fig. 1.7 Simulation model

Fig. 1.8 Cost curve

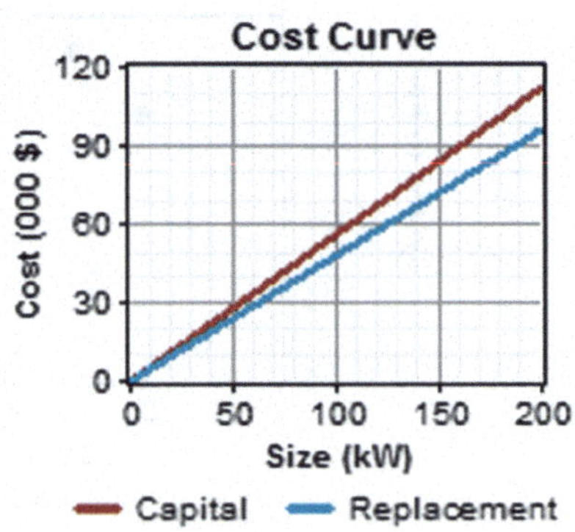

1.5.2 Grid

In order to increase system reliability, the PV system was grid-connected. Power management is key for the photovoltaic system if it is grid interfaced [10]. If the PV power is unavailable then the load is met from the grid power which is available every time. The purchase price and sell back price for each unit of energy was taken to be 0.120 $ and 0.100 $. Net purchasing was done on a monthly basis. The sale and purchase capacity for the grid was considered to be 200 kW each. Emission factors for CO_2, CO, SO_2, and NO_2 were taken into consideration for the calculation of emissions of overall system.

1.5.3 Battery and Converter

The role of the energy storage devices like the battery is to overcome the intermittent nature of the solar photovoltaic power (PV) that increases the overall system reliability [11]. The battery used in the simulation was Surrette 4KS25P, with nominal voltage and nominal capacity of 4 V and 1900 Ah (7.6 kWh). The battery bank consisted of 2 strings, each string consisted of 25 batteries. The capital cost for each battery was 800 $. Capacity curve and lifetime curve of the battery under consideration are given in Figs. 1.5 and 1.6.

At present most of the loads are of AC nature and the output of PV system is of DC nature, which makes it necessary to convert the DC power to AC power by using the converters like inverter (DC to AC) or rectifier (AC to DC). The size and capital for the converter used in the simulation model was 70 kW and 500 $/kW, respectively. Efficiency and lifetime for the converter were selected to be 90% and 15 years.

1.6 Results and Discussions

This section explains the results of the simulation. The analysis of the results is done on the basis of technical and economic factors. First, the system architecture and electrical section are analyzed, then it will be followed by a discussion over the cost summary.

1.6.1 Optimized System Architecture with Electrical Power Production

The optimized system architecture consists of 200 kW grid, 200 kW PV, 70 kW inverter, 70kw rectifier, and 50 number of Surrette 4ks25p batteries. The total net present cost (NPC) for this optimized combination of PV and grid was $275,077. The levelised cost of energy (COE) was$ 0.103/kwh and the operating cost was $6843/year.

Figure 1.9 shows the monthly average electrical production. The yellow portion of the bin shows the PV power production and blue one shows the grid power. From the Table 1.5, it is clear that 85% of the power is produced from the PV system and rest comes from the grid, in addition to it 18.3% of extra electricity can be fed back to the grid for sale. PV production depends on temperature and solar irradiance. PV output power increases with an increase in solar irradiance and decreases with increases in temperature. The region under consideration is having average temperature of 28 °C.

1.6.2 Economic Analysis

Figure 1.10 shows the complete cost summary of the project mentioning all the details like net present cost (NPC), operating cost, fuel cost, and salvage. The total NPC, operating cost and levelised cost of electricity (COE) were $ 275077, $ 6843/year, and $ 0.103/kWh, respectively. As per the statistical figures, levelised cost of per unit of energy for the optimal system is less than that of the purchase cost of energy unit from the grid. Table 1.3 shows the complete details of the monthly based energy exchange with the grid. From the energy exchange details it is clear that net purchases are positive only in few months like January, March, December, and August, the reasons for the positive net purchases are low irradiance levels in the months of December and January and high temperature in the month of August.

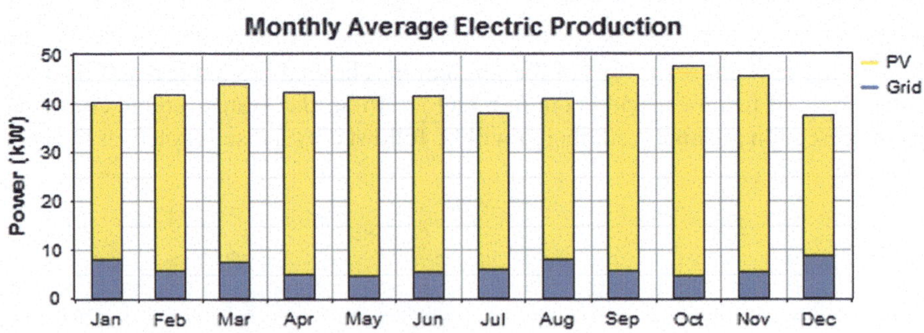

Fig. 1.9 Average monthly electric production

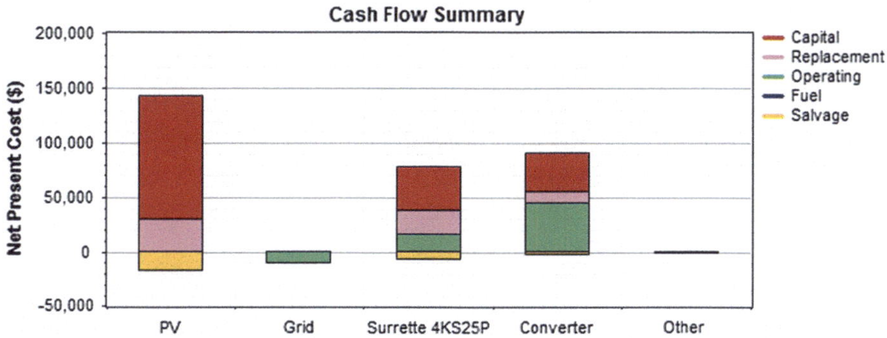

Fig. 1.10 Cash flow analysis

Table 1.3 Energy exchange with the grid

Month	Energy charge ($)	Demand charge ($)	Energy purchased (kWh)	Energy sold (kWh)	Net purchases (kWh)	Peak demand (kW)
Jan	187	0	5,825	4,270	1,555	109
Feb	−63	0	3,863	4,496	−634	92
Mar	12	0	5,457	5,353	104	93
Apr	−198	0	3,571	5,555	−1,983	102
May	−295	0	3,445	6,397	−2,952	83
Jun	−209	0	3,968	6,056	−2,088	98
Jul	−88	0	4,335	5,214	−879	107
Sep	−162	0	4,115	5,732	−1,616	98
Oct	−318	0	3,345	6,524	−3,180	71
Nov	−159	0	3,835	5,423	−1,587	93
Dec	395	0	6,542	3,254	3,288	111
Annual	−801	0	54,130	63,292	−9,162	111

Apart from the energy exchanges, statistical figures about the various emissions like CO_2, CO, SO_2, NO_2 should be given due attention as all the modern small and big integrated power systems have to satisfy the environmental policies. As per the optimized system architecture that is proposed for electrifying the BGSBU academic blocks, 85% of electricity is generated from the solar PV in Table 1.5. The decreased emissions that the proposed optimized system architecture can produce is given in Table 1.4.

Table 1.4 Emissions

Pollutant	Emissions (kg/year)
Carbon dioxide	−2.749
Carbon monoxide	−137
Particulate matter	0
Nitrogen oxides	−9.16
Unburned hydrocarbons	−45.8
Sulfur dioxide	−13.7

Table 1.5 Electrical production

Component	Production (kWh/year)	Fraction (%)
Grid purchases	54,130	15
PV array	316,179	85
Total	370,309	100

1.7 Conclusion

The purpose of the current study was to determine whether the dependency of BGSBU campus on grid power can be reduced by switching to PV power. By using a proper energy audit methodology, the existing load profile of the campus was shifted to a new scale. Based on the details of solar irradiance levels, new scale load demand, grid, battery, and converters simulation model was developed. It was concluded from the simulation results that BGSBU academic blocks can be electrified from the PV-based power plant since the integrated system is grid interconnected the excess power (18.3%) can be penetrated back into the system. The Levelized Cost of energy for the integrated system was calculated to be $0.103/kWh, which is less as compared to the purchase cost from the grid.

The work could be used in future to develop Renewable-based captive power plants for the college and university campuses. These captive power plants not only supply power to campus but can also act as a revenue source. Primarily the captive power plants are used to supply power to local demand, but in new college or university campuses where land is available the capacity of these captive power plants can be increased so that it can act as a revenue source for these new institutes.

References

1. Geth F, Tant J, Haesen E, Driesen J, Belmans R (2010) Integration of energy storage in distribution grids. IEEE PES Gen. Meet. PES 2010(1):1–6
2. Anand S, Fernandes BG (2010) Optimal voltage level for DC microgrids. In: IECON Proceedings of Industrial Electron. Conference, pp 3034–3039

3. Sen R, Bhattacharyya SC (2014) Off-grid electricity generation with renewable energy technologies in India: An application of HOMER. Renew Energy 62:388–398
4. Bekele G, Tadesse G (2012) Feasibility study of small Hydro/PV/Wind hybrid system for off-grid rural electrification in Ethiopia. Appl Energy 97:5–15
5. Ahmad F, Alam MS (2017) Feasibility study, design and implementation of smart polygeneration microgrid at AMU. Sustain Cities Soc
6. Swarnkar NM, Gidwani (2016) An Application of HOMER Pro in optimization of hybrid energy system for electrification of technical, pp 56–61
7. Data and statistics—IRENA resource. http://resourceirena.irena.org/gateway/dashboard/?topic=4&subTopic=17. Accessed 01 Jul 2018
8. Mohammad SS, Mishra S, Sinha SK Integration of renewable energy resources for rural electrification, pp 545–551 (2017)
9. NASA Prediction of Worldwide Energy Resources. https://power.larc.nasa.gov/#page-top. Accessed 10 Dec 2017
10. Anand S, Fernandes BG (2014) Power management control for solar photovoltaic based DC system. In: 40th Annual conference IEEE India electronic social IECON, pp 4365–4370
11. Grainger BM, Reed GF, Sparacino AR, Lewis PT (2014) Power electronics for grid-scale energy storage. In Proceedings of IEEE, vol 102, no 6

Chapter 2
Effect of Internal Customer Satisfaction Index on Revamping of ISO Implementation Structure in a R&D Organization

Sharma Ritu, Kaur Jatinder, Singh Jaspreet, and Srivastava Niraj

2.1 Introduction and Literature Review

The current worldwide competition is convincing R&D organizations to maximize the customer satisfaction and explore the quality management concepts in standardization of practices thereby enhancing the service quality. The worldwide commitment toward quality improvement curriculum is ongoing in all the areas of economical, scientific, and armed activities. The organizations around the world are greatly determined to adopt in-house management systems and improvement policies to enable them to produce latest technologies and strive in international market. The present paper is divided into five sections. Section 2.1 reports introduction and literature review on ISO, QMS, customer satisfaction, relation between QMS and customer satisfaction, influence of quality management principles on organization. Section 2.2 describes Research methodology in present study. Section 2.3 covers Results and discussions and is followed by conclusions in Sect. 2.4.

S. Ritu (✉) · K. Jatinder · S. Jaspreet · S. Niraj
Terminal Ballistics Research Laboratory, Chandigarh, India
e-mail: rits565@gmail.com

K. Jatinder
e-mail: jatinder05tbrl@gmail.com

S. Jaspreet
e-mail: dsinghjaspreet@gmail.com

S. Niraj
e-mail: telvastava@gmail.com

P. K. Kapur et al. (eds.), *Decision Analytics Applications in Industry*, Asset Analytics,
https://doi.org/10.1007/978-981-15-3643-4_2

2.1.1 ISO

ISO (the International Organization for Standardization) is a worldwide alliance of national standards bodies (ISO member bodies) founded on February 23, 1947. The work of preparing International Standards is done by ISO technical committees. ISO is the world's biggest developer of international standards with more than 150 countries as its members. ISO is a nongovernmental organization headquartered in Geneva with an objective to increase global business related to the change of goods and services [1]. ISO, a word originated from the Greek "iso" which means equal was selected as the short form as otherwise, International Organization for Standardization could have led to many acronyms in various languages [2]. As mentioned by Zharen [2], ISO came in order to aid the trade of goods and services by encouraging consistency and related activities throughout the world, and to develop collaboration in various sectors including science, technology, educational, and financial areas.

2.1.2 Quality Management System

A quality management system is defined as a system of procedures, processes, and resources that are formalized to achieve quality objectives. As stated by Dean et al. [3], "Quality Management (QM) defined as a 'philosophy or an approach to management' consisting of a 'set of equally supporting values, is maintained by a set of practice and methods'." In other words, it is a system of components which interact or are mutually connected and are organized in a way that makes it possible to consistently (and not randomly) attain a series of established intentions. As per ISO 9000, the quality management system is defined as a system "to direct and control an organization with regard to quality." The quality management system has to describe and manage a set of activities considered as processes, which are using resources for adding value to a customer's product or service. These sets of activities are described as processes, which have some inputs and outputs. In general, the output becomes a new input for the subsequent process. Implementing the quality management system is a purposeful decision of the organization. The requirements, objectives, output, employed processes and structure and size of the organization affect the planning of the quality management system. ISO 9000 standards series is the majorly adopted standard representing all international standards concerning to quality management systems. QMS provides the steadiness toward achieving organizational targets.

2.1.3 Customer Satisfaction

Customer satisfaction corresponds to a contemporary approach for quality in endeavors and organizations and serves in the increase of a truly customer-oriented management and culture. Kotler [4] mentioned customer satisfaction as the state experienced by an individual and related to evaluation of supposed features of products and expectations of that individual concerning those features. It is also defined as the levels of service quality performances that meet users' expectation. Customer satisfaction evaluation presents a significant and objective opinion about their choices and beliefs. Customer surveys can help management to resolve discrepancies between expectations and satisfaction.

2.1.4 Relation Between ISO 9001 QMS and Customer Satisfaction

The ISO 9000 family addresses various characteristics of quality management and contains several best-known standards of ISO. The standards provide direction and methods for organizations to ensure that their outputs time and again meet customer's requirements, and that quality is time and again improved. Clause 9.1.2 requires an organization to examine customer's view of the level to which their requirements and expectations have been met. The organization shall establish the methods for collecting and evaluating this information. ISO 9000 certifications are anticipated to enhance performance through customer satisfaction (Clause 5.1.2). Service organizations may determine service quality by using external information, by monitoring customer satisfaction in addition to their own internal statistics. Accomplishing high level of customer satisfaction desire that organizations continually monitor and study the experience, view, and recommendations of their customers and people who are possible customers (Clause 9.1.3.b).

2.1.5 Influence of Quality Management Principles (QMPs) on Organization

The principles of quality management are incorporated to improve the process performance and assist in process integration for betterment. The QMPs when practiced could be worked as a foundation to lead an organization's performance enhancement. The comparative significance of each principle could differ from organization to organization and is likely to vary over time. Sustained success is achieved when an organization retains and gains the confidence of customers and interested parties. These principles are of great importance for the success of leadership and they achieve full value when they work together in coherence. The understanding of system helps an organization to modify the performance of system and deliver steady outcome through a system of aligned processes. It enhances the capability to foresee and act in response to effect of internal and external uncertainty for improvement.

2.2 Research Methodology

The study was aimed at determining the "Effect of Internal Customer Satisfaction Index on revamping of ISO implementation structure in a R&D Organization.".The laboratory is ISO 9001 for the last 13 years. The development and implementation of QMS as per ISO 9001 requirements were started mainly with two aims: Standardization & Improvement. The first objective was to bring consistency and uniformity in documentation, record control, methods of executing technical and support activities; and the second was to bring a mechanism for upgradation and improvement in existing activities with a pro-active approach. The designed QMS has four centrally controlled documented procedures termed as Core Technical Procedures (CTP) for the technical activities carried out in laboratory, that are CTP for Test and Evaluation (CTP/TE), CTP for Manufacturing (CTP/MP), CTP for Technology Development (CTP/TD), and CTP for Design and Development (CTP/DD). A total of twenty-one technical divisions have been following these procedures and have been time and again working for improvement in their services. A five-point Likert scale rating has been selected to gather feedback to study the impact of QMS in their division from in-house customers of QMS in organization. Table 2.1 shows the format for customer feedback from in-house customers.

Table 2.1 Format for customer feedback

S. No.	Nature of Service	Bad Perception of Service Quality Excellent
1	Timeliness	1 2 3 4 5
2	Completeness	1 2 3 4 5
3	Courtesy	1 2 3 4 5
4	Accuracy	1 2 3 4 5
5	Ease of filling QMS formats	1 2 3 4 5
6	What in your opinion is the strongest aspect	
7	What in your opinion is the weakest area of service	

The customer feedback data (CFD) obtained from the technical divisions was taken and it was found that users following CTP/TD and CTP/DD were not satisfied that resulted in less customer satisfaction index (CSI) as compared to users of CTP/TE and CTP/MP. After thorough analysis, the deficient CSI value obtained indicated low score of 2 as average response from in-house users against Point 5 of nature of service that was ease of filling QMS format. The root cause analysis of the identified problem pinpointed toward ineffectiveness in implementation of CTP for Technology Development and CTP for Design and Development.

2.3 Results and Discussions

For effective ground-level implementation of ISO procedure, the management opinioned that understanding of expectations of users is very important. Interaction with customer gives a chance to add more value for customer. The role of quality management system is not limited to implementation as per ISO standard requirements but it should empower the interested parties to excel and perform in a better way. It also includes the management of risk that can affect the process output as well as the overall QMS output as the risk associated with noncompliance of procedure can adversely affect established QMS and can even lead to system collapse. The established procedures were reviewed at various levels to find out reasons for the ineffective implementation. It was found that users are not willing to follow and implement the established procedures (CTP/TD and CTP/DD). To facilitate the study, various attributes were identified and categorized to propose a solution for effective implementation.

Among all the attributes shown in Table 2.2, the majority of dissatisfaction was found to be in promptness in filling of QMS formats. Exhaustive interaction with users played a key role in tackling the problem along with management commitment with focus on improvement of organization's process performance. Table 2.3 shows comparison of old procedures (CTP/TD & CTP/DD) and revised procedure (CTP/TDD). Table 2.4 shows details of formats of old procedures (CTP/TD & CTP/DD) and revised procedure (CTP/TDD). The reason behind the hitch faced by users in filling the required formats was found to be complexity of procedures and duplicity of information sought through different formats making the Documentation cumbersome and time-consuming.

Table 2.2 Major attributes for ineffective implementation of CTP/TD & CTP/DD

S no.	Attribute	Nature
1.	Understanding/Clarity of procedure	Controllable
2.	Delay in activities due to procurement issues	Uncontrollable
4.	Promptness in filling QMS related formats	Controllable

Table 2.3 Comparison on number of formats between old CTP (CTP/TD and CTP/DD) and revised CTP (CTP/TDD)

Old core technical procedures		Revised core technical procedure	
Name of core technical procedure (CTP)	No of formats	Name of core technical procedure (CTP)	No of format
CTP/TD	05	CTP/TDD	01
CTP/DD	07		

Table 2.4 Details of formats between old CTP (CTP/TD and CTP/DD) and revised CTP (CTP/TDD)

Old procedures (CTP/TD & CTP/DD)				Revised procedure (CTP/TDD)	
Total: Twelve formats				Single format	
Format no.	(CTP/TD) Name of format	Format no.	(CTP/DD) Name of format	Part I Project details	Part II Review details
1	Statement of case (SOC) for new project	1	Design and Development Plan	Title of Project Project Type Statement of case Sanctioned Cost Probable date of completion Feasibility Report User Agency	Review Objective Review No. Review Date Agenda Points Discussed Decision taken Responsibility Target Date Review Minutes Reference No.
2	Action plan	2	Design and Development input		
3	Statement of case for sub project	3	Design and Development output		
4	Project review record	4	Design and Development review		
5	Project closure report	5	Design and Development verification		
		6	Design and Development validation		
		7	Design and Development changes		

For example, as depicted in Table 2.4, users had to repeatedly fill the information including the title of project, probable date of completion, sanctioned cost, project officer, project No. in all the twelve formats for CTP/TD and CTP/DD. In another example, details on input document, sub-system involved, name of the product/sub-system was required to be mentioned in all the formats of CTP/DD. The same data was replicated in multiple numbers of formats leading to improper record generation.

The users were not willing to complete the required documentation that has impacted the main goal of QMS implementation of streamlining the activities carried out in laboratory. After careful analysis of situation and its outcome on overall performance monitoring of laboratory, for better applicability in real scenario, it was decided to modify the procedures. Based on the work carried out by user divisions, it was decided to merge the two procedures into a more user-friendly simpler procedure.

Therefore, for ease of implementation, the procedures were reviewed, amended, and transformed into a single procedure as core technical procedure on Technology and Design Development (CTP/TDD) with minimum number of formats covering all the project-related parameters as mentioned in Tables 2.3 and 2.4.

The revised procedure (CTP/TDD) was implemented to carry out all activities as per planned arrangements to ensure systematic development of new products for better applicability in user divisions.

2.4 Conclusion

The objective of the paper was to improve the implementation of QMS in R&D organization and conducting this study was found to be very helpful. The problem of repeatability and duplicity of information was overcome and the core technical procedure got simpler and easy to implement. The average response from users improved from score 2 to score 4 against point number 5 in nature of service of customer feedback format. The findings indicated major improvement in the CSI obtained after the designing, development as well as implementation of new revised procedure (CTP/TDD) covering all the information related to activities. The feedback methodology has presented the means to capitalize on the advantages of ISO 9001 quality management system implementation. Effective QMS integrate customer point of view when defining their improvement initiatives. As ISO 9001 quality management system is known to improve organization's process, it can be inferred from the present study that it also contributes to corrective action procedure which is required to describe the source of weak efficiencies in processes, improve evidence in decision-making and create a culture of improvement in organization. It generates clarity and regulation in the system, which is influential in inspiring engagement of personnel at all levels within the organization.

2.5 Future Scope

- More data will be collected through structured survey from more respondents.
- After data collection, the data will be examined and inferred with the help of necessary statistical analysis; thereby, arriving at findings, suggestions, and conclusion based on analysis.

References

1. Psomas EL, Pantouvakis A, Kafetzopoulos DP (2013) The impact of ISO 9001 effectiveness on the performance of service companies. Manag Serv Qual Int J 23(2):149–164
2. Von Zharen WM (2001) ISO 14001: positioning your organization for environmental success
3. Dean JW, Bowen DE (1994) Management theory and total quality: improving research and practice through theory development. Acad Manag Rev 19(3):392–418
4. Kotler (1994) Marketing, analysis, planning, implementation and control

Chapter 3
An Empirical Research on the Role of Cloud-Based HRIS & HRM Functions in Organizational Performance

Rinku Sanjeev and Nidhi Shridhar Natrajan

3.1 Introduction

Digital transformations have changed the way business functions are conducted. Organizations have become ready for the real-time responses with respect to the ever-changing business environment. As per the report of IDC, the expenditure on Information Technology (IT) infrastructure across the globe was $1.3 trillion in 2017 and it is projected to cross $2.1 trillion by 2021. Gartner also reflects that 62% of organizations took initiative to make business digital [15]. Implementing new technologies provides competitive edge to the organizations. Cloud-based application hugely transformed business by giving numerous benefits such as reduced cost of IT infrastructure, scalability, collaboration, flexibility through on-demand service. Logic Monitor survey has highlighted that by 2020 83% of the organizations' workload will be mapped on cloud. Companies like Amazon, IBM, and Google are pioneers in this field and have provided cloud-based platforms [5]. The latest survey conducted by RightScale in January 2018 has emphasized that both private and public cloud adoption are increasing in the current year. More and more organizations are prioritizing public cloud, and trends indicate the popularity of Multi-cloud strategy. The revenue generated through cloud is not only generated by adoption of this technology but also from the number of workload deployed and usage of extended cloud service [4]. According to this the cloud service market with reference to India value up to 1353 million U.S. dollars as compared to 2015–2020 [18].

R. Sanjeev (✉) · N. S. Natrajan (✉)
Symbiosis Centre of Management Studies, Noida, India
e-mail: rinku.sanjeev@scmsnoida.ac.in

N. S. Natrajan
e-mail: nidhi.natrajan@scmsnoida.ac.in

P. K. Kapur et al. (eds.), *Decision Analytics Applications in Industry*, Asset Analytics,
https://doi.org/10.1007/978-981-15-3643-4_3

3.2 Background

Digitization in business and PC era in 1980s has transformed the way business and business data were managed. The technology is no more for competitive advantage but for the survival of the organization. The internet-based facilities have contributed to the globalization and enabled communication across the globe like never before. Cloud computing encompasses widespread adoption of virtualization, service-oriented architecture, autonomic, and utility computing. The use of cloud enables a company to store and access data, and runs the application without even investing in the IT infrastructure. Organization have recognized the potential of cloud computing and have started incorporating the cloud-based processes. The companies like Amazon and Microsoft Azure have become popular cloud providers and other companies especially small and medium organizations can benefit from the resources available virtually. Cloud not only supports low cost and smooth operations but also provides effective customer support. Customers need answers and the ability to purchase products 24 × 7 and around the world. The cloud makes this possible. The cloud also makes it easier to share information with customers. Another facility through cloud is flexible workplace. Since the organizations' information is stored in cloud the telecommuting worker can easily connect with the cloud with their own device and have greater flexibility. Companies like Zapier are strategizing for low-cost leadership by having fully virtual workforce. Startups and entrepreneurs can get hugely benefitted by the use of cloud-based services. Irrespective of the size and work of the company, the security of data and its availability makes cloud a suitable choice for every organization. The central and safe repository of data leads to higher productivity. Although cloud is a relatively new technology it is becoming highly popular due to its various benefits.

3.3 Review of Literature

Human resource (HR) is the most vital resource in the organization. HR heads and leaders support organizational decision-making process and help in enhancing the abilities of current workforce. They also enable forecasting the future needs regarding the workforce with the changing needs of the business. These specialists also help in minimizing the cost incurred due to attrition and replacement of the workforce. IT has enabled smooth and effective execution of various HR functions through enterprise resource planning (ERP). Human resources information systems (HRISs) like ADP workforce now have seamlessly combined the various functions of HR on a single platform while enabling smooth transaction of data. The traditional ERP is on-premise systems accessible within the intranet of the organization. The latest trends of on demand on payment-based cloud computing have extended this ERP support anywhere any time as per the needs of the organization. Whether the organization is large or small they all depend on data to carry out their routine activities and decision making. This large volume of data needs to be managed easily and should be accessible effortlessly and efficiently to the HR department. The cloud-based infrastructure provides facility to the managers to make their work effective. Software as a service (SaaS) supports cloud-based human resource management (HRM) software. The low-cost model of cloud and its scalability provide the organization an extra edge. The organizations must foster the culture of cloud-based HRM and provide adequate training for the same. Like every new technology, there are certain challenges associated with the adoption of cloud-based HRIS. The cloud-based applications are complex and require time to set up. It takes time for the organizations to adapt to a new working process and sometimes interfacing the applications with third party application becomes intricate. Being internet-based facility there are certain security-related issues to be taken care of, thus the current paper aims at understanding the role of cloud-based HRIS in enhancing the organizational performance [12, 36].

The organizations in the current era have recognized the importance of its workforce. A lot of importance is given to enhance HR policies and processes to keep their employees happy and satisfied. Technology has always taken business processes to a better execution level and provided organization a competitive edge [11, 17, 33, 38]. Globalization has not only provided opportunities but also poses a lot of challenges. But if the organizations strategize in a better way and incorporate the latest technology, there will be better outcome [7, 16, 31].

3.3.1 HRM and Organizational Performance

HRM has always been there in the core of the organizations working; however, technology has helped to evolve HRM practices in much better way [33]. Functions of HR like staffing, compensation, training, and planning, etc., lead to organizational development. According to Zakaria, apart from above-mentioned HR functions reward system, communication, and information sharing also result in better organizational performance and hence a competitive advantage. He also suggested innovative HRM practices for long-term survival of the organization [39].

3.3.2 Human Resource Information Technology (HRIT) and e-HRM

Information and Communication Technology (ICT) has become an integral part of the organization in the present day. Information systems provide a smooth functioning and better data management. Human resource information system (HRIS) and internet-based enterprise resource planning (ERP) have enabled e-HRM. HRIS supports collection, storing, maintenance, and retrieval of HR data and reports efficiently [8, 17, 21]. Most of the companies work on the global platform and thus require internet-based HRIS to maintain database of employee information related to recruitment, attrition, compensation, training, demographics, etc. e-HRM and HRIS enable smooth coordination between HR strategic and routine work creating value for the organization through technology [27].

3.3.3 Cloud Computing and Human Resources

Technology has been evolving at a very fast pace and organizations need to realign itself for better performance. The internet-based disruptive technology has a huge impact on the way various business functions are executed. HR department has always leveraged the technology for efficient HRM. The HR heads have been strategizing well and able to reduce the cost of HR operations by deploying the latest technological solutions [32].

Cloud computing has emerged as one of the new ways to utilize on-demand IT infrastructure. The distributed platform, application and storage accessible over internet, and different mobile devices have introduced a paradigm shift for organizational performance. All types of organization whether big or small are benefitted with this low-cost model [3, 14, 22]. According to the report of Deloitte [6] 68% of employees have highly recommended the use of technology and especially cloud computing for better HRM. In a similar report by Laurano has indicated the use of SaaS-based HR by 11% of the organization [19]. Thus there is a vast scope to deploy cloud-based technologies in the companies for low cost on demand use of IT applications, storage, and processing capabilities. Rader [24] talked about the importance of cloud-based computing service. According to the author this technology lowers the time taken to conduct a process, capability restrictions are overcome, reducing both fixed and variable cost of using the technology. Apart from this, the technology provides a platform for around the globe connectivity and high scalability.

According to him, this service not only minimizes the time taken to conduct process and capability limitations, and lowers costs by changing fixed costs to variable usage-based costs, but also provides anytime-anyplace connectivity and scalability without responsibility for maintenance or updates. This has to be penetrated in the culture of the organization and specifically HR personnel. There are several studies highlighting the business benefits of cloud computing [2, 9, 37]. Collaboration, innovation, low cost, and scalability are the most important among all the benefits. According to Sareen and Subramanian these benefits are also applicable for the HR functions; it just requires better technological planning [27]. The technology has its own set of challenges and complexities. The major area of concern is the adoption of cloud technology by the organizations. Both technical and behavioral factors are to be kept while changing the way organization transforms with the adoption of cloud-based process.

Fig. 3.1 Proposed research model

3.4 Methodology

The current study is empirically based on cross-sectional research design. The data for the current study is based on primary survey. A questionnaire was designed to collect the data from the top 10 IT companies in Delhi-NCR region. The instrument consisted of 42 items for eight distinct constructs. The constructs based on previous literature were modified to fit the needs of the study. The data were covered under two heads; demographic data and the eight distinct constructs that measure the benefit of the cloud-based HRIS [35].

Out of 450 questionnaires that were sent across to the respondents, we received 351 operational responses, yielding an overall response rate of 78%. The eight dimensions used in the questionnaire were adapted from the above-mentioned study.

The research methodology works in two ways; one through t-test to understand whether each variable shows above-average performance or not. Two regression analysis is applied to understand how much change in organizational performance is explained through changes in the cloud-based HRIS.

3.4.1 Proposed Research Model

See Fig. 3.1.

3.5 Data Analysis

The composite scores of the eight variables were calculated and the mean and standard deviation (SD) scores are shown in Table 3.1. To test the research hypothesis, the multiple linear regression technique was applied with all eight independent variables.

Table 3.1 Descriptive analysis

Descriptive statistics

	Mean	Std. deviation	*N*
Organizational performance	3.538	0.8996	351
Planning	0.0284998	0.98257234	351
Recruitment and selection practices	0.0075521	0.97392065	351
Training and development	0.0274575	1.01579197	351
Assigning and scheduling right people for job	0.0121462	0.99141039	351
Performance management	0.0244566	0.99661393	351
Retaining	0.0156253	1.01388526	351
Assessing	0.0156253	1.01388526	351

3.6 Result

Table 3.1 describes the mean and the standard deviation for $N = 351$. For all the independent variables mean and standard deviation is similar in nature.

In order to establish the significance of the independent variables one sample t-test was applied. The test was based on the comparison with average performance. Following is the set of hypothesis:

Ho: U = Uo and Ha: U ≠ Uo, where Uo = 2.5.

Ho represents the null hypothesis, Ha represents the alternative hypothesis, and Uo represents the hypothesized mean value (2.5). The importance of the particular variable was based on the statistical test based on mean.

The mean ranking of all the variables was compiled in order to analyze the significance of the decisions that the respondents expressed. For each variable, the null hypothesis was that, the variable at hand at a particular point in time is not significant which statistically indicated as Ho: U = Uo. The Uo is the critical rating above which the variable is considered significant or important.

The Likert scale was 1 = strongly agree, 2 = Agree, 3 = Neutral, 4 = Disagree, and 5 = strongly disagree. Under this section, the lower ratings of 1 and 2 were chosen for the rating scale as strongly agree and agree, respectively, while the Uo was set at 2.5, with 95% as the significance level [25].

Table 3.2 reveals that all the variables on human resource management measures with respect to the use of cloud-based HRIS were also assessed using a one sample t-test. All the statements are statistically significant at 0.000. It shows that in terms of cloud-based HRIS the mean is above average, reflecting better results in all the activities carried out related to HR.

Validity is done through extensive literature review; the citations are given under literature review section.

Table 3.3 indicates the reliability of the eight items Planning, Recruitment, Training, Assigning and Scheduling, Performance Management, Retaining giving value greater than 0.7 Nunnally [20].

Table 3.2 One sample T-Test HRM activities through cloud-based HRIS

One-sample test						
	Test value = 2.5					
	t	df	Sig. (2-tailed)	Mean difference	95% confidence interval of the difference	
					Lower	Upper
Organizational performance	17.980	350	0.000	0.634	0.70	0.56
Planning	23.730	350	0.000	0.791	0.86	0.73
Recruitment and selection practices	8.803	346	0.000	0.644	0.50	0.79
Training and development	20.268	350	0.000	0.717	0.79	0.65
Assigning and scheduling right people for job	−8.426	350	0.000	0.432	0.53	0.33
Performance management	10.976	344	0.000	0.799	0.66	0.94
Retaining	8.514	344	0.000	0.622	0.48	0.77
Assessing						

Table 3.3 Reliability test

Reliability statistics		
Cronbach's Alpha	Cronbach's Alpha based on standardized items	N of items
0.800	0.793	8

The multiple linear regressions require that the independent variables must have low correlation among themselves so that multi-collinearity is not present. Thus it is evident from Table 3.4 that the independent variables do not bring the multi-collinearity while studying them together.

H_1 There is a significant positive relationship between Planning and Organizational Performance.

H_2 There is a significant positive relationship between Recruitment and Selection and Organizational Performance.

H_3 There is a significant positive relationship between Training and Development and Organizational Performance

H_4 There is a significant positive relationship between Assigning and Scheduling right people for job and Organizational Performance

H_5 There is a significant positive relationship between Performance Management and Organizational Performance

Table 3.4 Interco-relation matrix

Correlations

		Organizational performance	Planning	Recruitment and selection practices	Training and development	Assigning and scheduling right people for job	Performance management	Retaining	Assessing
Pearson correlation	Organizational performance	1.000	0.352	0.407	0.437	0.106	0.300	0.172	0.172
	Planning	0.352	1.000	0.007	−0.031	0.021	0.011	−0.023	−0.023
	Recruitment and selection practices	0.407	0.007	1.000	−0.001	−0.020	0.036	0.037	0.037
	Training and development	0.437	−0.031	−0.001	1.000	−0.029	−0.004	0.030	0.030
	Assigning and scheduling right people for job	0.106	0.021	−0.020	−0.029	1.000	0.028	0.003	0.003
	Performance management	0.300	0.011	0.036	−0.004	0.028	1.000	0.045	0.045
	Retaining	0.172	−0.023	0.037	0.030	0.003	0.045	1.000	1.000
	Assessing	0.172	−0.023	0.037	0.030	0.003	0.045	1.000	1.000
Sig. (1-tailed)	Organizational performance	.	0.000	0.000	0.000	0.024	0.000	0.001	0.001
	Planning	0.000	.	0.448	0.282	0.351	0.419	0.336	0.336

(continued)

Table 3.4 (continued)

	Recruitment and selection Practices	0.000	0.448	.	0.494	0.355	0.253	0.244	0.244
	Training and development	0.000	0.282	0.494	.	0.293	0.473	0.286	0.286
	Assigning and scheduling right people for job	0.024	0.351	0.355	0.293	.	0.301	0.480	0.480
	Performance management	0.000	0.419	0.253	0.473	0.301	.	0.198	0.198
	Retaining	0.001	0.336	0.244	0.286	0.480	0.198	.	0.000
	Assessing	0.001	0.336	0.244	0.286	0.480	0.198	0.000	.
N		351	351	351	351	351	351	351	351

Table 3.5 Model significance

ANOVA[a]

Model		Sum of squares	df	Mean Square	F	Sig.
1	Regression	170.039	6	28.340	86.127	.000[b]
	Residual	113.192	344	0.329		
	Total	283.231	350			

[a]Dependent Variable: Organizational performance
[b]Predictors: (Constant), Planning, Recruitment, Training, Assigning & Scheduling, Performance Management, Retaining

Table 3.6 Regression model summary

Model summary[b]

Model	R	R square	Adjusted R square	Std. error of the estimate
1	0.775[a]	0.600	0.593	0.5736

[a]Predictors: (Constant), Planning, Recruitment, Training, Assigning & Scheduling, Performance Management, Retaining
[b]Dependent Variable: Organizational performance

H_6 There is a significant positive relationship between Retaining and Organizational Performance

H_7 There is a significant positive relationship between Assessing and Organizational Performance

The Sig value from Table 3.5 indicates that the proposed regression model is significant at $p < 0.05$ and hence further analysis is valid.

From Table 3.6, it is evident that the adjusted R square value is 0.593 which is above 0.5 (regression value ranges from 0 to 1) and hence the proposed model is good. The seven independent variables together are able to explain 59.3% variation caused in the dependent variable; Overall HCM practices.

Table 3.7 further establishes the acceptance of the proposed regression model. For all the seven variables, the significance value is less than 0.05 and hence all the independent variables are significant and cause a significant impact on the dependent variable organizational performance. The following table reflects the highest and lowest impacting variable.

Table 3.8 indicates the beta values and the corresponding order of effect of these variables on the dependent variables. Training and Development has the highest impact and Assigning & Scheduling right people for job has the least influence on the organizational performance. The cloud-based training and development process has highly impacted the performance of the organization. The entire seven hypotheses are accepted.

Table 3.7 Regression coefficient table

Coefficients[a]

Model		Unstandardized coefficients		Standardized coefficients	t	Sig.
		B	Std. Error	Beta		
1	(Constant)	3.563	0.031		116.208	0.000
	Planning	0.331	0.031	0.361	10.581	0.000
	Recruitment and selection practices	0.362	0.032	0.392	11.494	0.000
	Training and development	0.397	0.030	0.448	13.132	0.000
	Assigning and scheduling right people for job	0.101	0.031	0.111	3.255	0.001
	Performance management	0.247	0.031	0.274	8.020	0.000
	Retaining	0.124	0.030	0.274	4.079	0.000
	Assessing	0.124	0.030	0.139	4.079	0.000

[a]Dependent Variable: Organizational performance

Table 3.8 Impact rank

Independent variables	Beta coefficient	Impact rank
Planning	0.331	3
Recruitment and selection Practices	0.362	2
Training and development	0.397	1
Assigning and scheduling right people for job	0.101	6
Performance management	0.247	4
Retaining	0.124	5
Assessing	0.124	5

3.7 Managerial Implications

The outcomes of the study shown many benefits of cloud computing and are becoming highly popular for resource sharing. The practitioners, researchers, and top consultants agreed that cloud-based computing services is nowadays becoming a focus of design and act as a predominant method for positioning of HR systems and services with mobile fast [6, 16, 29, 30].

The study stated that most of the organizations prefer tailor-made HR information system though low success rates are being reflected in this scenario. While the success rate is much higher in cloud-based computing solution, it depends upon the maintenance and customization of the service provider. Many researchers and consultants reported that technology plays a critical role in human resource management [10, 13, 29]; adoption of cloud computing remains to forecast but lack of understanding about its importance by most of the organization [16].

The major challenge for cloud adoption is data security [10, 13, 14, 29]. Inclusive execution plan, communication, and training are the leading factors to meet up with the challenges of adoption and ensure the successful implementation of cloud-driven technology. For better adoption of end new technology, user backing and training are also important [28]. According to Babcock [1], better technology adoption may accomplish through strong and simplified communication with all stakeholders. Researcher have developed a model to address factors for e-business adoption based on framework which talks about the fusion of users and new technology [23, 34, 40].

The result of the study focuses on the factors supporting for effective method of cloud technology adoption. Other researches and consultants pointed that Cloud-Based commuting emphasized on effectiveness, efficiency, performance, and engagement [13, 29]. Organization growth and maturity challenges the organizational readiness toward adoption of technology [2, 13, 29], resulting in slow progressing through the maturity curve.

The technology has its own set of challenges and complexities. The major area of concern is the adoption of cloud technology by the organizations. Both technical and behavioral factors are to be kept while changing the way organization transforms with the adoption of cloud-based process. For smaller organizations the low-cost model may bring challenges associated with security of the company's data. Availability of strong network facility is a must otherwise the cloud services will not be available seamlessly.

3.8 Conclusion and Recommendations for Future Research

The current paper highlights the various factors responsible for slow implementation of cloud-based functionality specifically HR-based functions. It also tries to establish the HRM process re-engineering in context of cloud computing. The results further add to the existing body of knowledge in the area of cloud-based HRIS. The study through its extensive literature review and empirical result will be able to provide support to HR leaders while adopting and enhancing cloud-based HRIS. The managers and leaders will also be able to appreciate and implement the technology-driven strategy for the better performance of the organization.

Many organizations consider training as an overhead but the result of the study highlights the importance of training and development, it has been top ranked as per the result. This is very relevant since technology continually upgrades and employees need training to work on a new platform. Recruitment and planning are the next important parameters in the cloud-based organization. Thus it is clear that recruiting right people with the right skills including technology is very critical. In the digital era, the planning and strategy formulation is to be woven around the technology to be used in the organization. Planning and then fitting the cutting edge technology in various processes has become the old way of working. Thus technology drives the planning, this point has to be clearly understood by the HR leaders.

The adoption of cloud-based HRIS enables a company to store and access data, and runs the application without even investing in the IT infrastructure. Cloud-based technology is there from the last two decades but its adoption is quite slow. The leader must self-appreciate this technology and create a culture so as to build technology-driven workforce. Cloud-based HRIS is cost effective, flexible and scalable technology, thus without even investing huge amount in IT infrastructure the organization can have storage, computational facilities and various applications online 24 × 7. Use of cloud-based HRIS is also a sustainable way of working due to decreased paperwork, energy saving and cut down on the hardware requirements. Due to optimal IT infrastructure the commuter emissions are also low and thus an environmental friendly technology.

References

1. Babcock C (2010) Management strategies for the cloud revolution: How cloud computing is transforming business and why you can't afford to be left behind. McGraw-Hill, New York
2. Borgman, H. P., Bahli, B., Heier, H., & Schewski, F. (2013). Cloudrise: Exploring Cloud Computing Adoption and Governance with the TOE Framework. 2013 46th Hawaii International Conference on System Sciences. https://doi.org/10.1109/hicss.2013.132
3. Cegielski CG, Allison Jones-Farmer L, Wu Y, Hazen BT (2012) Adoption of cloud computing technologies in supply chains. The International Journal of Logistics Management 23(2):184–211. https://doi.org/10.1108/09574091211265350
4. Cloud Computing Trends (2018). State of the Cloud Survey, https://www.rightscale.com/blog/cloud-industry-insights/cloud-computing-trends-2018-state-cloud-survey
5. Columbus L. (2018). 83% Of Enterprise Workloads Will Be In The Cloud By 2020, https://www.forbes.com/sites/louiscolumbus/2018/01/07/83-of-enterprise-workloads-will-be-in-the-cloud-by-2020/#df35e976261a
6. Deloitte, (2011). HR in the Cloud: It's inevitable. Retrieved from http://www.deloitte.com/assets/Dcom-UnitedStates/Local%20Assets/Documents/us_consulting_HCTrends_HRintheClouds_060611.pdf,
7. Deloitte, (2012). HR Transformation Survey: A Case for Business Driven HR'. Retrieved from http://public.deloitte.com/media/0524/us_bnet_HRTransformationSurvey_1207pdf.pdf,
8. Dhamija, P. (2012). E-RECRUITMENT: A ROADMAP TOWARDS: E- HUMAN RESOURCE MANAGEMENT. Researcher's World, 3(2), 33–39. Retrieved from http://www.researchersworld.com/vol3/issue3/vol3_issue3_2/ Paper_05.pdf
9. Gangwar H, Date H, Ramaswamy R (2015) Understanding determinants of cloud computing adoption using an integrated TAM-TOE model. Journal of Enterprise Information Management 28(1):107–130. https://doi.org/10.1108/jeim-08-2013-0065
10. Geczy P, Izumi N, Hasida K (2012) Cloud sourcing: managing cloud adoption. Global Journal of Business Research 6(2):57–70
11. Gibson, W. (2012) HR Cloud: Enabling a path for growth in the technology industry. Available from https://www2.deloitte.com/content/dam/Deloitte/global/Documents/Technology/gxtech-sap-hr-cloud-technology-industry-growth.pdf
12. https://www.statista.com/statistics/485177/public-cloud-services-market-revenue-india/
13. IBM (2015). The Economist, & Intelligence Unit, 'Mapping the cloud maturity curve: Measuring organizational excellence in the new era of IT' [White

paper]. Retrieved from http://www-1.ibm.com/common/ssi/cgibin/ssialias?subtype= WH&infotype=SA&htmlfid=KUL12355USEN&attachment=KUL12355USEN.PDF-1.ibm.com/common/ssi/cgibin/ssialias?subtype = WH&infotype = SA&htmlfid = KUL12355USEN&attachment = KUL12355USEN.PDF,
14. Indranil B (2012) Cloud Computing and Its impact on Corporate HR practices. Advances in Management. Advances in Management 5(10):68
15. International Data Corporation (2017). IDC Forecasts Worldwide Spending on Digital Transformation Technologies to reach $1.3 Trillion in 2018Available at https://www.idc.com/getdoc.jsp?containerId=prUS43381817 [accessed on 15- Oct- 2018]
16. Kapoor B, Sherif J (2012) Global human resources (HR) information systems. Kybernetes 41(1/2):229–238. https://doi.org/10.1108/03684921211213052
17. Karimidizboni R (2013) Human resources information system. Interdisciplinary Journal of Contemporary Research in Business 4(10):1004–1017
18. Khan MA (2016) A survey of security issues for cloud computing. Journal of Network and Computer Applications 71:11–29. https://doi.org/10.1016/j.jnca.2016.05.010
19. Laurano, M. (2014). Three reasons to move your HR to the cloud. Analyst Insight by Aberdeen Group (white paper), Retrieved from http://aberdeen.com/research/8816/ai-cloud-human-resources/content.aspx
20. Nunnally JC (1978) Psychometric theory, 2nd edn. McGraw-Hill, New York
21. Obeidat, B. Y. (2012). The Relationship between Human Resource Information System (HRIS) Functions and Human Resource Management (HRM) Functionalities. Journal of Management Research, 4(4). https://doi.org/10.5296/jmr.v4i4.2262
22. Obeidat, M. A., & Turgay, T. (2012). Empirical Analysis for the Factors Affecting the Adoption of Cloud Computing Initiatives by Information Technology Executives. Journal of Management Research, 5(1). https://doi.org/10.5296/jmr.v5i1.2764
23. Patel, H. (2007). Factors Influencing Technology Adoption: A Review. Available from: https://www.researchgate.net/publication/273140050_Factors_Influencing_Technology_Adoption_A_Review, accessed Nov 15 2018
24. Rader D (2012) How cloud computing maximizes growth opportunities for a firm challenging established rivals. Strategy & Leadership 40(3):36–43. https://doi.org/10.1108/10878571211221202
25. Rensis (1932) A Technique for the Measurement of Attitudes. Archives of Psychology 140:1–55
26. Rogers, E.M. (2003), 'Diffusion of Innovations (5th ed.) New York. NY: Free Press
27. Sareen P, Subramanian KV (2012) e-HRM: A Strategic Reveiw. International Journal of Human Resource Studies 2(3):119. https://doi.org/10.5296/ijhrs.v2i3.2100
28. Schepers J, Wetzels M, de Ruyter K (2005) Leadership styles in technology acceptance: do followers practice what leaders preach? Managing Service Quality: An International Journal 15(6):496–508. https://doi.org/10.1108/09604520510633998
29. Sierra-Cedar. (2015). 2014–2015 HR Systems Survey: HR Technologies, Deployment Approaches, Integration, Metrics, and Value - 17th Annual Edition. Retrieved from http://www.sierracedar.com/wp-content/uploads/sites/12/2014/11/Sierra-Cedar_2014-2015_HRSystemsSurveyWhitePaper.pdf
30. Sivak, M. & Card, D. (2014). Human resources technology and service delivery trends in 2014. Executive Summary research report. Retrieved from https://www.slideshare.net/Gumennui/human-resources-technology-and-service-delivery-trends-in-2014
31. SHRM (2011). SHRM workplace forecast: the top workplace trends according to HR Professionals'. Retrieved from https://www.shrm.org/Research/FutureWorkplaceTrends/Documents/11-0014WPF_Posting_6.pdf,
32. SHRM (2014). Mission and history. Retrieved from https://www.shrm.org/about/history/Pages/default.aspx
33. Theriou GN, Chatzoglou PD (2008) Enhancing performance through best HRM practices, organizational learning and knowledge management. European Business Review 20(3):185–207. https://doi.org/10.1108/09555340810871400

34. Tornatzky LG, Fleischer M (1990) The Processes of Technological Innovation. Lexington Books, Lexington
35. TurBan, E. and Volonino, L. (2012) Information technology for management (international student version). John Wiley & Sons (Asia) Pvt Ltd. 8th ed
36. Wang XL, Wang L, Bi Z, Li YY, Xu Y (2016) Cloud computing in human resource management (HRM) system for small and medium enterprises (SMEs). The International Journal of Advanced Manufacturing Technology 84(1–4):485–496. https://doi.org/10.1007/s00170-016-8493-8
37. Willcocks LP, Venters W, Whitley EA (2013) Cloud sourcing and innovation: slow train coming? Strategic Outsourcing: An International Journal 6(2):184–202. https://doi.org/10.1108/so-04-2013-0004
38. Yahya FB, Kaur A (2010) Competition for foreign talent in Southeast Asia. J Asia Pac Econ 15(1):20–32. https://doi.org/10.1080/13547860903488203
39. Zakaria N (2013) Enhancing organizational performance of Malaysian SMEs through human resource management (HRM) practices and organizational innovative capability: A proposed framework. J Global Entrepreneurship 5(1):326–351
40. Zhu K, Kraemer KL, Xu S (2006) The process of innovation assimilation by firms in different countries: a technology diffusion perspective on e-business. Manage Sci 52(10):1557–1576. https://doi.org/10.1287/mnsc.1050.0487
41. Zikmund W (2014) Marketing Research. Cengage Learning. 3rd ed. Bib Id 6225219

Chapter 4
Synchronization of UPFC in Active Distribution Systems

Shiva Pujan Jaiswal and Vivek Shrivastava

4.1 Introduction

Power distribution network presently faces an issue of rapid growth in power utilization because of high living standard and heavy growth in industrialization. The old networks of power transmission and distribution are incapable of handling more power. So that production of power is shifting from centralized big units to small DGs. It is important to develop a better environment by utilizing the capabilities of renewable-based DG. DG can be classified on basis of purpose, MVA, technologies such as Small Hydro Plant (SHP), Photovoltaic (PV) cells, solar thermal, wind, fuel cell, tidal, wave energy, geothermal, biomass, or biogas [1]. Distributed generation system has many advantages over conventional centralized units, such as a reduction in transmission loss, good voltage control, shaving of peak demand, the congestion-free distribution system, and low emission of air pollutant. To optimize the size and location of DG mainly three objectives are considered in the literature: (i) control of losses, (ii) control of voltage, and (iii) economic benefits [2]. In 2010, Hung et al. [3] used the minimum loss as a factor for the optimization of size and site of DG. In 2013, an improved analytical technique was proposed for minimization of loss by DG at suitable location [4]. Soo and Jung [5] optimized multiple DG for improvement of system stability and performance. The authors of reference [1] improved the bus voltage and reduced losses of the power distribution system using modified PSO algorithms. Hung et al. [6] applied a cost-based objective function to locate renewable-based DG for reduction of distribution loss of the system. The integration of distributed generation in a system using PSO is proposed by Nasri et al. [7]. Genetic

S. P. Jaiswal (✉)
Depatrment of Electrical and Electronics Engineering SET, Sharda University, Noida, India
e-mail: shivajaiswal@gmail.com

V. Shrivastava
Department of Electrical Engineering, National Institute of Technology, New Delhi, India
e-mail: shvivek@gmail.com

P. K. Kapur et al. (eds.), *Decision Analytics Applications in Industry*, Asset Analytics,
https://doi.org/10.1007/978-981-15-3643-4_4

algorithm is employed by Ayodele et al. [8] to get the advantageous placement of DG for minimization of line loss as an intention. An intelligent algorithm was applied by Bhadoria et al. [9] to get the optimal location of DG. In literature, numbers of hybrid and heuristic algorithms such as cuckoo search, invasive weed, artificial bee colony, harmony search algorithms, and cloned SO are found [10]. Although DG has lots of advantages, it cannot control the variation of voltage with load. Watson et al. [11] used an electronics voltage regulator to counter overvoltage at light load with PV penetration in the distribution system. Agalgaonkar et al. [12] used tap changer and autonomous regulator to control the bus voltage after installation of DG in the network. Farng et al. [13] incorporated voltage regulator to provide protection against voltage vibrations in the distribution system.

Jaiswal et al. [14] discussed the application of SVC under contingency condition for stable power system operation. Velamuri et al. [15] controlled the power flow after integration of DG and FACTS controller. In reference [16], application and comparison of different optimization techniques were studied for optimization of UPFC. Garciá-Cerrada and Garcia-González [17] applied UPFC to enhance the stability and efficiency under the normal condition as well as under a fault condition. Kunwar and Bansal [18] studied the application of FACTS to enhance the transient stability of the system in the presence of a wind power plant. A genetic algorithm is used in reference [19] to optimize an objective function to get the most advantageous place of UPFC and fixed its parameter. UPFC is able to control all three parameters, i.e., bus voltage and its angle as well as line impedance of a power distribution system. UPFC is a set of two converters which are connected through a common DC link. Shunt converter controls the reactive power, whereas series converter mainly injects voltage with controlled magnitude and angle. UPFC is the most versatile FACTs device. To obtain maximum advantages from UPFC, it is most essential to the optimization of its parameters and location. It can improve stability margin, enhance power transfer capability of lines, reduce congestion, and prevent the blackout.

4.2 Problem Formulation

For stability and efficiency enhancement of the power distribution network, a mathematical equation is developed for optimizing the placement of distributed generation. To counter the issue of voltage variation with load change in the distribution system in the presence of DG, an objective function is proposed to obtain perfect parameters of UPFC at an optimal location.

4.2.1 Optimal Placement of DG

A multi-objective mathematical equation is proposed with three different indices considered as APLI, RPLI, and VPI_{DG}.

Active Power Loss Index (APLI):

$$APLI = \frac{P_L^{wdg}}{P_L^{wodg}} \tag{4.1}$$

Reactive Power Loss Index (RPLI):

$$RPLI = \frac{Q_L^{wdg}}{Q_L^{wodg}} \tag{4.2}$$

Voltage Profile Index (VPI_{DG}):

$$VPI = \left| \frac{V_1 - V_{\min}^{wodg}}{V_{\min}^{wodg}} \right| \tag{4.3}$$

where

P_L^{wdg} = Active power loss of system with DG
P_L^{wodg} = Active power loss of system without DG
$V_{\min}^{wodg}$ = Minimum bus voltage in system without DG
$V_{\min}^{wdg}$ = Minimum bus voltage in a system with DG
V_1 = Slack bus voltage.

Objective Function: By combining APLI, RPLI, and VPI_{DG} a multiobjective function is formulated as given by Eq. (4.4):

$$f_{DG} = w_1(APLI) + w_2(RPLI) + w_3(VIP_{DG}). \tag{4.4}$$

4.2.2 Modeling of UPFC

Distribution system faces unpredictable load variation, and DG alone is incapable of handling extreme corner loads. An optimal UPFC can make a system with DG more robust and efficient. Two converters' connection of UPFC is shown in Fig. 4.1.

4.2.3 Optimal Placement of UPFC

For controlling of voltage magnitude along with solar-based DG in the network, other objective functions with two different indices considered are Voltage Profile Index (VPI_{UPFC}) and Power Loss Index (PLI).

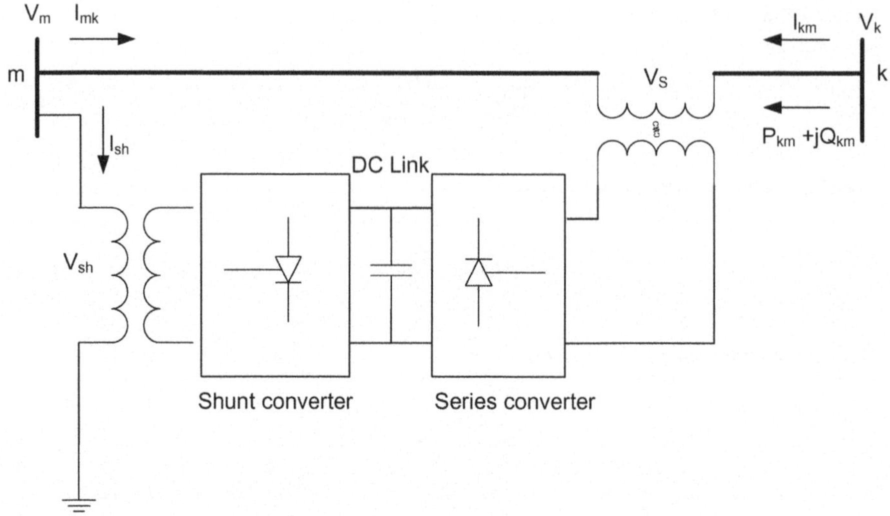

Fig. 4.1 Schematic representation two converters of UPFC

Power Loss Index (PLI):

$$PLI = \left| \frac{P_{WUPFC}}{P_{WOUPFC}} \right| \tag{4.5}$$

Voltage Profile Index (VPI_{UPFC}):

$$VPI_{UPFC} = \left| \frac{V_1 - V_{\min}^{WUPFC}}{V_{\min}^{WOUPFC}} \right| \tag{4.6}$$

where

P_{WUPFC}= Active power loss of system with UPFC
P_{WOUPFC}= Active power loss of system without UPFC
V_{MIN}^{WUPFC}= Minimum bus voltage in system with UPFC
V_{MIN}^{WOUPFC}= Minimum bus voltage in a system without UPFC

Objective Function: To achieve the desired benefit from it, the optimization of placement and parameters is essential. By combining TLI and $\mathrm{VPI_{UPFC}}$, a multiobjective function is formulated as Eq. 4.7.

$$f_{UPFC} = \{w_4(TLI) + w_5(VPI_{UPFC}). \tag{4.7}$$

4.3 Results and Discussion

The developed technique is validated on 33-Bus radial as well as 30-Bus interconnected system using PSO. The developed mathematical objective equation is optimized to get the optimal site and size of DG as well as UPFC in the presence of a solar PV unit. Its constraints are adjusted to counteract the consequences of high deviation in applied load, i.e., from the light load (60% of rated load) to heavy load (160% of the rated load).

4.3.1 The Case I IEEE 33-Bus System

The single line diagram is shown in Fig. 4.2; the real power demand is 3.73 MW and reactive load is 2.3 MVAR. The results of optimization are 2.1 MW DG at 7th and 27th bus is the optimized location with tuned parameters of UPFC in this system.

The profile of voltage magnitude of this system at different loading without solar-based DG and with no UPFC is presented in Fig. 4.3. The maximum magnitude of the bus voltage is 1.012 p.u. with a 60% load, whereas the minimum voltage is 0.701 p.u. at 160% load of base load. The bus voltage profile at different loads with optimal DG at 7th bus; the voltage goes beyond ±10 limits at extreme corner loads as publicized in Fig. 4.4. The ability of UPFC for enhancement of stability

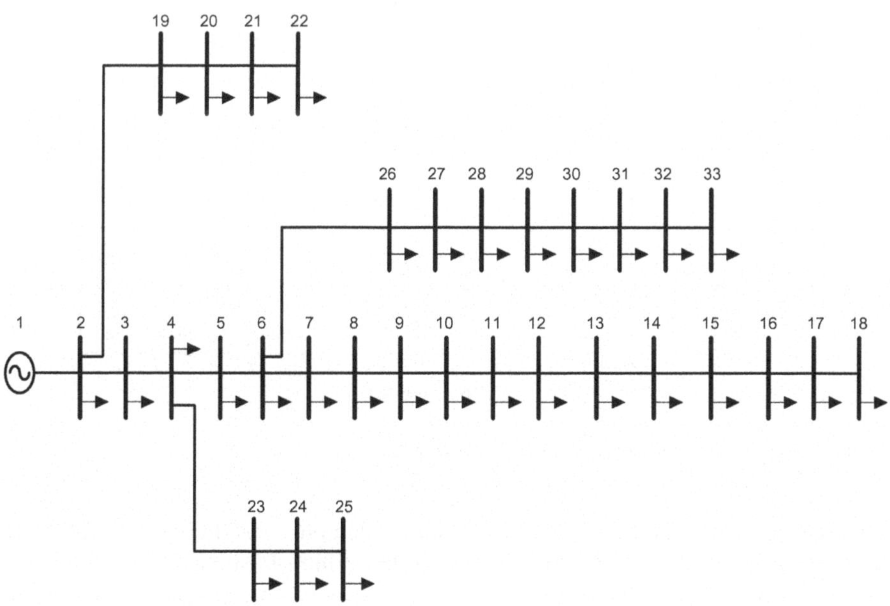

Fig. 4.2 IEEE. 33-Bus system

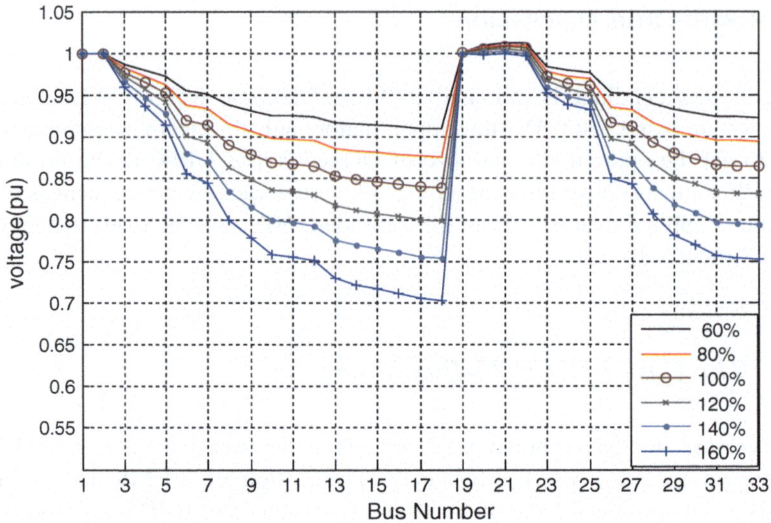

Fig. 4.3 Profile of bus voltage for IEEE. 33-Bus network

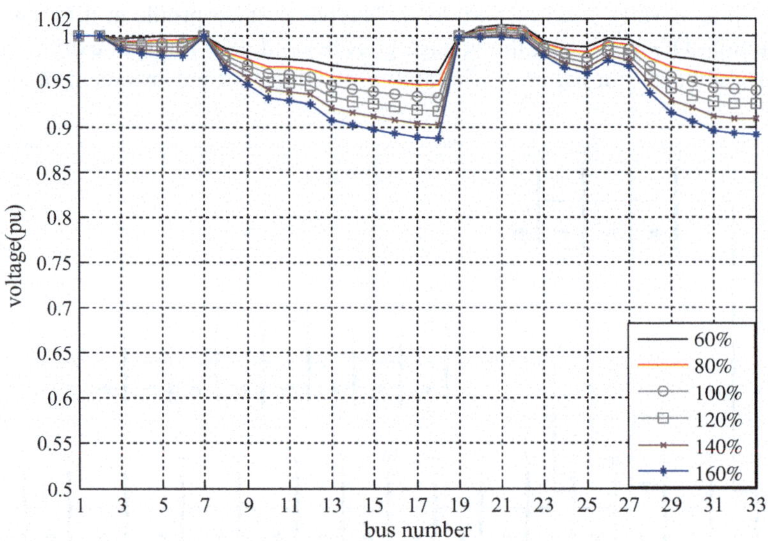

Fig. 4.4 Profile of bus voltage for 33-Bus network with DG

and efficiency with all possible load variations as demonstrated in Fig. 4.5. Table 4.1 shows the effectiveness of DG and UPFC at an optimal location. With best-tuned parameters, the UPFC reduces the loss in every condition and keeps the voltage within permissible limits. At rated load, UPFC in the active system also reduces the loss

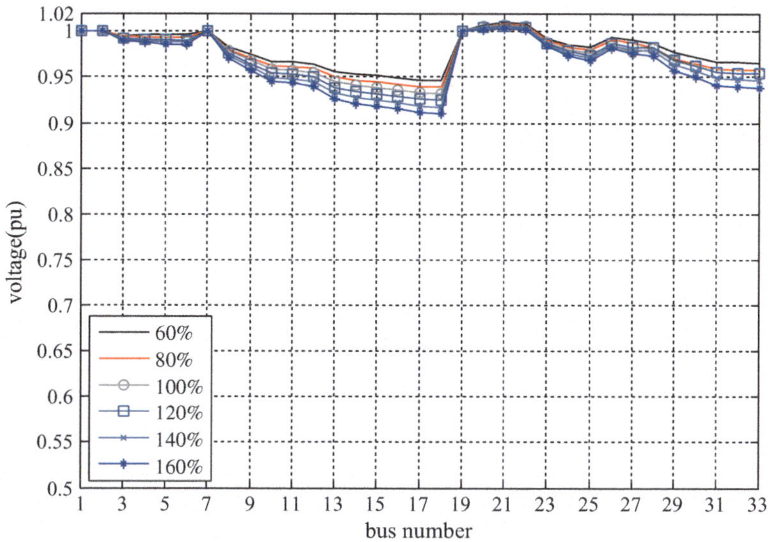

Fig. 4.5 Profile of bus voltage for IEEE. 33-Bus network with DG and UPFC

from 211.2 to 95 kW. The technique of optimization of placement and parameters of UPFC proves its effectiveness to reduce the loss with 160% loading.

4.3.2 Case II IEEE 30-Bus Interconnected Network

The developed technique is validated for 30-Bus interconnected power system as a line diagram is given in Fig. 4.6.

The profile of bus voltage magnitude of the system at various levels of load is shown in Fig. 4.7; with 160% loading 19th bus voltage crosses the permissible limit and at the light load, bus voltage goes beyond the 1.1 pu. Fig 4.8 displays the helplessness of DG to control voltage level at extreme loads. Figure 4.9 demonstrates a safe and secure voltage profile which is provided by UPFC. Table 4.2 proves that optimal UPFC can reduce the loss of the power system in the presence of DG with a range of load. Developed technique obtained 11th bus as an optimal location for 15 MW DG, and 10th line is an optimized allocation for UPFC. At rated load, UPFC in the active system also reduces the loss from 22.71 MW to 12.10 MW. The technique of optimization of placement and parameters of UPFC proves its effectiveness to reduce the loss with 160% loading, i.e., loss is reduced by more than 50%.

Table 4.1 Losses and voltage magnitude for 33-Bus network at different loading

% Loading	Loss in kW			w/o DG and UPFC		DG		DG and UPFC	
	w/o DG	DG	DG and UPFC	V_{min}	V_{max}	V_{min}	V_{max}	V_{min}	V_{max}
60	164	112	108.00	0.91	1.012	0.96	1.012	0.948	1.01
80	189	144	126.57	0.875	1.012	0.949	1.012	0.94	1.01
100	211.2	107	95.00	0.84	1.01	0.934	1.01	0.929	1.01
120	275	221	185.5	0.8	1.001	0.919	1.001	0.925	1.01
140	293	236	210.00	0.751	1	0.901	1	917	1.01
160	316	241	221.30	0.701	1	0.889	1	0.91	1.01

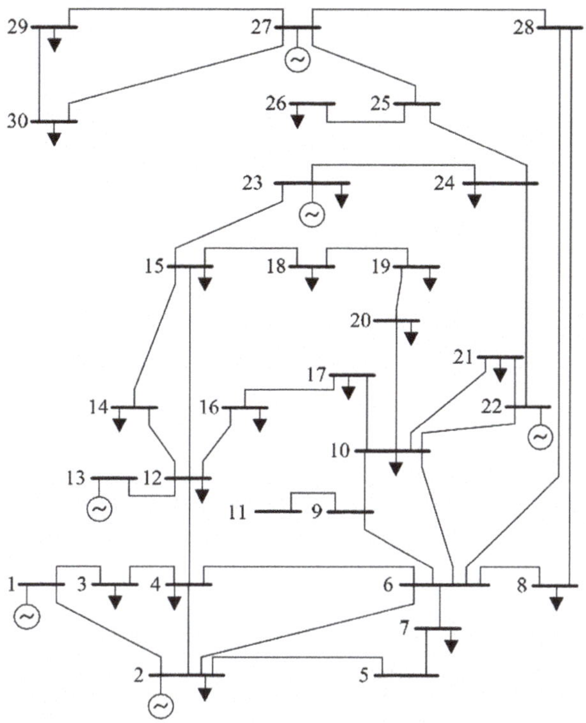

Fig. 4.6 IEEE. 30-Bus network

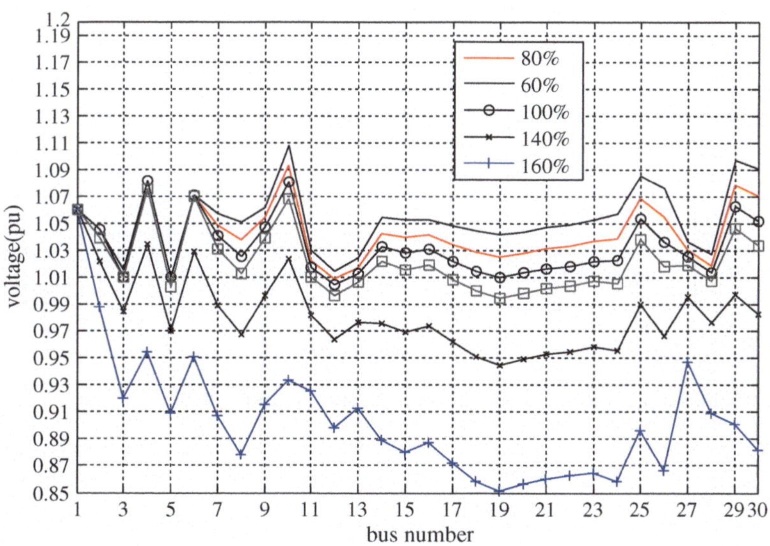

Fig. 4.7 Profile of bus voltage for IEEE. 30-Bus network

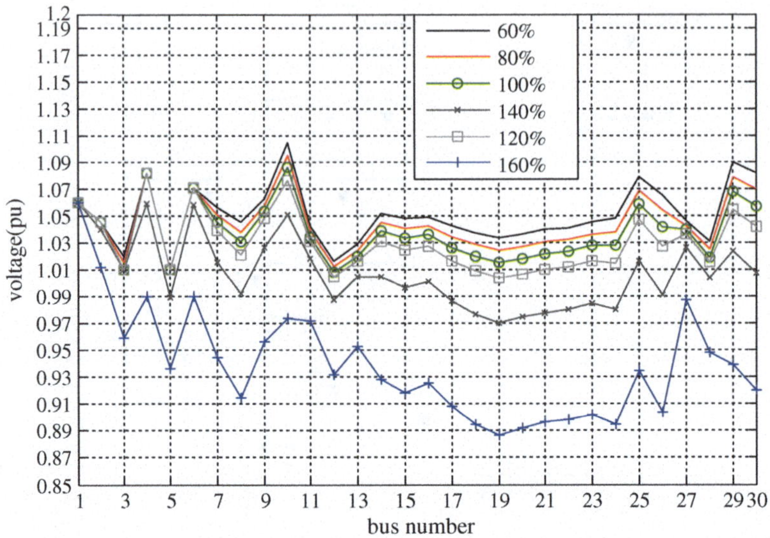

Fig. 4.8 Profile of bus voltage for IEEE. 30-Bus network with DG

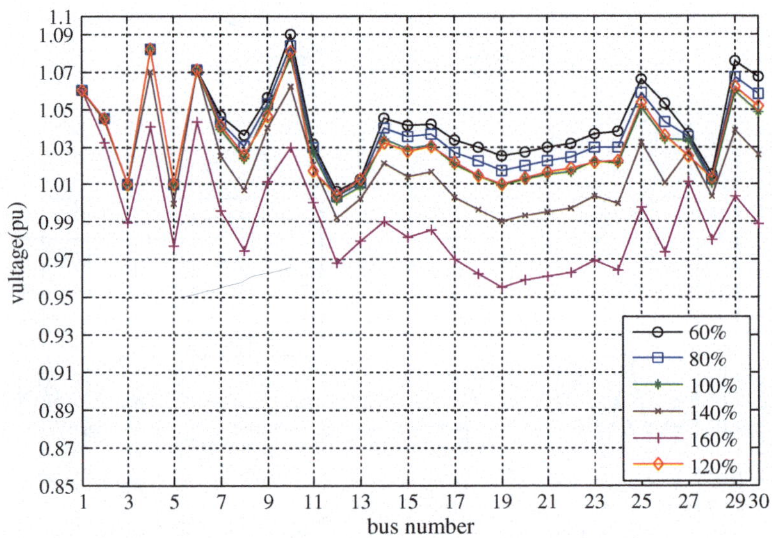

Fig. 4.9 Profile of bus voltage for 30-Bus network with DG and UPFC

Table 4.2 Losses and voltage magnitude for 33-Bus network at different loading

%Loading	Loss in MW			w/o DG and UPFC		DG		DG and UPFC	
	w/o DG	DG	DG and UPFC	V_{min}	V_{max}	V_{min}	V_{max}	V_{min}	V_{max}
60	61.08	59.09	50.50	1.01	1.103	1.01	1.10	1.01	1.08
80	43.03	39.53	31.00	1.01	1.099	1.01	1.09	1.01	1.08
100	22.71	14.20	12.10	1.00	1.08	1.00	1.08	1.00	1.08
120	33.44	26.80	21.67	0.99	1.07	1.01	1.07	1.00	1.08
140	49.23	41.10	27.16	0.94	1.06	0.96	1.06	0.99	1.07
160	82.72	66.60	40.68	0.85	1.06	0.88	1.06	0.95	1.06

4.4 Conclusion

This paper presented a technique to address the issue of high transmission and distribution loss, increasing power demand, moderate power quality of supplied power, and low life of electrical equipment. In this paper, the optimization of solar-based DG and UPFC has been carried out. DG minimizes the gap between demand and supply and reduces the T and D losses. The optimal UPFC is applied to enhance the power quality and to counter the effect of load variation on profile bus voltage magnitude. Results show that the proposed technique with the proper objective function enhanced the voltage stability and reduced the system losses at all possible levels of the applied load.

References

1. Bhadoria VS, Jaiswal SP, Shrivastava V, Pal NS. Optimal sitting and sizing of capacitor using iterative search method for enhancement of reliability of distribution system. In: Advances in energy and power systems, lecture notes in electrical engineering. https://doi.org/10.1007/978-981-13-0662-4_11
2. Ramavat SR, Jaiswal SP, Goel N, Shrivastava V Optimal location and sizing of DG in distribution system and its cost-benefit analysis. In: Advances in intelligent systems and computing. Springer. https://doi.org/10.1007/978-981-13-1819-1_11
3. Hung DQ, Mithulananthan N, Bansal RC (2010) Analytical expressions for DG allocation in primary distribution networks. IEEE Trans Energy Convers 25:814–820
4. Bhadoria VS, Pal NS, Shrivastava V (2013) A review on distributed generation definitions and DG impacts on distribution system. In: Proceedings of international conference on advanced computing and communication technologies (ICACCT™-2013), vol 7. pp 1–5
5. Soo HL, Jung WP (2013) Optimal placement and sizing of multiple DGs in a practical distribution system by considering power loss. Ind Appl IEEE Trans 49:2262–2270
6. Hung DQ, Mithulananthan N, Bansal RC (2013) Analytical strategies for renewable distributed generation integration considering energy loss minimization. Appl Energy 105:75–85
7. Nasri A, Hamedani Golshan M, Saghaian M, Nejad S (2014) Optimal planning of dispatchable and non-dispatchable distributed generation units for minimizing distribution system's energy loss using particle swarm optimization. Int Trans Electr Energy Syst 24:504–19
8. Ayodele T, Ogunjuyigbe A, Akinola O (2015) Optimal location, sizing, and appropriate technology selection of distributed generators for minimizing power loss using genetic algorithm. J Renew Energy
9. Bhadoria VS, Shrivastava V, Pal NS, Jaiswal SP (2017) Reliability improvement of distribution system by optimal sitting and sizing of disperse generation. Int J Reliab, Qual Saf Eng. https://doi.org/10.1142/S021853931740006X
10. Ramavt SR, Jaiswal SP, Goel N, Shrivastava V. Battery energy storage technology integrated for power system reliability. In: Improvement advances in energy and power systems, lecture notes in electrical engineering. https://doi.org/10.1007/978-981-13-0662-4_12
11. Watson JD, Watson NR, Bhaba D (2016) Effectiveness of power electronic voltage regulators in the distribution network. IET Gener Transm Distrib 10:3816–3823
12. Agalgaonkar YP, Pal BC, Jabr RA (2014) Distribution voltage control considering the impact of PV generation on tap changers and autonomous regulators. IEEE Trans Power Syst 29:182–192
13. Farag HE, Abdelaziz M, El-Saadany EF (2012) Incorporating voltage regulator and load models in unbalanced power flow studies of active distribution systems. In: IEEE Power and Energy Society General Meeting, San Diego, CA, pp 1–7

14. Jaiswal SP, Shrivastava V, Palwalia DK Impact of semiconductor devices on voltage stability of distribution system. Mater Today Proc. 3(3):581–589. https://doi.org/10.1016/j.matpr.2019.03.101
15. Velamuri S, Sreejith S (2017) Power flow analysis incorporating renewable energy sources and FACTS devices. Int J Renew Energy Res 7(1):452–458
16. Jaiswal SP, Shrivastava V (2014) Allocation of UPFC in distribution system to minimize the losses. In: IEEE international conference on innovative applications of computational intelligence on power, energy and controls with their impact on humanity. https://doi.org/10.1109/cipech.2014.7018212
17. Garciá-Cerrada, Garcia-González P, (2000) Control system for a UPFC in a transmission line. EPE J 10(3–4):54–59. https://doi.org/10.1080/09398368.2000.11463470
18. Kunwar, Bansal R (2013) Stability analysis of power system with doubly fed induction generator-based wind farm connected through series flexible AC transmission system device compensated line. Int J Ambient Energy 35(3):118–131. https://doi.org/10.1080/01430750.2013.789982
19. Jaiswal SP, Shrivastava V A PSO based Search for optimal tuning and fixing of UPFC to improve usefulness of distribution system. J Intell Fuzzy Syst IOS Press. https://doi.org/10.3233/jifs-169783

Chapter 5
Study of FM/FM(FM)/1/L Queue with Server Startup Under Threshold *N*-Policy Using Parametric Non-linear Programming Method

Anjali Ahuja and Anamika Jain

5.1 Introduction

Among the various concepts of the queueing theory, the concept of controlling queues has drawn notable attention these days. Numerous authors have devoted significant efforts to construct models of controllable queues which are extensively used in computer systems, production systems, communication systems, etc. The most popular control policy of the system of the queue, i.e., N-policy, was initially discussed by Yadin and Naor [1]. Ke et al. [2] analyzed the performance measures of the single server bulk queue where the server undergoes at least j vacations until he finds N jobs in the queue. Choudhary et al. [3] considered the queue wherein the service provider is unreliable and leaves for working vacation every time the queue becomes empty. The server waits for the count of jobs to attain level N in the queue and then he takes some time to start serving jobs again. Yang and Wu [4] examined the single unreliable server queueing system where the server takes vacation multiple times until he finds at least N jobs in the queue and solved the problem of cost optimization numerically. A finite state-dependent queue along with service halts and N-policy is investigated by Agrawal et al. [5]. Jayachitra [6] explored the unreliable server model with discouraged arrivals wherein the service provider operates only when there are at least N jobs present in the queue.

In actuality, we cannot know each parameter of waiting line accurately. The application of fuzzy logic is required in waiting line models for complex frameworks which is by all accounts significant due to its usefulness in routine life as well as in PC systems. Simultaneously, in viable situations, the framework planners are needing exact data for decision making. In the earlier writings, the times of inter-arrival and service

A. Ahuja · A. Jain (✉)
Manipal University Jaipur, Jaipur, Rajasthan, India
e-mail: anamikajain_02@rediffmail.com

A. Ahuja
e-mail: anjaliahuja070423@gmail.com

P. K. Kapur et al. (eds.), *Decision Analytics Applications in Industry*, Asset Analytics,
https://doi.org/10.1007/978-981-15-3643-4_5

are generally governed by distributions of probability. But in actual life, we tend to encounter the situations where the most suitable way of representing the patterns of arrival and service is by expressions given linguistically such as "frequent", "quick", "approximate", etc. The accuracy and applicability of fuzzified queues are substantially more than the queues taken as crisp. Zadeh [7] was the first one to establish the notion of fuzzy sets. Later, this notion turned out to be the foundation of a modeling approach exercised to handle uncertainty and ambiguity (see, Kaufmann [8], Zadeh [9], Zimmermann [10]). Many authors then studied the fuzzy sets in relation to queueing theory (see, Buckley [11] and Kao et al. [12]). Chen [13–15] applied the parametric non-linear programming technique for investigating the single server fuzzified queues with finite capacity, bulk service, and bulk arrival, respectively. Ke and Lin [16], Kalayanaraman [17], and Kumar [18] examined a queueing system subject to breakdown where the rate of arrival, service, breakdown, and repair followed a fuzzified exponential distribution. In Ke et al. [19], the α cut technique is exercised for constructing the function of membership of performance measure for the fuzzified queueing system where arrivals are in the form of batches. Ke et al. [20] and Khodadadi and Jolai [21] examined the retrial fuzzy queues. The mathematical programming approach for the queue with imperfect coverage and reboot is presented by Ke et al. [22]. The multi-server queueing system with fuzzified rates of arrival and service is analyzed by Aydin and Apaydin [23]. A fuzzy queue operating under F policy is studied by Yang and Chang [24]. Here, the principle of Zadeh's extension is used to convert the crisp queue into the fuzzy queue. Mary et al. [25] dealt with FM/M(a, b)/1 queueing system wherein the service provider is unreliable and goes for a vacation every time the system is empty. This problem is solved by utilizing a robust ranking technique. Kumar [26], Wang et al. [27], and Ritha and Menon [28] investigated the infinite queue under N-policy by making use of the mathematical programming approach. Recently, by exercising the α cut approach, Varadharajan and Susmitha [29] constructed the membership function from the system characteristic for the priority queue.

It is discovered from the related literature that not many scientists have given an effort toward the study of N-policy queue with customer discouragement and infinite capacity. Specifically, no examination work has shown up in past research work on queue in which both crisp and fuzzy parameters are taken into account to explore the finite buffer queue using N-policy and balking to the best of our knowledge. Thus, the absence of research on this theme propelled us to investigate such a queueing system. The remaining article is structured in the manner as explained ahead. Section 5.2 describes the crisp model and deduces the system characteristic "mean count of units in the system" at steady state. Section 5.3 defines the fuzzified exponential distribution. The model under the fuzzy environment, in brief, is explained in Section 5.4. In Sect. 5.5, α-cut method and principle of Zadeh's extension are employed for the transformation of the finite N-policy fuzzy queue into a finite N-policy crisp queue, and parametric non-linear programming technique is applied for developing the function of membership for the mean count of units in the system. Section 5.6 sketches out the step-by-step solution algorithm for the development of the membership function. A real-life problem is discussed, and analysis of the obtained results is done to demonstrate the proposed method in Sect. 5.7. Lastly, conclusions are sketched in Sect. 5.8.

5.2 The Description of the Crisp Model

Aimed at the analytical examination, we take into account the following assumptions in our model. The inter-arrival times of incoming jobs are distributed exponentially with parameter λ. The arrivals may balk with probability $1 - \beta$. The arriving units join a solitary waiting line and then get served according to their arrival order. The service time is distributed exponentially using parameter μ. The buffer size is finite (say L) so if a unit appears when there are already L units in the system, it is deprived of gaining access to the system. The service provider turns off and leaves to perform some other task when all the jobs are served. Once the count of jobs attains a threshold level, N, the service provider is quickly turned on yet is briefly inaccessible to the waiting units. It requires some exponential time for the startup using a parameter φ to finish the ongoing task which he was performing in his idle time. Immediately after the end of the startup period, the service provider starts serving the waiting units immediately and continues to serve them until the system becomes uninhabited again.

For the purpose of steady-state examination of the model under consideration, the below-mentioned notations are used:

$P_{I,k} \equiv$ The probability of k units in the system while the server is switched off and performing some other task where $k = 0, 1, 2,\ldots, L$.
$P_{B,k} \equiv$ The probability of k units in the system while the server is switched on and occupied in providing service where $k = 1, 2, 3,\ldots, L$.
$\Pi_k \equiv$ The probability of k units in the system where $k = 0, 1, 2,\ldots, L$.

The differential-difference equations of the above-described model are established as follows:

For $k = 0$

$$\frac{dP_{I,0}(t)}{\mathrm{dt}} = -\beta\lambda P_{I,0}(t) + \mu P_{B,1}(t) \tag{5.1}$$

For $k = 1$

$$\frac{dP_{I,1}(t)}{dt} = -P_{I,1}(t) + P_{I,0}(t) \tag{5.2}$$

$$\frac{dP_{B,1}(t)}{dt} = -(\beta\lambda + \mu)P_{B,1}(t) + \mu P_{B,2}(t) \tag{5.3}$$

For $1 \leq k \leq N - 1$

$$\frac{dP_{I,k}(t)}{dt} = -P_{I,k}(t) + P_{I,K-1}(t) \tag{5.4}$$

$$\frac{dP_{B,k}(t)}{dt} = -(\beta\lambda + \mu)P_{B,k}(t) + \beta\lambda P_{B,k-1}(t) + \mu P_{B,k+1}(t) \tag{5.5}$$

For $N \leq k \leq L - 1$

$$\frac{dP_{I,k}(t)}{dt} = -(\beta\lambda + \varphi)P_{I,k}(t) + \beta\lambda P_{I,k-1}(t) \tag{5.6}$$

$$\frac{dP_{B,k}(t)}{dt} = -(\beta\lambda + \mu)P_{B,k}(t) + \beta\lambda P_{B,k-1}(t) + \mu P_{B,k+1}(t) + \varphi P_{I,k}(t) \tag{5.7}$$

For $k = L$

$$\frac{dP_{I,L}(t)}{dt} = -\varphi P_{I,L}(t) + \beta\lambda P_{I,L-1}(t) \tag{5.8}$$

$$\frac{dP_{B,L}(t)}{dt} = -\mu P_{B,L}(t) + \beta\lambda P_{B,L-1}(t) + \varphi P_{I,L}(t) \tag{5.9}$$

In steady state $\underset{t\to\infty}{Lim}\ P_{I,k}(t) = P_{I,k}$ and $\underset{t\to\infty}{Lim}\ P_{B,k}(t) = P_{B,k}$. Therefore, the steady-state equations corresponding to Eqs. (5.1)–(5.9) are given below:

For $k = 0$

$$\beta\lambda P_{I,0} = \mu P_{B,1} \tag{5.10}$$

For $k = 1$

$$P_{I,1} = P_{I,0} \tag{5.11}$$

$$(\beta\lambda + \mu)P_{B,1} = \mu P_{B,2} \tag{5.12}$$

For $1 \leq k \leq N - 1$

$$P_{I,k} = P_{I,K-1} \tag{5.13}$$

$$(\beta\lambda + \mu)P_{B,k} = \beta\lambda P_{B,k-1} + \mu P_{B,k+1} \tag{5.14}$$

For $N \leq k \leq L - 1$

$$(\beta\lambda + \varphi)P_{I,k} = \beta\lambda P_{I,k-1} \tag{5.15}$$

$$(\beta\lambda + \mu)P_{B,k} = \beta\lambda P_{B,k-1} + \mu P_{B,k+1} + \varphi P_{I,k} \tag{5.16}$$

For $k = L$

$$\varphi P_{I,L} = \beta\lambda P_{I,L-1} \tag{5.17}$$

$$\mu P_{B,L} = \beta\lambda P_{B,L-1} + \varphi P_{I,L} \tag{5.18}$$

The boundary condition is given by

$$P_{I,0} + \sum_{k=1}^{L} P_{I,k} + \sum_{k=1}^{L} P_{B,k} = 1 \tag{5.19}$$

We solve the equations recursively to obtain the system's solutions at steady state.

$$P_{I,k} = P_{I,0}, \ k = 1, 2, \ldots, N-1 \tag{5.20}$$

$$P_{I,k} = P_{I,0} \times \omega^{k-N+1}, \ k = N, N+1, \ldots, L-1 \tag{5.21}$$

$$P_{I,L} = P_{I,0} \times \frac{\omega^{L-N+1}}{1-\omega} \tag{5.22}$$

where

$$\omega = \frac{\beta\lambda}{(\beta\lambda + \varphi)} \tag{5.23}$$

Also,

$$P_{B,k} = P_{I,0} \times \left\{ \frac{\kappa\left(1-\kappa^{k}\right)}{1-\kappa} \right\} k = 1, 2, \ldots, N-1 \tag{5.24}$$

$$P_{B,k} = P_{I,0} \times \left\{ \frac{\kappa^{k-N+2}(1-\omega)}{(1-\kappa)(\kappa-\omega)} - \frac{\kappa\omega^{k-N+1}}{(\kappa-\omega)} - \frac{\kappa^{k+1}}{(1-\kappa)} \right\} k = N, N+1, \ldots, L \tag{5.25}$$

where

$$\kappa = \frac{\beta\lambda}{\mu} \tag{5.26}$$

Solving for $P_{I,0}$ by putting the values of $P_{I,k}$ and $P_{B,k}$ in the boundary condition, we obtain

$$P_{I,0} = 1 \Big/ \left\{ \frac{\kappa^2\left(1-\kappa^{L-N+1}\right)(1-\omega)}{(1-\kappa)^2(\kappa-\omega)} - \frac{\omega^2\left(1-\kappa\omega^{L-N}\right)}{(\kappa-\omega)(1-\omega)} - \frac{\kappa^2\left(1-\kappa^L\right)}{(1-\omega)^2} + \frac{(N-\kappa)}{(1-\kappa)} \right\} \tag{5.27}$$

Further, we know

$$\Pi_k = P_{I,k} + P_{B,k} \tag{5.28}$$

where

$$P_{B,0} = 0 \tag{5.29}$$

Now putting the values of $P_{I,k}$ and $P_{B,k}$ in Eq. (5.28) from Eqs. (5.20)–(5.22), (5.24)–(5.25), and (5.27) we get

$$\Pi_k = \Pi_0 \times \left\{ \frac{1-\kappa^{k+1}}{1-\kappa} \right\} \quad k = 0, 1, \ldots, N-1 \tag{5.30}$$

$$\Pi_k = \Pi_0 \times \left\{ \frac{\kappa^{k-N+2}(1-\omega)}{(1-\kappa)(\kappa-\omega)} - \frac{\kappa^{L+1}}{(1-\kappa)} - \frac{(1-\kappa)\omega^{L-N+2}}{(1-\omega)(\kappa-\omega)} \right\} \quad k = N, N+1, \ldots, L-1 \tag{5.31}$$

$$\Pi_L = \Pi_0 \times \left\{ \frac{\kappa^{k-N+2}(1-\omega)}{(1-\kappa)(\kappa-\omega)} - \frac{\kappa^{k+1}}{(1-\kappa)} - \frac{\omega^{k-N+2}}{(\kappa-\omega)} \right\} \tag{5.32}$$

where

$$\Pi_0 = 1 \Big/ \left\{ \frac{\kappa^2\left(1-\kappa^{L-N+1}\right)(1-\omega)}{(1-\kappa)^2(\kappa-\omega)} - \frac{\kappa^2\left(1-\kappa^L\right)}{(1-\kappa)^2} + \frac{(N-\kappa)}{(1-\kappa)} - \frac{\omega^2\left(1-\kappa\omega^{L-N}\right)}{(1-\omega)(\kappa-\omega)} \right\} \tag{5.33}$$

Moreover, the expected customers' count in the system can be acquired as

$$E(L)_S = \sum_{k=0}^{L} k\Pi_i \tag{5.34}$$

After putting the values of Π_i from Eqs. (5.30)–(5.32) in Eq. (5.34) and then doing some manipulation, we deduce

$$E(L)_S = \Pi_0 \times \left\{ \begin{array}{l} \frac{(N-1)N}{2(1-\kappa)} + \frac{L\omega^{L-N}\left(\kappa\omega^2-\omega^2\right)}{(1-\omega)(\kappa-\omega)} - \frac{\kappa\left(\kappa-\kappa^{L+1}-L\kappa^{L+1}+L\kappa^{L+2}\right)}{(1-\kappa)^3} \\ + \frac{\kappa^{2-N}(1-\omega)\left[\kappa^{L+1}(L\kappa-L-1)-\kappa^N(N\kappa-N-\kappa)\right]}{(1-\kappa)^3(\kappa-\omega)} \\ - \frac{\omega^{2-N}\left[\omega^L(L\omega-L-\omega)-\omega^N(N\omega-N-\omega)\right]}{(1-\omega)^2(\kappa-\omega)} \end{array} \right\} \tag{5.35}$$

where ω, κ, and Π_0 are given in Eqs. (5.23), (5.26), and (5.33), respectively.

5.3 Fuzzified Exponential Distribution

The exponential distribution is a widely used distribution due to its simplicity. It is utilized to evaluate and study the traits of the units that possess a constant failure rate. An exponential distribution estimates the time interim between the events of occasions that happen freely at a consistent rate, as per Poisson distribution with parameter θ. Its probability density function is stated by

$$\phi(t) = \theta e^{-\theta t}, \ t > 0 \tag{5.36}$$

And its cumulative distribution function is

$$\Theta_\tau(t) = P(\tau \le t) = 1 - e^{-\theta t} \tag{5.37}$$

Let the estimates of fuzzified random variable $(\hat{\tau})$ are fuzzy number

$$\hat{t} = \{[0, \infty), \mu_{\hat{t}}\} \tag{5.38}$$

wherein $\hat{t} = \hat{s}t, \ t \in \tau, \hat{s}$ is a fuzzy number and $\mu_{\hat{s}}(x)$ is a membership function.

While the random variable (τ) possess a crisp exponential probability distribution $\{\text{Exp}(\theta)\}$, the resultant fuzzified random variable $(\hat{\tau})$ have fuzzified exponential probability distribution function $\{Exp(\hat{\theta})\}$. So, $\forall t \in [0, \infty)$, the cumulative fuzzy distribution function is

$$\Theta_{\hat{\tau}}(t) = 1 - e^{-\theta \hat{s} t} \tag{5.39}$$

5.4 *N*-Policy Queue in Fuzzy Environment

We know that according to the standard theory of sets, an element 'χ' in universal set U is either an individual from some crisp set X or not which is represented by an indicator function

$$Y_X = \begin{cases} 1 \ if \ \chi \in X \\ 0 \ if \ \chi \notin X \end{cases} \tag{5.40}$$

Zadeh [9], in contrast, defined the fuzzy set X by the membership function

$$\mu_{\hat{X}}(\chi) \in [0, 1] \tag{5.41}$$

via broadening the idea of binary membership to contain several membership degrees on the real continuous interim [0, 1].

We consider $FM/FM(FM)/1/L$ queue with threshold N-policy as well as server startup. Here, the first, second, and third FM symbol denotes fuzzified exponential times of inter-arrival, service, and startup, respectively. The incoming units are distributed exponentially with fuzzy arrival rate $\hat{\lambda}$. The arrivals may balk with probability $1 - \beta$. The arriving units get the service in FCFS order. The time of service is distributed exponentially with a fuzzified rate of service $\hat{\mu}$. The buffer size is finite (say L). The server leaves to perform some other task when the queue becomes empty and returns once the count of the unit in the system reaches "N". But it takes exponential startup time before serving units again with fuzzy startup rate $\hat{\varphi}$. So here, suppose the fuzzy sets $\hat{\lambda}$, $\hat{\mu}$, and $\hat{\varphi}$ having the membership functions $\mu_{\hat{\lambda}}(a)$, $\mu_{\hat{\mu}}(b)$, and $\mu_{\hat{\varphi}}(c)$, respectively, can be stated as follows.

$$\hat{\lambda} = \left\{\left(a, \mu_{\hat{\lambda}}(a)\right)|a \in A\right\} \tag{5.42}$$

$$\hat{\mu} = \left\{\left(b, \mu_{\hat{\mu}}(b)\right)|b \in B\right\} \tag{5.43}$$

and

$$\hat{\varphi} = \left\{\left(c, \mu_{\hat{\varphi}}(c)\right)|c \in C\right\} \tag{5.44}$$

where A, B, and C are the crisp universal sets of rates of arrival, service, and startup, respectively. It is to be noted that crisp model defined in Sect. 5.2 is based on the differential equations whereas rules are used to model fuzzy system. Also crisp system is exact, linear and leads to inaccuracy if information is not complete but fuzzy system is vague, nonlinear and is able to handle incomplete information to some extent.

The useful system characteristic in this investigation is the mean count of units in the system, i.e., $E(L)_S(a, b, c)$. Clearly, when $\hat{\lambda}$, $\hat{\mu}$, and $\hat{\varphi}$ are all fuzzy numbers, $E(L)_S\left(\hat{\lambda}, \hat{\mu}, \hat{\varphi}\right)$ is too a fuzzy number. As per the principle of Zadeh's extension (see, Zadeh [9]), the function of membership of the meaHere it can be easily seen that the n count of units in the system is obtained as shown:

$$\begin{aligned}\mu_{E(L)_S\left(\hat{\lambda},\hat{\mu},\hat{\varphi}\right)}(x) &= \mu_{E\hat{(L)}_S}(x)\\ &= \sup_{a\in A, b\in B, c\in C} \min\left\{\mu_{\hat{\lambda}}(a), \mu_{\hat{\mu}}(b), \mu_{\hat{\varphi}}(c)\right\}x = E(L)_S(a, b, c)\end{aligned} \tag{5.45}$$

From the result obtained in the Eq. (5.35), we have the mean count of units in the single server finite buffer crisp queue with threshold policy as follows:

$$E(L)_S(a,b,c) = \Pi_0 \times \left\{ \begin{array}{l} \frac{(N-1)N}{2(1-\kappa)} + \frac{L\omega^{L-N}\left(\kappa\omega^2-\omega^2\right)}{(1-\omega)(\kappa-\omega)} - \frac{\kappa\left(\kappa-\kappa^{L+1}-L\kappa^{L+1}+L\kappa^{L+2}\right)}{(1-\kappa)^3} \\ +\frac{\kappa^{2-N}(1-\omega)\left[\kappa^{L+1}(L\kappa-L-1)-\kappa^N(N\kappa-N-\kappa)\right]}{(1-\kappa)^3(\kappa-\omega)} \\ -\frac{\omega^{2-N}\left[\omega^L(L\omega-L-\omega)-\omega^N(N\omega-N-\omega)\right]}{(1-\omega)^2(\kappa-\omega)} \end{array} \right\} \quad (5.46)$$

where

$$\omega = \frac{\beta a}{(\beta a + c)}, \quad (5.47)$$

$$\kappa = \frac{\beta a}{b} \quad (5.48)$$

and Π_0 is given by the Eq. (5.33).,

Here it can be easily seen that the expression in the Eq. (5.46) is complicated. Thus, the inference of the shape of $\mu_{E(\hat{L})_S}(x)$ is very difficult. Therefore, we obtain the α cut of $E(L)_S\left(\hat{\lambda}, \hat{\mu}, \hat{\varphi}\right)$ according to the principle of Zadeh's extension by applying the parametric non-linear programming method.

5.5 The Parametric Non-linear Programming (NLP) Method

The parametric NLP method is employed for the construction of a membership function $\mu_{E(\hat{L})_S}(x)$ and to discover its shape. For this purpose, we find α-cuts of $E(\hat{L})_S$ by applying the principle of Zadeh's extension. First of all, let us state the α-cuts of fuzzified rates of arrival $\hat{\lambda}$, service $\hat{\mu}$ as well as startup $\hat{\varphi}$, as given below:

$$\lambda(\alpha) = \left\{a \in A | \mu_{\hat{\lambda}}(a) \geq \alpha\right\} \quad (5.49)$$

$$\mu(\alpha) = \left\{b \in B | \mu_{\hat{\mu}}(b) \geq \alpha\right\} \quad (5.50)$$

and

$$\varphi(\alpha) = \left\{c \in C | \mu_{\hat{\varphi}}(c) \geq \alpha\right\} \quad (5.51)$$

For the $FM/FM(FM)/1/L$ queue with server startup under threshold N-policy $\hat{\lambda}$, $\hat{\mu}$, and $\hat{\varphi}$ are fuzzy numbers. The α-cuts of $\hat{\lambda}$, $\hat{\mu}$, and $\hat{\varphi}$ defined above are crisp intervals bearing the forms as shown below:

$$\lambda(\alpha) = \left[\min_{a\in A}\{a|\mu_{\hat{\lambda}}(a) \geq \alpha\},\ \max_{a\in A}\{a|\mu_{\hat{\lambda}}(a) \geq \alpha\}\right] = \left[a_\alpha^L,\ a_\alpha^U\right] \tag{5.52}$$

$$\mu(\alpha) = \left[\min_{b\in B}\{b|\mu_{\hat{\mu}}(b) \geq \alpha\},\ \max_{b\in B}\{b|\mu_{\hat{\mu}}(b) \geq \alpha\}\right] = \left[b_\alpha^L,\ \mathrm{b}_\alpha^U\right] \tag{5.53}$$

and

$$\varphi(\alpha) = \left[\min_{c\in C}\{c|\mu_{\hat{\varphi}}(c) \geq \alpha\},\ \max_{c\in C}\{c|\mu_{\hat{\varphi}}(c) \geq \alpha\}\right] = \left[c_\alpha^L,\ \mathrm{c}_\alpha^U\right] \tag{5.54}$$

As per the Eqs. (5.52)–(5.54), the rates of arrival, service, and startup can be denoted by various interims, and these equations also signify wherein the rates of arrival, service, and startup lie at possibility levels α. Using the basic convexity property of fuzzy numbers (see, Zimmermann [10]), the bounds of $\hat{\lambda}$, $\hat{\mu}$, and $\hat{\varphi}$ may be denoted as a function of α in the following manner:

$$a_\alpha^L = \min\left(\frac{1}{\mu_{\hat{\lambda}}(\alpha)}\right) \text{and } a_\alpha^U = \max\left(\frac{1}{\mu_{\hat{\lambda}}(\alpha)}\right) \tag{5.55}$$

$$b_\alpha^L = \min\left(\frac{1}{\mu_{\hat{\mu}}(\alpha)}\right) \text{and } b_\alpha^U = \max\left(\frac{1}{\mu_{\hat{\mu}}(\alpha)}\right) \tag{5.56}$$

and

$$c_\alpha^L = \min\left(\frac{1}{\mu_{\hat{\varphi}}(\alpha)}\right) \text{and } c_\alpha^U = \max\left(\frac{1}{\mu_{\hat{\varphi}}(\alpha)}\right) \tag{5.57}$$

respectively. Thus, the membership function $\mu_{E(\hat{L})_S}(x)$ stated in Eq. (5.45) is similarly parameterized by α. As a result, the method of α-cuts may be employed for obtaining the function of membership $\mu_{E(\hat{L})_S}(x)$.

Also, $\mu_{E(\hat{L})_S}(x)$ is the minimum of $\mu_{\hat{\lambda}}(a)$, $\mu_{\hat{\mu}}(b)$, and $\mu_{\hat{\varphi}}(c)$ as per Eq. (5.45).

For the derivation of $\mu_{E(\hat{L})_S}(x)$, we require that at least one of the cases stated below satisfies such that $\mu_{E(\hat{L})_S}(x) = \alpha$.

Case 1 $\mu_{\hat{\lambda}}(a) = \alpha,\ \mu_{\hat{\mu}}(b) \geq \alpha,\ \mu_{\hat{\varphi}}(c) \geq \alpha$

Case 2 $\mu_{\hat{\lambda}}(a) \geq \alpha,\ \mu_{\hat{\mu}}(b) = \alpha,\ \mu_{\hat{\varphi}}(c) \geq \alpha$

Case 3 $\mu_{\hat{\lambda}}(a) \geq \alpha,\ \mu_{\hat{\mu}}(b) \geq \alpha,\ \mu_{\hat{\varphi}}(c) = \alpha$

$$x = \Pi_0 \times \left\{ \begin{array}{l} \frac{(N-1)N}{2(1-\kappa)} + \frac{L\omega^{L-N}\left(\kappa\omega^2-\omega^2\right)}{(1-\omega)(\kappa-\omega)} - \frac{\kappa\left(\kappa-\kappa^{L+1}-L\kappa^{L+1}+L\kappa^{L+2}\right)}{(1-\kappa)^3} \\ +\frac{\kappa^{2-N}(1-\omega)\left[\kappa^{L+1}(L\kappa-L-1)-\kappa^N(N\kappa-N-\kappa)\right]}{(1-\kappa)^3(\kappa-\omega)} \\ -\frac{\omega^{2-N}\left[\omega^L(L\omega-L-\omega)-\omega^N(N\omega-N-\omega)\right]}{(1-\omega)^2(\kappa-\omega)} \end{array} \right\} \tag{5.57}$$

where ω and κ are stated in Eqs. (5.47) and (5.48), respectively.

The parametric NLP method can be employed for this purpose. So, we can deduce the bounds of the α cut of $\mu_{\hat{E(L)_S}}(x)$ for Case 1 by means of this method as given below:

$$(E(L)_s)_\alpha^{L_1} = \min_{a\in A, b\in B, c\in C} \Pi_0 \times \left\{ \begin{array}{l} \frac{(N-1)N}{2(1-\kappa)} + \frac{L\omega^{L-N}\left(\kappa\omega^2-\omega^2\right)}{(1-\omega)(\kappa-\omega)} - \frac{\kappa\left(\kappa-\kappa^{L+1}-L\kappa^{L+1}+L\kappa^{L+2}\right)}{(1-\kappa)^3} \\ +\frac{\kappa^{2-N}(1-\omega)\left[\kappa^{L+1}(L\kappa-L-1)-\kappa^N(N\kappa-N-\kappa)\right]}{(1-\kappa)^3(\kappa-\omega)} \\ -\frac{\omega^{2-N}\left[\omega^L(L\omega-L-\omega)-\omega^N(N\omega-N-\omega)\right]}{(1-\omega)^2(\kappa-\omega)} \end{array} \right\} \tag{5.58}$$

such that $a_\alpha^L \le a \le a_\alpha^U$, $b \in \mu(\alpha)$, and $c \in \varphi(\alpha)$
and

$$(E(L)_s)_\alpha^{U_1} = \max_{a\in A, b\in B, c\in C} \Pi_0 \times \left\{ \begin{array}{l} \frac{(N-1)N}{2(1-\kappa)} + \frac{L\omega^{L-N}\left(\kappa\omega^2-\omega^2\right)}{(1-\omega)(\kappa-\omega)} - \frac{\kappa\left(\kappa-\kappa^{L+1}-L\kappa^{L+1}+L\kappa^{L+2}\right)}{(1-\kappa)^3} \\ +\frac{\kappa^{2-N}(1-\omega)\left[\kappa^{L+1}(L\kappa-L-1)-\kappa^N(N\kappa-N-\kappa)\right]}{(1-\kappa)^3(\kappa-\omega)} \\ -\frac{\omega^{2-N}\left[\omega^L(L\omega-L-\omega)-\omega^N(N\omega-N-\omega)\right]}{(1-\omega)^2(\kappa-\omega)} \end{array} \right\} \tag{5.59}$$

such that $a_\alpha^L \le a \le a_\alpha^U$, $b \in \mu(\alpha)$, and $c \in \varphi(\alpha)$.

The alpha cut bounds (lower and upper) of $\mu_{\hat{E(L)_S}}(x)$ in Case 2 can be deduced by the same method as shown:

$$(E(L)_s)_\alpha^{L_2} = \min_{a\in A, b\in B, c\in C} \Pi_0 \times \left\{ \begin{array}{l} \frac{(N-1)N}{2(1-\kappa)} + \frac{L\omega^{L-N}\left(\kappa\omega^2-\omega^2\right)}{(1-\omega)(\kappa-\omega)} - \frac{\kappa\left(\kappa-\kappa^{L+1}-L\kappa^{L+1}+L\kappa^{L+2}\right)}{(1-\kappa)^3} \\ +\frac{\kappa^{2-N}(1-\omega)\left[\kappa^{L+1}(L\kappa-L-1)-\kappa^N(N\kappa-N-\kappa)\right]}{(1-\kappa)^3(\kappa-\omega)} \\ -\frac{\omega^{2-N}\left[\omega^L(L\omega-L-\omega)-\omega^N(N\omega-N-\omega)\right]}{(1-\omega)^2(\kappa-\omega)} \end{array} \right\} \tag{5.60}$$

such that a $\in \lambda(\alpha)$, $b_\alpha^L \le b \le b_\alpha^U$, and $c \in \varphi(\alpha)$
and

$$(E(L)_s)_\alpha^{U_2} = \max_{a\in A, b\in B, c\in C} \Pi_0 \times \left\{ \begin{array}{l} \frac{(N-1)N}{2(1-\kappa)} + \frac{L\omega^{L-N}(\kappa\omega^2-\omega^2)}{(1-\omega)(\kappa-\omega)} - \frac{\kappa(\kappa-\kappa^{L+1}-L\kappa^{L+1}+L\kappa^{L+2})}{(1-\kappa)^3} \\ +\frac{\kappa^{2-N}(1-\omega)[\kappa^{L+1}(L\kappa-L-1)-\kappa^N(N\kappa-N-\kappa)]}{(1-\kappa)^3(\kappa-\omega)} \\ -\frac{\omega^{2-N}[\omega^L(L\omega-L-\omega)-\omega^N(N\omega-N-\omega)]}{(1-\omega)^2(\kappa-\omega)} \end{array} \right\} \quad (5.61)$$

such that a $\in \lambda(\alpha)$, $b_\alpha^L \le b \le b_\alpha^U$, and $c \in \varphi(\alpha)$.

We employ the same method to deduce the alpha cut bounds of $\mu_{\hat{E(L)_S}}(x)$ for Case 3 as follows:

$$(E(L)_s)_\alpha^{L_3} = \min_{a\in A, b\in B, c\in C} \Pi_0 \times \left\{ \begin{array}{l} \frac{(N-1)N}{2(1-\kappa)} + \frac{L\omega^{L-N}(\kappa\omega^2-\omega^2)}{(1-\omega)(\kappa-\omega)} - \frac{\kappa(\kappa-\kappa^{L+1}-L\kappa^{L+1}+L\kappa^{L+2})}{(1-\kappa)^3} \\ +\frac{\kappa^{2-N}(1-\omega)[\kappa^{L+1}(L\kappa-L-1)-\kappa^N(N\kappa-N-\kappa)]}{(1-\kappa)^3(\kappa-\omega)} \\ -\frac{\omega^{2-N}[\omega^L(L\omega-L-\omega)-\omega^N(N\omega-N-\omega)]}{(1-\omega)^2(\kappa-\omega)} \end{array} \right\} \quad (5.62)$$

such that a $\in \lambda(\alpha)$, $b \in \mu(\alpha)$, and $c_\alpha^L \le c \le c_\alpha^U$

and

$$(E(L)_s)_\alpha^{U_3} = \max_{a\in A, b\in B, c\in C} \Pi_0 \times \left\{ \begin{array}{l} \frac{(N-1)N}{2(1-\kappa)} + \frac{L\omega^{L-N}(\kappa\omega^2-\omega^2)}{(1-\omega)(\kappa-\omega)} - \frac{\kappa(\kappa-\kappa^{L+1}-L\kappa^{L+1}+L\kappa^{L+2})}{(1-\kappa)^3} \\ +\frac{\kappa^{2-N}(1-\omega)[\kappa^{L+1}(L\kappa-L-1)-\kappa^N(N\kappa-N-\kappa)]}{(1-\kappa)^3(\kappa-\omega)} \\ -\frac{\omega^{2-N}[\omega^L(L\omega-L-\omega)-\omega^N(N\omega-N-\omega)]}{(1-\omega)^2(\kappa-\omega)} \end{array} \right\} \quad (5.63)$$

such that a $\in \lambda(\alpha)$, $b \in \mu(\alpha)$, and $c_\alpha^L \le c \le c_\alpha^U$.

As per Eqs. (5.52)–(5.54) defining $\lambda(\alpha)$, $\mu(\alpha)$, and $\varphi(\alpha)$, one can substitute $a \in [a_\alpha^L, a_\alpha^U]$, $b \in [b_\alpha^L, b_\alpha^U]$, and $c \in [c_\alpha^L, c_\alpha^U]$ in the place of a $\in \lambda(\alpha)$, $b \in \mu(\alpha)$, and $c \in \varphi(\alpha)$, respectively. Going along with the proposition of Kaufmann [8] and Zimmermann [10], we represent the alpha-cuts of a, b, and c in nested manner. Taking into consideration the two levels of possibility α_1 and α_2 such that $0 < \alpha_2 < \alpha_1 \le 1$, we obtain $[a_{\alpha_2}^L, a_{\alpha_2}^U] \supseteq [a_{\alpha_1}^L, a_{\alpha_1}^U]$, $[b_{\alpha_2}^L, b_{\alpha_2}^U] \supseteq [b_{\alpha_1}^L, b_{\alpha_1}^U]$, and $[c_{\alpha_2}^L, c_{\alpha_2}^U] \supseteq [c_{\alpha_1}^L, c_{\alpha_1}^U]$. The determination of lower bound $(E(L)_s)_\alpha^L$ and upper bound $(E(L)_s)_\alpha^U$ is essential for the development of membership function $\mu_{\hat{E(L)_S}}(x)$. So, we have

$$(E(L)_s)_\alpha^L = \min_{a\in A, b\in B, c\in C} \Pi_0 \times \left\{ \begin{array}{l} \frac{(N-1)N}{2(1-\kappa)} + \frac{L\omega^{L-N}(\kappa\omega^2-\omega^2)}{(1-\omega)(\kappa-\omega)} - \frac{\kappa(\kappa-\kappa^{L+1}-L\kappa^{L+1}+L\kappa^{L+2})}{(1-\kappa)^3} \\ + \frac{\kappa^{2-N}(1-\omega)\left[\kappa^{L+1}(L\kappa-L-1)-\kappa^N(N\kappa-N-\kappa)\right]}{(1-\kappa)^3(\kappa-\omega)} \\ - \frac{\omega^{2-N}[\omega^L(L\omega-L-\omega)-\omega^N(N\omega-N-\omega)]}{(1-\omega)^2(\kappa-\omega)} \end{array} \right\} \quad (5.64)$$

such that $a_\alpha^L \le a \le a_\alpha^U$, $b_\alpha^L \le b \le b_\alpha^U$, and $c_\alpha^L \le c \le c_\alpha^U$ and

$$(E(L)_s)_\alpha^U = \max_{a\in A, b\in B, c\in C} \Pi_0 \times \left\{ \begin{array}{l} \frac{(N-1)N}{2(1-\kappa)} + \frac{L\omega^{L-N}(\kappa\omega^2-\omega^2)}{(1-\omega)(\kappa-\omega)} - \frac{\kappa(\kappa-\kappa^{L+1}-L\kappa^{L+1}+L\kappa^{L+2})}{(1-\kappa)^3} \\ + \frac{\kappa^{2-N}(1-\omega)\left[\kappa^{L+1}(L\kappa-L-1)-\kappa^N(N\kappa-N-\kappa)\right]}{(1-\kappa)^3(\kappa-\omega)} \\ - \frac{\omega^{2-N}[\omega^L(L\omega-L-\omega)-\omega^N(N\omega-N-\omega)]}{(1-\omega)^2(\kappa-\omega)} \end{array} \right\} \quad (5.65)$$

such that $a_\alpha^L \le a \le a_\alpha^U$, $b_\alpha^L \le b \le b_\alpha^U$, and $c_\alpha^L \le c \le c_\alpha^U$.

As per principle of Zadeh's extension, $E(\hat{L})_s$ stated in Eq. (5.46) is a fuzzy number that holds the property of convexity. The α-cuts of $E(\hat{L})_s$ is represented by a crisp interim $\left[(E(L)_s)^L, (E(L)_s)^U\right]$. In view of the two levels of possibility α_1 and α_2 such that $0 < \alpha_2 < \alpha_1 \le 1$, we get $(E(L)_s)_{\alpha_1}^U \le (E(L)_s)_{\alpha_2}^U$ and $(E(L)_s)_{\alpha_1}^L \ge (E(L)_s)_{\alpha_2}^L$. That is with an increase in α, $(E(L)_s)^U$ decreases and $(E(L)_s)^L$ increases. Thus $E(\hat{L})_s$ possess the property of convexity. Hence membership function $\mu_{E(\hat{L})_S}(x)$ can be deduced.

For developing the membership function $\mu_{E(\hat{L})_S}(x)$, we take a decreasing function and an increasing function given by $(E(L)_s)^U : \alpha \to (E(L)_s)^U$ and $(E(L)_s)^L : \alpha \to (E(L)_s)^L$, respectively. Now, with reference to α, if both the bounds (upper $(E(L)_s)^U$ and lower $(E(L)_s)^L$) are invertible then in that case the function of membership is established as shown:

$$\mu_{E(\hat{L})_S}(x) = \begin{cases} L(x) & (E(L)_s)_{\alpha=0}^L \le x \le (E(L)_s)_{\alpha=1}^L \\ 1 & (E(L)_s)_{\alpha=1}^L \le x \le (E(L)_s)_{\alpha=1}^U \\ R(x) & (E(L)_s)_{\alpha=1}^U \le x \le (E(L)_s)_{\alpha=0}^U \end{cases} \quad (5.66)$$

wherein

$$L(x) = 1/(E(L)_s)_\alpha^L \quad (5.67)$$

is a left shape function and

$$R(x) = 1/(E(L)_s)_\alpha^U \quad (5.68)$$

is a right shape function. However, in some cases, the analytical solution of $L(x)$ and $R(x)$ cannot be deduced. Then the numerical approximation of the shape of $\mu_{E(\hat{L})_S}(x)$ can be done by taking different possibility levels α.

5.6 Methodology to Compute the Membership Function Using Triangular as Well as Trapezoidal Fuzzy Numbers

5.6.1 Triangular Fuzzy Number

A triangular fuzzy number $\hat{X} = [x_1, x_2, x_3]$ is a fuzzy set whose membership function $\forall\, x \in R$ can be expressed as

$$\mu_{\hat{X}}(x) = \begin{cases} \frac{x-x_1}{x_2-x_1}, & \text{if } \mathrm{x}_1 \le x < x_2 \\ 1, & \text{if } x = x_2 \\ \frac{x_3-x}{x_3-x_2}, & \text{if } \mathrm{x}_2 < x \le x_3 \\ 0, & \text{otherwise} \end{cases} \tag{5.69}$$

Suppose that $\hat{\lambda}$, $\hat{\mu}$, and $\hat{\varphi}$ are triangular fuzzy numbers which may be specified as $\hat{\lambda} = [a_1, a_2, a_3]$, $\hat{\mu} = [b_1, b_2, b_3]$, and $\hat{\varphi} = [c_1, c_2, c_3]$, where $a_1 < a_2 < a_3$, $b_1 < b_2 < b_3$, and $c_1 < c_2 < c_3$.

The function of membership for triangular fuzzy numbers, $\hat{\lambda}$, $\hat{\mu}$, and $\hat{\varphi}$ are

$$\mu_{\hat{\lambda}}(x) = \begin{cases} \frac{x-a_1}{a_2-a_1}, & \text{if } a_1 \le x < a_2 \\ 1, & \text{if } x = a_2 \\ \frac{a_3-x}{a_3-a_2}, & \text{if } a_2 < x \le a_3 \\ 0, & \text{otherwise} \end{cases} \tag{5.70}$$

$$\mu_{\hat{\mu}}(x) = \begin{cases} \frac{x-b_1}{b_2-b_1}, & \text{if } \mathrm{b}_1 \le x < b_2 \\ 1, & \text{if } x = b_2 \\ \frac{b_3-x}{b_3-b_2}, & \text{if } \mathrm{b}_2 < x \le b_3 \\ 0, & \text{otherwise} \end{cases} \tag{5.71}$$

$$\mu_{\hat{\varphi}}(x) = \begin{cases} \frac{x-c_1}{c_2-c_1}, & \text{if } \mathrm{c}_1 \le x < c_2 \\ 1, & \text{if } x = c_2 \\ \frac{c_3-x}{c_3-c_2}, & \text{if } \mathrm{c}_2 < x \le c_3 \\ 0, & \text{otherwise} \end{cases} \tag{5.72}$$

5.6.2 Trapezoidal Fuzzy Number

A trapezoidal fuzzy number $\hat{X} = [x_1, x_2, x_3, x_4]$ is a fuzzy set whose membership function $\forall\, x \in R$ can be expressed as

$$\mu_{\hat{X}}(x) = \begin{cases} \frac{x-x_1}{x_2-x_1}, & \text{if } x_1 \le x < x_2 \\ 1, & \text{if } x_2 \le x \le x_3 \\ \frac{x_4-x}{x_4-x_3}, & \text{if } x_3 < x \le x_4 \\ 0, & \text{otherwise} \end{cases} \tag{5.73}$$

Suppose that $\hat{\lambda}$, $\hat{\mu}$, and $\hat{\varphi}$ are trapezoidal fuzzy numbers and can be represented as $\hat{\lambda} = [a_1, a_2, a_3, a_4]$, $\hat{\mu} = [b_1, b_2, b_3, b_4]$, and $\hat{\varphi} = [c_1, c_2, c_3, c_4]$, where $a_1 < a_2 < a_3 < a_4$, $b_1 < b_2 < b_3 < b_4$, and $c_1 < c_2 < c_3 < c_4$.

The function of membership for trapezoidal fuzzy numbers, $\hat{\lambda}$, $\hat{\mu}$, and $\hat{\varphi}$ are

$$\mu_{\hat{\lambda}}(x) = \begin{cases} \frac{x-a_1}{a_2-a_1}, & \text{if } a_1 \le x < a_2 \\ 1, & \text{if } a_2 \le x \le a_3 \\ \frac{a_4-x}{a_4-a_3}, & \text{if } a_3 < x \le a_4 \\ 0, & \text{otherwise} \end{cases} \tag{5.74}$$

$$\mu_{\hat{\mu}}(x) = \begin{cases} \frac{x-b_1}{b_2-b_1}, & \text{if } b_1 \le x < b_2 \\ 1, & \text{if } b_2 \le x \le b_3 \\ \frac{b_4-x}{b_4-b_3}, & \text{if } b_3 < x \le b_4 \\ 0, & \text{otherwise} \end{cases} \tag{5.75}$$

$$\mu_{\hat{\varphi}}(x) = \begin{cases} \frac{x-c_1}{c_2-c_1}, & \text{if } c_1 \le x < c_2 \\ 1, & \text{if } c_2 \le x \le c_3 \\ \frac{c_4-x}{c_4-c_3}, & \text{if } c_3 < x \le c_4 \\ 0, & \text{otherwise} \end{cases} \tag{5.76}$$

The methodology for constructing the function of membership $\mu_{E(\hat{L})_S}(x)$ consists of the following steps:

Step 1. Select α such that $0 \le \alpha \le 1$ and has a step size of 0.1.
Step 2. Find the bounds (upper and lower) when inputs $\hat{\lambda}$, $\hat{\mu}$, and $\hat{\varphi}$ are

a. Triangular fuzzy number

$$a_\alpha^L = (a_2 - a_1)\alpha + a_1,\, a_\alpha^U = a_3 - (a_3 - a_2)\alpha,$$

$$b_\alpha^L = (b_2 - b_1)\alpha + b_1,\, b_\alpha^U = b_3 - (b_3 - b_2)\alpha,$$

$$c_\alpha^L = (c_2 - c_1)\alpha + c_1,\, c_\alpha^U = c_3 - (c_3 - c_2)\alpha$$

b. Trapezoidal fuzzy number

$$a_\alpha^L = (a_2 - a_1)\alpha + a_1, a_\alpha^U = a_4 - (a_4 - a_3)\alpha,$$

$$b_\alpha^L = (b_2 - b_1)\alpha + b_1, b_\alpha^U = b_4 - (b_4 - b_3)\alpha,$$

$$c_\alpha^L = (c_2 - c_1)\alpha + c_1, c_\alpha^U = c_4 - (c_4 - c_4)\alpha$$

Step 3. For 'a' from a_α^L to a_α^U with step size $\Omega_1 = \frac{(a_\alpha^L - a_\alpha^U)}{100}$, for 'b' from b_α^L to b_α^U with step size $\Omega_2 = \frac{(b_\alpha^L - b_\alpha^U)}{100}$, and similarly in case of 'c' from c_α^L to c_α^U have $\Omega_3 = \frac{(c_\alpha^L - c_\alpha^U)}{100}$.

Step 4. Compute α-cut of output fuzzy number $E(\hat{L})_S$, $\left[(E(L)_s)_\alpha^L, (E(L)_s)_\alpha^U\right]$ by applying Eqs. (5.64) and (5.65).

Step 5. Approximate the shape of L(x) and R(x) by means of $(E(L)_s)_\alpha^L$ and $(E(L)_s)_\alpha^U$ described in Eq. (5.66), respectively, at different possibility levels α to develop $\mu_{E(\hat{L})_S}(x)$.

5.7 Application Illustration

The proposed procedure for FM/FM(FM)/1/L queue with threshold N-policy can be practically demonstrated taking an example of Courier and Delivery Service. The parcel delivery service is able to provide cheap yet reliable service by following the simple process of keeping the parcels to be dispatched to a particular region in a queue. According to the standard operating procedure of the company, the service provider dispatches the parcels to the region after they have a minimum of N parcels in the queue. The threshold limit of N is kept, keeping a number of factors into consideration say operating cost, size of the parcel delivering wagon, expected footfall, etc. This technique enables the service provider to deliver each parcel one after another within a geographic region at a reduced cost. After the company reaches the threshold limit of N parcels, some startup time is taken to assign delivery agents. However, the increased delivery time in this process of queuing leads to the unavoidable issue of balking. The service provider cannot exceed the capacity of L parcels, given the logistic constraints.

Since the number of parcels in the system is of great importance to analyze the performance of the system, it is necessary for the manager of the courier company to examine the impact of different threshold values on the mean count of parcels in the system. For this purpose, we take three different values of threshold levels $N = 5$, $N = 10$, and $N = 15$ and for each critical value "N" we compare two cases (i) $\beta = 0.5$ $(1 - \beta = 0.5)$ (ii) $\beta = 1$ $(1 - \beta = 0$, i.e., no balking) to see the impact of balking on the mean count of customers. Also, the capacity of the system is fixed as $L = 30$. This analysis can be done by constructing a membership function for the mean count of jobs in the system with the technique discussed in Sect. 5.5.

It will be interesting to study the variations in the optimum solution for various values of a_α^L, a_α^U, b_α^L, b_α^U, c_α^L, and c_α^U since the range of interval of α is (0, 1]. At distinct values of α, the closed-form solutions of crisp intervals $\left[(\mathrm{E(L)_s})_\alpha^{\mathrm{L}}, (\mathrm{E(L)_s})_\alpha^{\mathrm{U}}\right]$ are very difficult to determine for two main reasons (i) the number of fuzzy variables are three and (b) the mean count of parcels in the system has complex expression. So we use Matlab software for numerical analysis of $\mu_{E(\hat{L})_S}$ instead of doing it analytically.

5.7.1 *Triangular Fuzzy Number*

In this case, we suppose that the rates of arrival, service as well as startup for FM/FM/(FM)/1/30 queue are triangular fuzzy numbers and are represented by $\hat{\lambda} = [20\ \ 23\,26]$, $\hat{\mu} = [22\ \ 25\ \ 28]$, $\hat{\varphi} = [3\ \ 6\ \ 9]$ per hour, respectively, for the courier service system discussed above. One can easily determine

$$\left[\mathrm{a}_\alpha^{\mathrm{L}}, \mathrm{a}_\alpha^{\mathrm{U}}\right] = \left[\min\left\{1/\mu_{\hat{\lambda}}(\alpha)\right\},\ \max\left\{1/\mu_{\hat{\lambda}}(\alpha)\right\}\right] = [20 + 3\alpha, 26 - 3\alpha]$$

$$\left[b_\alpha^L, \mathrm{b}_\alpha^U\right] = \left[\min\left\{1/\mu_{\hat{\mu}}(\alpha)\right\}, \max\left\{1/\mu_{\hat{\mu}}(\alpha)\right\}\right] = [22 + 3\alpha, 28 - 3\alpha]$$

and

$$\left[c_\alpha^L,\ \mathrm{c}_\alpha^U\right] = \left[\min\left\{1/\mu_{\hat{\varphi}}(\alpha)\right\}, \max\left\{1/\mu_{\hat{\varphi}}(\alpha)\right\}\right] = [3 + 3\alpha, 9 - 3\alpha]$$

The results for membership function $\mu_{E(\hat{L})_S}(x)$ are displayed in Figs. 5.1, 5.2, 5.3, 5.4 and 5.5 and obtained numerically.

Furthermore, the triangular shape appears in figure on plotting membership function $\mu_{E(\hat{L})_S}(x)$ when the rates of arrival, service as well as startup are taken as triangular fuzzy numbers. In Figs. 5.1, 5.2, 5.3, the results of both $\beta = 1$ and $\beta = 0.5$ against $N = 5$, $N = 10$, and $N = 15$ are shown, and clearly $\beta = 1$ has wider range of membership function than $\beta = 0.5$. The largest $E(L)_S$ of $\beta = 0.5$ is lesser than smallest $E(L)_S$ of $\beta = 1$ when α goes above 0.089, 0.061, and 0.0384 in Figs. 5.1, 5.2, 5.3, respectively.

Also when α escalates beyond 0.933 and 0.232 in Figs. 5.4, 5.5, respectively, the highest $E(L)_S$ for $N = 5$ is lesser than lowest $E(L)_S$ for $N = 10$. In the same way, when α escalates beyond 0.921 and 0.147 in Figs. 5.4, 5.5, respectively, the highest $E(L)_S$ calculated for $N = 10$ is lesser than the lowest $E(L)_S$ calculated for $N = 15$. Moreover, $N = 5$ has broader range of membership function than $N = 10$ and $N = 15$ in Figs. 5.4 and 5.5.

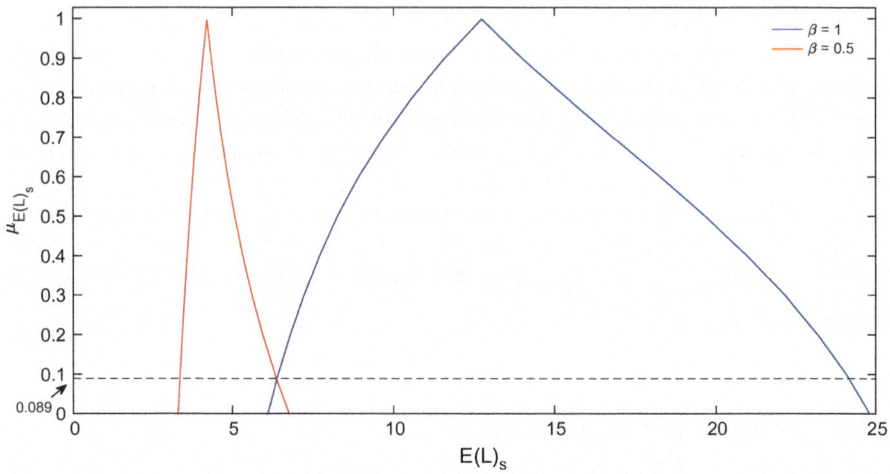

Fig. 5.1 The function of membership of the expected count of parcels in the system for FM/FM(FM)/1/30 queue when $N = 5$ if fuzzy numbers are taken as triangular

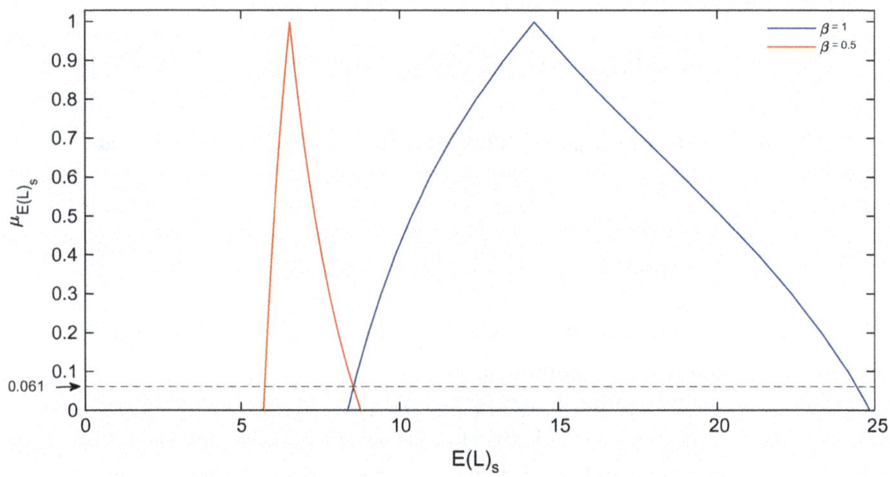

Fig. 5.2 The function of membershipof the expected count of parcels in the system for FM/FM(FM)/1/30 queue when $N = 10$ if fuzzy numbers are taken as triangular

The α-cut values of the rates of arrival, service, startup, and mean count of units in the system when $\beta = 1$ and $\beta = 0.5$ for triangular fuzzy numbers are reported in Table 5.1 and Table 5.2, respectively. If we take $N = 10$ for, instance, Table 5.1 depicts that at one boundary $\alpha = 1$, mean count of parcels in the system is 14.2695 and at other boundary $\alpha = 0$, impossible falls outside [8.3799, 24.8649]. Similarly Table 5.2 depicts that at one end $\alpha = 1$, mean count of parcels in the system is 6.5439 and at other end $\alpha = 0$, impossible falls outside [5.7167, 8.7971]. It should

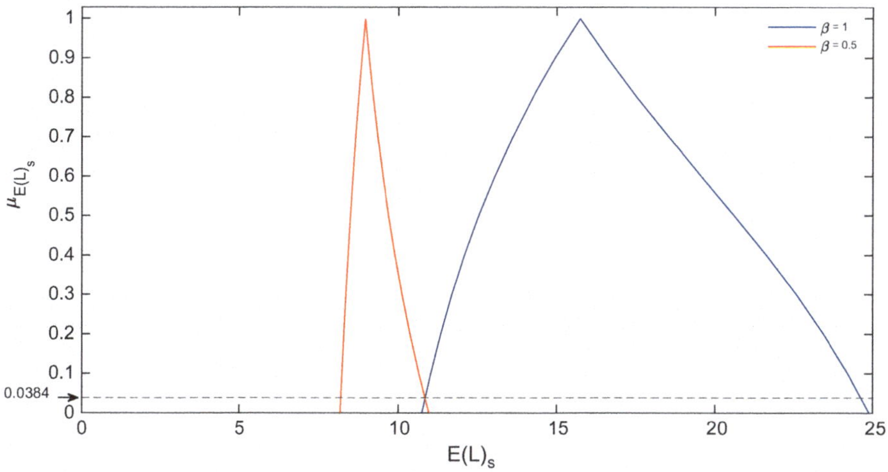

Fig. 5.3 The function of membership of the expected count of parcels in the system for FM/FM(FM)/1/30 queue when $N = 15$ if fuzzy numbers are taken as triangular

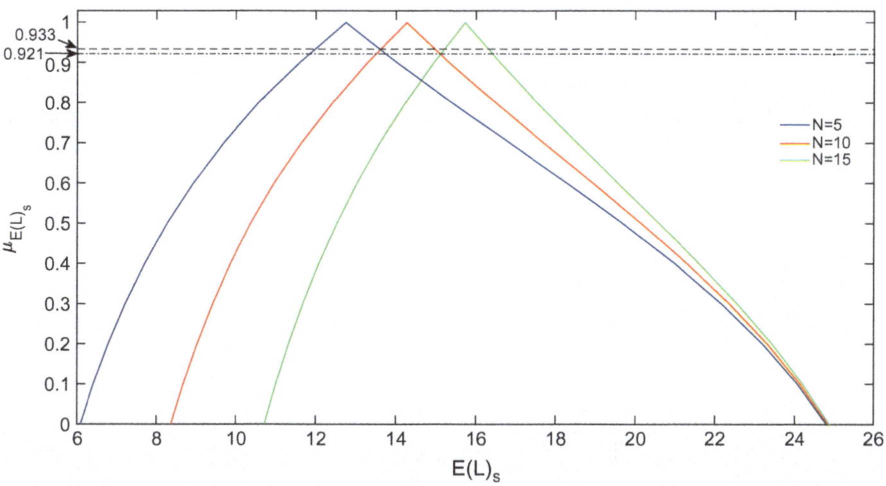

Fig. 5.4 The function of membership of the expected count of parcels in the system for FM/FM(FM)/1/30 queue if fuzzy numbers are taken as triangular and $\beta = 1$

be noted that the value is greater for $\beta = 1$ than $\beta = 0.5$ which should be the case. If the range of mean count of parcels is specified as [9.8750, 21.3160] by the manager when $\beta = 1$ then the corresponding value of $\alpha = 0.4$. So, the preferred service rate ranges $\left[b^L_{0.4}, b^U_{0.4}\right] = [23.2,\ 26.8]$ and the preferred startup rate ranges $\left[c^L_{0.4}, c^U_{0.4}\right] = [4.2,\ 7.8]$.

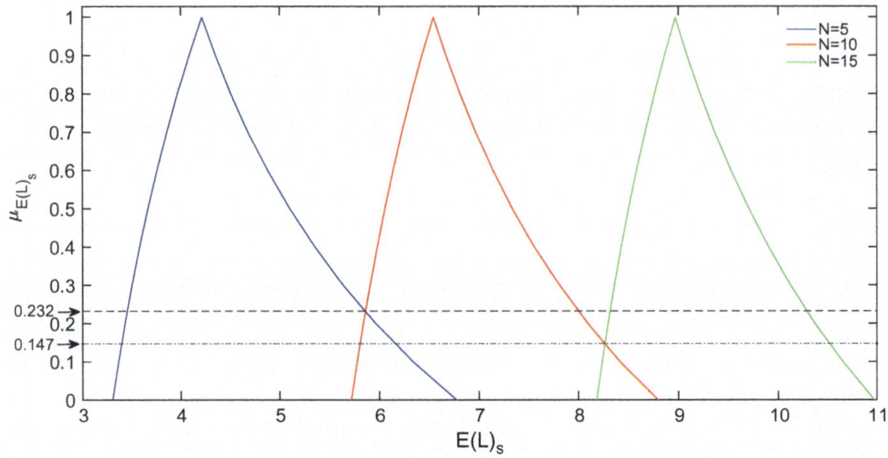

Fig. 5.5 The function of membership of the expected count of parcels in the system for FM/FM(FM)/1/30 queue if fuzzy numbers are taken as triangular and β = 0.5

5.7.2 *Trapezoidal Fuzzy Number*

In this case, we suppose that the rates of arrival, service as well as startup for FM/FM(FM)/1/30 queue are trapezoidal fuzzy numbers and are represented by $\hat{\lambda} = [16\ 17\ 18\ 19], \hat{\mu} = [17\ 18\ 19\ 20], \hat{\varphi} = [1\ 2\ 3\ 4]$ per hour, respectively, for the courier service system discussed above. One can easily determine

$$\left[a_\alpha^L, a_\alpha^U\right] = \left[\min\left\{1/\mu_{\hat{\lambda}}(\alpha)\right\}, \max\left\{1/\mu_{\hat{\lambda}}(\alpha)\right\}\right] = [16+\alpha,\ 19-\alpha]$$

$$\left[b_\alpha^L, b_\alpha^U\right] = \left[\min\left\{1/\mu_{\hat{\mu}}(\alpha)\right\},\ \max\left\{1/\mu_{\hat{\mu}}(\alpha)\right\}\right] = [17+\alpha,\ 20-\alpha]$$

and

$$\left[c_\alpha^L, c_\alpha^U\right] = \left[\min\left\{1/\mu_{\hat{\varphi}}(\alpha)\right\},\ \max\left\{1/\mu_{\hat{\varphi}}(\alpha)\right\}\right] = [1+\alpha,\ 4-\alpha]$$

The results for membership function $\mu_{E(\hat{L})_S}(x)$, obtained numerically, are displayed in Figs. 5.6, 5.7, 5.8, 5.9 and 5.10.

Furthermore, the trapezoidal shape appears in figure on plotting the function of membership $\mu_{E(\hat{L})_S}(x)$ when rates of arrival, service, and startup are taken as trapezoidal fuzzy numbers. In Figs. 5.6, 5.7, 5.8, the results of both β = 1 and β = 0.5 against $N = 5$, $N = 10$, and $N = 15$ are shown, and clearly β = 1 has wider range of membership function than β = 0.5. The largest $E(L)_S$ of β = 0.5 is lesser than smallest $E(L)_S$ of β = 1 when α goes above 0.158, 0.105, and 0.0555 in Figs. 5.6, 5.7, 5.8, respectively.

Table 5.1 The α-cuts of the arrival rate, service rate, startup rate and mean count of units in the system at distinct values of α for the triangular fuzzy number when $\beta = 1$ (i.e. balking probability is $1-\beta = 0$)

							$N = 5$		$N = 10$		$N = 15$	
α	a_α^L	a_α^U	b_α^L	b_α^U	c_α^L	c_α^U	$(E(L)_s)_\alpha^L$	$(E(L)_s)_\alpha^U$	$(E(L)_s)_\alpha^L$	$(E(L)_s)_\alpha^U$	$(E(L)_s)_\alpha^L$	$(E(L)_s)_\alpha^U$
0	20	26	22	28	3	9	6.0980	24.8242	8.3799	24.8649	10.7452	24.9018
0.1	20.3	25.7	22.3	27.7	3.3	8.7	6.4245	24.0802	8.6864	24.1531	11.0270	24.2201
0.2	20.6	25.4	22.6	27.4	3.6	8.4	6.7957	23.1994	9.0328	23.3244	11.3415	23.4407
0.3	20.9	25.1	22.9	27.1	3.9	8.1	7.2205	22.1715	9.4263	22.3761	11.6937	22.5680
0.4	21.2	24.8	23.2	26.8	4.2	7.8	7.7097	20.9975	9.8750	21.3160	12.0891	21.6159
0.5	21.5	24.5	23.5	26.5	4.5	7.5	8.2759	19.6939	10.3884	20.1639	12.5340	20.6076
0.6	21.8	24.2	23.8	26.2	4.8	7.2	8.9335	18.2944	10.9770	18.9513	13.0354	19.5727
0.7	22.1	23.9	24.1	25.9	5.1	6.9	9.6978	16.8475	11.6512	17.7174	13.6004	18.5426
0.8	22.4	23.6	24.4	25.6	5.4	6.6	10.5832	15.4085	12.4207	16.5033	14.2357	17.5465
0.9	22.7	23.3	24.7	25.3	5.7	6.3	11.6003	14.0299	13.2928	15.3450	14.9471	16.6066
1	23	23	25	25	6	6	12.7522	12.7522	14.2695	14.2695	15.7378	15.7378

Table 5.2 The α-cuts of the arrival rate, service rate, startup rate and mean count of units in the system at distinct values of α for the triangular fuzzy number when $\beta = 0.5$ (i.e. balking probability is $1-\beta = 0.5$)

							$N = 5$		$N = 10$		$N = 15$	
α	a_α^L	a_α^U	b_α^L	b_α^U	c_α^L	c_α^U	$(E(L)_s)_\alpha^L$	$(E(L)_s)_\alpha^U$	$(E(L)_s)_\alpha^L$	$(E(L)_s)_\alpha^U$	$(E(L)_s)_\alpha^L$	$(E(L)_s)_\alpha^U$
0	20	26	22	28	3	9	3.3030	6.7792	5.7167	8.7971	8.1838	10.9765
0.1	20.3	25.7	22.3	27.7	3.3	8.7	3.3666	6.3392	5.7749	8.4198	8.2397	10.6567
0.2	20.6	25.4	22.6	27.4	3.6	8.4	3.4346	5.9616	5.8370	8.0930	8.2994	10.3748
0.3	20.9	25.1	22.9	27.1	3.9	8.1	3.5075	5.6353	5.9035	7.8082	8.3631	10.1251
0.4	21.2	24.8	23.2	26.8	4.2	7.8	3.5859	5.3513	5.9749	7.5585	8.4314	9.9030
0.5	21.5	24.5	23.5	26.5	4.5	7.5	3.6703	5.1024	6.0517	7.3383	8.5047	9.7047
0.6	21.8	24.2	23.8	26.2	4.8	7.2	3.7615	4.8826	6.1345	7.1430	8.5837	9.5269
0.7	22.1	23.9	24.1	25.9	5.1	6.9	3.8603	4.6874	6.2242	6.9687	8.6690	9.3668
0.8	22.4	23.6	24.4	25.6	5.4	6.6	3.9679	4.5128	6.3216	6.8125	8.7613	9.2221
0.9	22.7	23.3	24.7	25.3	5.7	6.3	4.0853	4.3559	6.4278	6.6716	8.8617	9.0907
1	23	23	25	25	6	6	4.2141	4.2141	6.5439	6.5439	8.9711	8.9711

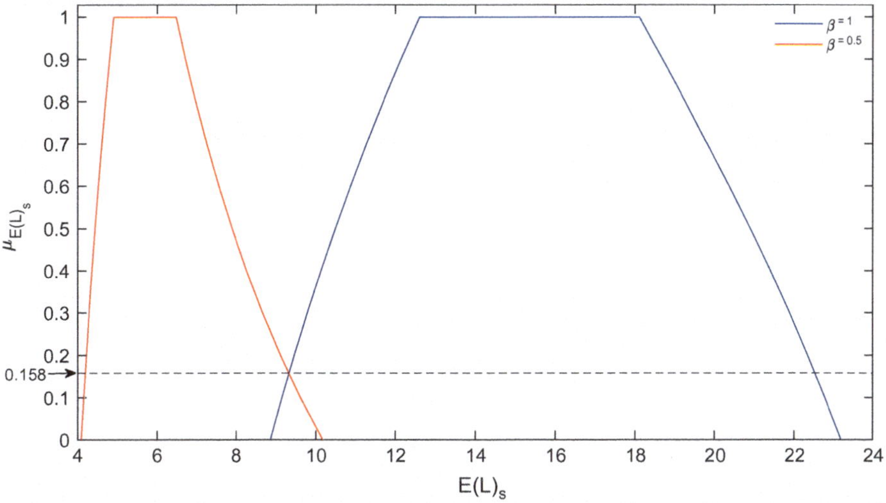

Fig. 5.6 The function of membership of the expected count of parcels in the system for FM/FM(FM)/1/30 queue when N = 5 if fuzzy numbers are taken as trapezoidal

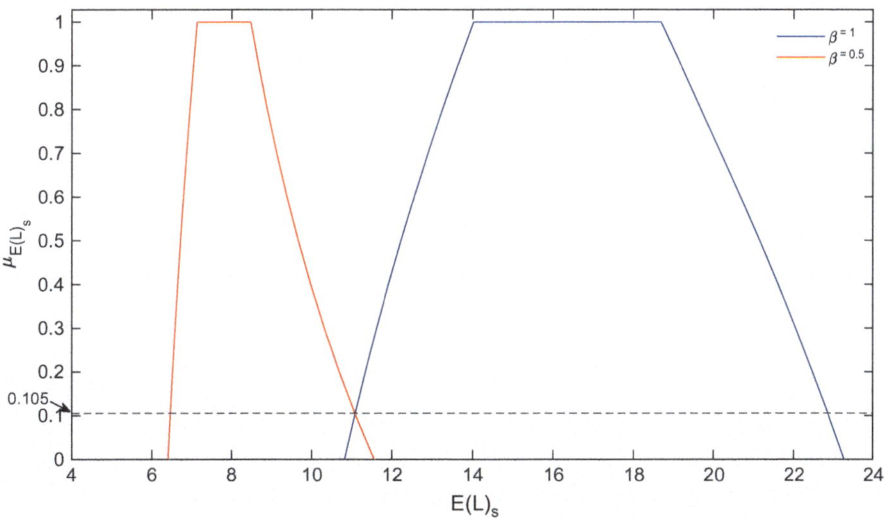

Fig. 5.7 The function of membership of the expected count of parcels in the system for FM/FM(FM)/1/30 queue when N = 10 if fuzzy numbers are taken as trapezoidal

The membership function for N = 5, N = 10, and N = 15 are plotted against β = 1 and β = 0.5 in Figs. 5.9–5.10. Also when α escalates beyond 0.805 in Fig. 5.9, the highest $E(L)_S$ for $N = 5$ is lesser than the lowest $E(L)_S$ for $N = 10$. In the same

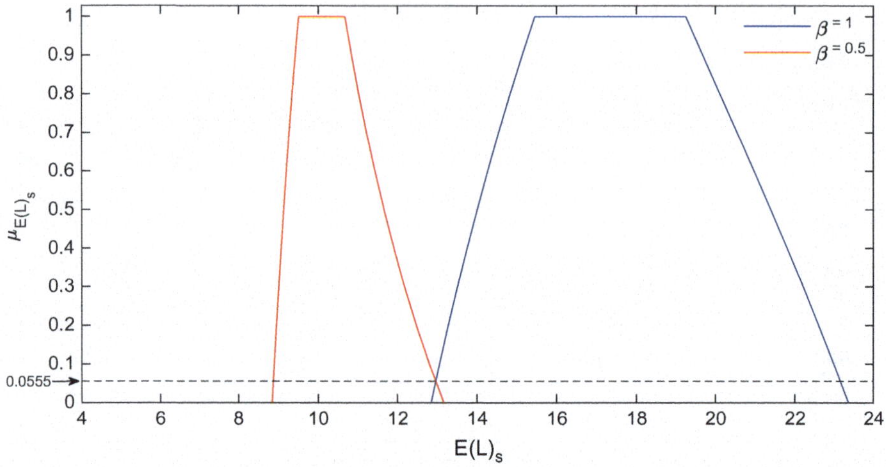

Fig. 5.8 The function of membership of the expected count of parcels in the system for FM/FM(FM)/1/30 queue when N = 15 if fuzzy numbers are taken as trapezoidal

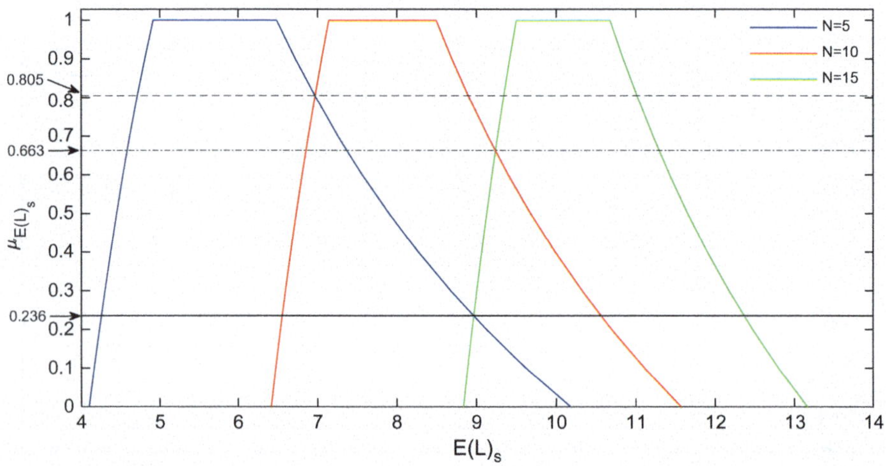

Fig. 5.9 The function of membership of the expected count of parcels in the system for FM/FM(FM)/1/30 queue if fuzzy numbers are taken as trapezoidal and β = 1

way, when α escalates beyond 0.663 in Fig. 5.9, the highest $E(L)_S$ for $N = 10$ is lesser than the lowest $E(L)_S$ for $N = 15$.

Likewise, as soon as α escalates beyond 0.236 in Fig. 5.9, the highest $E(L)_S$ for $N = 5$ is lesser than the lowest $E(L)_S$ for $N = 15$. Moreover, $N = 5$ has a broader range of membership function than $N = 10$ and $N = 15$ in Figs. 5.9 and 5.10.

The α-cut values of the rates of startup, arrival, service, and mean count of units in the system when β = 1 and β = 0.5 for trapezoidal fuzzy numbers are reported

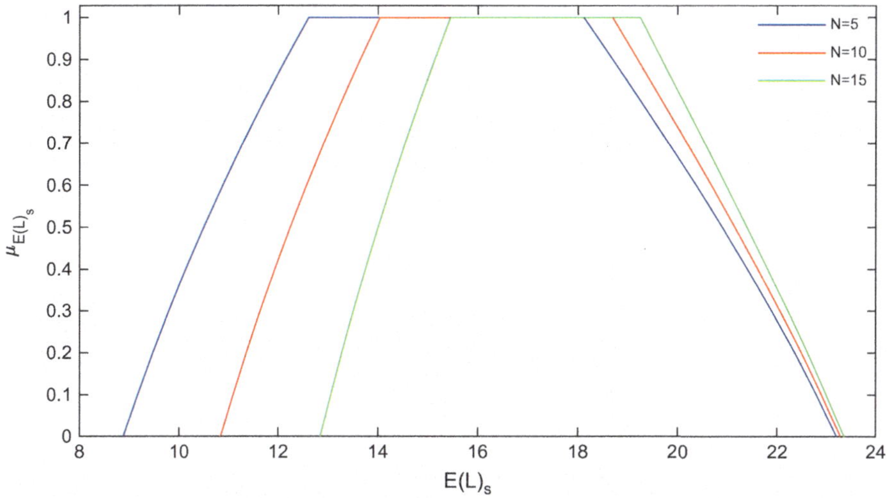

Fig. 5.10 The function of membership of the expected count of parcels in the system for FM/FM(FM)/1/30 queue if fuzzy numbers are taken as trapezoidal and $\beta = 0.5$

in Table 5.3 and Table 5.4, respectively. If we take $N = 5$, for example, Table 5.3 describes that at one boundary $\alpha = 1$, mean count of parcels in the system falls between [12.6074, 18.1136] and at other boundary $\alpha = 0$, impossible falls outside [8.8708, 23.2072]. On the contrary, Table 5.4 shows that at one end $\alpha = 1$, mean count of parcels in the system falls between [4.9157, 6.4870] and at other end $\alpha = 0$, impossible falls outside [4.0950, 10.1823]. It should be noted that the value is greater for $\beta = 1$ than $\beta = 0.5$ which should be the case. If the range of mean count of parcels is specified as [4.5316 7.5658] by the manager when $\beta = 0.5$ then the corresponding value of $\alpha = 0.6$. Here, $\left[b^L_{0.6}, b^U_{0.6}\right] = [17.6, 19.4]$ is the preferred service rate and $\left[c^L_{0.6}, c^U_{0.6}\right] = [1.6,\ 3.4]$ is the preferred startup rate.

Therefore the analysis done above for both triangular and trapezoidal fuzzy numbers aids system designers to design the desired queueing system and managers in handling decision and policymaking.

5.8 Conclusion

The set theory of fuzzy has evidenced its usefulness in a wide range of queueing frameworks. In this article, truncated *N*-policy queue with server startup and balking is studied under fuzzy environment. Firstly, the steady-state probability distribution, as well as the mean count of units in the system, is derived for the model. Then, the α-cut methodology and extension principle of Zadeh are employed for the transformation of the finite *N*-policy fuzzy queue with server startup and balking into a group of crisp queue with server startup and balking. Using parametric non-linear

Table 5.3 The α-cuts of the arrival rate, service rate, startup rate and mean count of units in the system at distinct values of α for the trapezoidal fuzzy number when $\beta = 1$ (i.e. balking probability is $1-\beta = 0$)

							$N = 5$		$N = 10$		$N = 15$	
α	a_α^L	a_α^U	b_α^L	b_α^U	c_α^L	c_α^U	$(E(L)_s)_\alpha^L$	$(E(L)_s)_\alpha^U$	$(E(L)_s)_\alpha^L$	$(E(L)_s)_\alpha^U$	$(E(L)_s)_\alpha^L$	$(E(L)_s)_\alpha^U$
0	16	19	17	20	1	4	8.8708	23.2072	10.8258	23.2869	12.8450	23.3628
0.1	16.1	18.9	17.1	19.9	1.1	3.9	9.1589	22.7928	11.0787	22.8961	13.0560	22.9950
0.2	16.2	18.8	17.2	19.8	1.2	3.8	9.4644	22.3538	11.3451	22.4857	13.2770	22.6127
0.3	16.3	18.7	17.3	19.7	1.3	3.7	9.7878	21.8907	11.6257	22.0567	13.5083	22.2172
0.4	16.4	18.6	17.4	19.6	1.4	3.6	10.1298	21.4046	11.9211	21.6105	13.7504	21.8105
0.5	16.5	18.5	17.5	19.5	1.5	3.5	10.4913	20.8966	12.2317	21.1489	14.0037	21.3945
0.6	16.6	18.4	17.6	19.4	1.6	3.4	10.8728	20.3688	12.5580	20.6738	14.2685	20.9715
0.7	16.7	18.3	17.7	19.3	1.7	3.3	11.2750	19.8235	12.9006	20.1876	14.5454	20.5438
0.8	16.8	18.2	17.8	19.2	1.8	3.2	11.6981	19.2636	13.2595	19.6929	14.8345	20.1137
0.9	16.9	18.1	17.9	19.1	1.9	3.1	12.1423	18.6920	13.6351	19.1925	15.1363	19.6837
1	17	18	18	19	2	3	12.6074	18.1136	14.0273	18.6853	15.4510	19.2524

Table 5.4 The α-cuts of the arrival rate, service rate, startup rate and mean count of units in the system at distinct values of α for the trapezoidal fuzzy number when β = 0.5 (i.e. balking probability is 1–β = 0.5)

							$N = 5$		$N = 10$		$N = 15$	
α	a_α^L	a_α^U	b_α^L	b_α^U	c_α^L	c_α^U	$(E(L)_s)_\alpha^L$	$(E(L)_s)_\alpha^U$	$(E(L)_s)_\alpha^L$	$(E(L)_s)_\alpha^U$	$(E(L)_s)_\alpha^L$	$(E(L)_s)_\alpha^U$
0	16	19	17	20	1	4	4.0950	10.1823	6.4158	11.5707	8.8389	13.1673
0.1	16.1	18.9	17.1	19.9	1.1	3.9	4.1588	9.6278	6.4724	11.1113	8.8916	12.8015
0.2	16.2	18.8	17.2	19.8	1.2	3.8	4.2257	9.1292	6.5318	10.6987	8.9467	12.4726
0.3	16.3	18.7	17.3	19.7	1.3	3.7	4.2962	8.6792	6.5943	10.3261	9.0045	12.1748
0.4	16.4	18.6	17.4	19.6	1.4	3.6	4.3705	8.2718	6.6600	9.9884	9.0652	11.9037
0.5	16.5	18.5	17.5	19.5	1.5	3.5	4.4488	7.9021	6.7292	9.6812	9.1289	11.6557
0.6	16.6	18.4	17.6	19.4	1.6	3.4	4.5316	7.5658	6.8023	9.4008	9.1959	11.4279
0.7	16.7	18.3	17.7	19.3	1.7	3.3	4.6193	7.2591	6.8795	9.1443	9.2664	11.2180
0.8	16.8	18.2	17.8	19.2	1.8	3.2	4.7122	6.9790	6.9612	8.9090	9.3407	11.0241
0.9	16.9	18.1	17.9	19.1	1.9	3.1	4.8108	6.7224	7.0478	8.6927	9.4192	10.8444
1	17	18	18	19	2	3	4.9157	6.4870	7.1398	8.4935	9.5020	10.6775

programming technique, the function of membership for the mean count of units in the system is constructed. Hereafter, the proposed solution algorithm is followed for getting the α-cuts of the function of membership numerically as well as bounds of their interim are rearranged to get the expression for the membership function in closed form. It is noteworthy that by this technique, more information is conveyed to the management since the mean count of units in the system is stated by means of membership function instead of crisp values. Lastly, to exemplify the relevance of the proposed methodology, a real-life application of courier and delivery system is studied and meaningful results are provided for system designers and practitioners. Furthermore, this study can be extended to more compound Poisson process.

References

1. Yadin M, Naor P (1963) Queueing systems with a removable service station. J Oper Res Soc 14:393–405
2. J C Ke, H I Huang, Y K Chu (2010) Batch arrival queue with N-policy and at most J vacations. Appl Math Modell 34:451–446
3. Choudhury G, Tadj L, Ke J (2011) A two-phase service system with Bernoulli vacation schedule, setup time and *N*-policy for an unreliable server with delaying repair. Qual Technol Quan Manage 8:271–284
4. D Y Yang, CHW (2015) Cost-minimization analysis of a working vacation queue with *N*-policy and server breakdowns. Comput Industr Eng 82:151–158
5. Jayachitra P (2018) Analysis of an N-policy policy M/M/1 queueing system with server start up, unreliable server and balking. Res J Math Stat Sci 6:1–9
6. Agrawal PK, Jain M, Singh A (2017) Optimal N-policy for finite queue with server breakdown and state-dependent rate. Int J Comput Eng Res 7:61–68
7. Zadeh LA (1965) Fuzzy sets. Inf Control 8:338–353
8. Kaufmann A (1975) Introduction to the theory of fuzzy subsets. Academic Press, New York
9. Zadeh LA (1978) Fuzzy sets as a basis for a theory of possibility. Fuzzy Sets Syst 1:3–28
10. H J Zimmermann (2001) Fuzzy set theory and its applications, 4th ed. Boston
11. Buckley JJ (1990) Elementary queueing theory based on possibility theory. Fuzzy Sets Syst 37:43–52
12. Kao C, Li CC, Chen SP (1999) Parametric programming to the analysis of fuzzy queues. Fuzzy Sets Syst 107:3–100
13. Chen SP (2004) Parametric nonlinear programming for analyzing fuzzy queues with finite capacity. Eur J Oper Res 157:429–438
14. Chen SP (2005) Parametric nonlinear programming approach to fuzzy queues with bulk service. Eur J Oper Res 163:434–444
15. Chen S (2006) A bulk arrival queueing model with fuzzy parameters and varying batch sizes. Appl Math Model 30:920–929
16. Ke JC, Lin CH (2006) Fuzzy analysis of queueing systems with an unreliable server: A nonlinear programming approach. Appl Math Comput 175:330–346
17. Kalayanaraman R, Thillaigovindan N, Kannadasan G (2010) A single server fuzzy queue with unreliable server. Int J Computat Cogn 8:1–4
18. Kumar VA (2012) A nonlinear programming approach for a fuzzy queue with an unreliable server. Bullet Soc Math Services Standards 1:69–89
19. Ke JC, Huang HI, Lin CH (2006) Parametric programming approach for batch arrival queues with vacation policies and fuzzy parameters. Appl Math Comput 180:217–232

20. Ke JC, Huang HI, Lin CH (2007) On retrial queueing model with fuzzy parameters. Phys A 374:272–280
21. Khodadadi SB, Jolai F (2012) A fuzzy based threshold policy for a single server retrial queue with vacations. CEJOR 20:281–297
22. Ke JC, Huang HI, Lin CH (2008) Parametric programming approach for a two-unit repairable system with imperfect coverage, reboot and fuzzy parameters. IEEE Trans Reliab 57:498–506
23. Aydin O, Apaydin A (2008) Multi-Channel fuzzy queueing systems and membership functions of related fuzzy services and fuzzy inter-arrival times. Asia-Pac J Oper Res 25:697–713
24. Yang D, Chang P (2015) A parametric programming solution to the F-policy queue with fuzzy parameters. Int J Syst Sci 46:90–598
25. Mary KJR, Rajalakshmi R, Pavithra J (2016) Analysis of FM/M(a, b)/1/MWV/Br queueing model. Int J Comput Appl 142:18–22
26. Kumar VA (2016) Parametric Non-linear programming approach for *N*-policy queues with infinite capacity. J Progr Res Math 10:1470–1483
27. Wang TY, Yang DY, Li MJ (2010) Fuzzy analysis for the *N*-Policy queues with infinite capacity. Int J Inf Manage Sci 21:41–55
28. Ritha W, Menon BS (2011) Fuzzy N policy queues with infinite capacity. J Phys Sci 15:73–82
29. Varadharajan R, Susmitha R (2018) Evaluation of performance measures of priority queues with fuzzy parameters using A—cut approach. ARPN J Eng Appl Sci 13:2636–2641

Chapter 6
NHPP-Based SRGM Using Time-Dependent Fault Reduction Factors (FRF) and Gompertz TEF

M. Jain, P. Agarwal, and R. Solanki

6.1 Introduction

In today's era, testing of the quality of software before launching in the market is a huge concern. Softwares are regularly updated in order to meet modern requisites by the software developer. Software failure is defined as the conveyed services which do not fulfil the requirements supposed by the developer/user and provides inadequate performance. From the last four decades, for upgrading the previous versions and improve the reliability, software reliability growth models (SRGMs) are recycled. Failure data collected during testing is implemented on SRGMs by which we get the stochastic performance of the software's behaviour.

6.2 Gompertz Testing Effort Function (GTEF)

Gompertz TE consumption curve has been used to fit the statistical data by many authors. Winsor [24] summarized that Gompertz and logistic curves both are having similar properties of representing the growth aspect. Berger [2] compared the logistic and Gompertz curves and found that Gompertz is a more reliable growth curve than the logistic growth curve. The software reliability was improved and optimized by Rafi and Akhtar [18].

M. Jain
Department of Mathematics, Indian Institute of Technology Roorkee, Roorkee, India
e-mail: madhufma@iitr.ernet.in

P. Agarwal (✉) · R. Solanki
Department of Applied Sciences, SRMIST, NCR Campus, Modinagar, India
e-mail: priyanka1354@gmail.com

R. Solanki
e-mail: rs.solanki78@gmail.com

P. K. Kapur et al. (eds.), *Decision Analytics Applications in Industry*, Asset Analytics,
https://doi.org/10.1007/978-981-15-3643-4_6

The cumulative TEF is defined by Weibull function as

$$W(t) = \alpha e^{-\beta e^{-ct}} \tag{6.1}$$

Current TEF is given by

$$w(t) = \alpha\beta c e^{-ct-\beta e^{-ct}} \tag{6.2}$$

where α and β are the total expenditure and consumption rate of the testing effort, respectively.

6.3 Literature Survey

SRGM based on NHPP under different environments have been developed with certain improvements by many researchers. Some of them have contributed toward the software reliability under the features of imperfect debugging and testing coverage. Such types of works can be found in Cao et al. [3], Pham and Zhang [17], Wang et al. [23], Wang and Wu [22], Li and Pham [11], Chatterjee and Shukla [27] and many more. Sagar et al. [20] presented inflection S-shaped SRGM by considering Weibull distribution. Relationship between software defect and defeat is expressed by fault reduction factor (FRF) that influences software reliability growth directly. Hsu et al. [6] explained FRF as both decreasing and increasing functions. Jain et al. [28] analysed the FRF under imperfect phenomenon with several change-points of testing. After that Pachauri et al. [15] extended previous model using FRF as inflection S-shaped function. Chatterjee and shukla [31] used a general Weibull type FRF incorporating change-point and imperfect debugging for both detection and correction processes.

Testing effort functions demonstrate the consumed resources or expenses during the software development. There is a testing effort curve between testing period and consumed resources. Some factors which influence testing efforts are software development process, CPU hours, quality, testability, man power, quality goals, testing strategy, etc. TEF can be explained by special functions like Exponential, Rayleigh, Weibull, logistic, log-logistic functions, Gompertz, etc. TEF has been used to develop SRGMs by many researchers from time to time. Huang and Kyo [8] and Rafi et al. [19] incorporated logistic testing effort, for the assessment of software reliability growth. Different TEFs have been studied with distinct scenarios such as change-point and release policies by Li et al. [12], Lin and Huang [13], Satoh [29]. Ohishi et al. [30] used Gompertz TEF to evaluate the performance of reliability and software failure data. Ahmad and Imran [1] compared the SRGMs with log-logistic TEF model. After Ahmad and Imran's work, Peng et al. [16] used log-logistic TEF to analyze both detection and correction phenomenon in software system. Recently, Jin and Jin [10] discussed the growth of software reliability by considering the testing effort function and used a new optimization technique based on Swarm intelligent application.

It has been noticed that a large number of research papers have been published to locate the ideal release time with different surroundings but the main goal of this problem is to reduce the total assets and make more robust model than before. Okumoto and Goel [14], Ehrlich et al. [5] and Huang and Lo [7], Wang et al. [21] discussed the optimization method during testing phase based on NHPP. Xie and Yang [25] developed the cost model under imperfect debugging environment. Some optimal release policies were suggested by Dohi et al. [4]. Jha et al. [9] developed NHPP-based model for the prediction of reliability growth under some important features, namely, debugging time lag, testing coverage and testing effort function, etc. Using Gompertz TEF, Rafi and Akhtar [18] examined the SRGM and optimized the total testing cost.

6.4 Methodology

In this section, three types of FRF with Gompertz TEF have been incorporated for the software reliability modelling. To address the reliability growth issue of SRGM based on these issues, certain assumptions are made. It is considered that occurence of failure is a random phenomenon and does not depend on each other. To analyze the fault detection/removal process, NHPP is considered. We assume that the mean value function (MVF) is proportional to the residual faults in the software and time-dependent function FRF is used for proportionality.

Usually, fault reduction factors are evaluated as the ratio of the total faults eliminated to the total failures maturity. Musa [26] evaluated the MVF and reliability considering constant FRF. But it is realized that some external or internal aspects such as environmental effects, imperfect debugging, testing coverage, system allocation, etc. affect the reliability growth. These aspects precisely influence the FRF. In this study, three patterns, i.e. constant, increasing, decreasing of FRFs are taken into account to obtain MVF.

Based on the above discussions, the equation for MVF is framed by

$$\frac{d}{dt}m(t) \times \frac{1}{w(t)} = r(t)[a - m(t)] \tag{3}$$

$$and \quad r(t) = r \times b(t), \tag{4}$$

where a is the initial fault content, m(t) is the MVF, r(t) is detection rate of the faults and b(t) is the FRF.

Case 1 When $b(t) = b_0$.

Here FRF b(t) is constant and $0 < b_0 \leq 1$. Thus

$$m(t) = a[1 - e^{-rb_0\alpha e^{-\beta e^{-ct}}}] \tag{5}$$

Case 2 When $b(t) = 1 - (1 - b_0)e^{-ct}$.

Here FRF is increasing curve so that $b_0 \leq b(t) \leq 1, \; b_0 > 0$ and c is the increasing parameter, $0 \leq c \leq 1$.

Now $m(t) = a[1 - e^{-r\alpha[1-(1-b_0)\{e^{-ct}+\frac{1}{\beta}\}]e^{-\beta e^{-ct}}} \times e^{r\alpha\left[1-(1-b_0)\{1+\frac{1}{\beta}\}\right]e^{-\beta}}]$ (6)

Case 3 When $b(t) = b_0e^{-ct}$.

Here FRF is decreasing curve with time and $0 \leq b(t) \leq b_0, \; b_0 \leq 1$, and c is the decreasing parameter, $0 \leq c \leq 1$. In this case

$$m(t) = a[1 - e^{-r\alpha b_0\{e^{-ct}+\frac{1}{\beta}\}e^{-\beta e^{-ct}}} \times e^{r\alpha b_0\{1+\frac{1}{\beta}\}e^{-\beta}}] \tag{7}$$

6.4.1 *Expected Total Cost*

Let C_1 (C_2) be the cost of removing an error during testing (operational) phase, ($0 < C_1 < C_2$), The per unit testing effort cost consumed is denoted by C_3 (> 0). Furthermore, T_{LC} denotes the s/w life cycle. The total cost incurred on the software using the development, operational and testing effort expenditures is obtained as follows:

$$EC(T) = C_1m(T) + C_2[m(T_{LC}) - m(T)] + C_3[\int_0^T w(t)dt] \tag{8}$$

6.4.1.1 Reliability Evaluation

The reliability can be evaluated for the software using the function

$$R\left({}^{\Delta T}/{}_{T}\right) = \exp[-\{m(T + x) - m(T)\}] \tag{9}$$

6.4.1.2 Optimization Problem

Here, we minimize the total cost of the s/w so as to determine the optimal release time of the s/w testing to achieve the desired reliability. Now, we frame the optimization problem as follows:

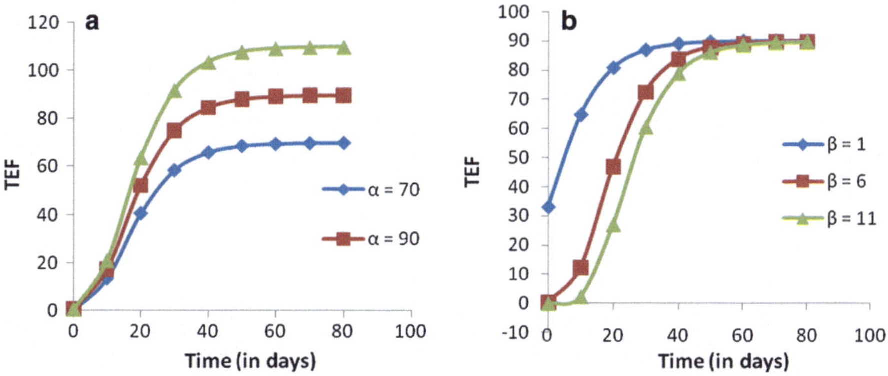

Fig. 6.1 Gompertz testing effort behaviour by varying α and β in (**a**) and (**b**) respectively

$$\text{Minimize} \quad EC(T)$$
$$\text{Subject to} \; R\left({}^{\Delta T}/_{T}\right) \geq R_{\min} \tag{10}$$

R_{min} is desired level of reliability (Fig. 6.1).

6.5 Numerical Results

In this segment, we review the applicability of our model by coding a program in MATLAB. We set some default parameters as $\alpha = 70.55$, $\beta = 3.304$, $c = 0.1109$, $r = 0.05$, $b_0 = 0.01$, $C_1 = 10\$$, $C_2 = 40\$$, $C_3 = 100\$$ and $T_{LC} = 100$. To analyze the impact of varying values of descriptors on the MVF, total expected cost and reliability for all three cases numerical results are obtained and displayed in Figs. 6.2, 6.3, 6.4 and 6.5.

In Fig. 6.2a–c, we depict the MVF for cases 1, 2 and 3, respectively. We see that MVF increases sharply at the beginning and after some time it becomes almost constant for case 1. For case 2, we notice that detected faults decreases with time (till $t = 10$) and then after the number of faults detected becomes constant. When FRF b(t) is an increasing function then MVF increases with time (till $t = 40$ CPU hours) and after that it becomes almost constant.

Reliability and total expected cost for three cases are depicted in Figs. 6.3a–b, 6.4a–b, 6.5a–b, respectively. We calculate the respective reliability and cost for the same parameters. For case 1, approx. till $t = 10$ (CPU hours), the cost decreases and after some time approx. ($t = 40$), the cost starts to increase and as time reaches 80 h, it attains almost constant.

Figures 6.4a, b depict the reliability and expected cost for case 2 when FRF b(t) is a decreasing function. MVF is very large initially, which reveals that the reliability is approximately zero. Further as the mean value function decreases, reliability

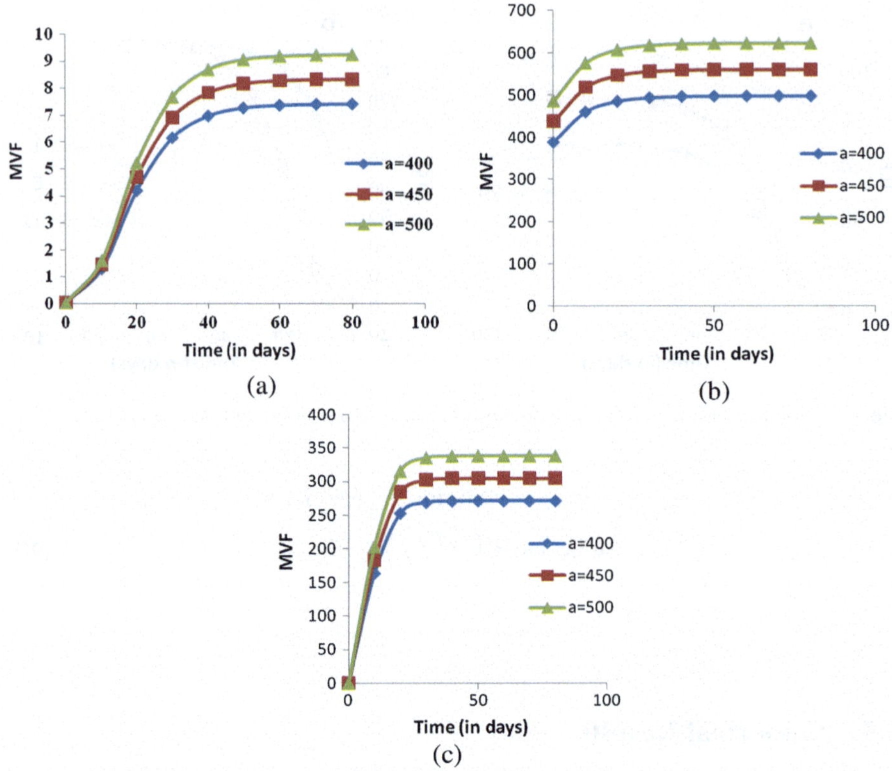

Fig. 6.2 Mean value function for **a** case 1 **b** case 2 **c** case 3, by varying **a**

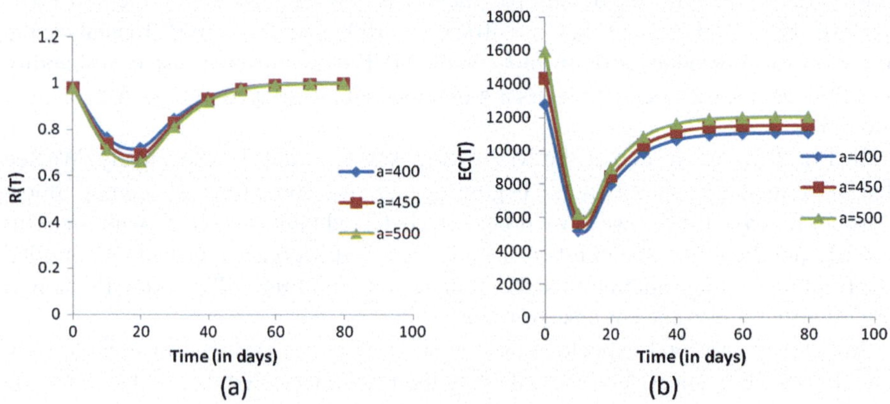

Fig. 6.3 For case 1 **a** Total Expected Cost **b** Reliability by varying α

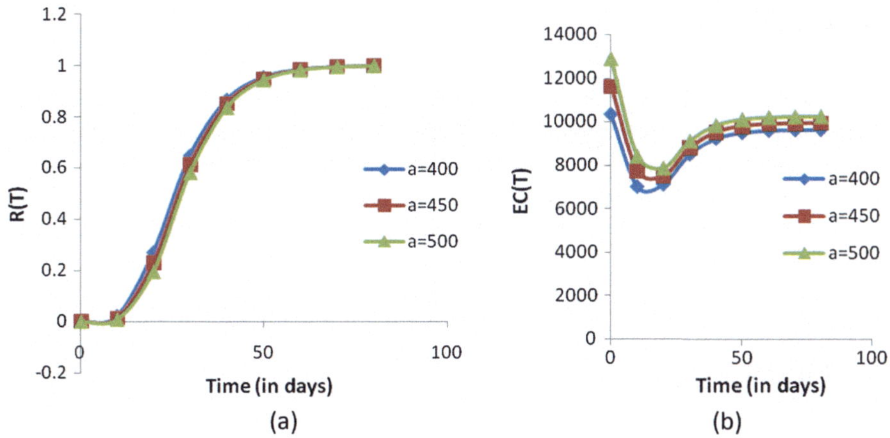

Fig. 6.4 For case 1 **a** Total Expected Cost **b** Reliability by varying α

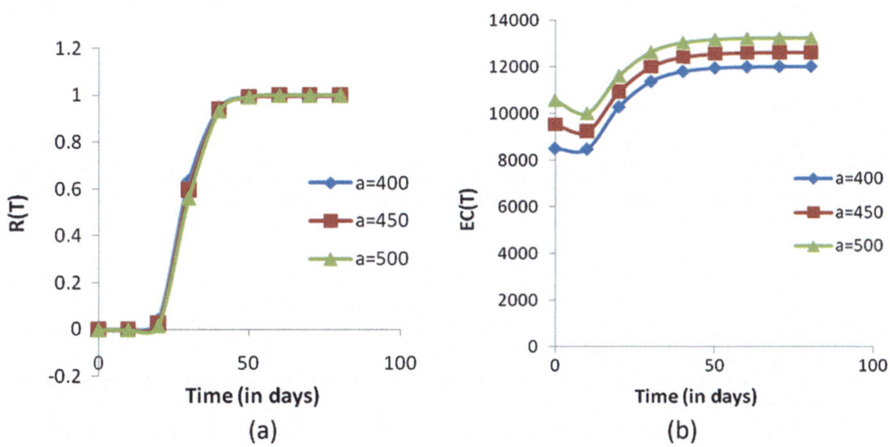

Fig. 6.5 For case 1 **a** Total Expected Cost **b** Reliability by varying α

increases and approx. at t = 50 CPU hours becomes constant. We get the optimal cost to release the software due to convexity of EC(T).

When FRF b(t) is an increasing function then reliability and cost are displayed in Fig. 6.5a, b, respectively. Reliability increases rapidly with time and total expected cost EC(T), decreases up to nearly t = 10 and then after increases and later on it becomes asymptotically constant. And finally, it is observed that nearly about t = 40 CPU hours we can release the software. This is the optimal release time. At that point (t = 40) MVF is about to be constant, reliability is maximum and total cost is minimum.

6.6 Conclusion

In this work, an SRGM with fault reduction factor (FRF) and Gompertz TEF is developed. We have incorporated the FRF for three cases by considering it as constant, decreasing and increasing. The optimal release times for all the three cases have been determined and validated by taking numerical illustration. It is noted that FRF and GTEF play a key role and should be included to find the higher level of reliability for minimum cost.

6.7 Future Scope

By using these models, software developers can take reliable decision for releasing their product at the correct time so that he (software developer) can gain the maximum profit and customer can gain more satisfaction.

References

1. Ahmad N, Imam Z (2013) Software reliability growth models with Log-logistic testing-effort function: A comparative study. International Journal of Computer Applications 75:126–11
2. Berger RD (1981) Comparison of the Gompertz and Logistic equations to describe plant disease progress. Ecology and Epidemiology 71:716–719
3. Cao P, Dong Z, Liu K, Cai KY (2013) Quantitative effects of software testing on reliability improvement in the presence of imperfect debugging. Inf Sci 218:119–132
4. Dohi T, Kaio N, Osaki S (1997) Optimal software release policies with debugging time lag. Int J Reliab Qual Saf Eng 4:241–255
5. Ehrlich W, Prasanna B, Stampfel J, Wu J (1993) Determining the cost of a stop-testing decision. IEEE Softw 10:33–42
6. Hsu CJ, Huang CY, Chang JR (2011) Enhancing software reliability modeling and prediction through the introduction of time-variable fault reduction factor. Appl Math Model 35:506–521
7. Huang CY, Lo JH (2006) Optimal resource allocation for cost and reliability of modular software systems in the testing phase. Journal of System and Software 79:653–664
8. Huang CY, Kyo SY (2002) Analysis of incorporating logistic testing effort function into software reliability modeling. IEEE Trans Reliab 51:261–270
9. Jha PC, Gupta D, Yang B, Kapur PK (2009) Optimal testing resource allocation during module testing considering cost, testing effort and reliability. Comput Ind Eng 57:1122–1130
10. Jin C, Jin SW (2016) Parameter optimization of software reliability growth model with S-shaped testing effort function using improved swarm intelligent optimization. Appl Soft Comput 40:283–291
11. Li Q, Pham H (2017) NHPP software reliability model considering the uncertainty of operating environments with imperfect debugging and testing coverage. Appl Math Model 51:65–68
12. Li X, Xie M, Ng SH (2010) Sensitivity analysis of release time of software reliability models incorporating testing effort with multiple change-points. Appl Math Model 34:3560–3570
13. Lin CT, Huang CY (2008) Enhancing and measuring the predictive capabilities of testing-effort dependent software reliability models. J Syst Softw 81:1025–1038
14. Okumoto K, Goel AL (1980) Optimum release time for software systems based on reliability and cost criteria. Journal of System and Software 1:315–318

15. Pachauri B, Dhar J, Kumar A (2015) Incorporating inflection S-shaped fault reduction factor to enhance software reliability growth. Appl Math Model 39:1463–1469
16. Peng R, Li YF, Zhang WJ, Hu QP (2014) Testing effort dependent software reliability model for imperfect debugging process considering both detection and correction. Reliability Engineering & System Safety 126:37–43
17. Pham H, Zhang X (2003) NHPP software reliability and cost models with testing coverage. Eur J Oper Res 145:443–454
18. Rafi SM, Akthar S (2010) Software reliability growth model with Gompertz TEF and optimal release time determination by improving the test efficiency. International Journal of Computer Applications 7:34–43
19. Rafi SM, Rao KN, Akthar S (2010) Software reliability growth model with logistic exponential test effort function and analysis of software release policy. International Journal on Computer Science and Engineering 02:387–399
20. Sagar BB, Saket RK, Singh CG (2016) Exponentiated Weibull distribution approach based inflection S-shaped software reliability growth model. Ain Shams Engineering Journal 7:973–991
21. Wang J, Wu Z, Shu Y, Zhang Z (2016) An optimized method for software reliability model based on nonhomogeneous poisson process. Appl Math Model 40:6324–6339
22. Wang J, Wu Z (2016) Study of the nonlinear imperfect software debugging model. Reliability Engineering and System Safety 153:180–192
23. Wang J, Wu Z, Shu Y, Zhang Z (2015) An imperfect software debugging model considering log-logistic distribution fault content function. J Syst Softw 100:167–181
24. Winsor CP (1932) The Gompertz curve as growth curve. Proceeding of National Academy of Sciences 18:1–8
25. Xie M, Yang B (2003) A study of the effect of imperfect debugging on software development cost model. IEEE Trans Software Eng 29:471–473
26. Musa JD (1975) A theory of software reliability and its applications. IEEE Transaction on Software Engineering 3:312–327
27. Chatterjee S, Shukla A (2016) Change point–based software reliability model under imperfect debugging with revised concept of fault dependency". Proceedings of the Institution of Mechanical Engineers Part O: Journal of Risk and Reliability, SAGE 230:579–597
28. Jain M, Manjula T, Gulati T (2014) Imperfect debugging study of SRGM with fault reduction factor and multiple change point. International Journal of Mathematics in Operational Research 6:155–175
29. Satoh D (2000) A discrete Gompertz equation and a software reliability growth model. IEICE Trans Informat Syst E83-D: 1508–1513
30. Ohishi K, Okamura H, Dohi T (2009) Gompertz software reliability model: estimation algorithm and empirical validation. J Syst Softw 82:535–543
31. Chatterjee S, Shukla A (2016) Modeling and analysis of software fault detection and correction process through Weibull-type fault reduction factor, change point and imperfect debugging. Arabian J Sci Eng 41:5009–5025

Chapter 7
Implementation of Low-Carbon Logistics in Indian SMEs—A Statistical Study

K. M. Ahamed Sheriff, M. Elango, and K. M. Mahaboob Sheriff

7.1 Introduction

India faces major development challenges—access to the fundamental amenities such as imbibing dihydrogen monoxide, electricity, sanitation, and clean cooking energy still remain a luxury for urban and rural dwellers homogeneously [1]. Groundwater, which has been the key resource for meeting the irrigation and consumption desiderata of the urban and rural population, is coming under tremendous pressure because of haphazard urban orchestrating and climate change [2]. The convey sector is a vital part of global climate change mitigation strategies, as it accounts for 23% of the energy cognate greenhouse gas emissions [3]. Considerable efficiency gains can be made cost efficaciously to set the convey sector on a sustainable development pathway. They can be achieved through already available technologies and practices, which will not only reduce greenhouse gas emissions significantly, but also generate social, environmental, and economic co-benefits. However, progress in the take-up of low-carbon mobility measures substantially lags behind the potential. A number of barriers contribute to this lack of uptake. This paper explores those barriers by focusing on vehicle fuel efficiency in particular, but will also touch on the wider policy framework to improve the efficiency of the transport sector and reduce emissions.

K. M. Ahamed Sheriff (✉)
Sethu Institute of Technology, Virudhunagar District, Virudhunagar, Tamil Nadu, India
e-mail: adsharif.09@gmail.com

M. Elango
Thiagarajar College of Engineering, Madurai, Tamil Nadu, India
e-mail: memech@tce.edu

K. M. Mahaboob Sheriff
Nawab Shah Alam Khan College of Engineering and Technology, Hyderabad, TS, India
e-mail: dr.mahaboobsheriff@gmail.com

P. K. Kapur et al. (eds.), *Decision Analytics Applications in Industry*, Asset Analytics,
https://doi.org/10.1007/978-981-15-3643-4_7

7.1.1 Low-Carbon Logistics in SMEs

SMEs play a key role in national economies around the world, generating employment and adding value and contributing to innovation. SMEs are central to the efforts to achieve environmental sustainability and more inclusive growth. However, these contributions vary widely across firms and across countries and sectors. Reducing the environmental impact of SMEs by achieving and going beyond environmental compliance with existing rules and regulations in both manufacturing and services is a key factor for success in the green transformation [4].

Currently, under the country's strong control, key energy-consuming enterprises gradually tapped the potential of energy saving and emission reduction. But the energy saving and emission reduction measures for SMEs are relatively weak, so it is an urgent need to further intensify related policies. With the implementation of industrial restructuring and rejuvenation program, supports for SMEs have intensified, so, many SMEs are entering a rapid development stage. But energy consumption and carbon emissions gradually increased. Obviously, SMEs will gradually become government's key management objects.

While the impact of individual SMEs on the environment is small, their collective impact is considerable—indicating that there are significant opportunities for the adoption of resource conserving measures which could potentially also contribute to business competitiveness in an increasingly low-carbon economy [1]. Commercial waste, as well as being an issue of concern by itself [5] is an important source of "embedded" emissions of greenhouse gases.

Albeit much effort has been made to reduce carbon emission in manufacturing activities, little has been done in the logistics industry. The kineticism toward reduced carbon intensity of logistics activities will engender both opportunities and imperils for logistics and conveyance operation in SMEs. As suppliers to the ecumenical market, it would be obligatory for SMEs to understand carbon management policies and the requirements of the overseas buyers on suppliers.

Low carbon SMEs are virtually twice as liable to have export customers as their counterparts in the non-low carbon businesses. Global statistics reveal that virtually two in five low-carbon SMEs are already exporting to a diverse set of countries and three-quarters of them plan to enter or expand exports to incipient markets in the next two to five years. In India, SMEs account for 60% of all private sector jobs and about 40% of low carbon business.

But obviously, Indian SMEs have their share of unique challenges in terms of labour-intensive and low-technology ecosystem. Given that smaller enterprises drive innovation, magnification and job engenderment and the strategic paramountcy they now relish, it seems inevitably ineluctable that SMEs additionally play a pivotal role in driving the ecumenical low-carbon economy.

7.2 Research Methodology

A typical research process consists the following steps [3]:

- Formulation of the problem
- Development of research hypotheses
- Planning the study
- Data collection and processing
- Analysis and Interpretation
- Presentation of results.

Each step of the research process has been explained in detail as follows.

7.2.1 Formulation of the Problem

The research problem has been identified by making thorough literature review.

Theiben et al. [6] applied the case study method for reducing carbon emission by operational activity [6]. Freis et al. [7] developed the energy interrelations within the system logistics center for decision-making during the orchestrating of energy-efficient and CO_2-neutral systems [7]. Nie et al. [8] developed a level, which has positive effect to energy efficiency [8]. Cheng and Zhang [9] analyzed that high energy consumption is a major feature of the logistics industry [9]. Qian et al. [10] proposed that the social corporate environmental responsibility of a low-carbon logistics provider has positive relationship with customer environmental requirement [10]. According to the United Nations, weather events are becoming more extreme and greenhouse gas emissions are now at their highest levels in history [11]. Herold and Lee [12] analyzed that there has been no research considering supply chain integration as a carbon management strategy of logistics service providers [12]. Low-carbon sectors are subset of both the low-carbon sectors strategically identified by India and BIS study [13]. It is well-kenned that the 3 TBL (Environmental, Economical, and Social) perspective framework was proposed by Elkington [14] for industrial sustainability [14].

In order to fill the identified research gaps in the literature, it is proposed to conduct qualitative case study research in Indian SMEs and also to find out the effect of those influential factors based on TBL concept.

7.2.2 Development of Research Hypotheses

Initially, field investigations, in-depth interviews, telephonic conversation with subordinates were conducted in the SMEs to have a broad idea about the real factors affecting the implementation of low-carbon logistics in Indian SMEs.

Based on this observation as well as from extensive literature review, it has been decided to frame initial research hypotheses that have to be analyzed in this qualitative statistical study. Then pilot study is conducted by sending these research hypotheses to the experts in this field to know the opinion of them on the suitability to the problem. Finally the following research hypotheses have been selected on the aspect of Triple Bottom Line (TBL) (Environmental, Economical and Social) principle.

RH.1: Lack of awareness affects the implementation of low-carbon logistics in SMEs (General)
RH. 2: It is necessary to collect logistics activity data regarding carbon emissions (Environmental)
RH. 3: Application of Reverse logistics activities (Reprocessing, Remanufacturing, Repairing, Reusing, Recycling) leads to reduce carbon emission in packaging (Environmental)
RH. 4: Usage of Radio Frequency Identification (RFID) technology can assist in reducing carbon emission of warehousing in SMEs (Economical)
RH. 5: Usage of biofuel in land transport leads to neutralize carbon emission (Environmental and Economical)
RH. 6: Multimodal transport leads to efficient implementation of low-carbon logistics in SMEs (Environmental and Economical)
RH. 7: Corporate social responsibility is one of the major motivations for low carbon logistics (Social)
RH. 8: All SMEs have to obtain ISO 14000 certificate to reduce environmental impact (Social).

7.2.3 *Planning the Study*

A survey was designed to explore stakeholder views on the suitability of broad application of low-carbon logistics in Indian SMEs. In this regard, it has been decided to focus on the SMEs located in southern parts of India and survey questionnaire was planned to be circulated among the various levels of stakeholders in those SMEs in order to analyze the factors affecting the implementation of LCL on the aspect of Environmental, Economical and Social issues.

7.2.4 *Data Collection and Processing*

Survey data has been collected by targeting wide range of stakeholders with different level of experience and knowledge in the logistics industry and are involved in the business of SMEs. Most of the respondents participated have been passed through the interview process. Survey questionnaire consists of 3 parts which are listed as follows:

Introductory section—Short notes on research problem was provided
Background section—Personal and company details of the respondent were asked
Hypotheses section—Designed research hypotheses questions (as explained in Sect. 7.2.2) with multiple-choice, 5-point Likert scale type were listed.

A sample of around 50 participants from various types of SMEs has been selected for conducting survey and they have been asked to express their opinion on the research hypotheses questions. In order to collect data from respondents, an online survey development cloud-based software called survey monkey (www.surveymonkey.com) which is an effective source to collect survey data has been used. it is clear that chosen SMEs belong to manufacturing, warehousing, service and other sectors and most of the SMEs participated in this survey are from manufacturing and service sector. Figure 7.1 provides the detail of types of SMEs and number of stakeholders from those SMEs involved in this survey.

Similarly, Fig. 7.2 provides the detail of size of each SMEs, i.e., number of employees working in those SMEs. There are five categories on the size of chosen SMEs, i.e., (1–10, 11–30, 31–50, 51–100 and above 100 employees). Most of the SMEs participated in this survey are having more than 100 employees.

Sample population for this survey consists of the following level of stakeholders in those chosen SMEs: Trainee, Junior level/Subordinate, Trainer level/Team leader, Intermediate level/Superior position, Top-level management. Most of the respondents belong to Trainer level/Team leader category.

Before making analysis, data collected through survey monkey were transferred to SPSS for further processing. Figure 7.3 shows the level of stakeholders in Likert scale.

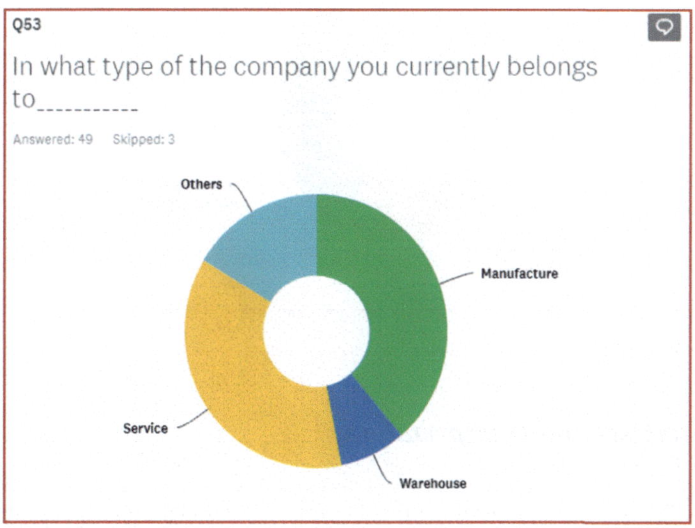

Fig. 7.1 Types of SMEs participated

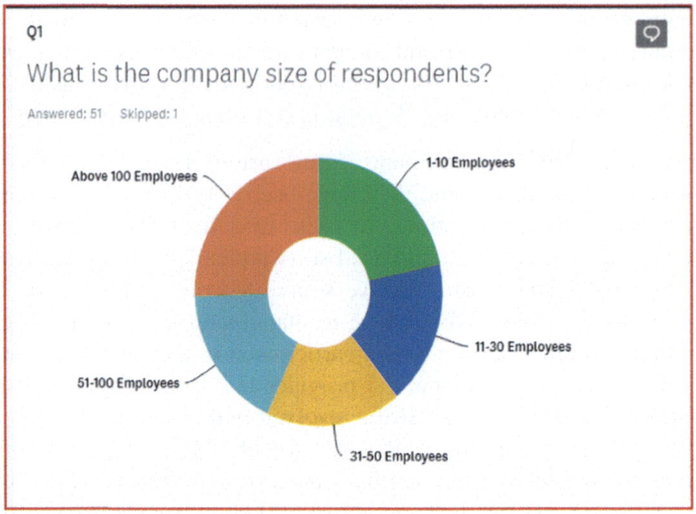

Fig. 7.2 Size of SMEs

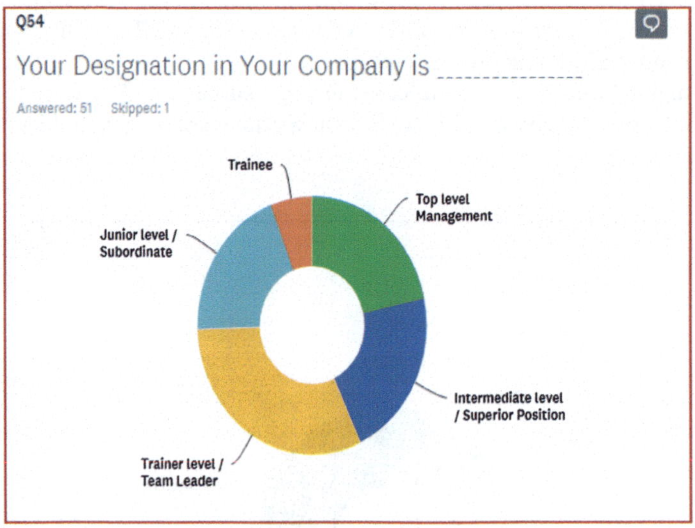

Fig. 7.3 Level of SMEs

7.2.5 Analysis and Interpretation

Tables 7.1 and 7.2 are the output results that are obtained by conducting one sample t-test on the collected data using SPSS. Table 7.1 provides the detail of sample statistics. To test the null hypotheses, two-tailed test was conducted by fixing $\alpha =$

Table 7.1 Sample statistics output

One sample statistics				
	N	Mean	Std. deviation	Std. error mean
Corporate social responsibility	52	3.98	1.111	0.154
Lack of awareness	52	4.15	1.211	0.168
ISO 14000 certification	52	3.73	1.223	0.170
Efficient vehicle usage	52	4.33	1.279	0.177
Collecting logistics activity	52	4.25	1.454	0.202
RFID technology usage	52	4.04	0.969	0.134
Biofuel usage	52	3.96	1.009	0.140
Reverse activities	52	3.04	1.521	0.211

Table 7.2 Sample test result

One sample test						
	Test value = 1.675					
	t	df	Sig. (2-tailed)	Mean diff.	95% confidence interval of the diff.	
					Lower	Upper
Corporate social responsibility	14.962	51	0.000	2.306	2.00	2.62
Lack of awareness	14.763	51	0.000	2.479	2.14	2.82
ISO 14000 certification	12.125	51	0.000	2.056	1.72	2.40
Efficient vehicle usage	14.950	51	0.000	2.652	2.30	3.01
Collecting logistics activity	12.775	51	0.000	2.575	2.17	2.98
RFID technology usage	17.582	51	0.000	2.363	2.09	2.63
Biofuel usage	16.341	51	0.000	2.287	2.01	2.57
Reverse activities	6.466	51	0.000	1.363	0.94	1.79

0.05, $t_{1-\alpha/2,\nu} = t_{0.975,51} = 2.008$ (from critical values of student's *t*-distribution table) for 95% confidence interval with $52 - 1 = 51$ degrees of freedom. Since the level of significance value for all research hypotheses is lesser than α Value (i.e., 0.05) as shown in Table 7.2, it was decided to reject null hypothesis. Thus all the designed research hypotheses were found acceptable.

7.2.6 Presentation of Results

In this section, details of responses from participants on selected research hypotheses are provided with its justification.

RH. 1: Lack of awareness affects the implementation of LCL-SMEs

Stakeholders from chosen SMEs were asked to assess to what extent lack of knowledge impacts in current Indian scenario. The answers for the respondents belonging to chosen SMEs were rated on a five-point Likert scale.

From our investigation, it has been found that among 52 respondents, this question was skipped by 4 participants. Among the 48 participated, 54% of the respondents replied that they were partially aware of Low-carbon logistics, 19% of the respondents accepted that they were not aware of LCL, 15% of the respondents replied that they understood the concept very well, 10 and 2% of the respondents replied that they have partially and well-implemented this LCL concept in their SMEs. Figure 7.4 shows the response of stakeholders in Likert scale.

From this survey, it is clear that stakeholders in Indian SMEs are not that much aware about LCL and benefits of its implementation and hence LCL has not effectively implemented in Indian scenario in comparison with other developed countries like US, UK, and Australia. In order to make awareness about the importance of LCL concept, Indian SMEs have to conduct training and awareness programs to their people by qualified and experienced logistics professionals.

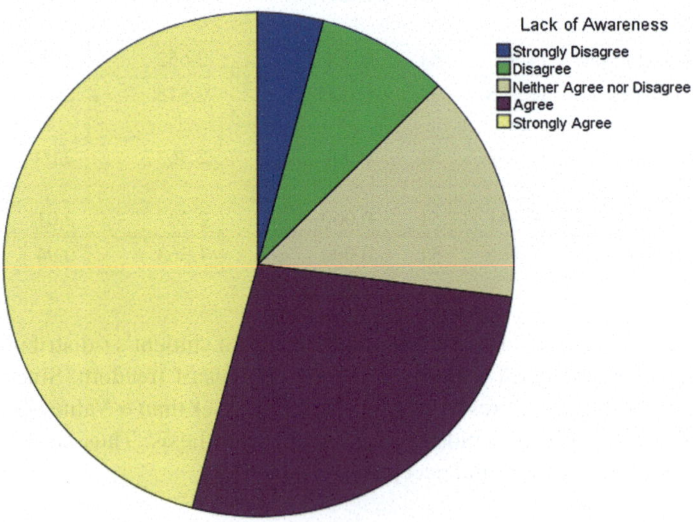

Fig. 7.4 Influence of lack of awareness

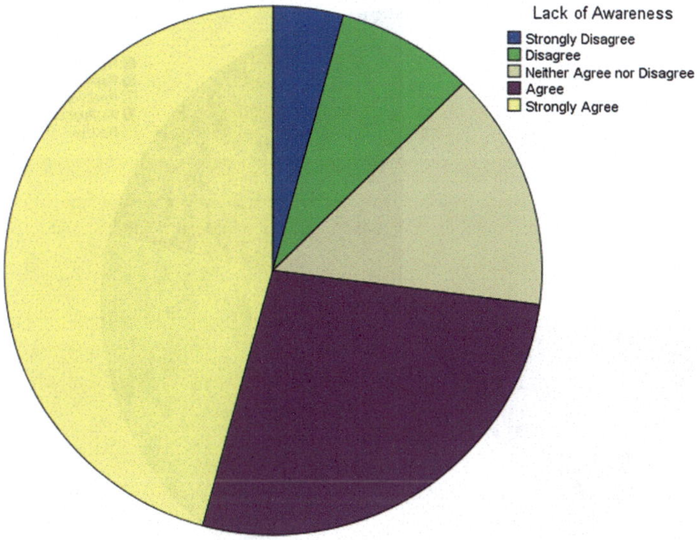

Fig. 7.5 Collecting logistics activity

RH. 2: It is necessary to collect logistics activity data regarding carbon emissions

Stakeholders from chosen SMEs were asked to express their opinion on the necessity of collecting logistics activity data regarding carbon emissions.

From our investigation, it has been found that among 52 respondents, this question was skipped by 4 participants. Among the 48 participated, 60% of the respondents strongly accepting that it is necessary to collect logistics activity data while 15% of the respondents strongly refusing to accept it. 18% of the respondents just accepted the hypotheses and 6% of the respondents have not given any opinion. Hence most of the respondents stressed the importance of collecting logistics activity data in order to identify emission sources of energy consumption and also to impact environmental performance. Figure 7.5 shows the response of stakeholders in five-point Likert scale.

Fuel receipts/records for stationary sources consumption, vehicle fuel receipts/records, refrigeration and air-conditioning records, electricity bills, town gas bills, water supply bills, paper purchasing, and recycling records are some of the data sources to collect logistics activity data regarding carbon emissions.

RH. 3: Application of Reverse logistics (RL) activities (reprocessing, remanufacturing, repairing, reusing, and recycling) leads to reduce carbon emission in packaging

Research hypothesis with 5 point multiple-choice options was given to the stakeholders of chosen SMEs in order to identify their preference on the application of reverse logistics activities to reduce carbon emission in packaging.

Most of the respondents have opted the reverse logistic activity "Reuse" as shown in Fig. 7.6, which involves collection of used packaging materials for further packag-

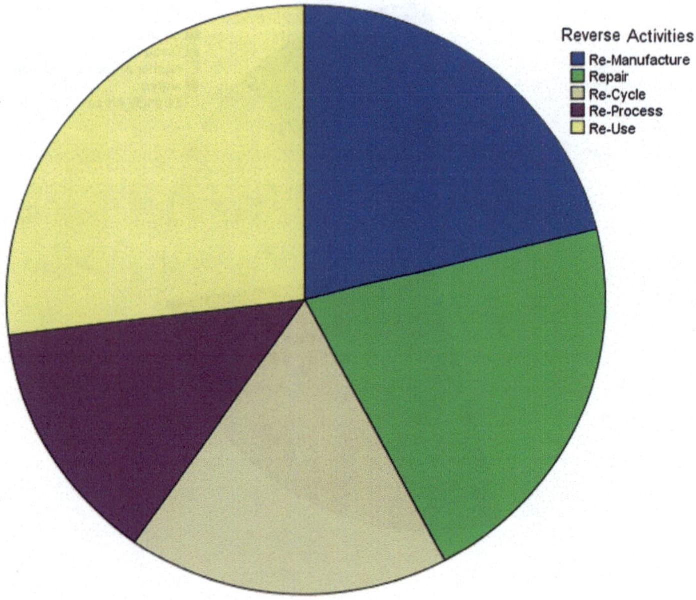

Fig. 7.6 Reverse activities

ing purpose that will reduce the carbon emission due to new packaging production. "Repair" and "Remanufacturing" are the next preferred RL activities by the respondents for packaging. "Recycling" and "Reprocessing" are the least preferred RL activities by the respondents since these will be expensive as it involves mechanical or chemical disassembly processes that require additional resources. All this RL activities will impact economic performance of the packaging process.

RH. 4: Usage of Radio Frequency Identification (RFID) can assist in reducing carbon emission of warehousing in SMEs

The influence of RFID in reducing carbon emission of warehousing was tested by getting opinions from the stakeholders of chosen SMEs through survey. The output of survey responses was shown in Fig. 7.7.

From this survey it was found that among 48(out of 52) respondents, 46% of the respondents accepted that RFID can assist in reducing carbon emission of warehousing, 23% of the respondents strongly accepting RFID's usage, 27% of the respondents have not provided any opinion on this issue, and 2% of the respondents refused to accept RFID usage. The cause behind the selection of RFID by number of respondents was its contribution to act as an extension of Warehouse Management System which may provide complete visibility of pallets, cases, and products as well as it may give clear updated inventory information while receiving and storage.

RH. 5: Usage of biofuel in land transport leads to neutralized carbon emission

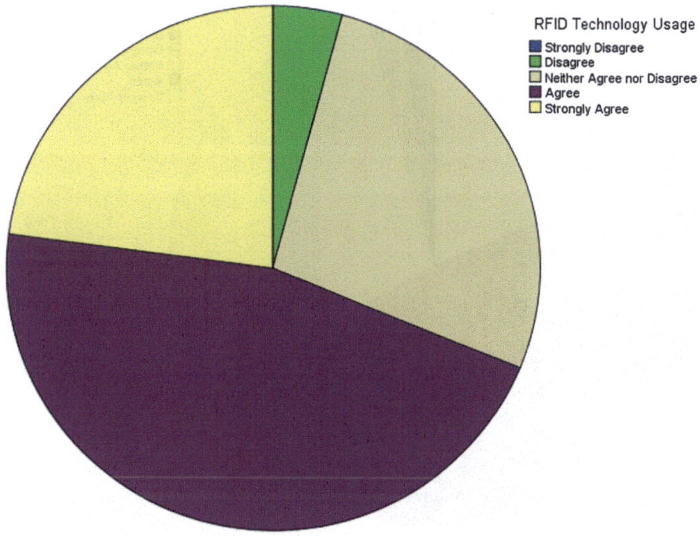

Fig. 7.7 RFID Tech. usage

From survey output, it was found that among 48(out of 52) respondents, 19% of the respondents strongly accepting that the usage of biofuel in land transport will reduce carbon emission, 48% of the respondents just accepting it, 29% of the respondents have not given any opinion on this issue and 2% of the respondents have refused to accept the biofuel usage as one of the way to neutralize carbon emission.

It was clearly evident from this study that most of the stakeholders in SMEs have chosen this biofuel as an alternate choice to avoid the scarcity of conventional fuel such as petrol, diesel, etc. Since the biofuel is made of food by-products such as sugarcane ethanol, algae, etc.

Developing countries like India and China have to explore more on the application of this biofuel in order to meet high future demand and scarcity in conventional fuels as well as to make it an alternate option for conventional fuel.

The biofuel usage survey output is shown in Fig. 7.8.

RH. 6: Multimodal transport leads to efficient implementation of low-carbon logistics in SMEs

In order to collect opinion from stakeholders of chosen SMEs on the choice of multimodal transport for efficient implementation of low-carbon logistics, survey questionnaire was being circulated to the respondents via online source.

Figure 7.9 shows the output on the various modes of transportation. From the output, it was found that most of the respondents (80%) have accepted that multi-modal transport leads to efficient implementation of low-carbon logistics in Indian SMEs. It is also clearly evident that carbon emission can be reduced by choosing various modes of transportation. It is always better to go for low emission modes of transportation (i.e., water and rail) instead of high emission modes of transportation

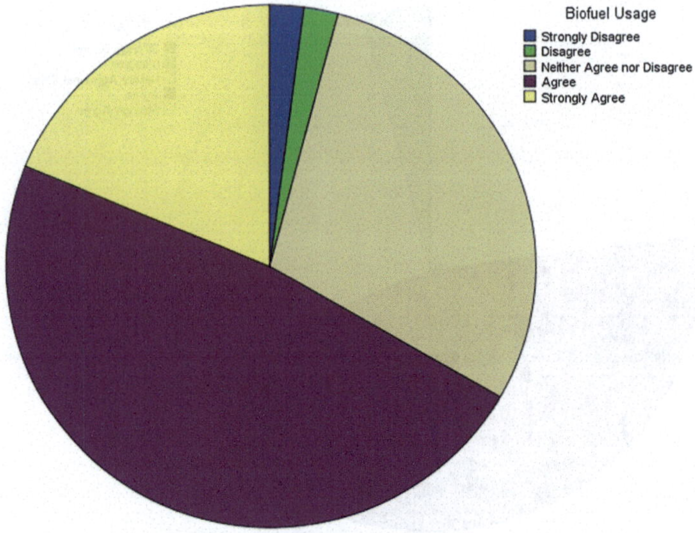

Fig. 7.8 Bio fuel usage

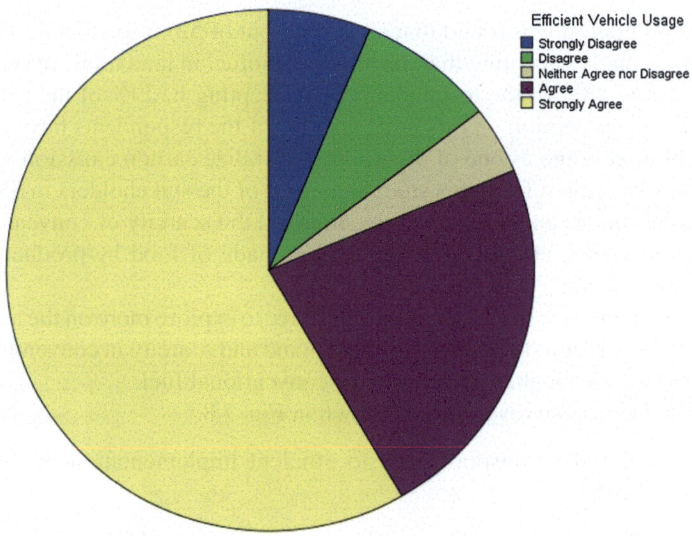

Fig. 7.9 Efficient vehicle usage

(i.e., truck and air) whenever possible. It was also found that choosing combination of transport modes is one of the best options to reduce carbon emission instead of choosing single transport mode. It has to be decided by the SMEs on the optimal choice of transport mode by making an analysis with due consideration of triple bottom line (Environmental, Economical, Social factors) principle.

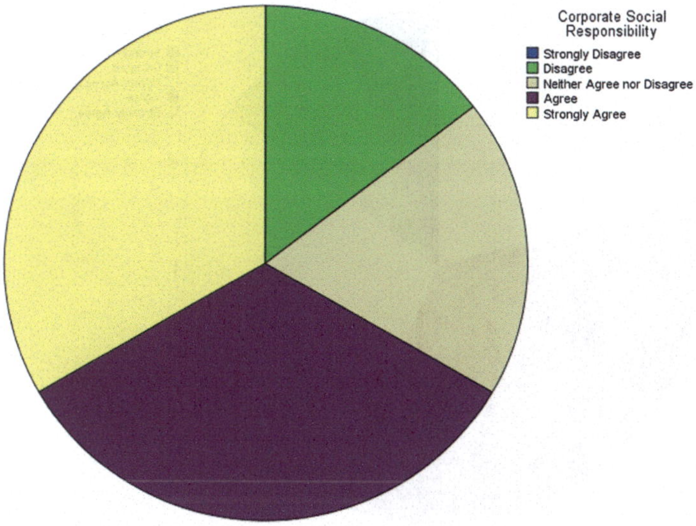

Fig. 7.10 Corporate social responsibility

RH. 7: Corporate social responsibility is one of the major motivations for Low-Carbon Logistics

From Fig. 7.10, it was found that among 50 (Out of 52) respondents, 68% of the respondents have accepted that Corporate Social Responsibility is one of the major motivation for low-carbon logistics in Indian SMEs. It was also found that CSR plays an important role in developed countries like US, EU, and Australia as their government made it as a mandatory policy for all SMEs. But in Indian scenario, CSR has not become mandatory like a policy as like developed countries. Hence policies and regulations on CSR are to be designed by Indian government in order to make the SMEs to have the societal welfare interest.

RH. 8: All SMEs have to obtain ISO 14000 certificate to reduce environmental impact

To assess the importance of ISO 14000 certification, a survey was conducted in Indian SMEs.

The output results are shown in Fig. 7.11. From the investigation, it was found that around 60% of the respondents have accepted that SMEs have to obtain ISO 14000 certification to reduce environmental impact. Around 30% of the respondents have not given any opinion on this hypothesis. It was also clearly evident from current industrial scenario, some of the customers who are dealing with low carbon products, use ISO 14000 certification as one of the essential criteria for supplier selection.

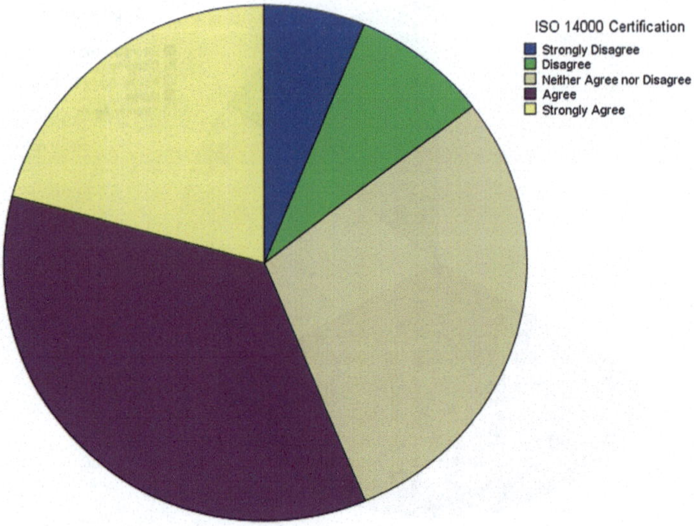

Fig. 7.11 ISO 14000 certification

7.3 Conclusion

In this paper, a research has been conducted on the analysis of factors affecting the implementation of low-carbon logistics in Indian SMEs. For this purpose, research hypotheses were designed according to triple bottom line (TBL) principle (i.e., with due consideration of environmental, economical, and social issues). In order to test these research hypotheses, a survey questionnaire was prepared and was circulated among the stakeholders of SMEs located in South India. After getting opinion from those stakeholders, the responses were analyzed using SPSS. Then based on the result analysis, each designed research hypothesis was proved with its detail justification.

From this research, it was found that low-carbon logistics is still in an early stage of implementation in developing countries like India. In order to meet the future demand of low carbon economy as well as to establish carbon-free, sustainable transportation, Indian government has to form more policies for controlling carbon emission. Apart from that more awareness has to be created among the stakeholders of Indian SMEs and effective training has to be provided to them for implementing LCL in Indian scenario. In addition to that more extensive research has to be conducted on the implementation of LCL in Indian SMEs by making in-depth survey and analysis.

References

1. Environment Agency., Spotlight on business environmental performance in 2006, Environment Agency
2. Mall RK, Gupta A, Singh R, Singh RS, Rathore LS (2006) Water resources and climate change: an Indian perspective
3. Maylor H, Blackmon K (2005) Researching business and management. Palgrave Macmillan, Houndmills
4. OECD (2013a) Green entrepreneurship, eco-innovation and SMEs. Working Party on SMEs and Entrepreneurship
5. House of Lords (2008) Waste Reduction. Science and Technology Committee, House of Lords, The Stationary Office, London, vol 1
6. Theiben S, Spinler S, Huchzermeier A (2014) Reducing the carbon footprint within fast-moving consumer goods supply chains through collaboration: the manufacturers' perspective
7. Freis J, Vohlidka P, Günthner WA (2016) Low-carbon warehousing: examining impacts of building and intra-logistics design options on energy demand and the CO_2 emissions of logistics centers
8. Nie Y, Xiao Z, Cao B, Zhou G (2016) Analysis of influence factors for low-carbon logistics
9. Cheng D, Zhang X (2017) Overview of low carbon logistics development in china and foreign countries
10. Qian C, Wang S, Liu X, Zhang X (2019) Low-carbon initiatives of logistics service providers: the perspective of supply chain integration
11. United Nation. Goal 13: take urgent action to combat climate change and its impacts
12. Herold DM, Lee K-H (2019) The influence of internal and external pressures on carbon management practices and disclosure strategies
13. https://www.teriin.org/blog/indian-industries-voice-support-creation-coalitions-affect-long-term-sustainable-growth, Accessed 6 Sep 2019
14. Elkington J (1998) Partnerships from cannibals with forks: the triple bottom line of 21st-century business

Chapter 8
A Game-Theoretic Model of Deceptive Ambush as Counter Measure for Habitat Selection in Cross-Border Infiltration

Shivam Sharma and Ashok Kumar Sinha

8.1 Introduction

This paper addresses the important subject of cross-border infiltration and seeks to add to the work that has been done over the past decade on applying game-theoretic models of conflict to effective border control. The most fundamental concept in our model of cross-border infiltration control is the notion of infiltration routes. Specifically, our game-theoretic model is centered on human settlements that are often found to exist close to the international border on the defending side. Research on the individuals apprehended at the border has revealed that these human settlements figure prominently in the conscious conflict decision processes of infiltrators to attempt illegal crossings with the motive of carrying out terrorist and subversive activities. The infiltration routes almost always pass through some of these human settlements. The reasons become clear once we understand the factors influencing the choice of infiltration routes. An inefficient infiltration route may incur many costs associated with several factors, such as food and water requirement, physical and psychological stress, higher chances of discovery by the defending forces, significant exposure to disease and the risk of losing one's way and not reaching the destination. In fact, the choice of infiltration routes may be constrained more by the need for transportation of supplies to the rest of the infiltrators along the trail than by the stress placed on the infiltrator. Human settlements along such trails, therefore, act as more or less permanent waypoints and serve as logistics bases, rest areas, and even hiding places.

S. Sharma (✉)
Indraprastha Institute of Information Technology, Okhla Industrial Estate,
Phase III, New Delhi 110020, India
e-mail: shivam17147@iiitd.ac.in

A. K. Sinha
Defence Terrain Research Laboratory, Defence Research and Development Organisation,
New Delhi 110054, India
e-mail: sinha23ak@gmail.com

P. K. Kapur et al. (eds.), *Decision Analytics Applications in Industry*, Asset Analytics,
https://doi.org/10.1007/978-981-15-3643-4_8

Various trails converge at these settlements before diverging to different destinations. On the other hand, as counter-measures, these semi-permanent settlements also serve as optimal points or choke points for constructing infiltration barriers, setting up sensors, and carrying out ambush attacks.

In this paper, we consider a game-theoretic model of deceptive ambush as an example of strong-point defense at the choke points. In Sect. 8.2, we describe the cognitive tactics of the players involved in cross-border infiltration. In Sect. 8.3, we begin by first investigating a 2×2 zero-sum game-theoretic model of habitat selection under the threat of ambush but no deception. The two players, namely, defender (Blue) who is the interdictor or ambusher and the infiltrator (Red) determine the tactical options available to them. The ambusher assigns a numerical value (Measure of Effectiveness, MOE) to each possible outcome by judging the potential gain or loss from an encounter. Both the players as intelligent opponents vary their tactics to appear unpredictable to the opponent. A player's overall strategy is determined by how often he chooses one option over another.

Players calculate all possible strategies and the net gain or loss to each one. The players adjust to each other's options over the play only to eventually settle on an optimum strategy. In a military scenario, the optimum strategy is not necessarily the most desirable outcome (winning the encounter) but the best action that one can do against an opponent of a given strength. With the proposed game-theoretic model, the players can check to see if the outcome of the encounter favors him or the opponent. If both players correctly ascertain the situation, then the losing player may decide not to participate in the encounter. In this game model, it is assumed that the players play rationally and has aspects of both simultaneous and sequential-move game, i.e., even though the players act simultaneously, each player has partial information about the other's progress and can respond accordingly. Such partial information may help the player to assign a value to success of an ambush and thereby calculate strategies, as shown in Sect. 8.3. In Sect. 8.4, the present game model is extended to a game of asymmetric information where one of the players, namely, the infiltrator is either denied information or fed with wrong information through deception [1]. This extended model emphasizes the value of deception achieved through a wrong projection of payoffs. By definition, asymmetric warfare revolves around the weaker player using its strengths to exploit the apparent weaknesses of the stronger opponent [2]. In this paper, however, the asymmetry considered is only in matters of information available to the players and the players may otherwise be evenly matched in the case of armed conflict. Section 8.5 models a game of deceptive ambush where deception is achieved through inverted signaling. The effect of deception is demonstrated through the outcome of a game where one player deceives the other player leading him to a waiting ambush. The paper concludes with a discussion of the proposed model's utility for strategic decision-making of security personnel.

8.2 Background

8.2.1 Ambush

Ambush is a tactic which has been widely employed both in recent and historical times. Ambush relies on leveraging the advantages of local domain knowledge and conducive environmental situations to present an element of surprise for an unsuspecting enemy. Concealed positions, such as, under a dense bush and other types of hiding in a forest or a hilly terrain are employed as intentional strategic attack points. In the case of an ambusher–infiltrator scenario, ambush plays a vital role in surprising and capturing the infiltrator. While the infiltrator may want to evade an ambush, the defending forces may attempt to carry out a successful trap. As mentioned, obstacles and environmental factors have a crucial influence on such interactions.

8.2.2 Deception

Deception is a prominent cognitive feature of various intelligent entities including human beings [3, 4]. Unlike other decisional abilities, deception involves the creation of situational dilemma to deceive the opponent to act in favor of the deceiver. It is a potential indicator of social behavior and theory of mind [5]. Deception conveys intentions at its core which involves utilization of theory of mind for manipulation of opponent beliefs. Agents with better knowledge of the mindset, beliefs, objectives, and other characteristics of an enemy can employ deception to gain various advantages [6, 7].

In military and defense operations, deception is a central parameter around which various strategies revolve [8, 9]. Deception involves propagating false or half-truth information. Lies, equivocations, concealment, exaggeration, and understatements are five primary types of deception. Interpersonal deception theory by Buller et al. [10] explains the conscious and subconscious behavior during a deceptive environment.

8.2.3 Habitat Selection

The issue of habitat selection in predator–prey interactions in the animal kingdom has been extensively addressed using game-theoretic models [11–13]. Predator–prey interactions are a major evolutionary driving force that determines the behavior of both predator and prey. Ideally, in order to maximize its fitness and survival, an animal would maximize the time spent in foraging for food while taking care not to become the meal while looking for one and has to evaluate a trade-off between feeding and survival. Avoiding overlap with potential predators is achieved by adjusting feeding behavior while remaining in a secure place during high-risk times and not risking any encounter. To the best of our knowledge, the present paper is the first application of game theory to human–human interaction similar to a predator–prey scenario for habitat selection.

In an infiltration scenario, people living in habitats close to international borders play a crucial role in the encounter between infiltrator and security personnel. An infiltrator on a covert mission suffers certain losses in different forms. These losses may either be regarding ammunition, health or any other kind of damage. Hence, an infiltrator requires human settlements where he can stay and recover his losses. There may be a single habitat or multiple habitats that he can choose depending on the situation. People residing in the habitat may be pro-infiltrator, anti-infiltrator, or neutral. It is possible that certain areas are pro-infiltrator because of socio-political reasons. Pro-infiltrator and neutral habitats are the most advantageous to the infiltrators.

On the other hand, the security personnel are fully aware of such habitats and try to use these habitats to capture or ambush the infiltrator. They try to win over the population of the habitat in case it is *pro-infiltrator*. Here, *win over* refers to convincing inhabitants to support the defense personnel and assist in the capture of the infiltrator. The infiltrator has to decide whether or not to visit a habitat, and the defender has to decide whether to *win over* the inhabitants or *penetrate* the habitat without winning over. Direct penetration has its risks but also saves resources of the defense that may have to be employed to win over the population and also gives them a chance to deceive the infiltrator.

It is evident that both the infiltrator and the ambusher have to choose from multiple strategies. The choice may be governed by the prevailing situation, future requirements, beliefs, or experiences on either side. The infiltrator also needs to take into account the possibility of the ambush and nature of the population of the habitat before he decides to visit.

8.3 Game-Theoretic Formulation of Habitat Selection for Ambush Operation

As introduced in the previous sections, a habitat presents an opportunity for the infiltrator to replenish its resources while on a mission. However, what is crucial is the decision-making process of an infiltrator who has to decide whether to visit a particular habitat or not given the risks involved. This decision-making process gets further complicated when multiple habitats are introduced. In this case, the infiltrator has to decide which all habitats may be visited and their corresponding order. In this paper, only a single habitat model is considered.

8.3.1 Single Habitat Model

Let us assume that

A. The *infiltrator* has two possible strategies

1. *To visit the habitat* and
2. *Not to visit the habitat*

B. The ambusher also has two possible strategies

1. To *penetrate* into the habitat. This involves the ambusher entering the habitat in order to encounter the infiltrator. The ambusher does not interact or take support of the inhabitants.
2. To *win over* the habitat. This involves the ambusher taking necessary measures to win over the local inhabitants in order to obtain covert support in his operation.

This problem can be formulated as 2×2 game as explained in Fig. 8.1. The corresponding payoffs for each of the rows and columns are represented by a_{ij}. The matrix represents the game from the ambusher's perspective, wherein a positive payoff value is a gain for the ambusher and an equivalent loss for the infiltrator, while a negative payoff value is a gain for the infiltrator and an equivalent loss for the ambusher.

Here, p, $(1-p)$ and q, $(1-q)$ denote the probabilities of strategy selection for Blue and Red, respectively. Value of the game (v) for the general game matrix of Fig. 8.1 is given by

$$v = pq(a_{11} - a_{12} - a_{21} + a_{22}) + p(a_{12} - a_{22}) + q(a_{21} - a_{22}) + a_{22} \tag{8.1}$$

$$\frac{\partial v}{\partial p} = 0 \Rightarrow q(a_{11} - a_{12} - a_{21} + a_{22}) + a_{12} - a_{22} = 0 \tag{8.2}$$

Probabilities for Red's strategies are

Ambusher (Blue) \ Infiltrator (Red)	Visit (q)	Not Visit $(1-q)$
Penetrate (p)	a_{11}	a_{12}
Win over $(1-p)$	a_{21}	a_{22}

Fig. 8.1 A two-player zero-sum game matrix for single habitat selection from ambusher's perspective is represented. Each a_{ij} corresponds to the quantitative measure of achievement received by the players

$$q = \frac{(a_{22} - a_{12})}{(a_{11} - a_{12} - a_{21} + a_{22})} \tag{8.3}$$

and

$$1 - q = \frac{(a_{11} - a_{21})}{(a_{11} - a_{12} - a_{21} + a_{22})} \tag{8.4}$$

Similarly,

$$\frac{\partial v}{\partial q} = 0 \Rightarrow p(a_{11} - a_{12} - a_{21} + a_{22}) - a_{22} + a_{21} = 0 \tag{8.5}$$

Probabilities for Blue's strategies are

$$p = \frac{a_{22} - a_{21}}{a_{11} - a_{12} - a_{21} + a_{22}} \tag{8.6}$$

$$1 - p = \frac{a_{11} - a_{12}}{a_{11} - a_{12} + a_{22} - a_{21}} \tag{8.7}$$

From Eqs. (8.1), (8.3), and (8.6) the value of the general game matrix can be obtained as

$$v = \frac{a_{11}a_{22} - a_{12}a_{21}}{a_{11} - a_{12} - a_{21} + a_{22}} \tag{8.8}$$

8.3.2 *Illustrative Scenarios for a Single Habitat Game*

In this section, we explain the formulation of a 2 × 2 game matrix under a variety of scenarios and their corresponding interpretations.

8.3.2.1 Scenario 1

Value judgment leading to figures for payoffs is carried out by the players before the game starts. The game matrix for this scenario is shown in Fig. 8.2. The Blue judges that he is worth 6 points. Thus, if Blue uses the first strategy, that is, penetrate without first winning over the population and Red visits the habitat then Blue gets killed and loses 6 points because the population is assumed to be pro-Red, and Red is able to incite the population against Blue, or Red is able to use the community as a shield (since Blue is not able to retaliate lest the civilians being used as shield get harmed) or the population informs Red beforehand about the whereabouts of Blue so that Blue gets killed. If Blue penetrates the population and does not visit

Fig. 8.2 The figure shows the game matrix formulated for scenario 1

Ambusher (Blue) \ Infiltrator (Red)	Visit (q)	Not Visit ($1-q$)
Penetrate (p)	−6	1
Win over ($1-p$)	6	0

the habitat, then Blue gets 1 point as he is able to take some steps towards winning over the population, eroding somewhat Red's defenses. Blue has an opportunity to begin a process of winning over the population in the absence of incitement by Red (a small but long-term gain). It can also be treated as −1 since there is a cost attached to making efforts for winning over which may go waste.

Conversely, Blue also judges that Red, if captured or killed, is also worth 6 points. If Blue attempts to win over the local population and Red visits the human habitat, then Blue will win 6 points because then Red will be captured or killed with the assistance of the local inhabitants who are now thoroughly won over. If Blue wins over local inhabitants and Red does not visit the habitat, then Blue will not gain anything because there will not be any encounter, so no immediate gain.

Blue may keep $a_{12} = 1$ but make $a_{22} = 2$ believing that winning over the population even if Red does not visit may pay a richer dividend in future. This has been observed to be true actually. The location of the habitat also plays a crucial role in deciding the payoffs. In the case of human habitat being near the point of entry, the likelihood of Red's health deteriorating due to various reasons such as extended travel, an encounter with obstacles or need to hide from Blue will be little. Red is, therefore, less likely to visit the habitat.

On the other hand, if the human habitat is far away from point of entry with many obstacles between point of entry and human habitat there is higher likelihood of Red needing the human habitat and therefore a higher likelihood of Red visiting human habitat resulting in a higher probability of kill for Blue (i.e., if Blue has been able to win over the population), an effect that Blue can take into account while deciding payoffs. It is more so if Red considers his mission very important. The high importance of Red's mission means Red must succeed in carrying out his mission. For this, Red must move cautiously so that he can survive for the sake of the mission. Red may be expected to visit the human habitat more often to ensure his survival.

Both Red and Blue will vary their strategies in order to appear unpredictable to the other. For example, Blue's overall strategy is determined by how often he chooses one strategy over another. All possible strategies and mathematical outcomes of each strategy for the given game are presented in Fig. 8.3, where, the color bands represent the outcomes of each combination from Blue's point of view. With this graph, Blue can understand how he may alter his strategy based on Red's action. For

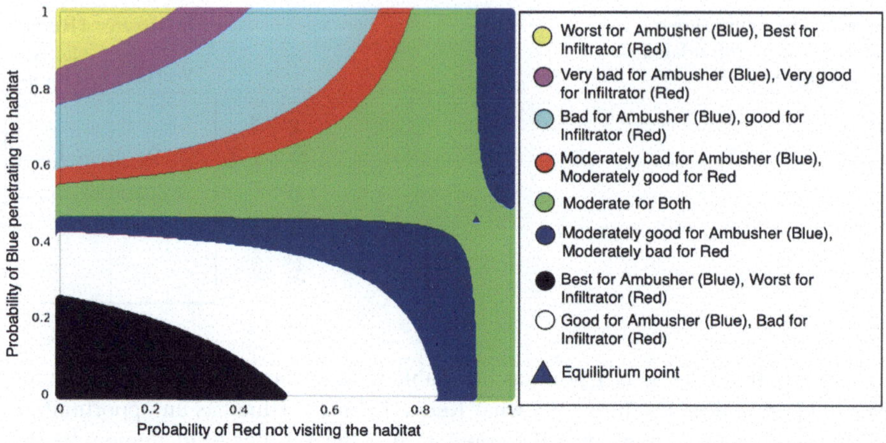

Fig. 8.3 Graph depicts the solution set for the game matrix described in Fig. 8.2 for the scenario 1. The x – axis represents Red's probability of not visiting the habitat and y – axis represents Blue's probability of penetration. The color bands represent the outcomes of each combination from Blue's point of view

instance, if Blue perceives that Red avoids the human population most of the time (this could be for many reasons, one reason being that he does not feel the need as described above; another, for instance, maybe that Red has been deceived by Blue into believing that the population is not pro-Red!), then Blue may choose to penetrate the population more frequently (upper right corner). This move will increase Blue's chances of catching Red because Blue will get more time to turn the population in its favor and therefore can capture Red in future. So, this is a good outcome for Blue. If Blue indeed does so then Red will turn the encounter in his favor by visiting the population more often (upper left corner of the plot). This readjustment takes place after Red learns through experience that he had been a victim of deception. This is a bad outcome for Blue because he will lose a high percentage of his vital resources. Blue may reverse the outcome by decreasing his effort to penetrate but increasing his effort to win over the local population most of the time (lower left corner). Based on this kind of analysis Blue can anticipate how Red is likely to respond to a given strategy.

The triangle in Fig. 8.3 represents that Blue's optimum strategy which is to penetrate the habitat with 0.46 probability, while Red's optimum strategy is to avoid the population with 0.077 probability. If Red visits more often, then Blue can move the outcome towards the lower left corner by increasing his time and efforts not in penetrating but gaining the support of the population (i.e., winning over). On the contrary, if Red visits less often then Blue can move toward the upper right corner by increasing his efforts to penetrate the local population. Having found each player's optimum strategy Blue may check to see whether the outcome of the encounter favors Blue or Red. Blue gets an average score of 0.46 even if red plays his optimum strategy. This is important for Blue because it indicates that even if both sides play intelligently,

Blue will have positive results. Moreover, it indicates that if Red correctly assesses the situation then he would not want to visit the habitat at all.

8.3.2.2 Scenario 2

The game matrix for scenario 2 is shown in Fig. 8.4. Figure 8.5 shows the solution set for this game.

The payoff value, $a_{12} = 100$, pertains to a situation where Blue is carrying vital sensors to be installed at the habitat urgently for surveillance and information gathering. So, it is important that Blue does not encounter Red as Blue penetrates the human habitat without winning over.

Ambusher (Blue) \ Infiltrator (Red)	Visit (q)	Not Visit ($1-q$)
Penetrate (p)	−5	100
Win over ($1-p$)	5	0

Fig. 8.4 The figure shows game matrix formulated for scenario 2

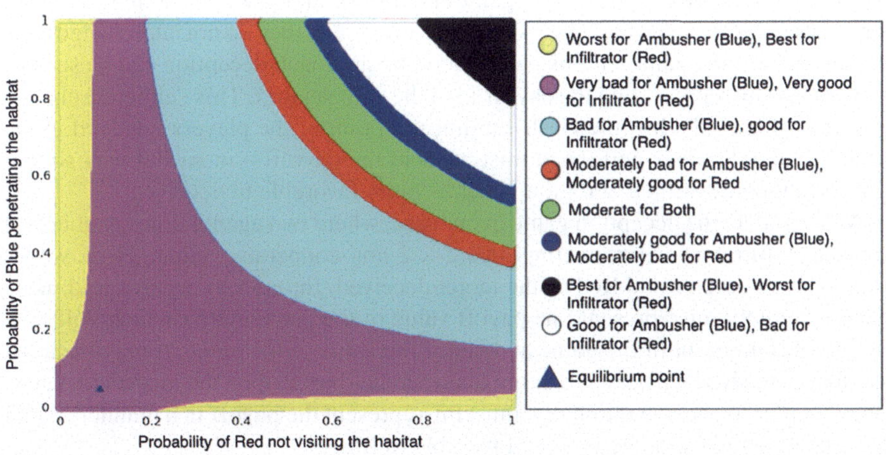

Fig. 8.5 Graph depicts the solution set for the game matrix in Fig. 8.4 for scenario 2. The $x-$axis represents the Red's probability of not visiting the habitat and $y-$axis represents Blue's probability of penetration. The color bands represent the outcomes of each combination from Blue's point of view

In other words, it is important that at the time of installation the probability of Red's visit is minimum. We see from Fig. 8.5 that the equilibrium solution is near the left corner of the graph. This means that Red should visit the habitat at a high rate, and Blue should respond by penetrating at a low rate. The average value of the game is 4.55, a very favorable outcome for Blue and a direct result of the effectiveness of Blue's low rate of penetration for a successful installation of the sensor.

8.4 Deception Strategy in an Ambush Operation: Manipulating Payoffs

Deception becomes unavoidable both in cooperative interactions and war-like situations since deception is based on the claim that people adjust their interactive behavior in response to their perception of a social situation's pattern of rewards and costs. The most effective element of ambush is surprise. However, deception, when used in an ambush operation that is staged (or is pretended to be staged but actually not staged) has the power to confound the enemy's combat decisions by several degrees at multiple levels. It has also been shown in the literature that deception always pays to the deceiver. Moreover, being rational players, each one knows that the opponent is trying to deceive him. The situation becomes more complex where players act deceptively while trying to avoid deception. Thus, each choice of behavior by an individual offers the likelihood of specific rewards and costs (also known as Measure of Effectiveness, MOU) after the interaction.

We first look at the deception purely from a game-theoretic viewpoint where two players are interacting in a zero-sum game and for simplicity, we also assume that one player is the deceiver and the other is being deceived, and they do not interchange their role throughout the game. In this game-theoretic approach deception can be seen as deliberately misrepresenting the payoff(s) of the game matrix. This deliberate change in the payoff(s) will shift the Nash equilibrium point of the players involved in the game. So, the deceiver will try to misrepresent the payoff(s) in such a way so that equilibrium point shifts to a region which is more favorable to deceiver.

We formulate the deception game in our case, where two agents Blue (Ambusher) and Red (Infiltrator) are interacting, in a 2×2 non-cooperative game. Here, we fix Blue to be a deceiver and Red as the target/deceived. In order to deceive Red, Blue will deliberately misrepresent one payoff value of the game matrix which will shift the equilibrium point of the game and hence the value of the game. It may be noted here that as the Red will be calculating his strategies based upon the modified matrix, Blue, in order to deceive rationally, must misrepresent the payoff in a manner so that Red's new optimal strategy is more favorable to Blue.

To understand the deception game, let's analyze the game in Fig. 8.6.

Consider that $1 \geq \alpha \geq \beta \geq \gamma \geq \delta$ in Fig. 8.6a. Here the deceiver will choose to play for strategy R1, while deceived will choose to play strategy S2. So, the game has an equilibrium point (R1, S2). In order to deceive, the deceiver changes the payoff

(a)

Deceived (Red)	Deceiver (Blue)	R1 (p)	R2 ($1-p$)
	S2 (q)	α	β
	S1 ($1-q$)	γ	δ

(b)

Deceived (Red)	Deceiver (Blue)	R1 (p)	R2 ($1-p$)
	S2 (q)	α	$\alpha+\zeta$
	S1 ($1-q$)	γ	δ

Fig. 8.6 Game matrix formulated for deception. **a** represents the original game matrix formulated with respect to Blue's (deceiver's) point of view. **b** represents the intended deception game matrix

of (R2, S2) to $\alpha + \zeta$ where $\zeta \geq 0$. The deceiver will not change his strategy as he knows the actual matrix. But with the misrepresented matrix, the deceived agent will recalculate his strategy and will play S1. Thus, misrepresentation of the game gives more payoff to the deceiver as shown in Fig. 8.6b. Now, in order to make it more realistic, we argue that if deceiver deceives with probability p, i.e., the deceived agent is getting two different matrices at different frequencies and unable to decide which one is correct, the target chooses left column and right column with probability $(1-p)$ and p, respectively. These strategies will be in equilibrium if choosing the first row is individually rational for the deceiver. This can be achieved by making p satisfy $(1-p)\alpha + p\beta > (1-p)\gamma + \delta p$. This condition guarantees that the deceiver has a higher payoff in the deception game. Since deceiver always chooses strategy R1, in a deception game the value of the game will be $(1-p)\alpha + p\beta$. While in a game with no deception the value of the game will be $(1-q)\alpha + q\beta = (1-q)\gamma + q\delta$.

Based upon the above ideas, if we look at our game, we find that the major factor deciding Red's decision about how often to visit is the apparent value of Blue's penetration (as made to appear to Red) when Red does not visit. If the value of Blue's penetration when Red does not visit is made to appear to be lower, in particular, lower than an attack on Red, then Red will engage less often. It is intuitively clear that placing a high value on Red will make Red more conservative about himself. But it is less intuitive that placing a high value on Blue will make Red conservative. Yet, increasing the value of Blue drives Blue to devote more assets and/or take more precautions for its protection, thereby making an actual encounter by Blue more dangerous for Red. Red's benefit is that he has been able to force Blue to devote more assets, i.e., he has deceived Blue to devote resources without any actual encounter.

The matrix in Fig. 8.7a gives the ambush matrix or habitat selection matrix for Red–Blue interaction. Based upon the above rationale we deduce that in order to deceive Red, Blue must change the payoff (Win over, Not Visit) from -5 to $5 + \zeta$ (as shown in Fig. 8.7b), thereby, forcing Red to visit most of the time. Blue might

(a)

Blue \ Red	Visit (q)	Not Visit ($1-q$)
Penetrate (p)	−5	1
Win over ($1-p$)	5	0

(b)

Blue \ Red	Visit (q)	Not Visit ($1-q$)
Penetrate (p)	−5	1
Win over ($1-p$)	5	$5+\zeta$

Fig. 8.7 Deception game matrix. **a** represents original game matrix formulated with respect to Blue's (Deceiver) point of view. **b** represents the intended deception game matrix

Blue \ Red	Visit (q)	Not Visit ($1-q$)
Penetrate (p)	−5	$5+\zeta$
Win over ($1-p$)	5	0

Fig. 8.8 Deception game matrix formulated with respect to Blue's (Deceiver) point of view

also choose to alter the payoff of the cell (Penetrate, Not Visit) and make it $5+\zeta$ resulting in the matrix shown in Fig. 8.8.

It can be easily calculated that this kind of manipulation will force Red to visit the habitat with the probability $q = (5+\zeta)/(15+\zeta)$ and in turn it will make Blue to penetrate with probability $p = 5/(15+\zeta)$ which shifts the value of the game to $V = (5+\zeta)/(15+\zeta)$. With these expressions, it is easy to observe that by changing the payoff for (Penetrate, Not Visit), Blue is able to force Red more often while Blue wins over the habitat population with higher probability and more the value of ζ, more is the value of the game for Blue. In any case, Blue has a higher value after deception. Thus, Blue must deceive.

8.5 Deception Strategy in an Ambush Operation: Signaling

"Information Superiority" plays a crucial role in all forms of combat operations. Knowledge of an opponent's likely moves greatly influences the strategies of the player. For example, if Red has reasons to believe that Blue has information about

Red's plan of action then Red will generally play the equilibrium strategy, thereby negating Blue's information superiority (unless deception by Blue makes Red compute with wrong MOEs, a situation dealt with in the previous section). Value of information in military game models is discussed in [14–17]. Reference [18] explores several aspects of deception, from robustness to value of information in detecting deception in adversarial modeling for opponent intent inferencing. Success of an ambush depends on several factors, information superiority being one of them. Factors, such as, terrain, weather conditions, players' experiences of staging and evading ambush, military judgment, and the opponent's beliefs, which influence the success of ambush are difficult to quantify.

In this section, we consider a game where information (true or false) is used to achieve deception in the game of ambush at a border habitat. Our analysis is largely based on the concepts dealt with in reference [14]. Lies are spread to mislead the opponent. There are two possible motives for bluffing (also called inverted signaling), namely,

a. to give false impression of the strength in the case of real weakness. This kind of bluff succeeds if the opponent believes this.
b. to give a false impression of weakness in the case of real strength. This fails if the opponent sees through it.

In this 2×2 zero-sum game, Blue represents the security personnel and Red denotes the infiltrators who are spread over two habitats—Hab1 and Hab2. Blue is the superior force and Red is the weaker force but has the advantage of habitats. Blue can attack Red either at Hab1 or at Hab2 (penetration, no winning over). Red's success in avoiding defeat lies in stopping Blue's attack.

Strategies of Red are

a. Divide his defenses between Hab1 and Hab2 and face almost certain defeat.
b. Concentrate his forces at one of the habitats and have a chance to defeat Blue for which he must choose correctly. If he concentrates his forces on Hab1 and Blue attacks Hab2 or vice versa he will be routed.

So Red would want to know beforehand Blue's likely target of attack. For this, he has to gain information superiority over Blue.

Strategies of Blue:

a. Attack Hab1
b. Attack Hab2

So Blue plans for an attack of Hab2 and tries to deceive Red to concentrate his forces on Hab1 or vice versa. If Blue succeeds with his deception, he accomplishes his goals with far less cost. Let us assume that Blue sends a signal which says that Blue forces are approaching Hab1.

Red, who now is the ambusher, has, on receiving the information, the following choices

a. Assume the information to be true (i.e., direct) and prepare for a defense at Hab1.

Fig. 8.9 Deception game matrix

	Blue: Send a true information (*D*)	Blue: Send a false information (*I*)
Red: Assume information true (*D'*)	α a_{11}	$-\beta-\psi$ a_{12}
Red: Assume information false (*I'*)	$-\beta$ a_{21}	$\alpha+\psi'$ a_{22}

b. Assume the information to be a lie (i.e., inverted) and prepare for an encounter (ambush) at Hab2.

The game matrix from Red's perspective is shown in Fig. 8.9.
The entries in the matrix have the following interpretations:

α value to Red's best defense. Encounter takes place at Hab1.
$-\beta$ value to Red of Blue achieving an unanticipated surprise. Blue gains from the encounter. Unanticipated because it is an unplanned, unexpected surprise for Blue. Blue had not intended a deception.
$-\beta-\psi$ value to Red of blue achieving an anticipated surprise. Blue's bluff was believed as true by the Red. Blue planned for this.
$\alpha+\psi'$ value to Red of an ambush of Blue. Actual encounter takes place at Hab2. Red correctly anticipates Blue's intentions and is able to plan an ambush.
α, β, ψ' are positive numbers. ψ may be positive or negative

Based on historical data of actual combat it is assumed that

$$\beta \gg \alpha, \beta+\psi \gg \alpha, \psi' \gg \alpha$$

It may be noted that α is a value to Red's best defense. $(\alpha+\psi')$ is payoff for ambush. Blue does not expect Red to see through his bluff, so does not expect an encounter and is surprised by an ambush.α, $-\beta$ and $\alpha+\psi'$ are payoffs that come with a surprise.

As $\alpha+\psi' \gg \alpha$, it is obvious that, notwithstanding surprise, an encounter at Hab2 is considered preferable to encounter at Hab1.

Recommended strategies for Blue and Red are mixed and are obtained using standard methods as

For Red

$$P_R(D') = \frac{\alpha+\beta+\psi'}{2\alpha+2\beta+\psi+\psi'} \tag{8.9a}$$

$$P_R(I') = \frac{\alpha + \beta + \psi}{2\alpha + 2\beta + \psi + \psi'} \tag{8.9b}$$

For Blue

$$P_B(D) = \frac{\alpha + \beta + \psi + \psi'}{2\alpha + 2\beta + \psi + \psi'} \tag{8.10a}$$

$$P_B(I) = \frac{\alpha + \beta}{2\alpha + 2\beta + \psi + \psi'} \tag{8.10b}$$

It is rational for the recipient of the signal to be more suspicious of the information than would be concluded from an observed frequency of lies.

The value of this game to Red is

$$v = -\frac{(\alpha + \beta)(\beta + \psi) - \alpha(\alpha + \beta + \psi + \psi')}{2\alpha + 2\beta + \psi + \psi'} \tag{8.11}$$

By definition, counter-deception requires the ability to detect deception, i.e., it assumes the player knows his opponent is attempting to deceive. He must be aware of adversary deception activities so that he can plan informed and coordinate responses. One way to avoid direct aggression or surprise attacks is to anticipate surprise through warnings, collection and analysis of other indicators and evidences. Any event with a surprise element is an event to be anticipated [7]. Accordingly, we simplify Eqs. (8.9a, 8.9b) to (8.11) having a cognitive factor '*A*' (called value of anticipation)

$$\beta = \beta + \psi = \alpha + \psi' = A\alpha \tag{8.12}$$

The value of the cognitive factor A denoting anticipation represents how much more effective an operation would be if anticipation of acts (such as ambush) that lead to surprise is achieved than if it is not. An anticipating entity is an entity whose current state is determined by a perception of future state. Players in a game of conflict, such as, in a deceptive ambush game, must employ perception-based anticipative control to guide decision-making on goal-oriented stage decision processes. References [19–22] discuss anticipation in a wide spectrum of applications.

With Eq. (8.12), the probabilities of Red and Blue players become identical. For Red we have

$$P_R(D') = \frac{2A}{1 + 3A} \tag{8.13a}$$

and

$$P_R(I') = \frac{1 + A}{1 + 3A} \tag{8.13b}$$

Value of the game becomes

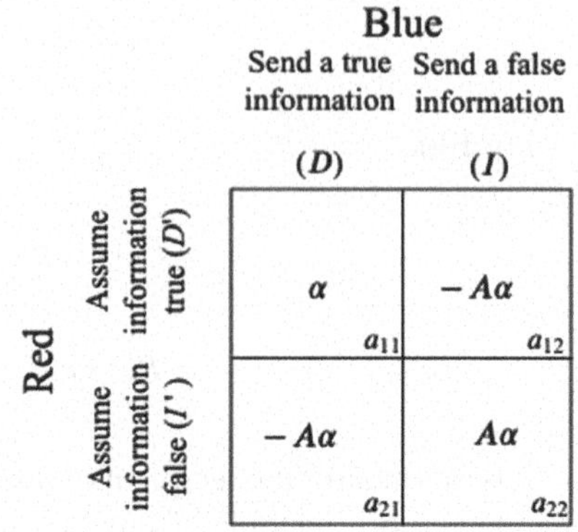

Fig. 8.10 Deception matrix after introduction of value of anticipation

$$v = \frac{A\alpha(A-1)}{1+3A} \tag{8.14}$$

The deception matrix reduces to Fig. 8.10.

Ambush is a surprise to the target when it is not expected. After an ambush has taken place, it is no longer a surprise. Surprise is replaced by a state of shock for the victim.

An analysis of historical data of successful ambushes suggests that A lies between 3 and 5 so that Blue and Red should play the Direct option about 60% of the time. In other words, Blue should not engage in deception the majority of the time. The game is unfair to Red as the value of the game to him is of –0.6α for $A = 3$ and –1.58 for $A = 5$. This shows why Blue must sometimes use bluffing. If Red could discount this possibility (i.e., correctly anticipate that there is deception) the value of game would be α in favor of Red.

Occasional practice of deception despite the risks, converts Blue from a loser (on average) to a winner on average. The probabilities associated with direct and inverted signaling result from finding a rational balance between the reward of anticipating surprise and the risk of being ambushed.

Let

$$\beta = \beta + \psi = A\alpha \tag{8.15a}$$

and

$$\alpha + \psi' = kA\alpha \tag{8.15b}$$

where k is a positive number. Here, we assign a higher value to true ambush (a loss to Blue).

Then,

$$P_R(D') = \frac{(1+k)A}{1+(2+k)A} \tag{8.16a}$$

and

$$P_R(I') = \frac{1 + A}{1 + (2 + k)A} \tag{8.16b}$$

For $A = 3$ and $k = 2$, inverted signaling should occur in just $\frac{4}{13}$ of the cases (down from $\frac{4}{10}$ when $k = 1$) and value to Red becomes -0.23α. Thus, as the value of ambush increases Blue is forced to lie less (for fear of ambush) and the game becomes more favorable to Red.

With $\alpha = 1$, the deception matrix of Fig. 8.10 becomes Fig. 8.11.

from Eq. (8.14) the value of the game is

$$v = \frac{-A(A - 1)}{3A + 1} \tag{8.17}$$

Deception matrix in Fig. 8.11 allows a quantitative understanding to be gained about the value of feedback information about the deceived to the deceiver. Since deception has a cost to the deceiver (here, Blue), feedback removes the penalty of ambush. Let us assume that Red has correctly anticipated Blue's bluff and has prepared for an ambush. Blue, on the other hand, has come to know about Red's plans through his intelligence gathering. If this feedback is acted upon, the deception matrix with feedback is recreated by converting "ambush" into a "best defense". So, the deception matrix in Fig. 8.11 is replaced by the matrix in Fig. 8.12.

Deception matrix in Fig. 8.12 has a value for Red as

$$v = \frac{-A - 1}{2} \tag{8.18}$$

That is, the value of this game with feedback to Blue is $\frac{3A+1}{2A}$ times the value of the same game without feedback. The advantage is even more if $k > 1$. For A = 3, the feedback game is $\frac{5}{3}$ as valuable to Blue as in the basic game without feedback. This increase in value comes because feedback removes the penalty of ambush, resulting

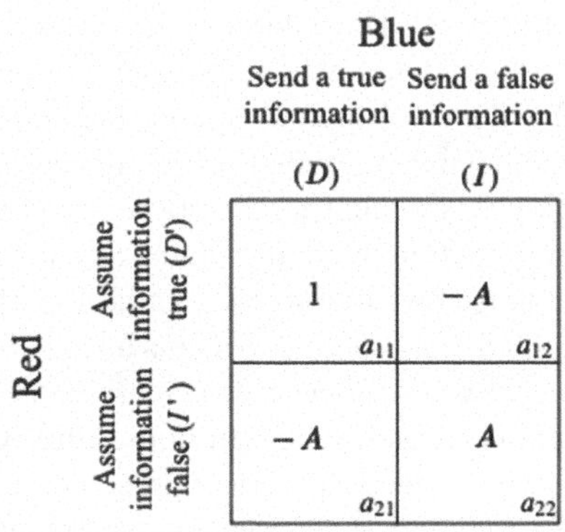

Fig. 8.11 Deception matrix with $\alpha = 1$

		Blue: Send a true information (*D*)	Blue: Send a false information (*I*)
Red	Assume information true (*D'*)	1 (a_{11})	$-A$ (a_{12})
	Assume information false (*I'*)	$-A$ (a_{21})	1 (a_{22})

Fig. 8.12 Deception matrix with feedback and $\alpha = 1$

in a game in which Blue has no fear of ambush and is free to lie exactly half the time. Thus, the possibility of feedback removes the penalty of ambush and promotes deception.

We have already noted that the deception came is unfair to Red. One way of increasing the fairness is to increase the penalty to Blue by increasing the value of ambush. Another way of reducing the unfairness is to delay the choice until more information is available. The benefit of waiting is that if the unfolding situation clarifies in time, ambush (surprise) can be avoided. The risk is that if the situation does not clarify in time to take proper action, ambush by the opponent (here, Red) is guaranteed. The player has to decide under what conditions and to what extent should he wait rather than commit himself.

8.6 Conclusion

This paper addresses the critical subject of cross-border infiltration and seeks to add to the work that has been done over the past decade on applying game-theoretic models of conflict to effective border control. With game-theoretic models, we demonstrated that deceptive ambush can be considered as a strong-point defense for ambushing infiltrators. The habitat selection model under the threat of ambush with deception has been modeled, where two agents Blue (Ambusher) and Red (Infiltrator) interact in a 2×2 non-cooperative game and emphasizes the value of deception in reducing one's predictability. It also shows, deception always pays to the deceiver when the target player is denied information or fed with wrong information. The paper introduces anticipation in another game as a cognitive factor and illustrates how anticipation of ambush and intelligence gathering affects both the attacker and the defender. Even though anticipatory systems have been widely studied in various contexts, the role of anticipation in deceptive search and ambush scenarios requires deeper exploration and is certain to benefit both military strategists and tacticians alike. The ideas and habitat selection models presented can be applied to devise optimal strategies that use and counteract deception in many other problems.

References

1. Whaley B (2016) Practise to deceive: learning curves of military deception planners. Naval Institute Press
2. Wagner AR (2009) Creating and using matrix representations of social interaction. In: Proceedings of the 4th ACM/IEEE international conference on Human robot interaction, pp 125–132. ACM
3. Bond CF, Robinson M (1988) The evolution of deception. J Nonverbal Behav 12(4):295–307
4. de Waal FBM (1992) Intentional deception in primates. Evolution Anthropol Issues News Rev 1(3):86–92
5. Nijholt A, Arkin RC, Brault S, Kulpa R, Multon F, Bideau B, Traum D, Hung H, Santos E, Li D, Yu F (2012) Computational deception and noncooperation. IEEE Intell Syst 27:60–75
6. Whaley B (1969) Stratagem: deception and surprise in war. Cambridge, Mass.: Center for International Studies, Massachusetts Institute of Technology, 1969
7. Grabo CM (2012) Anticipating surprise: analysis for strategic warning. Lulu.com
8. Dewar M (1989) The art of deception in warfare. David & Charles Publishers
9. Army, U. S. (1988) Field Manual 90–2, Battlefield Deception. Washington, DC: Department of the Army 3
10. Buller David B, Burgoon Judee K (1996) Interpersonal deception theory. Commun Theory 6(3):203–242
11. Brown JS, Laundré JW, Gurung M (1999) The ecology of fear: optimal foraging, game theory and tropic interactions. J Mamal 80(2):385–389
12. Bouskila A (2001) A habitat selection game of interactions between rodents and their predators. Ann Zool Fenn 38(1):55–70
13. Cressman R, Křivan V, Brown JS, Garay J (2014) Game-theoretic methods for functional response and optimal foraging behavior. PLoS One 9(2):1–13
14. Bracken J, Richard D (1998) Information superiority and game theory: the value of information in four games. Phalanx: 6–7, 33–34
15. Perry Walter, Signori David, Boon John (2004) Exploring information superiority: a methodology for measuring the quality of information and its impact on shared awareness. Rand National Defense Research Institute, Santa Monica, CA
16. Jormakka J, Jarmo VEM (2005) Modelling information warfare as a game. J Informat Warfare 4(2): 12–25
17. Daniel Donald C, Herbig Katherine L (eds) (1982) Strategic military deception. Pergamon Press, New York
18. Kott Alexander, McEneany William M (eds) (2007) Adversarial reasoning: computational approaches to reading the opponent's mind. Chapman and Hall/CRC, Boca Raton, FL
19. Nadin Mihai (ed) (2016) Anticipation across disciplines, cognition systems monographs, vol 29. Springer International Publishing, Switzerland
20. Wenke D, Rico F (2013) (Eds) Anticipation and the control of voluntary actions. Front Psychol. www.frontier.org
21. Piunti M, Castelfranchi C, Falcone R (2007) Surprise as short-cut for anticipation; clustering mental states in reasoning. In: Proceedings of the international joint conference on artificial intelligence, pp 507–512
22. Riegler A (2001) The role of anticipation in cognition. In: Dubois DM (Ed) Computing anticipatory systems. In: Proceedings of American Institute of Physics

Chapter 9
Can We Hold on To Our Heritage Monuments for Long? Relooking the Existing Challenges In Heritage Management Through "Alpha & Omega Yog Model of Mind" (Series-1)

Nitin Arora

9.1 Introduction: Reflections on the Context

> It shall be the duty of every citizen of India to value and preserve the rich heritage of our composite culture. *Part IVA Fundamental Duties Article, 45(51A) Fundamental Duties, Constitution of India.*

9.1.1 Fear embedded in Heritage Management

The management and conservation of heritage monument or sites is a challenge requiring much greater attention, awareness, and not just training the stakeholders in order to preserve them for present and future generations. Once these heritage monuments are preserved and conserved from getting lost in ruins of time only then can culture feel authentic in near future. Else converting desert into oasis is possible but it will be the feat of technology and not the soft power of culture for which all heritage sites stand for in the eyes of all tourists and people living in that region for centuries. Apart from the conservationist, good-hearted citizens, and NGOs, the role of the private individuals and government bodies is also of grave importance in the act of conservation. Heritage conservation is a living symbol of cultural heritage that once gave meaning to the life of the community in the past. There is a definite need to address the reasons behind why the challenges [1] exist in heritage management ranging from better interpretation of heritage sites, better control, appropriate use, appropriate conservation techniques, the contribution to the

N. Arora (✉)
Amity International Business School, Amity University Uttar Pradesh, Noida, India
e-mail: narora4@amity.edu

P. K. Kapur et al. (eds.), *Decision Analytics Applications in Industry*, Asset Analytics,
https://doi.org/10.1007/978-981-15-3643-4_9

surrounding environment, and the local community's continuum. The fear of losing these heritage monuments is grave and can send chill down to the spine of anyone who knows that culture comes alive only via monuments, and tourists get attracted toward these alive symbols of culture or civilization [2–5]. Can we let go the heritage in the winds of time while struggling to stand in the dilemma of priorities? Is it so simple to let go one's past?

9.2 Objectives

The objective of this research paper is to study the existing approaches of heritage management, which leads to preservation of heritage and proposes alpha and omega model of mind to relook the (mind)gaps in heritage management from the perspective of both non-engaged ordinary citizens and fully engaged stakeholders (tourist, conservationists, government, and locals).

9.3 Methodology

The scope of this paper is limited to discussing the two approaches and proposing the perception framework of the model in Series-1 of this paper. The model will be highlighted from a theoretical perspective integrating the diverse actions taken so far by stakeholders toward either promoting heritage tourism or heritage preservation. This research paper is based on the published literature on heritage management. It includes the published articles, International conservationist and government documents, books, reports, and websites related to heritage management. In the next theoretical Series-2 of this research paper, the authors will attempt to elaborate the transactional perspectives of the "*alpha and omega perception model*" between two or more stakeholders to explain the breaking of rules as desired by the model and response of the other stakeholders from the perspective the model. The model can be related to the transactional analysis of Eric Berne for layman existing knowledge that can be used to explain the proposed model highlighting the ***outcomes of clash of perceptions*** within the context of heritage management.

9.4 Theoretical Base for the Research

9.4.1 Heritage Management

What does the past mean? That's the question once in a while the contemporary person asks oneself in one's ongoing life when one visits any heritage monument on a holiday or when one looks on it while crossing the street near to it. The fundamental question that always arises in the mind of any conservationist or an onlooker is "*which heritage is important*," "to whom," "why," and "how" it should be conserved and "*for whom it should be conserved.*" That's the question a conservationist asks from different perspectives. Ultimately, the management of the heritage sites is actually the management of a *country's past* or also called as *the common past*. The past existed in all shapes and colors due to unstoppable changes caused by socially, technologically, politically, and environmentally influenced factors. Past connects with our present moment definitely and that archeological site or heritage monument or anything from historical heritage perspective was the witness to that special milepost of past.

There is a wide variety of cultural heritage [6]. It can be tangible, such as buildings, landscapes, and artefacts; and intangible, such as language, music, and customary practice. It is not just old things, pretty things, or physical things, and it often involves powerful human emotions. Is that building near you or in your village or your city just a brick and wall or is there something more hidden behind the wrinkles of time visible on the decaying monument? *Who owns that past, that culture? Any relevance to someone or is it sufficient to just click a picture with that monument in one's background and smile and upload it in social media!* Heritage—*Do we connect to it? How do we cherish the connection?*

There are majorly 27 world heritage sites in India, 5 are natural and 22 are cultural. The cultural properties range from the early cave paintings in Bimbedka, Madhya Pradesh to the nineteenth century Nilgiri Mountain Railways in Tamil Nadu; from Mughal architecture in Delhi and Agra to the ancient capital of Vijayanagara in Hampi, Karnataka.

Does one connect to them just intellectually or emotionally or spiritually or existentially? Management cum Maintenance of White House of USA or the toilets of world-class airports or shopping malls or in the Rashtrapati Bhawan of India are far better than the management of our ancient cultural or natural heritages. The former gets all the investment and resources for maintenance unlike the latter. And this is the most common challenge faced by heritage management. The other equally important challenges cum risks are like neglect, inappropriate renovation, public apathy, lack of understanding, and even necessary demolition for economic, political gain. A lot was done by various conservationist actively engaged in preserving the cultural symbols of a country following a few standard approaches for heritage management.

9.4.2 Challenges

The heritage sector is standing at a deaf and dumb crossroad. Government budgets are decreasing with every passing day and so is participation in traditional cultural activities. Urbanization, globalization, and technological change are diversifying potential audiences. High focus on tourism is invoking environmental and physical pressures. Digitization and online accessibility of cultural content are having mixed influence on tourism thus transforming value chains. These challenges ask for new approaches and perceptions to address the importance of our heritage [1]. Trafficking of cultural artefacts remains a silent yet difficult issue requiring action at helm of government affairs both nationally and internationally. Debatable global warming and climate change, in particular rising sea levels and extreme weather events, are putting heritage sites and cities at risk [7].

These above challenges are addressed since ages using two broad approaches of heritage management.

(a) Conventional approach and (b) Value-based approach.

9.4.3 The Existing Two Approaches for Heritage Management

The various approaches laid down to conserve heritage are actually an attempt to answer the old three questions that help in formulating any government policy or movement of conservationists which is cultural heritage monument or site oriented.

1. *Are the resources to be used for managing heritage scarce?*
2. *Has proper communication to the stakeholders for the site's designation been done effectively?*
3. *Has the site's heritage value been taken into account in decision-making?*

9.4.3.1 Conventional Approach

The '*conventional*' approach refers to the ways adopted by the conservation professionals in the Western world. The main focus is on the conservation of the monuments and sites to be preserved for future generations to wonder. Conservation experts started identifying and defining what needs to be protected. *It basically includes (a) Define (identify) (significance implied), (b) Documentation, (c) Assessing conditions, and (d) Planning for conservation interventions.*

9.4.3.2 Values-Led Approach

This is a positive way toward increased acknowledgment of multi-dimensionality of sites [8]. It got evolved in various parts of the world and became better known through the Burra Charter, first developed by ICOMOS Australia in 1979. The Charter promoted the assessment of the significance of a place—based on the values attributed by all stakeholders and the use of a Statement of Significance—as a basis for developing conservation and management strategies. It got furthered by Conservation Plans of James Kerr emphasizing cultural significance of a heritage place in society.

This approach adopts the premise that people in society ascribe various values to heritage. It essentially includes.

(a) Collecting data, (b) Evaluating significance, (c) Evaluating settings, and (d) Scheduling for preservation.

9.5 The Proposed Model

Both the models are in one way or the other addressing collectively the questions (Table 9.1) on the deteriorating state of heritage sites or monuments in a *politically correct manner that suits the people of the time*. Both approaches have common framework addressed below in tabular format with slight change on emphasis in evaluation of heritage significance only.

The conventional approach is a comparative layman approach with the first come first serve basis following the rule of the law if compared with value-based approach. Value-based approach is like a normative approach on what ought to be done on value-based reasons generated from the region of heritage. But value itself is debatable and limited by space. There exists a conflict on the importance of values within a region and within a country and varies as we go farther from the site of heritage. This approach makes people who value heritage to join from far off places to declare all heritages as universal heritage and universal responsibility and not just local responsibility. Value-based approach in short is an evasion of local responsibility and calling for responsible citizens from all across the world for sites that managed to capture their attention over a period of time of deterioration. Further, the next common problem in both approaches mentioned—values-led and conventional approach—is that the significance of the monument/site which should meet heritage criteria fixed by the concerned conservation bodies in order to be considered for preservation or not. This raises many concerns as the criteria can be diluted or crystalized suiting the needs of conservation bodies.

The authors propose a framework model "*Alpha & Omega Yog Model of Perception of Heritage*" (Fig. 9.1) below that includes the four dimensions of Mind as a kind of filter for perception of stakeholders to all concerns for heritage management

Table 9.1 Two old approaches for addressing heritage management

Planning stage	Historic preservation planning (some key activities)	Heritage tourism planning (some key activities)
Stage one	Historic preservation inventory	Community vision; develop heritage tourism inventory
Stage two	Evaluation of historical assets	Env. scanning and assessment
Stage three	Developing preservation plans	Develop goals and strategies
Stage four	Curatorial management	Implementation of strategies
Stage five	Presentation and interpretation	Evaluate, monitor, adjust

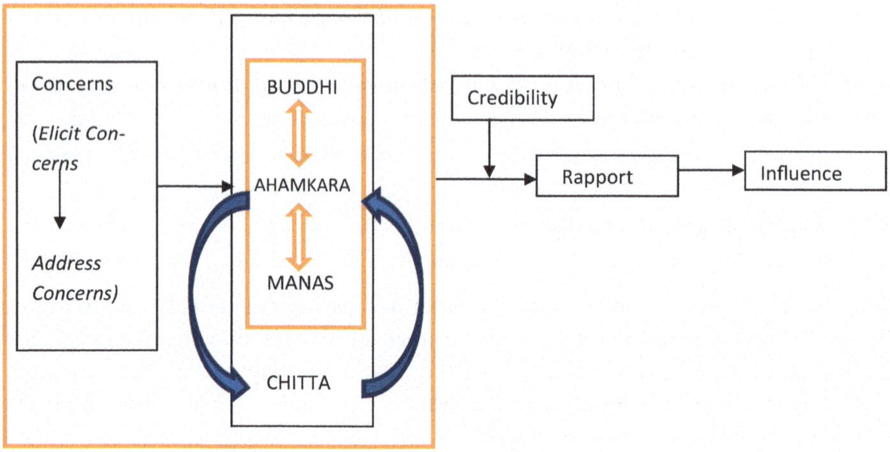

Fig. 9.1 Alpha & Omega Yog model of perception of heritage

leading to building rapport with the locals and the tourist thereby creating influence on preserving the site for many next generations to feel proud of.

The Model: The antecedents and consequences to the human mind for generating insights: creating influence on heritage management practices via allocating roles functioning of human mind

9.5.1 The Four Dimensions of Mind

The Mind is famously divided into four dimensions as not only declared by Lord Krishna in Bhagavad Gita (Scriptures of Hindus) but also a sensible way to understand the approximate functioning of mind:

(a) **Buddhi (Intellect)**,
(b) **Ahamkara (Ego, Identity)**,
(c) **Manas (Silos of Memory)**,
(d) **Chitta (Unsullied Memory, Pure Intelligence)**.

9.5.1.1 Buddhi

Buddhi is the other name of intellect which works in knowing in detail by dividing the whole into parts. It needs proof and is willing to go deeper and deeper into the object of knowledge like a surgeon with a knife or a biologist under the microscope or a physicist examining the approximate inferences of subatomic particles. The finesse of physicist is far sharper than that of surgeon but is incomprehensible for ordinary citizens. Surgeons and doctors are considered God in society because of the identification of humans as body rather spirit. Hence physicists' great work is not visibly important in the eyes of patients having fractured bone(s) or brain hemorrhage or other physical damage. He is concerned with gross and not subtle. Doctors can appreciate the importance of medical technological equipments used during such operations but even they want to use it just like a tool and have not much interest in the roots of the technological innovation processes.

9.5.1.2 Ahamkara

Essential for social identity and collective purpose, heritage enriches us through remembered precursors and prospective heirs. But these enduring benefits blind us to a mounting backlash too. The Spirit-led approach for heritage management assumes that all existing methods that necessitate the documentation and procedures to reinforce management practices are like replacing the electrical wires of house, whenever one's home electric bulb gets dysfunctional. Or like someone searching for camel on the roof of Burj Khalifa whenever the camel is lost. But due to wrong identification (**Ahamkara**), the multitude of problems are arising from different angles. This wrong identification toward what is called heritage needs to be changed. Initially, the royal sites were included in heritage list and all the other more culturally relevant sites and archeology were ignored.

9.5.1.3 Manas

On a negative side, we had brought from the past not so many promising situations but rather doubts. Heritage offers neither solace for present angst nor guidance to avoiding future perils. On positive side we are nothing but an outcome of our past culture only. If we consider our society into the baleful dilemmas crossroad then the correction can be made by reading the past lines (like astrological human hand lines) standing still in the form of heritage sites. No doubt seeing heritage destruction as integral to heritage preservation to some extent too due to inability to understand the best ways to preserve using feasible tools in the hands of the available experts and recommended policies [9].

Education depends on building **Manas (silos of facts, memories)**, theories and conclusions from the past experiments and discoveries. Without the function of memory in society of human, the education impact, the social norms, so called culture can't keep hold itself from one generation to next generation intact. **For example**, The neglect and destruction of the remaining concentration camps built in Italy during WW-II is largely due to the denial of this almost-forgotten memory of Italy's recent past. Denial by neglect is leading to similar disappearances of various sites worldwide. Some sites are attached with painful memories while some with the pride of the past.

9.5.1.4 Chitta

The only witnessing dimensions of mind called as Soul in ancient terminology that is like a detached observer to whatsoever is happening but is always present behind the ongoing actions like a monitor, watcher of all whose presence affects all actions. This subtle aspect of human is like subtle inference of physicist for sub-atomic particles that exists but can't be seen in literal sense. They can be seen only indirectly by inference. **For example**, in 1998 researchers at the Weizmann Institute of Science have conducted a highly controlled experiment demonstrating how a beam of electrons was affected by the act of being observed. The experiment revealed that the greater the amount of "***watching***," the greater the observer's influence on what actually takes place.

> "What does it mean to see an object". I.e. what do you see, or what does seeing mean, when you say "I can see that building of that ball" In that sense we are constantly seeing "electrons" since they are the ones doing the work sending out photons in the visible spectrum. Protons and neutrons on the other hand are 'seen' by machines by how other particles scatter of them (maybe this could be compared to a blind or blind-folded person, building an image of an object by feeling it) Electrons on the other hand are point particles so they do not show any structure no matter how fine-grained you can feel them. Now you may state that you see some real life object by the light they reflect, and as such you may imagine a this is also possible for sub-atomic particles, however, the resolution required is very high which means that the wavelength of the light you use needs to be less than that…this is very hard Gamma radiation, so if you were able to produce it, and see it, you would technically be able to see these subatomic particles. But again, in this discussion the question: "what does seeing mean" is equally important as the question posed here, and their answers will be linked.

Similarly, watching is already done by all citizens of earth. They all are seeing what's being lost whenever a site is demolished or breaks apart due to apathy. Fall in ticket sales; lack of promotion for tourist sites; distortion of history by deleting any relevance of site in history books, or movies or museum; ignoring the mind disturbing discoveries of sites like the Baby disposal lake in Israel, or scientific important ones like Baghdad battery. It is like hiding facts and keeping them under the carpet for creating a better tomorrow. Inaction after watching the facts have multiple reasons behind it that may be political, economical or technological and social too. Fall in ticket sales and lack of promotion of sites is like keeping balance between the costs of maintenance of the site. Recently, Thailand has stopped tourist visiting few beaches as too much plastic waste has collected at the site making the site unsustainable for future. The rulers of the time prefer to hide some brutal experiments done like the baby disposal or screaming mummies in Egypt. Some technological sites like Baghdad battery were a great revelation of the advancement of the society to make galvanic cells of the past in Iraq that common citizens knew recently only.

So Chitta is like having infinite memories of past and action or inaction depends on the willingness of the people of that time and situation to act on it. The actions can vary from hiding, ignoring, preserving, and making it known to the world in times to come. The more the evolved Chitta the higher the acceptance of the actions of time and moving forward in positive direction quicker than unevolved Chitta ones. It's the presence of the Chitta that matters in the direction of evolution quicker or slower or even standstill. It's the one which can take laidback approach and believe in any intellectual theories (biased toward the vested interest) of evolution over the spiritual perception of the world at large and not just humanity. If Chitta is highly evolved, there would have been less ruined heritage sites than as of now due to apathy. Apathy is nothing but inaction of Chitta or unconcerned to the importance of the site. The locals and the non-local tourists prefer to spend just good time on the site rather than mulling over the hurdles that can be crossed to preserve that site.

The four dimensions of mind are explained below in the table (Table 9.2) comparatively on their roles that can be used accordingly.

9.5.2 Rapport

All the recommendation and advocacies by leading heritage experts, government officials, conservation architects, heritage NGOs, professionals, and academicians in favor of heritage management is nothing but an interplay of (a) Individual or Collective intellect, (b) Cultural memories Manas of locals and pro-heritage volunteers and professionals, (c) and Ahamkara of the people w.r.t. the site. This correct interplay is what build rapport with the past, with the culture, with the pride of past that has recorded our journey till today.

Even if the local and government intends to connect (*Ahamkara*) with their past (*manas*) with pride via preserving these heritage monuments and archetypes, the more they have to deploy the comprehensive willingness to make it feasible (*intellect*). Heritage management is a luxury that only few can afford considering the ongoing emphasis on upcoming grand urban structures and artificial structures for entertainment, namely, **Atomium** in Brussels, the **Hallgrimur church** in Reykjavik, the **Cristo Rei** in Lisbon, the **Palace Bubbles** near Cannes or the **Cubic House** in Rotterdam. Among Burj Khalifa in Dubai or the upcoming further taller, Kingdom Tower of Jeddah which one will be included in the world heritage monument in next 100 years no one knows.

On a small family scale there exists lack of rapport between parents and children due to their different priorities. One can imagine how difficult it will be to connect the whole society for the common ancestral past which can be thousands of years old. The priorities are simply not matching. The success of building rapport between the present and the past can happen only when one can align spiritual side of human rather than the economics side of it. The tree is tree and not a log of wood! Tree can give oxygen and but not that dead log of wood being transformed in your study room as your study table! Animated birds or animals how appealing they may look but seeing them alive in natural landscapes with one's grandchildren should be a great joy for all.

The **concerns** need to be understood and credibly addressed. The **intelligence level** (intellect + Ahamkara + Manas in presence of Chitta) has the capacity to value the understanding. The credibility is all the more if the thing one valued is shown care. ***Caring is credibility.*** Caring means healing the wounds caring for the past while

Table 9.2 Roles of the four dimension of mind

Mind dimension	Approximate reference in english language	Feature	Organizational reference	Positive/negative dimensions
Chitta (Level 4)	Soul, Atman, Bliss body	Pure Intelligence, chooses yet Choiceless Consciousness, Witness, Observer, Watcher, No Judgment, Observation, Mirror, Accepts all decisions, Subtle, Spiritual Principles, Universality, Intelligence, Highest Will power to act (4 Level)	Spiritual leadership and crystallized-ethical followership (Ex: King Janaka)	+Sustainable, Self-Driven, Leading Examples/Heroic deeds/Enlightened actions, individual—Can't be communicated in normal language, can only be transmitted
Manas (Level 3)	Memory	Foundation, Culture, Glue, Accumulation, Contradictory possibilities, Higher Will power to Act (3 Level)	Team, Organization culture, climate, Policies	+Factual-Distorted, Implanted facts
Ahamkara (Level 2)	Identity,	Identification to some society, community, nation, Moderate Will power to Act (2 Level)	Team Culture, community of practice, Habit Formation,	+True Skill Values-False Skill Values
Buddhi (Level 1)	Intellect	Action mode, Skilled Worker in Action, Least will power to Act (1 Level)	Individual Divisive actions, At Leadership level	+ Can be Communicated Strongly-Can be Communicated Weakly

actively embracing the present via protection, planning and management (stewardship) and intervention measures taken in that direction. Caring cannot be done on long term basis unless something or someone is really valued. No half-hearted protection, planning, and interventions will be of sincere use from the perspective of heritage site. And it is better to integrate the citizens along with the experts too because heritage experts are similarly knowledgeable and armed like the others—they too are not so rational, or glued with their culture.

And finally, the **rapport** is build up by adding elements that break the gap between past and present in an interesting manner. For example, The Archaeological Survey of India (ASI) and Google jointly launched 360° panoramic imagery of 30 out of 100 "***nationally-important monuments***" for building engaging virtual feeling by global audience. The images of sites can be accessed on Google Maps through Street [10].

The reason for making digitization is to increase world attention to world heritage. To think of cultural heritage was to think of art objects, archeological sites, and historic monuments.

9.5.3 Credibility of Advise or in Actions Monitoring Any Plan

All the recommendations of the galaxy of experts to build rapport with the past will become credible if there is an effective and credible management plan too. The credibility acts as a catalyst to diffused interplay of dimensions of mind. It can strengthen the weakness of mind visible in policies and recommendations by more specific, site-specific plans [2]. The credibility is again a crystalized form of presence of Chitta in the form of strong intellect. **For example**, If you have to give work to an American you have to be very detailed in the first step itself to avoid confusion later. But dealing with Japanese workers is much different. It is not mandatory to give the experts the detailed plan in the first step. Just guidelines and framework is enough and the expert can layout the detailed map. One can rely on the intelligence of the subordinates. This adds credibility factor that even while implementing the recommended policy, the minor tweaks and a major overhaul is still possible at the site-level operators and managers. It is solely the question of how Chitta responds to the operator while executing the orders. His Chitta, presence, will not let his conscience allow leaving the gaps in policy even if it costs him whatever. The various factors that can be added and thus making the recommendation for addressing the various concerns that can make it credible are mentioned below:

1. Focus globally and with local support.
2. Save values of the heritage realistically.
3. Produce outcomes to protect values associated with the property.

9.5.3.1 Concerns to Be Addressed to Get Credibility

Some of the concerns that are widely known among the scholars and professional occupied with one intention heritage management that includes

1. Encroachment of sites in and around the sites of importance [11].

2. Threats from the Skies-Noise pollution of planes creating cracks in monuments.
3. Monumental blunders-negligence of Archaeological Survey of India (ASI).
4. Promotion of tourism.
5. Little success in getting listed the sites under world heritage.
6. Implementation of Laws related to monument and archeological sites preservation, conservation, and maintenance.
7. Irresponsible visitors inscribing names, scratching on monument, and others.

How many of these elicit concerns have got addressed by the ASI, or the state or central government is yet another barrier to cross and it needs to pass from the four roles of minds collectively for both citizens and lawmakers and stakeholders with spiritual perceptions.

All concerns for heritage management [12] are presented to the intellect part of the mind. The questions remain the same as asked before *What does the past means? Do we connect to it? How do we use it? Who owns the past?* If the site disappears, does it has any relevance and impact to our current life? Is there any sense of attachment to our past be it violent or sacred. Does that connect to our past is unconditionally accepted? Will the site get necessary funds for its management? Many such questions stand before the intellect that needs an answer. The Sites are financed by subsidies, public support, besides resources, and patronage. These own resources are generated by the tourist-based activities and on the strength (how old or rare) of its site artefacts. A site which is not maintained will eventually be lost to all but the archeologists. It is as equal to the future generation not even knowing who we are now and our collection of great stupidities and achievements. Hence the concerns that are being observed may or may not get addressed depending on the awareness level of the society at large. Sharper the intellect, more crystallized the exuberance of Chitta is and the positive circle goes on reinforcing each other roles.

The very same intellect with less evolved Chitta degrades to using itself for vested interests only or for limited time. Hence the Alpha and the Omega Yog model, the joining of cause and the effect impairing perceptions toward heritage sites. Identification needs to be shifted from swiftly changing objects of pride to the objects with deep roots in our culture. Interplay of memories deludes the identification of each generation as it requires the clear discrimination power of Chitta to direct intellect to choose between what is truth to that which is false. Culture is truth, the real roots of any individual or nation. Imposed culture is false and it recedes with time.

For example, in the first place the world war was itself a sign of weak Chitta of world at large. The intellect deployed revolves accordingly and hence arms manufacturing, weapons of mass destruction and designing concentration camps serve only vested interest ignoring the large base of population. For example, during the Second World War the occupying Italian army used Mamula as a concentration camp which is now transformed into a Luxury resort. They were facing two options: to let the site in the hands of wastage or get sponsors for their restoration. Italy preferred the latter.

All options are presented before are known to Chitta with the power to discriminate based on the sharpness of intellect, degree of identification with the sites, and the power of past memories, if the sites are worth saving or not. Different interplay of Chitta, Buddhi, Ahamkara, and Manas will create different responses in terms of saving the site or letting it go in the ruins of time.

Unaddressed Heritage Concerns

In India, the unaddressed heritage concerns resulted in deterioration of natural or man-made sites with the sands of time. The few are highlighted

1. Dal lake in Srinagar shrunk from 34 to 9 km^2 (excluding the other channels and floating gardens) in the last four decades [13].
2. The Union government stated that Gujarat was ranked fourth in the country in the list of states with maximum number of encroachments. The state was behind Tamil Nadu (133 monuments), Uttar Pradesh (57 monuments) and Karnataka (46 monuments). Last year, between August and September, ASI's official Arif Agariya filed 121 complaints with the city police and reported illegal constructions near protected monuments across the city. But police did not take any action.
3. Light and Sound shows produce vibrations that create cracks, and its heat harms the plaster in few heritage sites.
4. Out of 70,000 monuments and sites listed by ASI, no caretaker is there for 60,000. ASI administers and manages 3687 ancient monuments and archeological sites and remains of national importance that too India is known for 23 famous sites for tourist purpose and 6 national parks. Over and above when India stands with thumping chest of having the first civilization in the world no city of India is listed on world heritage cities. We just lost Harappa and Mohenjo-Daro cities due to reasons unknown.
5. Absence of National policy and stringent legislations despite having laws ranging from Bengal Regulation XIX of 1810 till Ancient and Historical Monuments and Archeological sites and Remains Act of 1951, 1958, 1985
6. Constraints of [9]. Accessible Tourism for the disabled [14].
7. Pay scales of professionals in the organization failed to attract superior brains in the field of history and archeology.
8. Placing a mechanism to collect feedback from visitors on guiding activities within the centrally protected monuments or sites maintained/administered/managed by the Archaeological Survey of India.
9. The AMASR Act has been in place since 1904 and has been consequently amended in 1958 and 2010 but not once have there been provisions to rethink and reframe the notions of what heritage stands for in a multi-cultural and complex society like India. The ASI has been criticized on several occasions for following such a restricted act for conservation which doesn't consider the rich and diverse cultural heritage of the country.

10. The term conservation has multiple meanings. It ranges from signifying the entire field or realm of cultural heritage preservation, to policy making to planning to technical intervention.

Despite having constellation of heritage experts, government officials, conservation architects, heritage NGOs and professionals and academicians working to get attention for these concerns, the major constraint in the development of tourism [15] in India is absence of adequate tourism infrastructure. This leads to bad tourist experiences and untoward situations for tourists.

Tourists carry along new memories to known memories about that site and its history after visiting the site. Next that memory silos gets enlarged w.r.t. that site and presents to enlarged social Chitta awaiting either positive action or ongoing apathy or at the worst gets ignored. One can't expect much from the tourists who have come to that promoted site to only enjoy their holidays on personal family vacation or business tours.

Some negativities have emerged due to non-addressed concerns of tourists rampage on sites, namely,

(a) ***Undesirable Social and Cultural Change***: Hippy culture of Goa in-turn promoted use of drugs, prostitution, and human trafficking.
(b) ***Increase Tension and Hostility toward tourist*** from localities due to suspicion and lack of respect and understanding for each other's culture and way of life.
(c) **Sense of Antipathy** due to benefits of tourism going to the airlines, hotels, and other international companies, not to local businessmen and workers.
(d) **Adverse Effects** on Environment and Ecology due to increased pressure on the carrying capacity of the ecosystem leading to large-scale deforestation and solid waste dumping as well as depletion of water and fuel resources. Forests suffer negative impacts due to tourism in the form of deforestation and land clearing for trekking in the Himalayan region, Sikkim and Assam.
(e) **Depletion of Natural Resources**, namely, Water resources and also Local resources (energy, food, and other raw materials).

Some of the Concerns that Got Addressed in India Are Making Credible Positive Steps in the Preservation of Sites

1. Graduation in history, besides knowledge of one or more languages, has been made mandatory. Earlier the criteria were just graduation and knowledge of one or more languages. This modification has not been received well by guides, most of who do not have a degree in the subject. There are 456 regional tourist guides affiliated with the Union ministry of tourism, and around 600 UP tourism-approved guides [16].
2. ASI has limited the number of photographers at Taj Mahal.
3. Launch of India Heritage Passport programme with UNESCO in 2006 in few pilot state government of Madhya Pradesh, Karnataka, West Bengal, Rajasthan, and Tamil Nadu.

4. With launch of smart city initiative of the new central government of India emphasis is given on integrating heritage management with smart aspects of that city. For example, Jaipur Smart City aspires to leverage its Heritage and Tourism, and through Innovative and Inclusive solutions like Smart Multi-Modal Mobility and Smart Solid Waste Management.
5. In 2010, National Monuments Authority (NMA) has been set up under the Ministry of Culture as per the provisions of AMASR.
6. Another step in addressing this problem was initiated by the Indian National Trust for Arts Heritage & Crafts (INTACH) to preserve buildings and precincts of architectural importance and not just historical monuments.
7. A total of 18 monuments have been placed under the care of PSUs like GAIL, SAIL, BHEL, Shipping Corporation of India, and others. PSUs lag behind in spending under the CSR. These PSUs will spend part of their corporate social responsibility (CSR) funds on maintenance of these monuments under the aegis of National Culture Fund (NCF), a trust under the Union ministry of culture.

Tourism is also the key factor that forces the government to address concerns as healthy revenue comes from the tourism sectors too. The positive impact cum actions of tourism on heritage management and in general, namely, are

(a) **Generating Income and Employment** alleviates poverty and promotes sustainable human development [17].
(b) **Increased Foreign Exchange reserves and** favorable balance of payment.
(c) **Preservation of National Heritage and Environment** by Tourism Department.
(d) **Infrastructure development has** benefitted the local community like building means of transports, health care facilities, and sports centers, high-end restaurants that cater to foreign visitors.
(e) **Promoting Peace and Stability** becomes possible because of tourism in developing country like India as the local community gets new jobs.

9.5.4 Influence

Influence is the effectiveness of implementation, evaluation, reporting and communication done to increase longevity of the heritage in question be it tangible or intangible. The Influence is not always red carpet in nature but positive too. Logically, inheritance is outside the control of those who inherit. We don't control our biological or genetic inheritance. Inheritance of property is determined by law. We influence our heritage irresponsibly by neglecting it, but we cannot completely escape it, just as we cannot escape our biological inheritance and just as we have to manage, any property left to us. Yet the analogy between cultural heritage and heritage in the primary sense of inheritance has its limitations. Heritage is not always something already present in a culture. It is, on the contrary, selected, negotiated, and perhaps even constructed by the heirs.

So the ***question of influence***, i.e., impact of existing level of rapport with the past is still away from the present generation of locals and so-called global heirs. Why? Because they are not legally bound to maintain those heritage sites! So do we need to execute the legal binding (intellect + refreshing pride of locals for heritage) on the locals or the heirs, if any, like exercised on forts of Rajasthan, etc., by the heirs of the old kings to maintain the heritage with utmost responsibility.

Influence in normal sense assumes direct control of reins over something by something like cause and effect. But this is strangely absent in the case of heritage management as it is owned by none and yet is called social heritage. Does entrance fee and free market logic is sufficient to answer this question of influence or let whatsoever the market or society wants, let that be the destiny of that heritage site or archeological site!

This **question of extent of influence** can't be answered just from this perspective of legal binding to locals or the legal heirs alone. Trust between the heritage conservationists' bodies and the locals is not high and hence the support of legal route. Maintenance of heritage sites where there are not actual heirs, or heritage sites or designs or landscapes facing dangers of attack from civil conflict and wars are also equally important. Tourism attraction impact factor, the strong story associated with that site, the relevance or need for that site to be managed or not, conversion/transformation possibility to serve another tourist function are also the key issues for getting the maintenance financed!.

9.6 Recommendations

> The future of Present is in the past, The future of Past is in the future, The future of Future is in the present, Ultimately, Present makes both past and future [18]. The future is in the now.

This is the Last Mile Problem in the sustainable heritage management after all raising awareness, community engagement, building entertainment in heritage sites in various ways, and legal rules for heritage management! What the authors foresee is that the last mile problem is always open ended and will remain as a big gap forever as in no time no civilization as a whole was being raised to equal levels of economic prosperity, or spiritual heights. There remains some sense of apathy and some sense of absence of urgency among sections of society toward the accelerated ruins of heritage sites. One should not feel negative with this perspective but should work to close the gap as much as possible with raising consciousness pride for the past along with relevance and capacity of the present society. Today's generation has all the more intellect, all the more resources, and technologies than all past generations. The question of influence remains as the question in which direction the consciousness goes via the Mind route of refreshing significant memories that created the present! The hope is if consciousness goes higher and higher, the memories will not hide in unconscious storehouse for long and will direct the present consciousness to value it. If stakeholder's memory remains stagnant or degrades in growth with time, the chances of accelerated or delayed ruins of heritage are a certainty!

As memory defines our identity which further drives intellect to address concerns of importance, it becomes imperative to remove the silos of dirt over the mind that hides the old memories. This dirt cuts attachment from the past. It can be seen both positively and negatively from different contexts but from the heritage protection context it is obviously not positive and hence needs to refresh to relive the significance attached with each heritage for the locals' at least or even for the far off ones.

9.7 Questions to Be Further Addressed in Future Research

Manas have no limits. One can consider the whole world heritage as ours but that will be too much of asking from the government or locals to worry for. **Ahamkara limits the boundary of the influence** of any nation or tribe or individual. **Intellect then only becomes sharp when it works within the boundary**. In boundary-less problems it fails to address them. It's only in the boundary of cultural heritage of particular nation or tribe or geography the Buddhi (intellect) with time can develop the cultural heritage knowledge base, understand cultural heritage risks, undertake cultural heritage assessments easily which can influence others and can generate the strong feeling of culture via heritage.

So this explains easily that despite having great Indian heritage to be proud of in one's **Manas** (memory), one can't stop wondering then why the Indian Museum in Kolkata be as appealing as the British Museum in London? What went wrong? Why suddenly Madam Tussauds museum is so popular against the old museums? What went right? Why can't our strong presence of rich heritage able to exert influence on the better conditions of the heritage sites? Do we need to add more consciousness to our actions? Does that alone solve the problem? Does absence of spiritual consciousness perspectives is creating havoc on our actions and making us hide all mishaps in ruins so that we just forget them? Do we need to minimize excessive use of intellect in dealing with heritage and start valuing emotions too? Do we need to increase attachment and pride with past at the cost of present ongoing turmoil in various countries like Syria, Turkey, Iraq, Lebanon, Afghanistan, Pakistan, etc.? Will it be worth doing it? Do we change the direction of mind to stop acting so passionately attached toward the painful past that may create havoc on present? There are many questions that remain answered and should be attempted from spiritual perspective or the interplay of four dimensions of mind on how to influence the conservation and preservation of monuments and sites of national or world heritage.

1.7.1. The difference between USA or India or China is the pride of cultural past that keeps aloft the country despite economical challenges. Absence of this cultural pride is what is keeping Americans falling apart on multiple levels including occasional economical grounds too. So do we need to digest the fact that the countries who sprung recently only (300 years or more, new countries like USA) who strongly emphasizes on maintaining heritage is a fad and will mellow down like it is happening in India which has more than 10,000 years of strong cultural and social past and is witnessing apathy to monuments and sites? Is it ok to forget the past heritage symbols with time and focus only on economics of present?

1.7.2. Does Heritage management practices happen on stand alone model? There must exist strong links between Tourists, Heritage, and Reasons for Visiting Heritage Sites [19, 20]. A detailed study of those reasons and links will shed light on emphasis on Spirit side of tourists like World Heritage Commons! The motivations of tourists to travel is just for the sake of holidaying or also to satisfy deeper quests of Chitta and awakening manas that subconsciously connects them with the sites the deeper roots of existence.

1.7.3. We need to find to what extent conservation efforts going nationwide impacted positively on peaceful citizens versus negatively impacted on the civil wars. The ratio will give further insight into how much efforts are needed more in that direction.

1.7.4. On April 26, 2016, the National Museum of Natural History, New Delhi and its valuable collection of animal fossils and stuffed animals were destroyed by fire. And such incidences are not the first of its kind. An insight into the extent of loss due to negligence of caretakers or natural disasters is needed to know what percentage of loss can be prevented by minimizing such gross negligence. Forest fires are sometimes seasonal and sometimes the revenge of the local forest dwellers. A study might be needed on what behavioral issues of caretakers led to such unbearable accidents?

References

1. World Heritage Challenges for the Millennium (2007). http://whc.unesco.org/documents/publi_millennium_en.pdf
2. McKercher B, Du Cros H (2002) Cultural tourism-the partnership between tourism and cultural heritage management. The Hawrth Hospitality Press, New York
3. Hall C, McArthur S (eds) (1993) Heritage management in New Zealand and Australia. Oxford University Press, Oxford
4. Hall M, McArthur MS (1998) Integrated Heritage management, principles and practice. UK, Stationary office
5. Orbasli A (2000) Tourists in historic towns-Urban conservation and heritage management. E & FN Spon
6. Why cultural heritage matters: a resource guide for integrating cultural heritage management into communities WORK at Rio Tinto. https://www.riotinto.com/documents/ReportsPublications/Rio_Tinto_Cultural_Heritage_Guide.pdf
7. Communication from the commission to the European parliament, the council, the European economic and social committee and the committee of the regions Towards an integrated approach to cultural heritage for Europe. http://ec.europa.eu/assets/eac/culture/library/publications/2014-heritage-communication_en.pdf
8. Clark K (2016) Preserving what matters: value-led planning for cultural heritage sites. http://www.getty.edu/conservation/publications_resources/newsletters/16_3/feature.html
9. Dilawari V (1996) An approach to heritage conservation policies. In: Ghosh S (ed) Architectural and Urban conservation, centre for built environment. Center for Built Environment, Bombay, India, pp 95–10

10. Google Street View. https://www.google.com/culturalinstitute/beta/search/streetview
11. Kaushik H, 14% ASI monuments in Gujarat encroached. http://timesofindia.indiatimes.com/city/ahmedabad/14-ASI-monuments-inGujaratencroached/articleshow/55898408.cms
12. Goral A (2011) Research on cultural tourism development in sacral and spiritual sites from the UNESCO world heritage List. Int J Herit Sustain Dev 1:49–59
13. Wani MA, Dutta A, Wani MA, Wani UJ (2014) Towards conservation of world famous Dal Lake—A need of hour. Int Res J Eng Technol 1
14. Problems and Prospects Of Accessible Tourism In India (2010) A Study Report Commissioned by Government of India, Ministry of Tourism
15. Binoy TA (2011) Archeological and heritage tourism interpretation: a study. South Asian J Tour Herit 4:1
16. Prakash M, Chowdhary N (2010) What are we training tour guides for? (India). TURIZAM 14:253–65
17. Making Tourism Sustainable. A Guide for Policy Makers (2005) A report by UNEPand World Tourism Organization
18. Rincon L (2009) Intellectual property and protection of cultural heritage: the case of the national museums of arts and civilizations in France. Study carried out for the World Intellectual Property Organization (WIPO). https://www.wipo.int/export/sites/www/tk/en/resources/pdf/rincon_report.pdf
19. Poria Y, Butler R, Airey D (2016) Links between tourists, heritage, and reasons for visiting heritage sites. Unpublished paper. http://epubs.surrey.ac.uk/1135/1/fulltext.pdf
20. Singh RPB (1991) Panchakroshi Yatra, Varanasi. Sacred Journey, Ecology of Place and Faithscape. Tara Book Agency, Varanasi. #Expanded form book published in 2002

Chapter 10
Rendering Blockchain Immutability in Chatserver: A Node.js Approach

Umesh Kumar Chopra, Ayush Kumar Rathore, and Rajiv Pandey

10.1 Introduction

Volumes of digital data are growing at exponential rates; some reports even claim that in recent years, 90% or greater of the data has been generated [1]. Information and data sets are constantly produced, recorded, and processed in the age of AI, ML, and Big Data; the information being at the heart of any decision making process. The storage of data leads to severe information breaches. Centralized organizations collect huge amounts of sensitive personal data and information, both from the public and private sectors. Data has recently been an inseparable component of worldwide news, with the Facebook–Cambridge Analytica Scam being the most famous [2]. The data breach scam triggered a worldwide outcry and individuals asked how much they regulate their delicate information. Another popular event widely reported by the public media, WannaCry ransomware, has corrupted millions of PCs worldwide [3]. Cybercriminals are becoming more active and competent by the day, while the odds have never been higher for individuals and organizations. Today information has become one of the economy's most precious assets [4] and it is a desire/need to protect it. In most organizations, the information trail is of paramount importance, so organizations are looking for solutions that can make it immutable. Blockchain by the very nature of its architecture ensures that the information once created, validated through appropriate consensus algorithms are then committed as history in the various blocks of the Blockchain. The security of data even in chat trails need to incorporate immutability, as the use case in defense or security forces may apply,

U. K. Chopra · A. K. Rathore · R. Pandey (✉)
Amity University Uttar Pradesh, Lucknow Campus, Noida 226028, India
e-mail: rpandey@lko.amity.edu

U. K. Chopra
e-mail: ukchopra@lko.amity.edu

A. K. Rathore
e-mail: ayushkumarrathore@gmail.com

P. K. Kapur et al. (eds.), *Decision Analytics Applications in Industry*, Asset Analytics,
https://doi.org/10.1007/978-981-15-3643-4_10

where the trails are valuable when debriefs are carried out. Since these information trails are replicated at multiple nodes, change in information at a specific point shall be highlighted during the course of the validation.

This paper contains seven sections: Sect. 10.1 being "Introduction" describes the organization of the paper. Section 10.2, "Blockchain" section, describes features and presents a brief analysis of the blockchain. Section 10.3, "Blockchain Security," highlights the security features with some demonstration. Section 10.4, "Node.js" section, explores Node.js as the platform for blockchain implementation. Section 10.5, "Chatserver" section, demonstrates how blockchain can be used in implementation of the Chatserver. Section 10.6, "Blockchain Use Case": Chatserver, evaluates the relevance of immutability during communication between Client and Server and Sect. 10.7 the "Conclusion".

10.2 Blockchain

Blockchain is primarily classified as public and private, permissioned and permissionless; however, there isn't a firm agreement in differentiating between types of blockchains. Various researchers have attempted to differentiate blockchains on the basis following parameters.

- Speed
- Security aspects
- Identity
- Consensus algorithms
- Level of anonymity.

Thus resulting in the underlying matrix

Permissionless and Public	Permissioned and Public
Permissionless and Private	Permissioned and Private

The generalized definition of a blockchain is as under; however, the architecture, consensus, and validations may vary on need basis:

> Blockchain is a shared and trusted, public ledger of transactions that each user can inspect but no single user controls, it is a distributed database that maintains a continuously growing list of transactions data records(committed to blocks), cryptographically secured from tampering and revision. [5]

Features of a Blockchain (selectively may be visible as per the blockchain architecture and rules opted):

- Each node receives a replica of all transactions.

- Nodes broadcast transaction history to other nodes in order to synchronize with the blockchain.
- Every node can race to evolve a consensus for its block particularly in permissionless/public blockchains. Rule may vary in private versions.
- In Bitcoin blockchain, the mining nodes evolve a consensus to commit the data to the block and the validator is appropriately rewarded. The same is updated to every block on the network.
- All transactions are evaluated for their validity and subsequently committed (Fig. 10.7).
- Upon validation of the hash values the nodes render their acceptance by appending it as the next block.

10.3 Blockchain Security

Blockchain uses the concept of distributed ledger to increase its security. Third parties and Central Organizations validate transactions through their own servers with only a small knowledge of legitimacy but in a blockchain environment, a P2P network of nodes functioning on the same blockchain protocol do all the legitimate work and through consensus or majority vote. Consensus uses Smart Contracts to maintain legitimacy.

> The content of the ledger demonstrates the current and the historical state of all transactions maintained by the blockchain. It being replicated at participating node, all updates must be agreed upon by all parties [6]

Consensus mechanism or algorithms are primarily:

1. Proof of Work (POW)
2. Proof of Stake (PoS)
3. Proof of Elapsed Time (PoET)
4. Delegated Proof of Stake (DPoS)
5. Proof of Burn
6. Proof of Authority (PoA)
7. Voting or MultiParty Consensus Algorithms (Prevalent in Consortium and Private Blockchains).

Therefore, pre-defining consensus rules for approving transactions on peer to peer network enforces legitimate transactions by all the nodes/users in the network rendering the blockchain immutable.

Figure 10.1 clearly highlights the various components of a block, namely:

1. Previous Hash: Hash value of the previous block.
2. Timestamp: It ensures immutability via the clause of stating that the block existed at the given time T.

Fig. 10.1 Bitcoin block data [7, 8]

3. Tx_Root: It is the Merkel tree of all the transactions in the block.
4. Nonce: It is the actual data.

The above components are elucidated in detail in the Node.js code snippet.

Blockchain technology enables customers to keep their database cooperative, safe, and decentralized. Therefore, the database is dispersed and maintains an increasingly secured list of documents from manipulation [5]. It is also scattered. A typical blockchain system is made up of a number of blocks connected in a certain order. Blockchain originates its name from this linked depiction. The system users, also known as "nodes," carry out the validation and data storage task in blocks. The agreement algorithm used for confirming operations and adding fresh bricks to the loop is called the algorithm Proof of Work (PoW). This cyclical comportment assures and maintains the integrity of the previous block, which examines its history trail backward that is the initial block or the genesis block [9]. Every block has a mathematical value that is the "Hash value" of the previous block generated by the cryptographical function, in which transaction and timestamp related information is embedded [10].

The various security advantages of blockchain are listed below [11]:

- **Privacy Protection**: In order to eliminate the need for central database storages for confidential information, blockchain adopts a networking system based on peer to peer paradigm. This removes centralized items that a hacker could target and steal valuable information through. Likewise, a blockchain has no key fault point, rendering it more powerful than distributed networking technologies. Blockchain uses asymmetric encryption, with every person having two components: a public key that is noticeable to everyone on the network and is used for message encryption for the customer; and a private key for the decryption. A public key of a user is in no way related to his public address, and it is technically very difficult to find the private key of a user from his public key. Blockchain therefore maintains anonymity and confidentiality.
- **Crash Recovery**: The blockchain information is transmitted across peer nodes in contrast to a conventional database in which all the information is stored at central

storage. Every blockchain user subject to the blockchain being permissioned or permissionless may be entitled to generate and store a complete copy of the data. While leading to data redundancy, it greatly increases the network's efficiency and defect/crash tolerance. In the event of some nodes being assaulted or damaged, the remainder of the network can still be helpful in reconstruction and validation of the blockchain.

- **Immutability**: Blockchain has a special feature of composing information that prohibits the information from being changed as soon as it is recorded. This feature requires a timeline/time stamp to be generated when a new record is created [12]. Data alteration is no longer permitted. Furthermore, the recording of a new business is decided by means of a consensus mechanism requiring the mutual consent of more than the pre-configured number of network users.

10.4 Node.js

Node.js is a server-side scripting option built over Google Chrome's V8 JavaScript runtime. "Node.js is an asynchronous event driven JavaScript runtime, designed to build scalable network applications" [13]. It can be used to develop network applications fast and scalable. Node.js offers a wealth of JavaScript modules libraries. For blockchain growth, some of these modules can be used. For implementing a blockchain Chatserver, this paper uses crypto-js, express, http, and socket.io modules.

10.5 Chatserver

This section demonstrates a Blockchain Chatserver approach and not a complete code. Prerequisites for a local server:

- Ubuntu 16.04 or above
- Node.js
- Npm and a text editor (Fig. 10.2).

This above code represents a block in the blockchain. It has two member functions namely constructor and calculate hash. Constructor has index, timestamp, data, and previous hash as parameters. Index represents the address of the block. The timestamp contains information related to when the block was created. Data can contain any information a user wishes to secure by uploading it on the blockchain. The historical hash parameter stores the previous block's hash value. This may not be true for genesis block where the hash value is NULL. The calculate hash function calculates the hash value for the current block using SHA256 encryption module.

Figure 10.3 illustrates, blocks in a blockchain. The first block, namely, Genesis block has an index value of '0' implying that it to be the first in the chain of blocks.

```
class Block
{
  constructor(index, timestamp, data, previousHash = '')
  {
    this.index = index;
    this.timestamp = timestamp;
    this.data = data;
    this.previousHash = previousHash;
    this.hash = this.calculateHash();
  }

  calculateHash()
  {
    return SHA256(this.index+this.previousHash+this.timestamp+JSON.stringify(this.data)).toString();
  }
}
```

Fig. 10.2 Block in the blockchain

```
constructor()
{
    this.chain = [this.createGenesisBlock()];
    fs.appendFile('blocks.txt', JSON.stringify(this.getLatestBlock(),null,4), (err) =>
        {
            if (err) throw err;
    });
}

createGenesisBlock()
{
    return new Block(0,Math.floor(Date.now() / 1000),"Main","0000000000000000000000");
}
```

Fig. 10.3 Code for creating a genesis block

The previous hash of the Genesis block is "0000" further asserting the fact that it is indeed the first block as the previous hash cannot be zero except in the case of the first block in the blockchain. The Genesis block also contains a hash value that can be verified by everyone present on the blockchain. The next block has an "index" value equal to '1' implying that it is the second block in the chain. Its previous hash parameter contains exactly the same hash value of the previous Genesis block.

The code snippet in Fig. 10.4 demonstrates the consensus aspect of blockchain and distributed ledgers, this code shall be executed to verify the validity of the block. It is added to the chain only if found valid. For detailed implementation of blockchain the Consensus Algorithm of either POW or POS is naturally incorporated.

Figure 10.5 code snippet of a block creation demonstrates the creation of genesis and other blocks. The addBlock function is called only when the receiver is receiving the sender's message. The code will be changed in fully fledged application as per the need of the application.

```
chainIsValid()
{
  for(var i=0;i<this.chain.length;i++)
  {
    if(this.chain[i].hash !== this.chain[i].calculateHash())
      return false;

    if(i > 0 && this.chain[i].previousHash !== this.chain[i-1].hash)
      return false;
  }
  return true;
}
```

Fig. 10.4 Chain validity (Immutability)

```
addBlock(data)
{
    let id = this.chain.length;
    let time = Math.floor(Date.now() / 1000);
    let newBlock = new Block(id,time,data);
    if(this.chainIsValid())
    {
        newBlock.previousHash = this.getLatestBlock().hash;
        newBlock.hash = newBlock.calculateHash();
        this.chain.push(newBlock);
        fs.appendFile('blocks.txt', "\n"+JSON.stringify(newBlock,null,4), (err) =>
        {
            if (err) throw err;
        });
    }
}
```

Fig. 10.5 Adding a new block

10.6 Blockchain Use Case

This section of the paper emphasizes that the Chatserver approach through blockchain and be effectively used in any confidential environments, typically where the trail of communication is determined to be immutable. The trail modified at any future representation should lead to an alarm and the associating management bodies should take cognizance (Figs. 10.6 and 10.7).

10.7 Conclusion

This paper has explored the architecture of a typical blockchain whereby the components of a block have been elaborated with a purpose of implementing the same through the Node.js approach. The immutability aspect of a blockchain is the key

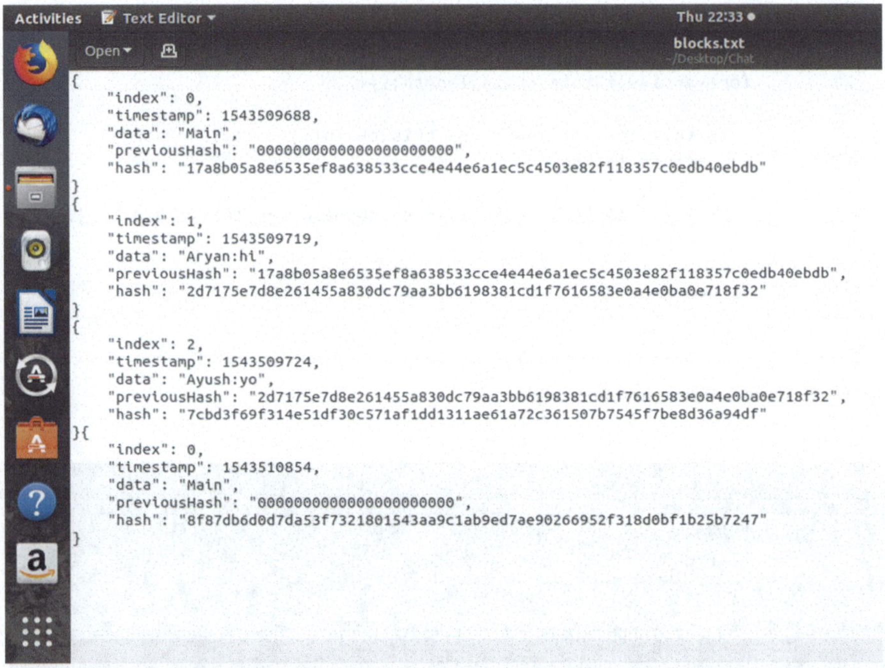

Fig. 10.6 Blockchain of chats

```
socket.on('usersend', function(msg){
  for (var i in socketInfo)
  {
      if(socketInfo[i].user!=socketInfo[socket.id].user)
      {
          socketInfo[i].socket.emit('mesrec', socketInfo[socket.id].user+":"+msg);
          blockch.addBlock(""+socketInfo[socket.id].user+":"+msg+"");
      }
  }
  socket.emit('mesrecuser', "You"+":"+msg);
});
```

Fig. 10.7 Block creation process

feature for acceptability in all business environments where security and privacy is the main concern. The aspect of immutability has been proposed for a chat server where the confidential trail of conversations is of prime importance [14, 15].

The future scope of extendibility of the said Node.js applications shall demonstrate the integration with DAPPs and other web applications. The paper presents the demonstrative approach of a Blockchain Chatserver and not a complete code by itself. The code needs to be extended for complete implementation of the blockchain and is beyond the scope of this paper [16, 17].

References

1. Brandtzæg PB (2013) Big Data, for better or worse: 90% of world's data generated over last two years. https://www.sciencedaily.com/releases/2013/05/130522085217.htm
2. Solon O (2018) Facebook says Cambridge Analytica may have gained 37 m more users'. https://www.theguardian.com/technology/2018/apr/04/facebook-cambridge-analytica-user-data-latest-more-than-thought
3. Jones S (2017) Global alert to prepare for fresh cyber-attacks. Accessed Financial times. https://www.ft.com/content/bb4dda38-389f-11e7-821a-6027b8a20f23
4. Economist T (2017) The world's most valuable resource is no longer oil, but data. Accessed The economist. https://www.economist.com/leaders/2017/05/06/the-worlds-most-valuable-resource-is-no-longer-oil-but-data
5. Voshmgir S (2017) What is Blockchain? fetched from Blockchain Hub. https://blockchainhub.net/blockchain-intro/
6. Dinh TTA et al (2017) Untangling blockchain: a data processing view of blockchain systems. IEEE
7. licensed under CC Attribution-Share-Alike 3.0, retrieved from Wikipedia
8. Tapscott D (2016) Blockchain revolution: how the technology behind Bitcoin is changing money, Business and the World
9. Nirupama Devi Bhaskar DL (2015) Handbook of digital currency. Elsevier
10. Nakamoto S, Bitcoin: a peer to Peer electronic cash system. www.bitcoin.org
11. What is Blockchain Technology? A Step-by-Step Guide For Beginners (n.d.) Accessed blockgeeks.com. https://blockgeeks.com/guides/what-is-blockchain-technology/
12. Gipp B, Meuschke N (2015) Decentralized trusted timestamping using the crypto currency bitcoin. In: iConference. Newport Beach, CA
13. https://nodejs.org/en/. The site of Node.js
14. Telegraph C (n.d.) How blockchain technology works. guide for beginners. Accessed Coin telegraph. https://cointelegraph.com/bitcoin-for-beginners/how-blockchain-technology-works-guide-for-beginners#nodes
15. Cachin C, Vukolic M (2017) Blockchain consensus protocols in the wild, IBM Research
16. Comparsion of Ethreum, Hyper Ledger Fabric And Corda, Martin Valenta, Philipp Sandner. www.fs-blockchain.de
17. Hyperledger-fabricdocs documentation, Release Master, Accessed 06 Mar 2018

Chapter 11
Comparative Study and Analysis of Various Facial Emotion Recognition Techniques

Naveen Kumari and Rekha Bhatia

11.1 Introduction

Facial emotion is a fundamental means to communicate with people. The ability to understand the facial emotion of others is a key to successful communication. Universal consistency is revealed by humans in order to recognize the emotions of a different individual. Emotion becomes an intermediate between human beings to assist interactions [1]. So, it often appears to be very challenging in order to understand emotional communication in the social environment. Recognition of emotion can be done through a variety of means like body language, voice modulation, and several other methods. However, the more practical technique is to observe the facial expressions. The human emotions that are recognized universally are as follows: anger, disgust, happiness, fear, neutral, surprise, and sadness. Sometimes, in order to express a complex emotion, an individual comes up with a more complex expression which needs to be acknowledged properly.

In Facial Expressions Recognition System, the image is processed to extract such information from it, which can help in recognizing six universal emotions that are neutral, happy, sad, angry, disgust, and surprise. This processing is done in several phases including image pre-processing, features extraction, and finally emotions classification using different techniques.

Facial Emotion Recognition is the one that recognizes the state of emotion of the individual personality by utilizing the facial image that is captured by various means like high-resolution cameras, surveillance cameras, and several other means that

N. Kumari (✉) · R. Bhatia
Department of Computer Science, Punjabi University Regional Centre for IT and Management, Mohali, India
e-mail: naveencse2k4@gmail.com

R. Bhatia
e-mail: r.bhatia71@gmail.com

P. K. Kapur et al. (eds.), *Decision Analytics Applications in Industry*, Asset Analytics,
https://doi.org/10.1007/978-981-15-3643-4_11

capture the face images. Facial expression, being a fundamental mode of communicating human emotions, finds its applications in human–computer interaction (HCI), health care, surveillance, driver safety, deceit detection, etc. [2]. Generally, face provides various kinds of signals to convey the message. Sometimes, the wrong signal is grabbed in order to recognize the emotion of an individual. When humans percept any type of emotion of other individuals they are almost correct though many emotions when combined can form a complex signal which will be difficult to understand. For recognizing facial expression, the focus should be entirely on signal to learn a particular type of message. Some facial expressions [3] are regularly used in the social environment during any conversation or for reactions during the conversation such as anger, surprise, sadness, happiness, fear, surprise, and disgust. Parallel face reading gives the major information about the person as the individual has to look at the rapid temporary changes on the face of the person. Humans express emotions in regular day to day communications. An individual can know how to react to people's expressions by understanding emotion which greatly enriches the interaction; face generally delivers three types of signs, i.e., static, slow, and rapid. The static signs are not entirely irreversible. The facial changes with growth such as a change in face structure, features, and skin pigmentation are the signs that is called static signs of a face. These facial signs are considered to transform far less than the slow signs. The slow signs include facial expression which gradually changes with age such as skin color, shape of face, the location of different facial features [4]. The rapid signs are generated by the movement of facial muscle which results in temporary variations in facial appearance, location shift, and temporary wrinkles. Individual facial features such as face structure, age, and skin type do not tell that a person is happy or angry but these features may affect face impression.

This paper is divided into various sections. Section 11.2 describes the facial emotion recognition system. Section 11.3 presents facial emotion recognition techniques. Section 11.4 represents result analysis. Section 11.5 describes research gaps and need of study. The paper is concluded in Sect. 11.6.

11.1.1 Facial Emotion Recognition System

The primary need for facial emotion recognition system is image pre-processing, feature extraction, and emotion classification. In Fig. 11.1 initially, the facial image is acquired from the particular database. Then, the acquired facial image is pre-processed to remove the noises and enhance the quality of the facial image. In the next step, the features are extracted from the image, i.e., geometric features and appearance features, with the assistance of suitable feature extraction technique. Later, a classifier is employed in order to apply classification on the set of acquired features. Subsequently, on the basis of classification, the emotions are recognized and the final output is delivered recognizing the exact type of the feature.

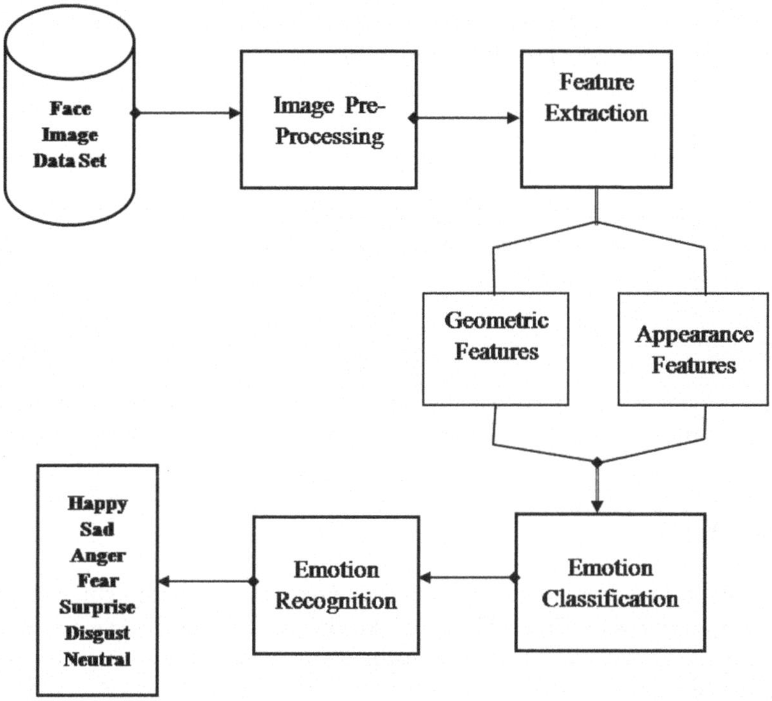

Fig. 11.1 Facial emotion recognition system

11.1.2 Types of Emotions

There are different types of emotions that can be easily detected on the human face [5]. The following are the main types of emotions:

11.1.2.1 Surprise

Facial feature creates an impression of total change such as eyes looks much bigger than normal expressions, jaw drop occurs and the brows go a little higher. The experience of surprise differs in intensity as the extreme surprise may lead to more wide eyes, extent jaw-drop and higher brows.

11.1.2.2 Fear

Fear expression is generally joined by the fear in eyes and brows. In the beginning, the expression is of worry which is referred to as momentary feeling. Near eye region, upper eyelids are raised as high as possible, eyebrows are raised and comes together, and eyes focus at one straight point. Whereas, near mouth region lips are stretched horizontally and there is a possibility of jaw drop.

11.1.2.3 Happy

Happy expression is symbolized by joy with a smile, raised cheeks, tightened eyelids, drawn down eyebrows and mouth can be open or closed.

11.1.2.4 Sad

Sad expression is generally showed by the corner lip down movement, dropped jaw and upper eyelids are dropped with the inner corner of eyebrow brought together.

11.1.2.5 Anger

Angry expression is symbolized by the raised upper eyelids with tightened up area around eyes, eyebrows are lowered and joined together, jaw tightly clenched, thinning of lips, forward lower jaw. In case of teeth closed position, mouth tends to get a rectangular shape.

11.1.2.6 Neutral

Neutral expression [6] is relaxed facial muscles. Eyelids are tangent to the iris. The mouth is closed and lips are in contact.

11.1.2.7 Disgust

Eyebrow and eyelids are relaxed. Also, there is raised and curled upper lip, frequently asymmetrically.

11.1.3 Common Dataset

A dataset is mainly generated for the determination of training and testing of the classifiers that are utilized in the research work. There are various types of common datasets [7] that are available to be employed in the research work related to the facial emotion recognition. Images used for facial expression recognition are static images or image sequences. An image sequence potentially contains more information than a still image, because the former also depicts the temporal information. The usage of these databases is restricted for research purpose only. Most commonly used databases include

11.1.3.1 Japanese Female Facial Expression Database (JAFFE)

JAFFE database [8] generally contains approximately 219 grayscale images with different facial expressions, i.e., happiness, sadness, anger, disgust, fear, surprise, and neutral face posed by 10 female subjects. Each of the subjects posed with three to four examples per expression to make a total of 219 images. Each image has been rated on different emotion adjective and the expression images are labeled as per the predominant expression in that image.

11.1.3.2 FERET Database

Face Recognition Technology program (FERET) [9] is basically a large database of facial images gathered independently from different algorithm developers. This database is the collaborative effort of Dr. Wechsler and Dr. Philips. The goal of the FERET database is to develop automatic face recognition abilities that could be used to support security, intelligence and law enforcement recruits in the enactment of their responsibilities. The database consists of 1564 sets of images from 14,126 total images which include 1199 individual images and 365 duplicate set of images that are taken over the lapse of time. This time-lapse is essential to study as it tells the changes that have occurred over a year.

11.1.3.3 Surveillance Camera Face Database (SCface)

The SCface database [10] primarily comprises the static images of human faces. The images were captured in an uncontrolled environment with 5 video surveillance cameras of variant qualities. The database is comprised of around 4160 static images which imitate the real-world circumstances and are utilized for testing and other use case scenarios. It is available free for the research works requiring facial images.

11.1.3.4 The Yale Face Database (Yale B)

The Yale B database [11] mainly delivers high-resolution images completely with the four ground truth locations of the four fiducially points, namely, two eyes, nose, and mouth points, to contribute extra probabilities to improve the performance of recognition. The database contains approximately 5760 images of around 10 subjects, each observed under 576 illumination conditions. Each image with ambient illumination pose was captured for every subject.

11.1.3.5 CK Facial Expression Database

The CK database [12] consists of 100 university students who at the time of their inclusion were between 18 and 30 years old, 65% were female, 15% were African-American, and 3% were Asian or Latino.

11.1.3.6 PIE

PIE database [13] primarily comprised nearly 41,368 images of around 68 individual subjects. The face images were captured by 13 synchronized cameras and 21 flashes, under varying pose, illumination, and expression. Each subject includes 13 different poses, under 43 different viewing conditions and with 4 distinct expressions.

11.1.3.7 MSRA Database

MSRA database [14] was collected at Microsoft Research Asia. This database contains face images that are posed by 12 individual subjects. These images were apprehended in two different sessions with changed backgrounds and lighting conditions. Around 64–80 facial images were collected for each individual in each session. All the images of faces are frontal.

11.2 Facial Emotion Recognition Techniques

There are three different types of techniques that are used to detect the facial emotions of a human face.

11.2.1 Image Pre-processing Technique

Image pre-processing plays an important role in emotion recognition. Image pre-processing stage enhances the quality of an input image and locates the data of interest by removing noise and filtering the image. Pre-processing is divided into 3 steps: face detection, normalization, and filtering of an image which produces uniform and enhanced images.

11.2.2 Feature Extraction Technique

Facial feature extraction [15] is the process of translating the input data into some set of features. Feature points such as nose, eyes, mouth are extracted and then used as input data to an application. The use of feature extraction can help reduce a huge amount of data to a relatively small set which is computationally faster. It is influenced by many complications like a difference in different pictures of the same facial expression, the light directions of imaging, and the variety of posture, size, and angle. The prime focus of the feature extraction techniques [16] is to extract the unique feature sets from the input facial image. The uniqueness of the extracted feature sets is directly proportional to the recognition rate.

A number of researches [6] have been conducted using different approaches in different areas of analysis for feature extraction. For instance, some use information dependent on geometric features in 2-D and 3-D facial images while others use static image information obtained by filtering the image. Several filter images that have been of use include Principal Component Analysis (PCA), Independent Component Analysis (ICA), Discrete Cosine Transform (DCT), Gabor Filters, Fast Fourier Transform (FFT), Singular Value Decomposition (SVD), Haar Wavelet transforms, etc. Sometimes a combination of multiple techniques and filtering is used for better performance analysis.

11.2.3 Emotion Classification Technique

Classification [17] is mainly employed for categorizing the facial emotions which are done by utilizing a classifier that usually comprised of pattern distribution merged with decision procedure. There is a wide variety of the classifier that is being opted, comprising the parametric as well as non-parametric techniques, in order to recognize the facial emotions problems. The classification approach is one of the most essential steps in the facial expression recognition system, which in turn ensures the reliability of the entire emotion recognition process, and also minimizes the probability of false positives that are being caused by miscalculations. Once, the set of a feature is extracted [18] from the facial image encompassing displaying the emotion, the optimization task is applied to the set of features. Finally, the classification is performed in order to recognize the specific emotion from the facial image. As supervised training is mandatory for this classification, therefore the labeled data is required to be enclosed in the training set. As soon as the training of the classifier is accomplished, the input images can be easily recognized by allocating them a precise class label. Generally, the facial expressions classification is performed in terms of Action Units (AU's) as well as in terms of seven common emotions: happy, sad, anger, surprise, disgust, neutral, and fear. There are two main categories of Feature Classification approach [6]:

1. Statistical non-machine learning approaches such as Euclidean and linear discrimination analysis.
2. Machine learning approach such as Feed forward Neural Network, Hidden Markov Model, Multilayer Perceptron, Radial Basis Function Network, etc.

11.3 Literature Survey

A study of about 35 papers which were represented in the field has been done in this research. All researchers have tried to improve the performance of the facial emotion recognition system by enhancing feature extraction techniques and classification techniques. Tables 11.1 and 11.2 show the summary of research work.

11.4 Research Gaps and Need of Study

Facial Emotion Recognition system is the one that recognizes the state of emotion of the individual personality by utilizing the facial image that is captured by various means like high-resolution cameras, surveillance cameras and several other means that capture the face images [10]. Since the advent of various face capturing devices, there has been a lot of work done by different researchers till today regarding the recognition rate of the facial emotion. Every proposed technique demonstrates certain variation in the recognition rate and has different accuracy due to the lack of appropriate feature descriptor [19]. There are still several factors that are required to be kept in focus while proposing an efficient facial emotion recognition technique.

The following points have been analyzed on the basis of which there is a need to elaborate the existing work:

(1) The feature extraction techniques [20] can be optimized to make the process more precise.
(2) Training and classification mechanism [21] may need more than one technique for better efficiency.
(3) Efficient emotion recognition technique can be developed for identifying the facial features under 'uneven lighting' [6].

11.5 Conclusion

The main objectives of this paper were to analyze the various techniques adopted in the facial emotion recognition system. Facial expressions are a very effective way to convey human emotions. It plays a vital role in non-verbal communication. There

Table 11.1 Comparative study and analysis of various facial emotion recognition techniques based on machine learning

S.No.	References	Dataset used	Recognition techniques used	Recognition rate
1	Chang and Ryoo [22]	Cohn JAFFE Self	Feature extraction: principal component analysis, linear discriminant analysis classification: SVM	90.1% 81.5% 85.1%
2	Jain et al. [23]	JAFFE MMI	Combination of CNN and RNN	94.91% JAFFE 92.07% MMI
3	Swati et al. [24]	JAFFE Yale Ck+	Feature extraction: discrete wavelet transform, HOG Classification: Multi-Class SVM	90% 88.3% 90%
4	Majumder et al. [25]	MMI extended cohn-Kanade	Feature extraction: fusion of geometric features and appearance feature using autoencoders classification: SOM based classifier	97.55% MMI 98.95% ck+
5	Priya and Muralidhar [26]	JAFEE	Classification: decision tree	78.4%
6	Tarnowski et al. [27]	Self-made	MLP classifier and "Natural" division of data	96 and 73% for natural division of data
7	Shan et al. [28]	JAFFE CK+	A deep convolution neural network (CNN), K-nearest Neighbor	76.7442% JAFFE 80.303% CK+
8	Qayyum et al. [29]	JAFFE CK+	Feed forward neural network, discrete cosine transform	98.83% JAFFE 96.61%CK+

(continued)

Table 11.1 (continued)

S.No.	References	Dataset used	Recognition techniques used	Recognition rate
9	Yanga et al. [30]	JAFFE	Feature extraction: HAAR cascade method, sobel Method followed by feature extraction neural network for classification	Sad 78.54% Surprise 93.26% Happy 95.25% Anger 91.22% Disgust 84.32% Fear 82.58%
10	Wang et al. [31]	Self-made	Feature extraction: stationary wavelet entropy Classification: feed forward neural network	96.80%
11	Al-Sumaidaee et al. [32]	CK	Feature extraction: novel feature extraction method Classification: Multi-class SVM	99.36%
12	Kamal et al. [16]	JAFFE CohnKanade	2D-LDA, 2D-PCA, support vector machines (SVM), K-nearest neighborhood (K-NN) classifiers	97.63% JAFFE 94.8% CK
13	Suja et al. [33]	CMU multiple	Haar cascade, active shape model (ASM), Adaboost Classifier	94%
14	Mert et al. [34]	DEAP	Feature extraction: multivariate empirical mode deposition Classification: k-NN and A-NN classifier	67% k-NN 72.87% A-NN

(continued)

Table 11.1 (continued)

S.No.	References	Dataset used	Recognition techniques used	Recognition rate
15	Mehta and Jadhav [15]	JAFFE	Log-Gabor filter + PCA	93.7%
16	Dixit et al. [35]	JAFFE	Feature extraction: Zernike moments Classification: Naive Bayesian classifier	81.66%
17	Basu et al. [36]	KTFE	Fusion of thermal image with histogram statistics	87.50%
18	Hsieh et al. [37]	Cohn-Kanade	Feature extraction: Gabor filter and Laplician of Gaussian Classification: Multi-class SVM	94.7%
19	De et al. [38]	Generic Dataset used for training	Feature extraction: PCA Classification: euclidean distance	Happiness: 93.1% Surprise, Anger: 91 and 86.2% Sorrow and fear: 78.9 and 77.7%
20	Agrawal et al. [21]	JAFFE	Feature extraction: face localization and feature detection process classification: PCA + SVM	94%
21	Owusu et al. [39]	JAFFE Yale	Viola-Jones descriptors, bessel transform, gabor feature extraction techniques, AdaBoost-based hypothesis, neural network classifier	JAFFE 96.83% Yale 92.22%

(continued)

Table 11.1 (continued)

S.No.	References	Dataset used	Recognition techniques used	Recognition rate
22	Lorincz et al. [20]	(CK+)	Feature extraction: 3D constrained local model Classification: time-series with SVM and Kernel method	99%
23	Gu et al. [40]	JAFFE CK database	Feature extraction: Gabor filter, PCA recursive LDA Classification: modified K-NN algorithm	89.67% JAFFE 91.51% CK
24	Liu and Wang [41]	PIE	SVM (Support vector machine) classifier-based Adaboost algorithm used detection and correction	95% or more
25	Bashyal and Venayagamoorthy [42]	JAFEE	Gabor filter + learning vector quantization	90.22%

Table 11.2 Comparative study and analysis of various facial emotion recognition techniques based on deep learning

S.No.	References	Dataset used	Recognition techniques used	Recognition rate
1	Zhang et al. [43]	CK+	Convolutional neural network (CNN), multitask network	98.9
2	Li et al. [44]	MMI	CNN, SVM (Support vector machine)	78.46
3	Pramerdorfer and Kampel [45]	FER	CNN + network Ensemble	75.2
4	• Huynh et al. [46]	BU-3DFE	CNN	92%
5	Guo et al. [47]	FER	CNN, k-NN	71.33
6	Hamester et al. [48]	JAFFE	CNN, CAE	95.8
7	Jung et al. [49]	Ck+	CNN DNN (Deep neural network)	98.75% 97.59
8	Liu et al. [50]	CK+MMI SFEW	Au-inspired deep networks	93.70% 75.85% 30.14%
9	Liu et al. [52]	Ck+JAFFE	Boosted Deep Belief Network	96.7% 93.0%
10	Ouellet et al. [51]	Ck+	Deep convolutional neural network + SVM	94.4%

has been a lot of work done by different researchers but still, there is a need to achieve a high level of recognition rate by reducing the computation time. Every proposed technique demonstrates certain variation in the recognition rate and has different accuracy due to the lack of appropriate feature descriptor. The research is still going on to increase the recognition rate of predicting facial emotions.

11.6 Future Scope

In future, researchers should try to hybridize the classification and feature extraction technique to improve the recognition accuracy/rate. The recognition rate can be enhanced by a combination of existing facial emotion recognition techniques. An efficient emotion recognition technique can be developed for identifying the facial feature under 'uneven lighting'.

References

1. Fasel B, Luettin J (2003) Automatic facial expression analysis a survey. Pattern Recognit 36(1):259–275
2. Happy SL, Routray A (2015) Automatic facial expression recognition using features of salient facial patches. IEEE Trans Affect Comput 6(1):1–12
3. Bai Y, Tang X, Zhu Y, Cai D (2014) A sentiment analysis method for facial expression generation in human-robot interactive communication. In: IEEE International Conference on Virtual Reality and Visualization (ICVRV), Shenyang, China, pp 97–102
4. Szwoch M, Pieniazek P (2015) Facial emotion recognition using depth data. In: IEEE 8th international conference on human system interactions (HSI), Warsaw, Poland, pp 271–277
5. Majumdar J, Avabhrith R (2014) Human face expression recognition. Int J Emerg Technol Adv Eng (IJETAE) 4(7):559–565
6. Khan NU (2013) A comparative analysis of facial expression recognition techniques. In: 3rd IEEE international advance computing conference (IACC), pp 1262–1268. https://doi.org/10.1109/iadcc.2013.6514409
7. Anitha C, Venkatesha MK, Narayana Adiga BS (2010) A survey on facial expression databases. Int J Eng Sci Technol 2(10):5158–5174
8. Huang Hung-Fu, Tai Shen-Chuan (2012) Facial expression recognition using new feature extraction algorithm. ELCVIA: Electron Lett Comput Vis Image Anal 11(1):0041–54
9. Hsu R-L, Abdel-Mottaleb M, Jain AK (2002) Face detection in color images. IEEE Trans Pattern Anal Mach Intell 24(5):696–706
10. Grgic M, Delac K, Grgic S (2011) SCface–surveillance cameras face database. In: Multimedia tools and applications, vol 51, no 3, pp 863–879
11. Priyanka MS, Jalan A, Nimbarte MS (2014) A survey on facial feature extraction to recognize facial expressions. Int J Eng Res Technol 3(2):539–544
12. Jabid T, Kabir MH, Chae O (2010) Robust facial expression recognition based on local directional pattern. ETRI J 32(5):784–794
13. Baker S, Sim T, Bsat M (2003) The CMU pose, illumination, and expression database. IEEE Trans Pattern Anal Mach Intell 25(12):1615–1618
14. He X, Yan S, Yuxiao H, Niyogi P, Zhang H-J (2005) Face recognition using laplacianfaces. IEEE Trans Pattern Anal Mach Intell 27(3):328–340
15. Mehta N, Jadhav S (2016) Facial emotion recognition using log gabor filter and PCA. In: International conference on computing communication control and automation (ICCUBEA), Pune, pp 1–5
16. Kamal S, Sayeed F, Rafeeq M (2016) Facial emotion recognition for human-computer interactions using hybrid feature extraction technique. In: IEEE international conference on data mining and advanced computing, Ernakulam, India, pp 180–184
17. Jun Ou (2012) Classification algorithms research on facial expression recognition. Phys Procedia 25:1241–1244
18. Geng C, Jian X (2009) Face recognition using sift features. In: IEEE international conference on image processing (ICIP), Cairo, Egypt, pp 3313–3316
19. Kumari J, Rajesh R, Pooja KM (2015) Facial expression recognition: a survey. Procedia Comput Sci 58:486–491
20. Lorincz A, Jeni L, Szabo Z, Cohn J, Kanade T (2013) Emotional expression classification using time-series kernels. In: Proceedings of the IEEE conference on computer vision and pattern recognition workshops, Portland, OR, USA, pp 889–895
21. Agrawal DD, Dubey SR, Jalal AS (2014) Emotion recognition from facial expressions based on multi-level classification. Int J Comput Vis Robot 4(4):365–389
22. Chang J, Ryoo S (2018) Implementation of an improved facial emotion retrieval method in multimedia system. In: Multimedia tools and applications, vol 77, no 4, pp 5059–5065
23. Jain N, Kumar S, Kumar A, Shamsolmoali P, Zareapoor M (2018) Hybrid deep neural networks for face emotion recognition. Pattern Recognit Lett. https://doi.org/10.1016/j.patrec.2018.04.010

24. Nigam S, Singh R, Misral AK (2018) Efficient facial expression recognition using histogram of oriented gradients in wavelet domain. In: Multimedia tools and applications, pp 1–23. https://doi.org/10.1007/s11042-018-6040-3
25. Majumder Anima, Behera Laxmidhar, Subramanian Venkatesh K (2018) Automatic facial expression recognition system using deep network-based data fusion. IEEE Trans Cybernet 48(1):103–114
26. Priya V, Muralidhar (2017) Facial emotion recognition using eye. Int J Appl Eng Res 12(16):5655–5659
27. Tarnowski P, Kolodziej M, Majkowski A, Rak RJ (2017) Emotion recognition using facial expressions. Procedia Comput Sci (Zurich, Switzerland) 108C:1175–1184
28. Shan K, Guo J, You W, Lu D, Bie R (2017) Automatic facial expression recognition based on a deep convolutional-neural-network structure. In: IEEE 15th international conference on software engineering research, management and applications (SERA), London, UK, pp 123–128
29. Qayyum H, Majid M, Anwar SM, Khan B (2017) Facial expression recognition using stationary wavelet transform features. Math Probl Eng 2017:1–9, Article ID 9854050
30. Yanga D, Alsadoona A, Prasad PWC, Singh AK, Elchouemic A (2017) An emotion recognition model based on facial recognition in virtual learning environment. In: 6th international conference on smart computing and communications (ICSCC), vol 125. Procedia Computer Science, Kurukshetra, pp 1–10
31. Wang S-H, Phillips P, Dong Z-C, Zhang Y-D (2018) Intelligent facial emotion recognition based on stationary wavelet entropy and Jaya algorithm. Neurocomputing 272:668–676. https://doi.org/10.1016/j.neucom.2017.08.015
32. Al-Sumaidaee S, Abdullah MAM, Al-Nima R, Dlay SS, Chamber JA (2017) Multigradient features and elongated quinary pattern encoding for image-based facial expression recognition. Pattern Recognit. https://doi.org/10.1016/j.patcog.2017.06.007
33. Suja SP, Tripathi S (2016) Real-time emotion recognition from facial images using raspberry Pi II. In: IEEE 3rd international conference on signal processing and integrated networks (SPIN), Noida, India, pp 666–670
34. Mert A, Akan A (2016) Emotion recognition from EEG signals by using multivariate empirical mode decomposition. Springer, London. https://doi.org/10.1007/s10044-016-0567-6
35. Dixit BA, Gaikwad AN (2015) Statistical moments based facial expression analysis. In: IEEE international advance computing conference (IACC), Banglore, India, pp 552–557
36. Basu A, Routray A, Shit S, Deb AK (2015) Human emotion recognition from facial thermal image based on fused statistical feature and multi-class SVM. In: Annual IEEE India conference (INDICON), New Delhi, pp 1–5
37. Hsieh C-C, Hsih M-H, Jiang M-K, Cheng Y-M, Liang E-H (2015) Effective semantic features for facial expressions recognition using SVM. In: Multimedia tools and applications, vol 75, no 11, New York, pp 6663–6682
38. De A, Saha A, Pal MC (2015) A human facial expression recognition model based on Eigen face approach. In: Procedia computer science, international conference on advanced computing technologies and applications (ICACTA), vol 45, pp 282–289
39. Owusu E, Zhan Y, Mao QR (2014) A neural-AdaBoost based facial expression recognition system. Expert Syst Appl 41(7):3383–3390
40. Gu W, Xiang C, Venkatesh YV, Huang D, Lin H (2012) Facial expression recognition using radial encoding of local Gabor features and classifier synthesis. Elsevier J Pattern Recognit, vol 45, pp 80–91
41. Liu S, Wang W (2010) The application study of learner's face detection and location in the teaching network system based on emotion recognition. Second Int Conf Netw Secur Wirel Commun Trust Comput 01:394–397
42. Bashyal S, Venayagamoorthy GK (2008) Recognition of facial expressions using Gabor wavelets and learning vector quantization. Eng Appl Artif Intell 21:1056–1064
43. Zhang Z, Luo P, Chen CL, Tang X (2018) From facial expression recognition to interpersonal relation prediction. Int J Comput Vision 126(5):1–20

44. Li S, Deng W, Du J (2017) Reliable crowdsourcing and deep localitypreserving learning for expression recognition in the wild. In: 2017 IEEE conference on computer vision and pattern recognition (CVPR). IEEE, pp 2584–2593
45. Pramerdorfer C, Kampel M (2016) Facial expression recognition using convolutional neural networks: State of the art. arXiv preprint arXiv:1612.02903
46. Huynh X-P, Tran T-D et al (2016) Convolutional neural network models for facial expression recognition using BU-3DFE database. In: Information science and applications (ICISA), vol 376. Springer, Singapore, pp 441–450. Print ISBN 978-981-10-0556-5
47. Guo Y, Tao D, Yu J, Xiong H, Li Y, Tao D (2016) Deep neural networks with relativity learning for facial expression recognition. In: 2016 IEEE international conference on multimedia & expo workshops (ICMEW). IEEE, pp 1–6
48. Hamester D, Barros P, Wermter S (2015) Face expression recognition with a 2-channel convolutional neural network. In: 2015 international joint conference on neural networks (IJCNN). IEEE, pp 1–8
49. Jung H, Lee S et al (2015) Development of deep learning-based facial expression recognition system. In: IEEE 21st Korea-Japan joint workshop on frontiers of computer vision (FCV), Mokpo, South Korea. ISBN: 978-1-4799-1720-4
50. Liu M, Li S et al (2015) Au-inspired deep networks for facial expression feature learning. Neurocomputing 159:126–136
51. Ouellet S (2014) Real-time emotion recognition for gaming using deep convolutional network features. arXiv:1408.3750
52. Liu P, Han S, Meng Z, Tong Y (2014) Facial expression recognition via a boosted deep belief network. In: The IEEE conference on computer vision and pattern recognition (CVPR), pp 1805–1812

Chapter 12
A Novel Block Hashing-Based Template Security Scheme for Multimodal Biometric System

Mahapara Khurshid and Arvind Selwal

12.1 Introduction

The biometric systems are automated pattern recognition systems that are used to verify/identify an individual based on different attributes like biological, behavioral, or chemical [1]. The various traits which can be used include fingerprint, iris, hand geometry, palmprint, face, gait, voice, etc. The paradigm of authenticating an individual has been shifted from traditional systems to biometric-based systems which are based on something we inherit, as the traditional systems suffer a lot of security breaches. Biometric systems have an extra layer of security as the person to be verified must be present at that time. Biometric systems are gaining importance in almost every field like national ID registration, border security, education, etc. [2]. The unimodal systems operating on one biometric trait suffers from many limitations like noise, non-universality, spoofing attacks, large intra-user variations, etc. [3, 4] In order to counter the limitations and improve the accuracy, a system consisting of more than one trait, referred to as multimodal systems [5–8] are being developed. The traits can be fused at different levels [9].

A fishbone model can be used to demonstrate the different causes of vulnerabilities present in a biometric system [10]. A variety of threats have also been identified by Ratha et al. [2] which if left unattended may cause a serious effect on the security of a system.

To mitigate the attacks and ensure the confidentiality of the system, the development of efficient template security schemes is of prime importance and an open field of research. Any biometrics template protection scheme must satisfy the ideal properties like diversity, revocability, reversibility, and performance [10]. In literature,

M. Khurshid (✉) · A. Selwal
Central University of Jammu, J&K, Samba, India
e-mail: mahaparakhurshid@gmail.com

A. Selwal
e-mail: arvind.cuj@gmail.com

P. K. Kapur et al. (eds.), *Decision Analytics Applications in Industry*, Asset Analytics,
https://doi.org/10.1007/978-981-15-3643-4_12

the template protection schemes have been classified as biometric cryptosystems and feature transformation-based systems [11, 12].

- Feature transformation approach: In this approach, the transformed template is produced by the application of a transformation function which is stored in the database. The different parameters that are used by the function are either derived from a password or a random key. At the time of matching, the query template is generated by applying the same transformation function and is matched against the stored template. Salting and non-invertible transforms are two classes of these types of systems categorized on the basis of the function used. In salting, the transformation function is invertible and the secrecy of the key determines the security of the system. If the key is disclosed to an attacker, the security of the system is compromised. Whereas in the case of non-invertible transform, the function used is one-way, i.e., it is very hard to get the original input template even if the secret key is known [10].
- Biometric cryptosystems: In such systems, only some public information/helper data other than the biometric template is stored about the biometric template. Matching is done indirectly by extracting the stored key from helper data and then verifying its validity. Key-binding and key-generation systems are the two variants of these systems which are categorized on the basis of the way of extracting the helper data. When a key is bound with the template and helper data is obtained from it, we say it is a key-binding scheme. Whereas, if we get the helper data directly from the template and also the key is generated from it and the query template, such systems are referred to as key-generation systems [10].

12.2 Literature Survey

A wide variety of template protection schemes have been developed by different researchers. These techniques have been evaluated based upon performance metrics and security parameters as per ISO Standard. The template security schemes for multimodal biometric systems are mainly based on the integration of feature vector information at various levels. Various analytical studies have also been presented to show various issues in designing the template security systems [13, 14].

Jeng et al. [15] proposed a template security scheme for a biometric system consisting of a face, iris, and palm as traits. The authors presented a technique called shuffle coding feature level fusion, to fuse the traits either from the same feature space or different feature space. The authors use this scheme with scaling and hashing. The results obtained shows better performance than the existing serial and parallel methods.

Gupta et al. [2] presented an NN-based hybrid scheme for the multimodal system to improve the storage. The authors use feature level fusion and apply bio hashing. In order to reduce the size, the authors apply their octet indexing technique. After experimentation, the results show that their proposed scheme performed better both in terms of security and storage. Also, this scheme shows a recognition rate of 98.4% and EER = 0.48%.

Aravinth et al. [16] presented the concept of remote biometric authentication. The authors use multiple classifiers and score level fusion. The authors use fingerprint, face, and iris as biometrics traits. After experimentation, it was revealed that the maximum accuracy shown by the system was 95%.

Sojan et al. [17] presented a template security scheme using trait as a fingerprint. The authors extracted one instance as minutia points and the other as orientation points. The two were mixed to get a single template. Correlation-based methods were used for matching purpose. The results demonstrated that the system shows good identification accuracy.

Ramalho et al. [18] proposed a multimodal biometrics system based on palmprint, hand geometry, and finger surface as biometrics trait. The authors have used error-correcting code, a hash function, and binarization process for designing the template security scheme. The results show that the system has a good recognition rate and is also fast as compared to others.

Selwal et al. [19] presented a scheme for the template security multi-instance system having fingerprint as a trait. The authors segment the ROI and assign a 3-bit code to each based on the angle. The weighted sum rule was used to combine the match scores. After experimentation, the results reveal that the error rates were significantly reduced.

Jagadiswary et al. [20] proposed an authentication scheme for multimodal BMS using fingerprint, retina, and finger vein. The authors use feature level fusion and RSA to protect the templates. The results show that the systems using RSA have better performance and overall performance was increased with GAR = 95.3% and FAR = 0.01%.

Rathgeb et al. [12] proposed a TSS for multi-BMS. The authors have used bloom filters to get iris codes. This technique is alignment-free and also the original size was reduced up to 20%. The performance evaluation shows that the EER was below 0.5%.

The background of various template security schemes exhibits clear domination of fingerprint as a biometrical trait. It is pertinent to mention that the information fusion and securing the biometric information is a vital design issue for a multimodal biometrics system. This is mainly because of underlying variation in the size and dimension of various feature vectors generated by respective biometrical traits in the multimodal biometric systems. For example, the size of a typical fingerprint feature vector is of the order of $n \times 3$, whereas the size of a hand geometry feature vector is only a one-dimensional vector of size m. In the case of iris, the feature vector is of the size of 512 bytes in length. Therefore, for designing a multimodal biometric system based upon various modalities, one of the challenges is to integrate the information and secure the resultant templates. The challenge for the research community is to design an effective template security scheme which satisfies all the ideal characteristics as well as maintaining better identification accuracy.

12.3 Methodology

This section explains the framework and algorithms for the proposed template security scheme.

12.3.1 Proposed Framework

Figure 12.1 shows the general framework of our proposed system. It comprises two phases: enrolment phase and verification phase. Both phases include the application of the corresponding transformation processes to get the new secured templates. During the enrolment process, the fingerprint and hand geometry are captured and features are extracted as per the algorithms discussed below. The fingerprint vector undergoes a block-based hashing technique to get the secured template which is then stored in the database.

The feature vector of hand geometry undergoes a transformation process to get the new secured transformed vector which is stored in the template database. At the time of verification, the corresponding feature templates are matched using two matchers resulting in two-match scores. The weighted match score level fusion is used to combine the two scores to get the final decision regarding the user.

Feature extraction of fingerprint: The fingerprint is one of the most widely used biometric traits. It exhibits high identification and verification capability. The irregularities present in the ridges are unique to an individual and also persistent. The extraction of the feature points called minutia can be done by using the following three steps [19].

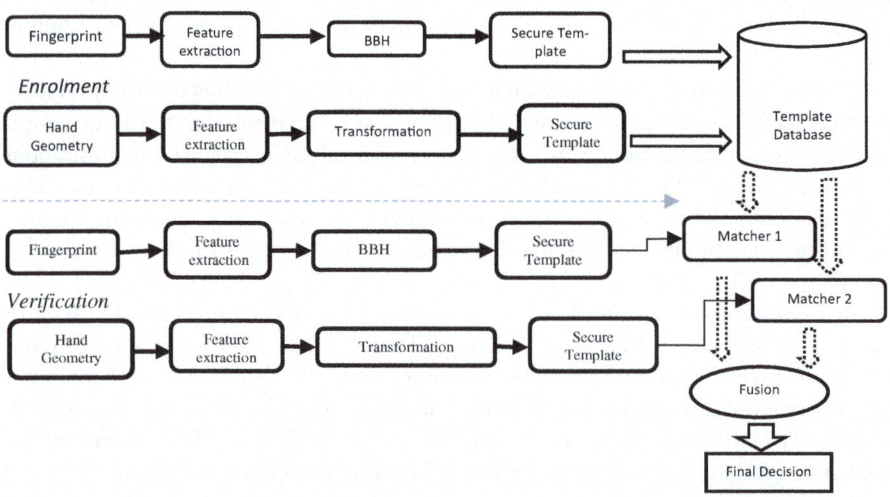

Fig. 12.1 Framework for the proposed template security scheme

1. *Orientation field estimation*: Let the fingerprint image sensed be F(y, z). The captured image is first divided into blocks having size $N \times N$. Then at each pixel (c, d), the local orientation M_y and M_z are calculated as per the following Eqs. (12.1–12.3):

$$M_y(c,d) = \sum_{k=c-N/2}^{m+N/2} \cdot \sum_{l=d-N/2}^{n+N/2} 2g_y(k,l)g_z(k,l) \tag{12.1}$$

$$M_z(c,d) = \sum_{k=c-N/2}^{m+N/2} \sum_{l=d-N/2}^{n+N/2} 2(g_y^2(k,l) - g_{yz}^2(k,l)) \tag{12.2}$$

$$\Theta(c,d) = \frac{1}{2}\tan^{-1}\left(\frac{M_Y(c,d)}{M_z(c,d)}\right) \tag{12.3}$$

The consistency levels in the local neighborhood of some block say (p, q) are calculated for the orientation field using the Eq. (12.4):

$$CL(c,d) = \frac{1}{T}\sqrt{\sum_{(c',d')\in H} \left|\Theta'(c',d') - \Theta(c,d)\right|^2} \tag{12.4}$$

$$\left|\Theta - \Theta'\right| = \left\{\begin{matrix} e & (if\ e = \left(\Theta - \Theta' + 360\right) \bmod 360\right) < 180 \\ e - 180\ otherwise & \end{matrix}\right\} \tag{12.5}$$

Here, H is the neighborhood of the given block, and T the total number of blocks in the neighborhood of Q. Θ (c′, d′) and Θ (c, d) are the local ridge orientations of the respective blocks.

- *Identifying the ridges*: There is an important property regarding ridges that the gray-level values are maximum in the direction that is normal to the local ridge orientation. The convolution process is used consisting of two masks. The value at each pixel is compared against a threshold to get the ridge.
- *Extraction of the minutia points*: Let us consider a pixel on a thinned ridge and its eight neighbors are H_0–H_7. Then the type of ridge can be identified by using Eq. (12.6)

$$\text{Ridge} = \left\{\begin{matrix} (\sum_{i=0}^{7} H_i) = 1 & \text{(x,y) is } ridge\ ending \\ (\sum_{i=0}^{7} H_i) > 2 & \text{(x,y) is } ridge\ bifurcation \end{matrix}\right\} \tag{12.6}$$

Feature extraction of hand geometry: The hand geometry identifies an individual based on various measurements of the hand like the length of the fingers and the width of the fingers. The extraction of the feature vector for hand geometry is done by the following steps.

- *Binarisation*: The input hand undergoes a binarization process to convert the RGB image to BW image so as to lower the computation complexity.
- *Border tracing*: For measuring the various lengths, only the boundary of the hand is of importance. So in order to get the contour of the hand, a proper border tracing algorithm is used. For example, square tracing algorithm can be used to find the contour.
- *Salient point detection*: The first order differential is calculated at each pixel. The point where the sign changes from $+1$ to -1 is considered as the fingertip and the point where the sign changes from -1 to $+1$ is considered as fingerroot.
- *Get the feature vector*: The feature vector of the hand is generated comprising 5 lengths, 10 widths, and width between point v1 and v2. The resulting feature vector is stored in the database.

12.3.2 Proposed Security Algorithm

Figure 12.2 shows the proposed algorithm for the template security scheme. Let the feature vector (FV) of fingerprint and hand geometry of the ith user be X_{fi} and X_{hgi}, respectively. The algorithm uses a block-based hashing (BBH) technique to be applied on the fingerprint vector which results in a transformed vector as S_{fi}. The feature vector of hand geometry undergoes a transformation process to get the new vector, say S_{hgi}. The algorithm works by dividing the fingerprint region into four blocks based on the angle of orientation of each minutia, i.e., 0–90, 90–180, 180–270, and 270–360. Each block undergoes a hashing process so as to get the new transformed blocks to say b'_1, b'_2, b'_3, b'_4. These four blocks are concatenated vertically to get the final transformed feature vector of the fingerprint. The feature vector of hand geometry undergoes a transformation function to get the new feature vector. The final result regarding the user is evaluated on the basis of Threshold (T).

12.3.3 Proposed Matching Algorithm

Figure 12.3 shows the matching algorithm for the proposed template security scheme. During testing, the two extracted transformed feature vectors of fingerprint and hand geometry are compared with the corresponding feature vectors previously stored at the time of enrolment in the template database. The similarity score for each comparison is found and let it be denoted by D and d. The two match scores are fused using weighted sum rule by assigning a weight ‘λ’ to the modalities. In order

Algorithm 1: Proposed template security algorithm
Input: *feature vectors X_{fi} and X_{hgi} of ith user*
Output: *Secured templates S_{fi} and S_{hgi}*
1. Start
2. Let $P=|X_{fi}|$ be the total minutia points extracted from the fingerprint.
3. For k=1 to P, perform the steps 4 to 11.
4. Read k^{th} minutia point, say $M_K(x, y, \Theta)$, read the value of Θ.
5. Divide the feature vector into four blocks based on the value of Θ using the following cases.
6. Case 1: $0< \Theta <=90$
$Block1 = M_k (x, y, \Theta)$
7. Case 2: $90< \Theta <=180$
$Block2 = M_k (x, y, \Theta)$
8. Case 3: $180< \Theta <=270$
$Block3 = M_k (x, y, \Theta)$
9. Case 4: $270< \Theta <=360$
$Block4 = M_k (x, y, \Theta)$
10. Perform hashing on each block resulting in B_1', B_2', B_3', B_4'.
11. Concatenate the hashed blocks to get the final transformed feature vector of the fingerprint and store it as S_{fi}.
12. Let X_{hgi} be the feature vector of the hand geometry of ith user.
13. Find the mean of the vector.
14. Transform the vector X_{hgi} by using the result of step 13 to get the transformed vector S_{hgi} of hand geometry for the ith user.
15. Stop

Fig. 12.2 Proposed template security algorithm

Algorithm 2: Matching algorithm
Input: *Query templates of fingerprint and hand geometry of Ith user.*
Output: *Final match score and Final decision regarding the user.*
1. Start
2. Let S_{fi} and S_{hgi} be the stored templates of fingerprint and hand geometry used for verification.
3. Calculate the similarity score between the corresponding feature vectors.
4. Let the two match scores be denoted by D and d.
5. By using the weighted sum rule, calculate the final score as
$\text{Final score} = \lambda * D + (1-\lambda) * d$
Here, λ is the assigned weight of each modality.
6. Set the threshold 'T', depending on the application
7. Perform a comparison of final score with 'T' to get the final decision regarding the user.
8. Stop

Fig. 12.3 Proposed template matching algorithm

to check the validity of the user, the final score obtained after combining the two scores is compared with a threshold, T, whose value depends on the sensitivity of the system. If the system is highly secure, a high value of the threshold is set otherwise moderate value can also be used.

12.4 Experimental Results

The dataset for experimentation has been collected by using high-definition sensor. The fingerprint and hand geometry images of 50 subjects with different variations were captured. The images of fingerprint trait were captured at a dpi of 300. The hand geometry images were captured with a local setup with the installation of a digital camera on a stand. The subjects were chosen carefully to include samples from a heterogeneous population like males, females, kids, old people, laborer, etc. The images were pre-processed in order to improve quality.

12.4.1 Performance Metrics

In order to assess the performance of any biometrics-based application, a variety of performance metrics have been used by various authors. A typical biometrics-based system must result in low error rates and high identification accuracy during the testing phase. The proposed algorithm has been evaluated using false accept rate, false reject rate, genuine accept rate, and identification accuracy.

- FAR (False accept rate) = (number of imposters given access)/total number of imposter accesses.
- FRR(False Reject Rate) = (number of genuine users rejected)/total number of users
- $\text{GAR} = 1 - \text{FRR}$
- $\text{IA} = 100 - (\text{FRR} + \text{FRR})/2$.

Table 12.1 shows the various error rates resulting from varying T and λ and is represented in Fig. 12.4. It is clearly shown that if we increase their values, the performance goes on decreasing and the identification accuracy is maximum at $T = 0.70$ and $\lambda = 0.55$. Table 12.2 shows the error rates computed at varying subjects and it is clearly shown that if we increase the subjects, the error rates are relatively increasing.

Table 12.1 Error rates at varying thresholds and weights

Threshold (T)/Weightage (λ)	FAR (%)	FRR (%)	GAR (%)	IA (%)
T = 0 .50 λ = 0.50	−3	−3	−97	−97
T = 0.60 λ = 0.50	−4	−3	−97	−96.5
T = 0.70 λ = 0.55	−3	−2	−98	−98.5
T = 0.75 λ = 0.60	−4	−4	−96	−96
T = 0.80 λ = 0.70	−6	−4	−96	−95

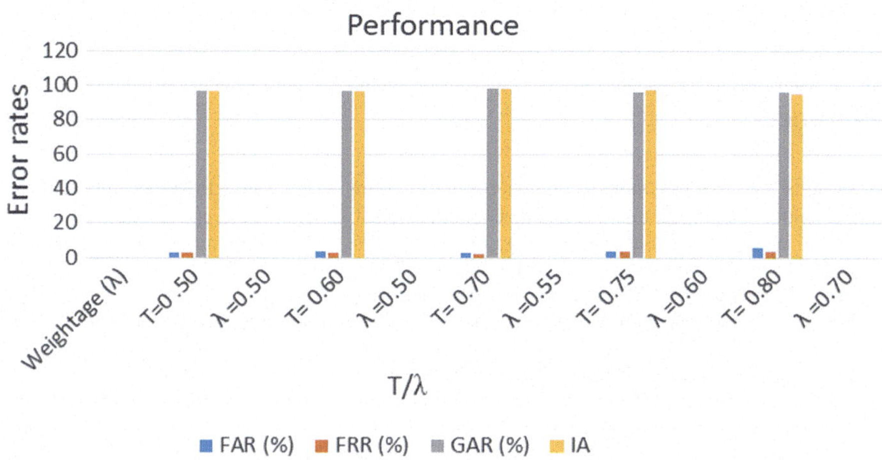

Fig. 12.4 Proposed template matching algorithm

Table 12.2 Error rates at varying subjects

Subjects	FAR (%)	FRR (%)	GAR (%)
10	0	0	100
15	0	0	100
20	0.001	0.001	99.999
25	0.001	0.002	99.9985
30	0.001	0.003	99.998
35	0.002	0.003	99.9975
40	0.002	0.003	99.9975
50	0.003	0.004	99.993

12.5 Conclusion

In this paper, a new transformation-based approach for template security has been presented for the multimodal system based on fingerprint and hand geometry traits. The proposed scheme uses a BBH on the fingerprint and average-based transformation on the hand geometry to get the secured templates. The weighted match score level fusion has been used to check the validity of the user during verification. The experimental results have shown that the proposed scheme exhibits better identification accuracy and also lower error rates. However, the proposed scheme can be verified on standard databases like NIST, FVC1, FVC2, etc. The performance of the proposed scheme may be further improved by using the BBH for the hand geometry trait. Furthermore, the security analysis of the proposed scheme may be carried out to test the ideal characteristics specified as per ISO standards.

References

1. Brindha VE (2011) Biometric template security using fuzzy vault. In: Proceedings of the International Symposium on Consumer Electronics ISCE, vol 3, no 6, pp 384–387
2. Selwal A, Gupta SK, Surender (2017) Low overhead octet indexed template security scheme for multi-modal biometric system. J Intell Fuzzy Syst 32(5):3325–3337
3. Mai G, Lim M, Yuen PC (2016) AC NU. IMAVIS
4. Kanade S, Dorizzi B, Multi-biometrics based cryptographic key regeneration scheme
5. Chin YJ, Ong TS, Teoh ABJ, Goh KOM (2013) Integrated biometrics template protection technique based on fingerprint and palmprint feature-level fusion. Inf FUSION
6. Gawande U, Zaveri M, Kapur A (2013) Bimodal biometric system: feature level fusion of iris and fingerprint. Biometric Technol Today 2013(2):7–8
7. Khellat-Kihel S, Abrishambaf R, Monteiro JL, Benyettou M (2016) Multimodal fusion of the finger vein, fingerprint and the finger-knuckle-print using Kernel Fisher analysis. Appl Soft Comput J 42:439–447
8. Miao D, Zhang M, Zhenan H, TieniuS, Bin-based weak classifier fusion of iris and face biometrics
9. Sree SRS, Radha N (2014) A survey on fusion techniques for multimodal biometric identification, pp 7493–7497
10. Nandakumar K, Jain AK, Nagar A (2008) Biometric template security. EURASIP J Adv Signal Process 2008(January):1–20
11. Belguechi R, Cherrier E, Rosenberger C, Ait-aoudia S (2013) Operational bio-hash to preserve privacy of fingerprint minutiae templates. IET Biom 2(February):76–84
12. Bringer J, Morel C, Rathgeb C (2015) Security analysis of Bloom filter-based iris biometric template protection. In: Proceedings of 2015 international conference on biometrics, ICB 2015, pp 527–534
13. Selwal A, Kumar S (2016) Template security analysis of multimodal biometric frameworks based on fingerprint and hand geometry. Perspect Sci 8:705–708
14. Selwal A, Kumar S (2016) Fuzzy analytic hierarchy process based template data analysis of multimodal biometric conceptual designs. Procedia Comput Sci 85(Cms):899–905
15. Jeng R, Chen W (2016) Two feature-level fusion methods with feature scaling and hashing for multimodal biometrics. 4602

16. Aravinth J, Valarmathy S, Aravinth J, Valarmathy S (2016) Modal biometric recognition and its application to remote biometrics authentication multi classifier-based score level fusion of multi-modal biometric recognition and its application to remote biometrics authentication. 2199
17. Sojan S, Kulkarni RK (2016) Mixing fingerprint features for template security. (7):355–359
18. Ramalho MB, Correia PL, Soares LD (2012) Hand-based multimodal identification system with secure biometric template storage. IET Comput Vis 6(3):165
19. Selwal A, Gupta SK (2016) Region coded hashing based template security scheme for multi-biometric system. (3)
20. Jagadiswary D, Saraswady D (2016) Biometric authentication using fused multimodal biometric. Procedia Comput Sci 85(Cms):109–116

Chapter 13
Edge Computing in Smart Villages

Rahul Gupta and Archana Singh

13.1 Introduction

Edge computing is a system that functions through a distributed network of nodes spread across large geographic distances. It can be defined as the pole opposite of the cloud computing system that functions on single centralized cloud environment. It is called edge computing as its functional units are edge devices that form the Internet of Things. The aim of this system is to provide faster and more efficient data analysis, artificial intelligence (ambient intelligence), and server resources closer to the smart sensors or actuators. The ability of a machine to perceive or sense the presence of humans is termed as Ambient Intelligence (AmI); motion sensors are a clear example of this technology. Actuators are defined as the moving part of a machine or a system that perform its desired function through machine instructions. Automation technologies have actuators as their functional units. Depending upon the size and number of users connected to a data bank, edge computing can be classified into three broad divisions, i.e., local devices, localized data centers, and regional data centers [1]. Edge computing is part of many real-world applications and is showing significant growth every year. It will be more nominal and versatile in the foreseeable future. The reasons for its success are the following advantages it provides through its use.

R. Gupta (✉) · A. Singh
Department of IT, ASET, Amity University, Noida, UP, India
e-mail: rahul.gupta190998@gmail.com

A. Singh
e-mail: archana.elina@gmail.com

P. K. Kapur et al. (eds.), *Decision Analytics Applications in Industry*, Asset Analytics,
https://doi.org/10.1007/978-981-15-3643-4_13

Being a distributive network disruption at a single point leading to complete system failure can be avoided.

- It reduces the amount of data to be transferred and as a result makes the data sharing faster and more efficient.
- It reduces the latency and improves the user experience.
- Multiple checkpoints can be installed between the edge devices and the central nodes making it more secure and reliable.

However, being a relatively newer system edge computing faces numerous challenges that hinder its performance, limiting its scope of function. These challenges need to be addressed by the institutions in order to achieve a smoother workflow.

- Edge devices should be designed to work under extreme conditions and should withstand irregular power supply.
- Edge devices located in public areas can be exploited and compromised and hence secure solutions are needed.
- Consistency is a big concern when dealing with edge devices distributed over a large geographic area.

Government of India has been making great strides toward complete digitization of the government systems as well as economy since the establishment of the Digital India initiative. Edge computing is a technology that can help the government authorities reach out to vast geographical distances and provide technical support to areas that were previously disconnected from the mainstream economy of the country. Through this technology, we can achieve the goals of providing better services to the population, more accurate data maintenance, making schemes more suited for the population, etc.

This paper explores the proposed concept of introducing edge computing into the government legislature judiciary and executive machinery. In Sect. 13.2, the existing technologies based on the concept of edge computing are mentioned along with their future prospects. In Sect. 13.3, the proposed concept of Smart Village Edge System (SVES) is explained and discussed in detail. In Sect. 13.4, the scope and implementation are discussed. Section 13.5 contains the discussion and future prospects.

13.2 Existing Applications of Edge Computing

Mobile edge computing is a massively growing application based on the edge computing technology. It aims to reduce latency, provide highly efficient network operation and service delivery, and improves the overall experience of the end user [2]. This technology also aims to deliver better results to the service providers as well. Automation in the industries using edge devices is a relatively new concept. It aims at creating localized automated environments within the industry using edge devices instead of automating the function using a centralized cloud-based environment. This technology aims to improve the efficiency of the automation and provides dedicated services as required by the industry. This technology will be making direct use of the actuators in the machines and perform the desired functions through machine instruction programmed by the user [3]. In today's era, data sharing is of utmost importance; hence, openness in the network is highly crucial. The user expects the service provider to deliver all the relevant information demanded by the user; however, the information demanded may be in another domain making it difficult for the service provider itself to access the information, resulting in failure of service; hence, new scheme must be developed for cross-domain data sharing and networking in order to provide a better service to the end user. The most concerning problem in cross-domain networking is the security threat and hence must be addressed as a primary concern. Edge computing can provide a solution to this problem by connecting all the domains through the cloud and using edge devices as nodes connected to the cloud [4]. Other future prospects of edge computing can be real-time traffic control, Remote surveillance, etc. Edge computing also has various applications in the Power Internet of Things (PIoT), such as in smart home, transmission line monitoring, intelligent substation, etc. The edge computing can support PIoT in edge network and reduce the dependence on the cloud center, to achieve the terminal level of transaction processing, improve system efficiency and real time [5]. LAVEA [6], a system based on the concept of edge computing provides low latency video analytics by lowering the computation tasks between the client and the server. Edge computing is a highly versatile technology and can be combined with various other technologies. Embedded deep learning for vehicular edge computing [7] is such an example where a low powered processor is used for capable and fast inference in vehicles. Table 13.1 provides a clearer difference between the three systems; it compares the three types of data centers on the bases of their uses and sizes.

Table 13.1 Comparative analysis of three typed of edge systems

Local devices	Localized data centers	Regional data centers
These systems are of small scale and are made for a specific purpose only. These provide solutions for homes, small businesses, and individual needs	These systems are medium-scale solution to the needs of businesses and solutions that need to store large amount of localized data that needs to be accessed within the premises only	These are high storage systems that provide data access to larger distances. Although the latency of data transfer depends upon the distance between the user and the data bank

13.3 Proposed Model of Edge Computing

Smart Village Edge System is an application of edge computing to be implemented in the government judiciary, legislature, and executive machinery. According to the Office of the Registrar General & Census Commissioner, India, about 70% of Indian population live in rural areas [8] most of which fall below the poverty line. Most of these villagers lack the basic resources and amenities that are fundamental to any living person. Government of India despite many efforts is not able to reach out to such areas because of the vast geographical landscape of India leading to decline of these areas and its residents. The proposed application of edge computing will provide a digital bridge to join these areas to the center of power. This system will consist of four sub-systems explained as follows: Local data and grievance centre, District data centre, State data centre, and Central database. Their roles and responsibilities are discussed as follows.

- Local data and grievance centre: This will be located in every village and will consist of the necessary edge devices (fingerprint sensor, retina sensor, etc.). Its purpose would be to collect and maintain a record of the data of all the local residents of the village. It can also act as a grievance helpdesk where the villagers can directly interact with the authorities and convey their problems.
- District data centre: They will be located one in each district and will be common for all the villages located in that district and will be under direct command of the District Magistrate (DM). It will be responsible for collecting all the grievances from the local data centers and convey them to the authorities. It will also act as a data bank for the District Courts and help the workflow of the local judiciary system.
- State data centers: These will be one in the capital of every state and will be under the direct watch of state authorities. It will collect the data from all the district data centers and help the state in making schemes according to the needs of the people. It will also help the High Courts in keeping the track of repetitive defaulters. It will also be responsible in monitoring the actions of the district and local data centers reducing the chances of failure in services.
- Central database: This central data bank will be directly connected to all levels of the system. It will reduce the need of manual census and will be able to collect data digitally from all the local data centers. It will also help the government in recording the demographic changes in different areas over the period of time. It will also reduce the need of manual census digitizing both the counting as well as the assessment process and as a result improving the accuracy of the data.

These components will be corresponding to each other in real time through the medium of the Internet. Edge devices such as mobile phones, surveillance cameras, and desktop computers can be used to monitor the functioning of different centers. With all the data centers in communication with each other, record maintenance can be done with ease in real time. Also through edge devices like, fingerprint sensors and iris scanners, services of Aadhar card registration can be done easily providing the

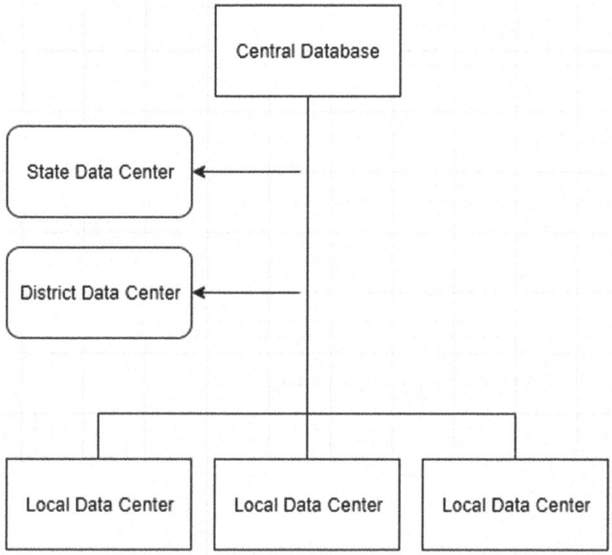

Fig. 13.1 Schematic of the smart village edge system

users access to all government schemes accessible by the Aadhar users. It will also reduce cases of fraudulent identities and other forms of corruption. The following proposed diagram shows the basic schematic of this system, it depicts how the local data centers are linked to a district data center that then provides data to the state data center. All state data centers are then connected to the central database (Fig. 13.1).

13.3.1 Local Data and Grievance Centers

Local data centers are the building blocks of this entire system. It is the only system that will be directly interacting with the people and hence is of the greatest importance. The main functions of this setup would be Aadhar Card registration and assistance, PAN Card registration and assistance, Ration Card registration and assistance, and Voter's Id registration and assistance. It will also function as an information centre about latest government schemes. It would act like an interface directly between the general public and the government authorities. The major edge devices used at this level will be biometric scanners like fingerprint scanners and retina scanners to create and maintain the database of the local residents. The digital components required by this system are stated in the following table; all these devices act as edge devices providing service to the end user (Table 13.2).

Table 13.2 Basic requirements of a local data centre

S no.	Devices
1	Fingerprint sensor
2	Retina sensor
3	Desktop computer
4	Internet connection
5	Local server system
6	Computer peripheries
7	Mobile application
8	Surveillance cameras

13.3.2 Goals of the Proposed Model

Through the application of edge computing and Internet of Things, this system aims to achieve the following goal:

- Providing knowledge and aids at the grass root level.
- Connecting the underprivileged population to the mainstream economy.
- Elimination of corruption at the execution and management level.
- Faster implementation of government schemes.
- Efficiency in the national and local record maintenance system.
- Better awareness among the local residents of the villages about the laws of the country.
- Easier access to resources.
- Connecting the underprivileged to the Aadhar network making them eligible to access government schemes.

13.4 Scope and Implementation of the Proposed Model

This paper proposes a model that is designed to introduce the technology of edge computing into the already existing government legislature and executive systems. With this project in implementation the authorities can achieve their goals faster and more efficiently. Also it will be an aid to the general public in understanding and benefiting from the government schemes. With this system in motion we can expect better and faster results from government schemes resulting in overall growth of the Indian population and development of the country. The following proposed diagram shows a hierarchical order of the different levels of the proposed model; it depicts how the central database is at the apex of the model, while the local data and grievance centers are the basic functional unit of the model (Fig. 13.2).

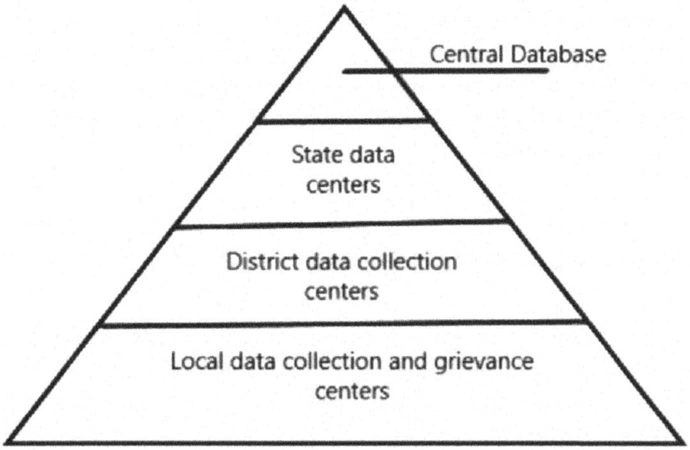

Fig. 13.2 A hierarchical representation of the model with the central database at the apex

13.4.1 Challenges of the Proposed Model

The proposed model aims to ease the workflow of the government machinery and make it more accessible to the local residents of the villages of India. However, to do so, cooperation from all different levels of national and state government departments is needed. Also, the staff to be appointed for operating the edge devices at the data centers must be trained accordingly.

Major concerns to be primarily addressed are as follows.

- The most important step is the identification of all the villages that lack the basic amenities and need to be connected to the network.
- Providing the necessary technical and financial support to the local data centers.
- Training the operating staff properly.
- Devising coping mechanisms for system failure.
- Keeping a backup of all the information at safe places.
- Keeping private and confidential data safe.
- Training the staff in public relations.
- Raising awareness among the local villagers about the uses and benefits of the local data centers.

After addressing all these issues the authorities will be able to run the establishments without any failure of services or failure of system for a foreseeable future. The cost of constructing and maintaining any system is a big factor in the overall feasibility of the system. The major components that will cost the most are discussed as follows. Construction of the data centers will be decided by how many villages in India are disconnected from the mainstream. The more the number of such villages the higher will be the construction cost. Hence, a minimalist approach should be used in order to cut the costs. Purchase and maintenance of the data servers is another major component that will be most expensive; however, given that the entire system is based on the communication between the servers it should not be compromised and should be given utmost importance. Staff salary, bills, and consumption costs at different data centers are other minor expenses that can be kept in check in order to avoid needless expenditure.

13.5 Discussion and Future Prospects

Edge computing opens up vast unexplored domains of data sharing and networking that were previously out of our reach. With this technology, every person can act as an edge node becoming part of a much larger network making everyone equal in the digital realm providing them access to all the relevant information. Edge devices like wearable technology, mobile phone, and personal computers empower us all to become a part of the digital network that joins the entire globe. Implementation of such a versatile technology in the government mechanism of India will increase the workflow and processing speed of the system manifold. Accepting this technology with an open mind and exploring its further prospects and future improvements will greatly benefit both the government and the citizens of India.

References

1. https://www.microsoft.com/en-us/research/project/edge-computing/
2. Brecher C, Pallasch C, Waltl J, Obdenbusch M, Hoffmann N, Wein S, Buchner T (2018) Edge powered industrial control: concept for combining cloud and automation technologies. In: IEEE international conference on edge computing (EDGE)
3. Hochstetler J, Chen Q, Yang Q, Padidela R, Fu S (2018) 2018 IEEE/ACM symposium on edge computing (SEC)
4. Li H, Wang J, Fan K, Pan Q, Liu T, Yang Y (2018) Cross-domain based data sharing scheme in cooperative edge computing. In: IEEE international conference on edge computing (EDGE)
5. Yi S, Li Q, Hao Z (2017) IEEE 37th international conference on distributed computing systems (ICDCS)
6. Zhang Q, Li Q, Zhang Q, Yi S, Shi W, Hao Z (2017) 2017 IEEE conference on energy internet and energy system integration (EI2)
7. Sabella D, Patel M, Sprecher N, Young V, Hu YC (2015) Mobile edge computing a key technology towards 5G. ETSI (European Telecommunications Standards Institute)
8. http://censusindia.gov.in/Census_And_You/area_and_population.aspx

Chapter 14
Time Series Analysis: A Machine Learning Approach

Kunal Mehta, Ekta Singhal, and Avinash K. Shrivastava

14.1 Introduction

Time series analysis is a dynamic research area which has attracted the attention of researchers and academicians over last few years. In today's world where technology has disrupted every business, it has helped newer and smaller organizations to fight the biggest of corporations. The largest hotel chain of the world does not really own any hotel, Airbnb; largest taxi firm does not own any car, Uber; the biggest content platform of the world, does not create any content, Facebook. All this begs the question about how bigger organizations can keep up with the changing landscape of business, consumer preferences and agility shown by these 'smaller' players? One of the biggest reasons for the success of these new-age organizations has been their proclivity to keep data at the center of their every decision. Marketing data sets have increased drastically over the last few years and are available in variety of forms like rank data, choice data, etc. [1]. One such form is time-series data, in the form of sales per month, market shares per quarter, the demand evolution over the last three years, or pattern of advertising-spending data over last few years. The main characteristic of time-series data is that the observations are ordered over time, and it is likely that earlier observation data have predictive content for future observations. Penpece and Elma [2] developed two frameworks by leveraging machine learning techniques that help fashion retailers in forecasting demand for new products. They further claimed

K. Mehta
Sapient Corporation, New Delhi, India
e-mail: kunalkmehta08@gmail.com

E. Singhal (✉)
Fortune Institute of International Business, New Delhi, India
e-mail: Ekta.singhal@fiib.edu.in

A. K. Shrivastava (✉)
International Management Institute, Kolkata, West Bengal, India
e-mail: kavinash1987@gmail.com

P. K. Kapur et al. (eds.), *Decision Analytics Applications in Industry*, Asset Analytics,
https://doi.org/10.1007/978-981-15-3643-4_14

that their model offers visibility into the underlying factors that impact demand, with insights into the importance of the different predictor variables and their influence on forecast accuracy [3]. Beheshti-Kashi et al. [4] presented state-of-the-art methods in the sales forecasting research with a focus on fashion and new product forecasting and reviewed different strategies to the predictive value of user-generated content and search queries.

Makridakis et al. [5] discussed, why the accuracy of machine learning models is below that of statistical ones and proposes some possible ways forward. Their results stressed on the need for objective and unbiased ways to test the performance of forecasting methods that can be achieved through sizable and open competitions allowing meaningful comparisons and definite conclusions. Rangapuram et al. [6] developed a novel approach to probabilistic time series forecasting that combines state-space models with deep learning. They claimed that their method retains desired properties of state-space models such as data efficiency and interpretability, while making use of the ability to learn complex patterns from raw data offered by deep learning approaches.

Kuremoto et al. [7] applied deep belief net (DBN) methodology stacked by multiple restricted Boltzmann machines (RBMs) to realize time series forecasting. They further introduced a state-of-the-art time series forecasting system that combines RBMs and multilayer perceptron (MLP) and uses SGA training algorithm. Petneházi [8] presented a recurrent neural network-based time series forecasting framework covering feature engineering, feature importances, point and interval predictions, and forecast evaluation. Pavlyshenko [9] studied the usage of machine learning models for sales predictive analytics. They considered the effect of machine learning generalization and used stacking approach for building regression ensemble of single models to show that it can improve the performance of predictive models for sales time series forecasting.

Considering most of the new age organizations are digital based, and are catering to the consumer base which is more comfortable with digital medium than ever before, the amount of data that they collect from their diverse platforms is not just humungous in size and pretty accurate, but it is also insightful about the behavioral aspects of users, not just transactional in nature. The main driver for the increasing popularity of time series technique in marketing is the shift from a period of scarce and aggregated data to huge, disaggregated data [10]. This means for the first time, if data analytics techniques are applied with caution and efficiency, can lead to see not just what's going on, but also predict what is about to happen.

Time series analysis is one such advanced analytics concept, where some of the best statistical algorithms are used to determine the impact of time, season, and trend on various metrics like visits on the platform; offline or online, revenue, conversions; both minor and major, etc. The aim of time series analysis is to carefully collect and study the past observations of a time series to develop an appropriate model. This model is then used to predict future values for the series. To a casual observer, business results are an output of organization's ability to reach out to its customers and customers' requirements, but in real world there are many variables in play which heavily skew customer's buying decision. Some of these variables are seasonal and trend indices. With plethora of data points now available at organization's disposal, it becomes difficult to understand how to decipher trends hidden within these data points by using traditional methods of looking into graphical and tabular reports, time series analysis can address this issue in an efficient manner.

Every sequence of data tends to follow a certain trend; this data can be revenue numbers, conversions, purchases, logins, anything and everything. As a marketer if we are able to identify this pattern, it can help us accomplish a lot. If we can find out these trends, we shall be able to predict, with a certain degree of accuracy, about what is going to happen in future, for example, what is going to be the revenue that we shall earn in next 3 months, what would be the sales in next 1 year, so on and so forth. The question at this point is how exactly we can find these trends hidden in plain sight. This is where mathematics in conjunction with machine learning comes to our rescue. As a prerequisite we would require observations that are collected over equal intervals of time for some specific dimension/category. Maybe monthly prices of a specific commodity, weekly revenue earned, monthly sale by volume, etc.

14.1.1 An Analogy

At the same time the question arises, why do marketers need to find out these trends, and how much it can help them. To give you an example of the value of pattern recognition, along with time, think about the fact that human species has been around for 300,000 years now. But we started investing our resources in math and science only since the last 3000 years. In these 3000 years we have been able to understand the history of universe, that is what happened 13.8 billion years ago, and how this universe will probably die down in next 100 billion years. All this happened just because we were able to recognize patterns of the way stars, planets, galaxies moved around us, and then on the basis of these patterns we were able to predict and prove with utmost accuracy about the future events of the observable universe around us. Now that we have got a glimpse of the power, we need to understand the underlying mathematics for such analysis so that we can utilize the same within the marketing domain.

14.1.2 Using Statistics with Machine Learning

Data makes sense to us in smaller quantities, revenue for the last 1 year, 12 months, 6 months, and so on. But when we increase the number of variables affecting our revenue numbers, and at the same time increase the observations from 1–2 to maybe 40, 50, or even 100 it becomes more and more difficult for us to understand the underlying patterns within our data points, and effect of each of the variables on our observations. This is where mathematics, statistics to be specific, and the concepts of machine learning comes into the picture. Before we go any further, let's understand the nuances of both these subjects.

14.1.2.1 Statistical Formula

Every mathematical equation can be mapped on a graph. There is a curve associated with every mathematical equation, which means, if we plot our observations on a graph, the nearest curve that can be carved, can probably give us an equation which will help us understand the hidden pattern in our observations. So, in short we are trying to create an equation which looks something like this:

$$Y = Base \pm Seasonality \pm Trend \pm Noise$$

where Y is the variable that we want to predict upon, for example, revenue, orders, visits, visitors, etc. Components of this equation are explained in the subsequent sections of **Solution Approach**.

14.1.2.2 Machine Learning

Machine learning is actually an application of Artificial Intelligence that provides the ability to learn, to improve, and create automated algorithms to technological tools. Machine learning models have been developed and established in the last decade as against classical statistical models in the area of forecasting [11].

Fundamentally it is the same as we humans do learn, just like we learn by observing our environment, interacting with different people (variables), and experiencing different outcomes of situations. I myself am told the story of how I learned the concept of hot and cold. When I was a little kid I used to run towards hot cup of coffee and/or tea that used to come for my parents. My parents always tried to shield me away from these, because of obvious reasons, but their efforts to keep me away from hot beverages used to trigger me even further. So, one day my father decided to actually provide me experiential learning, he took my finger, and made it touch the hot cup. That was it, I screamed, cried, and learned the concept of hot beverages forever. My parents tell me that since that day I never ever ran towards their cup.

Think of machine learning algorithms in the same way, we feed data in our tools, ask them to create a specific equation, on the basis of input and output observations that we are feeding it, and the tools come up with their own equation.

There are various types of machine learning algorithms like supervised, unsupervised, semi-supervised, reinforced, etc. However, fundamentally their background working remains the same.

Typically in machine learning scenarios, analysts try to divide data set into two parts: (1) Training set and (2) Test set. Training data set is initially fed into the tools and the algorithm is specified, on the basis of which it tries to come up with the equation. Accuracy of this equation is then tested on the test data, by seeing how far the predicted values are from the actual values.

More often than not it is found that machine learning equations give better results with training data set than with test data set. However, if the difference is not too high, then analysts can accept the output given by machine learning algorithms, else you would need to refine your data and algorithms to find an equation which provides you as accurate data as possible.

Now that we have established the basics of time series analysis, we will try to explain the principles and advantages of this methodology with the help of a case study based on one of the biggest hotel groups across the globe, with more than 1400 properties under 4 brand names, spread across different geographies.

14.2 Client Introduction

Ashton Group of Hotels (name changed to protect client privacy and data), has around 1000+ properties across the globe. They operate all their hotels under five global brands. Four of these brands can be classified as business hotels, with most of the hotels near airports of major cities in the world. Ashton launched multiple campaigns and campaign programs for their properties, some at global level, some at regional, and some at property level.

In such complex matrix of marketing strategies, there was a need to create a media calendar for the central marketing team, based out of their corporate head office in Madrid. Initially, the marketing team used to get inputs from each of the regional heads, talk about their views on upcoming trends, major holiday calendars of the region to come up with their own media calendar. It was proposed that we should use time series analysis to come up with the seasonal indexes to create media calendar in a more scientific manner.

Before we talk about the solution approach, we need to understand the various industry challenges in this space, which affects our client as well.

14.2.1 Industry Challenges

In middle ages, it were Greeks who came up with the idea of building resting places in villages for visitors before Romans started building expansive mansions for the same purpose.

Like pretty much everything, religion played its own vital role in developing and flourishing this business. Since a lot of people were traveling for religious pilgrimages, staying overnight at an INN made a lot of sense to these folks. As transportation business flourished, there was a significant rise in the hotel industry as well; city hotels came up all across Europe and New York that turned into many world-renowned hotels like Radisson Hotel (Cooper et al. [12]).

14.2.1.1 Customer Expectations

Hotel industry, like any other industry, does not really cater to a monolithic customer base. They also address the needs and requirements of various segments of customers, that is why you have hotels like

a) Transit hotels
b) Resorts
c) Heritage Hotels
d) Commercial Hotels
e) BnB.

The fact is that a facility that was created to provide accommodation, isn't supposed to provide just an accommodation anymore. The customer expectations have changed enough in last so many years that hotels are not just a place to crash on the bed at the end of the day anymore. And this brings us to the second challenge in this fast-evolving industry.

14.2.1.2 Airbnb

What Uber did to taxi business, Airbnb is doing to hotel industry, changing the face of the industry in a way that we have begun doubting if the traditional business will remain sustainable in the next decade?

Research showed that in the 10 cities with largest Airbnb share in US, resulted in 1.3% fewer hotel nights booked and 1.5% loss in revenue. This problem has come into effect because Airbnb has impacted the biggest principle of economics, which is demand and supply. Now that you don't have to own and operate a huge machinery to maintain a hotel, you don't need to be a hotelier or a hotel management graduate to operate Airbnb accommodations, hence huge supply of nominal, affordable, and personalized offerings from host of people in some of the major cities across the globe, spiraling the prices down.

The Airbnb effect can be felt much more significant in case there are planned events, huge volume of expected travelers, which traditionally meant higher room rates for hotels, thereby increasing the profit margins, without any significant rise in operating costs. Like in case of graduation ceremonies for local colleges, sporting events in the city, upcoming business events in the city, pilgrimage events, etc.

14.2.1.3 The OTA Gauntlet

Let us do a small experiment, search for hotel in a specific location, chances are OTA (Online Travel Agent like Makemytrip, Yatra, Boking.com, etc.) results will beat the hotel's own website. The fact is that nimble organizations, which started as small-time aggregators, have become huge players in the hospitality sector at large, and hotel industry in particular.

They give a lot of options to end customers to select and book from one of the many available options, and at the same time, OTAs seem to have broken the code on digital platforms to understand how consumer behaves, what they search, what they want, and utilize this information to curate better offerings for their customers. To top it all, they have started their own loyalty programs in which end customers get points for every booking, which can be redeemed for bookings in subsequent travel plans. Where OTAs do provide a regular and larger set of bookings and hence revenue, it also means that hotels have to (i) provide larger discounts to these OTAs and at the same time (ii) risk losing end customers to their own competitors since essentially, they are aggregators.

14.3 Business Problem

Ashton Group of hotels had an idea that if they could get a view of when their customers are going to visit their properties at regional level, they would be able to create discounted offerings for lean period to entice customers to visit. At the same time, for the peak periods they could pre-empt a surge pricing strategy, which again would help them increase their profit margins.

Such seasonal view of customer footfall will also help them give customized discounts to their OTAs, which were providing as much as 20% of their bookings, and 22% of their revenue, higher for the lean period, and lesser for peak periods.

They also wanted to understand the 'True Lift' of their campaign programs, that is, if due to their campaigns they are getting 12% higher bookings, and in the seasonal forecast they were supposed to get 10% higher bookings anyways, their true lift was only 2%. This way they would be in a better position to gauge the performances of their agency partners.

14.4 Solution Approach

To understand the time-based pattern analysis we utilized one of the most widely used methodology: **Time Series Analysis**. This helped us (a) identify the pattern represented by sequence of data and (b) predict the future observations.

In time series analysis, it is considered that data is made up of four components:

1) Recent trends
2) Seasonality index
3) Base level
4) Noise or random behavior.

The first two components represent a general systematic, at times linear, and more often nonlinear components that change over time. Trend is something that may or may not repeat itself over period of time, whereas seasonality index will always repeat itself in systematic intervals of time.

Base level is the value for the series if it were a straight line. Think about it this way, that if there was no trend, positive or negative, no seasonal impact on the consumer behavior, what would have been the minimum value that we could have expected?

Fourth component is the random behavior, which cannot be predicted by any pattern, which can be dependent upon many factors, including external.

For example, trend can be a lot of Indian travelers going Europe for vacations, whereas seasonality index can help us predict that 25% of the Indian travelers actually take vacations during the months of May and June, coinciding with the school vacations. Now it can be easy to assume that people go on vacations during school holidays, but the same trend might not be there during Diwali vacations, since during that time Indian families tend to stay at home to celebrate the festivities. These are some of the nuances that can be predicted on the basis of trend and seasonality analysis. Considering noise/random behavior can't be predicted, so, we try to formulate a mathematical equation for time series analysis that can minimize the noise and maximize the coefficients of trend and seasonality indexes along with base level.

14.4.1 Data Requirements

To create time series analysis monthly room revenue data was requested from each of the regional heads. It's imperative to note that booking numbers could also have been taken but it might not be an accurate picture of the trends for 2 reasons:

1) In hotel industry number bookings of the room depends heavily upon spatial and temporal dimensions, which is location and time. Hotels in city centers get more bookings, and at the same time, bookings also fluctuate significantly on the basis of when booking was done, what is the lead time for booking; that is, booking date to check-in date; time of booking; morning or evening, etc.
2) Booking numbers also get significantly affected by the discounts that are offered through multiple channels, for example, online bookings, bookings through OTAs (Online Travel Agents), bookings through agencies and partner businesses.

For now we only focused on the digital data, which is the most accurate data readily available within hotel industry. Since the data is completely owned by the hotel groups, it was easier to collect and work upon, without any limitations, constraints or dependence on external factors/agencies/stakeholders. For these reasons it was decided that revenue numbers, coming from digital data sources like Google Analytics or Adobe Analytics would be a better indicator of purchase trends.

14.4.2 Feasibility Study

Now we need to understand that to deploy machine learning time series analysis, we need to ensure that our data is actually in the shape and form that we can see is showing periodic trends. When looked into the revenue numbers of 2015–2017, along with revenue and visits on the hotel's digital platform, we were able to observe the following trends (Figs. 14.1 and 14.2).

When revenue numbers were plotted on the periodic level, we were able to observe data following a nice little trend of its own:

The above graphs showcased that revenue trends of 2015, 2016, and 2017 seemed to follow a pattern. Now the challenge was to create a trend line which could actually fit the curve and help us ascertain the seasonality indices to act on our data.

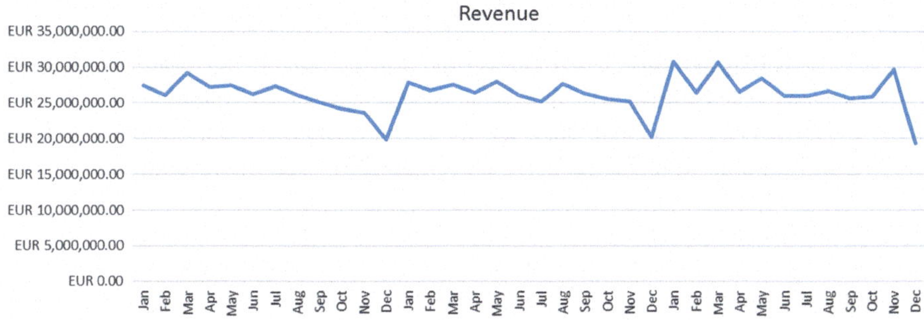

Fig. 14.1 Annual revenue trends for last 3 years

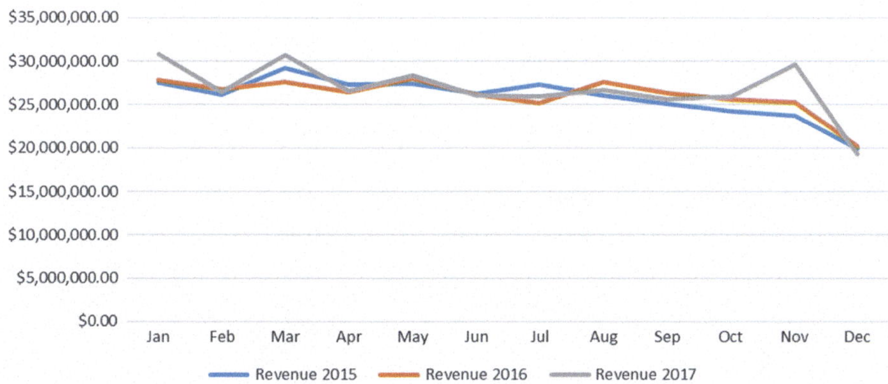

Fig. 14.2 Monthly revenue trends for last 3 years

14.4.3 The Time Series Analysis Algorithm

We used the ARIMA method introduced originally by Box and Jenkins [13], to understand the seasonality indexes. Following steps were taken to execute the algorithm on 30 months of data, that is, from January 2015 to June 2017:

a) Calculated 4 month moving average
b) Calculated centered average
c) Calculated seasonality index
d) Calculated de-seasonalized revenue numbers
e) Created linear regression trend line to come up with the regression equation
f) Calculated the future trend numbers with the trend equation
g) Included seasonality effect in the trend numbers.

By following these steps we were able to come up with the seasonal indexes as Table 14.1.

A seasonal chart like this clearly showcased how revenue in the month of December takes a nose dive by up to 20% of average, and bounces back in January to actually increase by 14% of the normal. Now taking a step further, we started predicting on the revenue numbers from July 2017 to December 2017 and observed the following trends:

Table 14.2 Revenue numbers from July 2017 to December 2017

It can be clearly see that prediction, at the overall level, had been 99.52% accurate, and similar was the case at monthly level as well.

Table 14.1 Seasonality index

Month	Seasonality index
January	1.14
February	0.98
March	1.04
April	0.97
May	1.02
June	0.97
July	0.99
August	1.02
September	1.00
October	1.01
November	1.00
December	0.80

Table 14.2 Revenue numbers from Jul 2017–Dec 2017

Actual	Forecast		
25,979,966.63	25,731,181.17	248785.46	0.96%
26,702,635.99	28,132,417.27	−1429781.27	−5.35%
25,612,703.88	26,832,819.21	−1220115.33	−4.76%
25,888,922.53	26,110,299.86	−221377.33	−0.86%
29,631,912.02	25,785,704.33	3846207.69	12.98%
19,313,287.68	20,719,324.73	−1406037.05	−7.28%
153,129,428.74	153,311,746.58	−182317.84	−0.12%
280,278,890.84	**280,892,311.98**	**(613,421.14)**	**−0.48%**

14.5 Managerial Implications

This methodology helped Ashton Group of Hotels to create their media calendar, understand the upcoming trends, and thereby come up with pricing and promotional strategies, and also gauge the 'True Lift' of the campaigns getting launched at various times of the year. The similar framework can be applied for different industries at regional levels, where seasonal trends can make a huge impact on the business metrics. It's important to note that this framework would be best placed to be applied at regional levels, since every region has their own trends and seasonal nuances. Doing this exercise on the global level can give misleading results.

Details of the algorithm and the underlying mathematical equations and proofs have been included in the sources section. The analysis explained in the paper can be used to answer several business questions which are given below:

- Create campaign calendars
- Identify true lifts for different campaigns and campaign programs
- Preparing inventory supply
- Creating pricing strategies
- Create discount policies for 3–6 months in advance
- Train employees to understand what might be coming next
- Better supply chain management
- Move toward just-in time from just-in case.

14.6 Conclusion

In this paper, we have considered a business example to show how machine learning combined with statistics can help in taking better business decision. We have further discussed about how it is different from the traditional way of forecasting using time series analysis. Our discussion with numerical results on proposed approach suggests that it is better to adopt machine learning method in time series analysis as it gives more accurate results. Finally, we have described the merits of this approach and its managerial implication. Currently, we have discussed the use of our approach from marketing point of view. In future, we can extend our approach for different sectors, viz., production, logistics, and supply chain.

References

1. Dekimpe M, Hanssens (2010) Time-series models in marketing: some recent developments. Neurosci Lett 32. https://doi.org/10.15358/0344-1369-2010-jrm-1-24
2. Penpece D, Elma OE (2014) Predicting sales revenue by using artificial neural network in grocery retailing industry: a case study in Turkey. Int J Trade, Econ Financ 5(5):435
3. Dekimpe MG, Franses PH, Hanssens DM, Naik, PA (2008) Time-series models in marketing. In: Handbook of marketing decision models. Springer, Boston, MA, pp 373–398
4. Beheshti-Kashi S, Karimi HR, Thoben KD, Lütjen M, Teucke M (2015) A survey on retail sales forecasting and prediction in fashion markets. Syst Sci Control Eng 3(1):154–161
5. Makridakis S, Spiliotis E, Assimakopoulos V (2018) Statistical and machine learning forecasting methods: concerns and ways forward. PLoS ONE 13(3):e0194889
6. Rangapuram SS, Seeger MW, Gasthaus J, Stella L, Wang Y, Januschowski T (2018) Deep state space models for time series forecasting. In: Advances in neural information processing systems, pp 7785–7794
7. Kuremoto T, Hirata T, Obayashi M, Mabu S, Kobayashi K (2019) Training deep neural networks with reinforcement learning for time series forecasting. In: Time series analysis, IntechOpen
8. Petneházi G (2019) Recurrent neural networks for time series forecasting. arXiv preprint arXiv: 1901.00069
9. Pavlyshenko BM (2019) Machine-learning models for sales time series forecasting. Data 4(1):15
10. Leeflang PS, Wittink DR (2000) Building models for marketing decisions: past, present and future. Int J Res Mark 17(2–3):105–126
11. Ahmed NK, Atiya AF, Gayar NE, El-Shishiny H (2010) An empirical comparison of machine learning models for time series forecasting. Econ Rev 29(5–6):594–621
12. Cooper C (2008) Tourism: Principles and practice. Pearson education
13. Box GEP, Jenkins GM (1976) Time Series Analysis: Forecasting and Control
14. Airbnb taking lessons from hospitality industry. https://www.theglobeandmail.com/life/travel/why-Airbnb-is-taking-some-lessons-in-hospitality-from-the-hotel-industry/article38139338/ (Accessed on 1 Aug 2018)
15. Airbnb effect. https://www.forbes.com/sites/hbsworkingknowledge/2018/02/27/the-Airbnb-effect-cheaper-rooms-for-travelers-less-revenue-for-hotels/#19dec141d672 (Accessed on 5 Aug 2018)
16. Creating ARIMA model on Excel. https://help.xlstat.com/customer/en/portal/articles/2062242-fit-an-arima-model-to-a-timeseries-in-excel?b_id = 9283 (Accessed on 10 Sep 2018)
17. Challenges in hotel industry. https://www.emeraldinsight.com/doi/abs/10.1108/17554211211217316 (Accessed on 20 July 2018)
18. Direct booking impact. https://www.forbes.com/sites/hbsworkingknowledge/2018/02/27/the-Airbnb-effect-cheaper-rooms-for-travelers-less-revenue-for-hotels/#19dec141d672 (Accessed on 21 Aug 2018)
19. How to identify and act of time series analysis. http://www.statsoft.com/Textbook/Time-Series-Analysis (Accessed on 2 July 2018)
20. Introduction to machine learning. https://www.expertsystem.com/machine-learning-definition/ (Accessed on 15 Sep 2018)
21. Time series forecasting. https://www.expertsystem.com/machine-learning-definition/ (Accessed on 10 June 2018)

Chapter 15
A Study of the Effectiveness of Online Marketing Strategies of Packaged Health Food Brands w.r.t. Gender

Amarnath Gupta and Pradnya Chitrao

15.1 Introduction

The dynamic business environment and the technological advancement have redefined the consumption, ways of doing business, and means of promotion for the Fast Moving Consumer Goods (FMCG) marketers. The twenty-first century marketing has touched an entirely new scale with the penetration of the Internet among almost every household in India and it has resulted in the emergence of online marketing. The prominence of digitization and online presence has led the companies getting involved into a very cut throat competition for seeking the customer's attention.

The increasing awareness of consumers regarding health and nutrition issues has greatly helped the packaged health food brands to drive their demand in the market. The marketers are trying new and innovative ways to take an advantage of this growing packaged healthy food market. With the rise in the competition, the FMCG companies have now moved their attention from conventional approaches to modern approaches like E-commerce, due to the increasing mobile Internet penetration globally.

Thus, keeping in view of all the above tools available in the body of knowledge of marketing are highly utilized and innovation is always looked forward to build the Brand Association which results in repeated buying and boosting the overall profits. The marketers focus on improving the quality of the packaged health food brands which in turn results in the Brand Loyalty and creating a Brand Equity.

A. Gupta (✉)
Research Scholar, Symbiosis International University, Pune, India
e-mail: amarg2460@gmail.com

P. Chitrao
Symbiosis Institute of Management Studies (SIMS), Pune, India
e-mail: pradnyac@sims.edu

P. K. Kapur et al. (eds.), *Decision Analytics Applications in Industry*, Asset Analytics,
https://doi.org/10.1007/978-981-15-3643-4_15

15.2 Literature Review

The past research studies done in this area have been highlighted in this chapter to understand the research gaps and for defining the research problem and the research objectives for the study. The researcher has evaluated the research work done in the past regarding Online Marketing Strategies of Packaged Health Food Brands.

15.2.1 Eric Allen, Jerry Fjermestad (2001)

Allen et al. [1] have investigated the online marketing strategies and explained its implication using the integrated model in the companies. It was concluded that the Internet has become a platform for the innovation of new products and also it is helping the consumer in choosing the right product for themselves. Internet was found to be a major source of information in the consumer's decision process. This study can be further extended toward study of effectiveness of online marketing strategies for packaged health food brands.

15.2.2 Mei-Fang Chen (2009)

Chen [2] has conducted the study on Organic Food Products and has tried to establish a relationship between the health consciousness levels of the consumer along with the environmental attitudes which can further affect the attitude of the consumers toward the Organic Food. It was found that the healthy lifestyle of the consumer promotes the positive attitude toward the Organic Food. The limitation of this research study is that it has considered only the Organic Food Products and therefore, future research studies could be extended toward study of online marketing strategies for FMCG sector also.

15.2.3 Sandra C. Jones Amanda Reid (2010)

Jones and Reid [3] have explored the Internet-based advertising techniques employed by Australian food organizations. It was found that majority of the meal groups have Internet websites or sections of websites that are being targeted toward children and/or teenagers. Future research studies could be extended toward packaged health food brands to generalize its findings.

15.2.4 Elyria Kemp My˜ Bui (2011)

Kemp and Bui [4] have examined the variables that area unit crucial within the method of brand-building that area unit is perceived as "healthy." An abstract model was further developed and tested on customers. The results of the study indicate that packaged believability, commitment, and affiliation area unit essential in developing stigmatization methods for "healthy brands." It was found that a reputable packaged food brand minimizes risk and will increase shopper's confidence.

15.2.5 Mahmud I. Nour, Mohammad Salamh Almahirah; Sultan "Mohammed Said," Sultan Freihat (2014)

Nour et al. [5] have carried out the research work to investigate how promotional mix elements are used by the Jordanian shareholding Ceramic and glass manufacturing businesses and to study the effect of promotional mix elements on buyers buying decision-making. The authors state that the main aim of promotional mix is to reach its target customers and make them buy and rebuy the product again and again to be specific inducing them for repeat purchases. It was found that out of the overall promotional mix elements being used by the companies under study, there was considerably more focus on advertising, personal selling, and promotional activities. This research study has a limited scope as it used to be carried out only for the Ceramic and glass manufacturing corporations in Jordan. The future research can consider the applicability of findings of this research in different international locations with other industries like FMCG, etc.

Research Gap–Out of the available literature survey, it has been observed that very few research studies have been conducted in past on online marketing strategies of packaged health food brands. Therefore, this research study aims to fill the gap by studying the effectiveness of online marketing strategies of packaged health food brands. The study will help the marketers to design and develop more effective online marketing strategies for the packaged health food brands which will help in generating more sales and revenue for the companies engaged in marketing of packaged health food brands.

15.3 Research Objectives and Scope

15.3.1 The Objectives for This Research Study Are as Below

1. To explore the effectiveness of Online Marketing Strategies of Packaged Health Food Brands w.r.t. Gender of the Customers.

2. To investigate if there exists any relationship between the awareness of packaged health food brands and the Gender of the customers.

15.3.2 Scope of the Study

The Research is mainly focusing on the Effectiveness of Online Marketing Strategies of Packaged Health Food Brands. The Products considered for this Study are Packaged Health Drinks, Packaged Healthy Soups, Packaged Healthy Breakfast Cereals and Meals, Packaged Healthy Food Supplements for Adults and Children, Packaged Dry Fruits, and Packaged Sugar Substitutes and Packaged Healthy Oils.

15.4 Research Methodology

Based on the outcome of the Pilot Study, the researcher has proposed the following research design:

15.4.1 Research Design

The researcher has used the Descriptive Research Design for conducting the research.

15.4.2 Population

The Population for this research was the people of Pune City who are above 18 years age group. (As per provisional reports of Census India, population of Pune in 2011 was 3,124,458; as per the estimates the population of Pune City in 2016 is 5,926,606 approx.).

15.4.3 Sources of Data

The researchers have used both Primary and Secondary data for the research. Primary data was mainly collected from the consumers of packaged health food brands from Pune region using the tools like personal observation, personal interviews and through the questionnaire. The secondary data was collected from the available sources like the print and electronic media along with the company's reports, research papers, and Internet websites.

15.4.4 Sample Design

For the Research, the researchers have used the Non-probabilistic Convenience Sampling technique.

15.4.5 Sampling Unit

The sampling unit for the research was the customers buying food (FMCG) products from various retailers or grocery shops or through ordering online in Pune City.

15.4.6 Sample Size

Around 150 plus questionnaire were administered but only 105 responses were found to be completely filled for the purpose of research study. Therefore, the researchers have considered a sample size of 105 for the research study.

15.4.7 Sampling Area

The researcher has collected the samples from the Pune City between May and June 2018.

15.5 Data Analysis

Since a self-designed questionnaire was used to collect the data, it was important to check its reliability. To check the reliability of the scale Cronbach's alpha was calculated. The value is shown in the Table 15.1.

From Table 15.2, it was found that since the Cronbach's alpha value is higher than 7; the scale used can be said reliable and can be used for further data collection process.

Table 15.1 Showing the percentage of valid responses

Case processing summary:			
		N	%
Casess	Valid	105	100.0
	Excluded[a]	0	0.0
	Total	105	100.0

[a]List wise deletion based on all variables in the procedure

Table 15.2 Showing the Cronbach's alpha value

Reliability statistics:	
Cronbach's Alpha	N of items
0.942	15

15.5.1 Descriptive Statistics

From Table 15.3 descriptive statistics it can be said that **Online Videos** of packaged health food provokes the customer to buy the packaged health food brands. Furthermore, it was also found that **Official and appealing website** of a packaged health food brand creates a good brand image and thus leads to positioning of the packaged health food brands. It was also observed that the respondents tend to buy packaged health food products if it is available on particular online retail stores' (Amazon, Flipkart, etc.). It was also observed that for the kids the respondents are not so much dependant on online stores like Amazon, Flipkart, etc., for buying the packaged health food brands. Further, One Sample t-test is further done to know statistical difference between sample mean and a hypothesized value of the mean in the population. The t-test here is run to know the difference between satisfaction and recommendation.

From the Tables 15.4 and 15.5 we get $p < .001$, which says that the satisfied consumers will further recommend the product to the other buyers. Further, a correlation test is run between the variable, which again shows that the variables are positively and significantly correlated. It means that if, satisfaction level increases, the recommendation for the packaged foods will also increase.

15.6 Testing of Hypothesis

The following Hypotheses are proposed:

H_0: **There is no relationship between the awareness of packaged health food brands and the Gender.**

H_1: **There is a significant relationship between the awareness of packaged health food brands and the Gender.**

In the Chi-Square Table 15.7, we see that the Pearson Chi-Square score is 19.502;

Table 15.3 The descriptive statistics of the variables

	N	Min.	Max.	Mean	Std. deviation
I am more likely to buy a particular packaged health food brand if I come across by a promotion **email** of that brand	105	1	5	3.21	0.8955
I am more likely to recommend a particular packaged health food brand if I come across by its testimonial through **social media**	105	2	5	3.13	0.6660
I am more likely to buy a particular packaged health food brand if I come across by its testimonial through **social media**	105	1	5	3.13	0.8994
I am more likely to recommend a particular packaged health food brand if I come across by a promotion **email** of that brand	105	1	5	3.12	0.9374
Online videos of packaged health food provoke me to buy that brand	105	1	5	4.11	0.9639
I buy (would buy) only those packaged health food brands for my kids for which there are **good reviews available online**	105	1	5	3.10	1.0277
Official and appealing website of a packaged health food brand creates a good brand image in my mind and thus I find that brand worth buying	105	2	5	4.05	0.9028
I am more likely to buy a particular packaged health food brand if it comes as one of the **online search results**	105	1	5	3.00	0.9094
Pop-up advertisement of packaged health food while browsing Internet make me buy that particular brand	105	1	5	2.97	1.1966
I would recommend packaged health food brand if it is listed in an online retailer's (amazon, Flipkart, etc.) website or mobile application	105	1	5	2.97	1.2046
I would more likely to recommend a particular packaged health food brand if it comes as one of the online search results	105	1	5	2.91	1.0295
I am more likely to buy particular packaged health food brand if it is available on particular online retail stores' (Amazon, Flipkart etc.) product portfolio	105	1	5	4.01	1.1131
I buy (would buy) only those packaged health food brands for my kids which are available on online stores	105	1	5	2.83	1.1804
I buy (would buy) only those packaged health food brands for my kids for which an official website is available	105	1	5	4.1	0.8933
Valid N (list-wise)	105				

Table 15.4 Showing the mean, Std. deviation and Std. error mean values of variables as calculated from one sample t-test

One-sample statistics				
	N	Mean	Std. deviation	Std. error mean
I am satisfied with the packaged health food brand which I am using currently	105	4.6857	0.95388	0.09309
I will recommend the brand of packaged health food to others	105	4.6190	1.00366	0.09795
I am ready to pay little higher price for good quality packaged health food brands as compared to local heath food (non-branded)	105	4.8095	1.02933	0.10045

Table 15.5 The values from running one sample t-test

One-sample test						
	Test value = 4					
	T	df	Sig. (2-tailed)	Mean difference	95% confidence interval of the difference	
					Lower	Upper
I am satisfied with the packaged health food brand which I am using currently	7.366	104	0.000	0.68571	0.5011	0.8703
I will recommend the brand of packaged health food to others	6.320	104	0.000	0.61905	0.4248	0.8133
I am ready to pay little higher price for good quality packaged health food brands as compared to local heath food (non-branded)	8.059	104	0.000	0.80952	0.6103	1.0087

it shows that the two variables are associated with each other (Table 15.6).

Again, If the p value is less than or equal to the significance level, we reject the null hypothesis and conclude that there is a statistical significant relationship between the variables. Hence, H_0 is rejected

***H*0: There is no relationship between Frequency of beingOnline and Gender.**

H_1: There is a significant relationship between Frequency of being Online and Gender.

Table 15.6 The pearson chi-square values

Chi-square tests			
	Value	df	Asymp. Sig. (2-sided)
Pearson chi-square	19.502[a]	4	0.001
Likelihood ratio	24.701	4	0.000
Linear-by-linear association	17.356	1	0.000
N of valid cases	105		

[a]3 cells (30.0%) have expected count less than 5. The minimum expected count is 1.37

Table 15.7 The pearson chi-square values

Chi-square tests			
	Value	df	Asymp. sig. (2-sided)
Pearson chi-square	3.427[a]	4	0.489
Likelihood ratio	4.681	4	0.322
Linear-by-linear association	1.751	1	0.186
N of valid cases	105		

[a]2 cells (20.0%) have expected count less than 5. The minimum expected count is 1.37

In line with the first hypothesis tested, the researchers have further tested the frequency of being online and Gender. The results here show that, since the p value is less than.05, the null hypothesis (H_0) is rejected and the alternative hypothesis (H_1) is accepted. Here, the Pearson Chi-Square is 3.427 with a df of 4 so we can say that there is a significant relationship between Frequency of being Online and Gender.

15.7 Findings

It was found that **Online Videos** of packaged health food provokes the customer to buy the packaged health food brands. Furthermore, it was also found that **Official and appealing website** of a packaged health food brand creates a positive brand image and thus leads to positioning of the packaged health food brands because the customers are more attracted toward such brands which have an official website and have a positive brand image. With the growing penetration of Internet, more and more number of the customers are buying the products mostly online. Therefore, the marketers are now giving more importance to the online marketing strategies which have been more effective in terms of customer satisfaction for the packaged health food brands. It was also observed that the respondents tend to buy packaged health food products if it is available on particular online retail stores' (Amazon, Flipkart, etc.). It has been also observed that for the kids, the respondents tend to purchase the packaged health food products whose official website is available online since, the parents do not want to take any risk with the health of their kids. It was also witnessed that, for the kids, the respondents are not so much dependant on online stores like Amazon, Flipkart, etc. for buying the packaged health food brands online as they generally buy such products from the market by physically checking the brands at the stores. It was found that with the increase in the satisfaction level of the customers, the loyalty toward the brand increases and therefore, such loyal customers get positively associated toward the brand and they do recommend such brands to others. It was found that the customers are ready to pay little higher price for good quality packaged health food brands as compared to local heath food as they don't want to take any risk of their health while buying such brands of which they are totally unaware.

15.8 Conclusion

It has been concluded that males are generally more aware about the packaged health food brands as compared to females as the former tend to be more health conscious as compared to the latter ones. It has been concluded that the online videos of packaged health food brands provoke the customers to buy that brand. Furthermore, it has been also concluded that there is a significant relationship between Frequency of Online Buying and Gender. It has been further concluded that males tend to prefer online buying of the packaged health foods more frequently as compared to females because the males are more frequently online as compared to females.

Research Implications and Limitations:
The findings of this study could be useful for the marketers in crafting more effective online marketing strategies to promote the specific Gender-based packaged health food brands which can further add value to the overall offerings made to the consumers. The research study highlights the growing importance of packaged health food brands among the consumers and how packaged health food brands should amplify their online marketing efforts in crafting more consumer-centric online marketing mix strategies that could deliver more value to consumers. One limitation of this study is that it has been conducted on Online Marketing Strategies of Packaged Health Food Brands only and other FMCG Products are not being considered under this study. Therefore, future studies could be extended to other FMCG Products to generalize the findings of this study to other FMCG Categories.

15.9 Discussion

The emergence and advancement of technology has directed the marketers to explore new possibilities to attract the customers of packaged health food brands. As the consumers are spending more time on the Internet, packaged health food brands have amplified their online marketing efforts in crafting more consumer-centric content that could deliver more value to consumers. This research study attempts to study the different types of online marketing strategies which the marketers are implementing to attract the customers who are buying packaged health food brands online and to determine the gaps that exist in the previous research being done so as to help the marketers to subsequently design the online marketing mix strategies that can deliver the value to the customers at a profit. The study has helped in identifying the factors that contribute toward the effectiveness of online marketing strategies of packaged health food brands w.r.t. Gender.

References

1. Allen E, Fjermestad J (2001) E-commerce marketing strategies: an integrated framework and case analysis. Logist Inf Manag 14(1/2):14–23. https://doi.org/10.1108/09576050110360106
2. Chen MF (2009) Attitude toward organic foods among Taiwanese as related to health consciousness, environmental attitudes, and the mediating effects of a healthy lifestyle. Br Food J 111(2):165–178
3. Jones S, Reid A (2010) Children's magazines: reading resources or food marketing tools? Public Health Nutr 13(3):393–399. https://doi.org/10.1017/S1368980009991133
4. Kemp E, Bui M (2011) Healthy brands: establishing brand credibility, commitment and connection among consumers. J Consum Mark 28(6):429–437. https://doi.org/10.1108/07363761111165949
5. Nour MI, Almahirah MS, Said SM, Freihat S (2014) The impact of promotional mix elements on consumers purchasing decisions. Int Bus Manag 8(2):143–151. http://www.cscanada.net/index.php/ibm/article/view/4800. http://dx.doi.org/10.3968/4800

Chapter 16
A Proposal for Dual Data Selection Using Parallel Genetic Algorithm

Seema Rathee and Jyoti Ahuja

16.1 Introduction

Nowadays, data is breeding unexpectedly fast in almost all the areas of data sciences. Researchers are recognizing to facilitate successful data mining, and instance and feature selection is a requisite component. It is a procedure to select a subset of dataset in both the dimensions called as dual selection. It removes irrelevant, redundant, or noisy data, and takes about discernible possessions for mining algorithms and also improves predictive accuracy, therefore, leading to superior classifier comprehensibility.

Searching optimal subset of features and instances is identified as an NP-hard problem and the number of local minima can be quite large. The search space includes all the possible subsets of instances as well as features. For any dataset with n features and m instances, comprehensive evaluation of all feasible subsets (2^{m+n}) is generally infeasible in carried out due to the huge amount of effort required [1–4].

The evolutionary algorithms (EAs) paradigms especially genetic algorithms (GAs) have many benefits for optimization problems in data mining [5]. But, traditional GA needs all the data should be in memory and replicate the iterations of the algorithm. Therefore, GAs performance is not satisfactory for very large databases because of exponential increment in computational time. Thus, there is a need of parallel processing based on genetic algorithm also called PGA. Parallel processing is capable of addressing the problem of scalability efficiently by quick convergence. In addition, PGAs are known to find better optimal solutions. One can state that the PGAs outperform the simple GAs in terms of efficiency as well as efficacy.

S. Rathee (✉)
Guru Jambheshwar University of Science & Technology, Hisar, India
e-mail: seema27rathee@gmail.com

J. Ahuja
Government Post Graduate College for Women, Rohtak, India
e-mail: kwatra.jyoti@gmail.com

P. K. Kapur et al. (eds.), *Decision Analytics Applications in Industry*, Asset Analytics,
https://doi.org/10.1007/978-981-15-3643-4_16

In parallel model, several GAs are run in parallel independently with their own subpopulations. The individuals relocate from one subpopulation to another. This process is called migration and the individuals that relocate are called emigrants or migrants. The better individuals of each subpopulation are sent to another subpopulation, so that they can spread in all subpopulation. The most important advantage for parallelization is reducing the completion time of an algorithm. Moreover, parallel GA provides advanced generalization accuracy. Thus, a GA is very suitable to deal with the parallel computing environment [6].

In this article, we recommend a parallel genetic algorithm (PGA) for dual data reduction. The suggested parallel approach will considerably reduce the size of dataset with reasonable accuracy. The algorithm performance will be measured in terms of reduction rates and predictive accuracy obtained by KNN classifier on the reduced dataset.

The following paper is structured as follows: Sect. 16.2 describes the background details and related work. The next Sect. 16.3 discusses the proposed approach and the design of the PGA. Section 16.4 gives the implementation details and experimental results. The last Sect. 16.5 concludes the paper.

16.2 Background and Related Work

16.2.1 Dual Selection

The most important aim of data reduction is to select a subset of instances and features that provide higher accuracy with least generalization error. Dual selection techniques are used to select the most valuable instances and features simultaneously from the dataset. The dual selection technique works as shown in Fig. 16.1. We are here using a simple example to understand the working of dual data reduction. The original dataset is reduced by extracting some rows and columns. The variables A1, A2, A3, and A4 indicate attributes (columns), whereas I1, I2, I3, I4, and I5 signify instances (rows) of the dataset. The row number I2 and I5 have been deleted. Similarly, column

Original Dataset

#	A1	A2	A3	A4
I1	p	q	r	s
I2	e	f	g	h
I3	t	u	v	w
I4	m	n	o	p
I5	w	x	y	z

Reduced dataset

#	A1	A2	A4
I1	p	q	s
I3	t	u	w
I4	m	n	p

Fig. 16.1 Dual data reduction process

number A3 has been deleted while retaining only A1, A2, and A4. This procedure will simultaneously reduce both instance and features (see in Fig. 16.1).

Authors have fruitfully applied EAs to crack the problem of dual data selection. Kuncheva and Jain have adopted genetic algorithm to select a feature and reference set simultaneously [7]. Ros et al. have given a hybrid genetic algorithm for dual selection [8]. A cooperative co-evolutionary model has been proposed by Garcia et al. [9]. Herrera et al. have designed an integrated method for numerous correlated techniques such as Instance selection (IS), Instance weighting (IW), and Feature weighting (FW) into the structure of a co-evolutionary model [10]. A unified criterion for both instance and feature selection (UFI) based on the optimum experimental design (OED) is proposed by Zhang et al. [11]. Ratnoo et al. have proposed a clustering-based dual data selection approach using CHC genetic algorithm. They have worked on two stages. The earlier stage is based on clustering approach and the dual data reduction algorithm is applied on the second stage [12].

16.2.2 PGA for Dual Data Reduction

Parallel GA is an attractive approach for dual selection. PGA avoids the premature convergence to maintain the diversity. PGAs are better than the sequential GAs due to migration of individuals. Relocation of genetic substance provides several types of new chromosomes in the further population. Hence, there are three basic approaches available to execute PGA, i.e., (a) single population and parallel evaluation of fitness, (b) division of core population into several subpopulations and evaluate them parallel, and (c) parallelize only GA operators with universal population [6, 13].

Some authors have successfully applied parallel GAs for instance and feature selection individually. Tahseen implemented a new global control parallel GA to solve the problem of only instance reduction [13]. This method improves the classification accuracy while reducing the computational time. Othman et al. have purposed a wrapper model, a web-based FS (Feature Selection) tool which is combined with a random search strategy [14]. A distributed GA is designed for core feature subset selection by Lee et al. in [6]. Chen et al. [15] applied a CGPGA (coarse-grained parallel genetic algorithm) for optimizing the parameters for SVM classifier as well as feature subset concurrently. Parallel evolutionary algorithms (PEAs) have been applied to high-dimensional datasets to improve the accuracy of classification also reduces the execution time [16]. Inspite of lots of work in dimension reduction, there is hardly any research work in the field of dual data reduction using PGAs. Due to the intensifying success of PGAs, there is a need to explore them for dual reduction of data.

16.3 The Proposed Approach: PGA for Dual Selection

A Parallel GA has hardly been applied earlier for the aim of dual reduction. Thus, we purpose a dual data reduction algorithm based on parallel genetic algorithm (PGA). We first divide the original dataset in numerous disjoint subsets in this paper. These subsets of the dataset are dispersed among the different processing nodes. Every node obtains a same size subset in which instances and features are randomly chosen. Then, the algorithm of data selection is applied to all subset concurrently. Subsequently, the outcomes of all the nodes are joined by a union operator to obtain the ultimate result.

The entire population is divided into numerous subpopulations based on ring network topology. In each subpopulation, GA is executed independently. After a predetermined number of generations, one or two individuals migrate from a subpopulation to its neighbor subpopulation. According to the quality of the subpopulations, the individuals are chosen for migration to maintain the diversity in subpopulations. The higher fitness individuals than the average fitness individuals of the subpopulation, will migrate in the proposed PGA. These individuals replace the inferior individuals in the target populations [6].

The suggested PGA is executed on parallel architecture; we expect a considerable reduction in size of the dataset. The primary thing which we must obtain is that we necessitate numerous partitions of the dataset. Each partition represents a diverse problem of optimization for the instance/feature selection algorithm.

16.4 Experimental Analysis and Results

The initial population will be randomly initialized according to the population size. The chromosome comprises instances and features by representing them as binary, i.e., by 1 and 0. The bit 1 represents the presence and 0 represents absence of instances or a feature.

The fitness function used to evaluate chromosomes will be the weighted percentage of accuracy and reduction rate.

$$\text{Fitness function: - } \left\{ f1 = \text{Accuracy}; f2 = \left(1 - \left(R_f / T_f\right) * R_i / T_i\right) * 100 \right\} \tag{16.1}$$

In fitness function, the terms R_f and T_f indicate the reduced and total number of features in given dataset. Similarly, R_i and T_i denote the reduced and total number of instances.

Including crossover and mutation operators, PGA consists of some different parameters such as number of subpopulations, migration policy, number of migrants, migration probability, the network topology, the replacement and selection methods of migrants [6]. Therefore, the convergence criterion is also an important factor. The

Table 16.1 The parameter specification for PGA

Sr. No.	Name of the parameters	Value
1	Population size	20
2	#Subpopulation	4
3	Chromosome length	According to the size of the dataset (#Fetures(m) + #Instances(n))
4	Total #Generations	50
5	Crossover probability	0.7
6	Mutation probability	0.8
7	#Migrants	2
8	Stopping criteria	10

algorithm stops when generations reach to a certain predefined value. The values for various parameters are given in Table 16.1.

We expect the following benefits of employing a parallel GA:

1. It will maintain diversity in the population to avoid premature convergence to sub-optimal reduced dataset.
2. Investigation of different search space areas simultaneously with exchange of information will lead to overall a better optimal solution, i.e., a better optimal reduced dataset that contains the relevant instances and attributes
3. We can give diverse weights to different phase of the fitness function in the various subpopulations. This can give different reduced datasets with respect to weights assigned to different criteria like reduction rate and accuracy of the classifier.
4. Several experiments can be conducted with different sets of GA parameters for different subpopulations.

For experimental analysis, we have used Matlab tool. Here we have taken ten UCI machine learning datasets. The depiction of these datasets is given in Table 16.2.

The accuracy and reduction rate results are encountered from the reduced dataset, as well as the accuracy of KNN classifier is also measured on whole dataset. The reduction rate signifies both instance and feature reduction rate together (see Table 16.3).

Results from Table 16.3 show that the proposed PGA for dual data selection achieves higher reduction rate without any compromise on accuracy when classified with KNN.

Table 16.2 Sumarry of datasets

Sr. No.	Datasets	#Features	#Instances	#Classes
1	Phoneme	5	5404	2
2	Page-block	10	5473	5
3	Texture	40	5500	11
4	Optdigit	64	5620	10
5	Mushroom	22	5644	2
6	Coil2000	85	9822	2
7	Nursery	8	12960	5
8	Letter	16	20000	26
9	Shuttle	9	58000	7
10	Connect4	42	67557	3

Table 16.3 Accuracy and reduction rate results of dual PGA

Datasets	KNN accuracy on whole dataset	KNN accuracy on reduced dataset by PGA	
		Accuracy	Reduction rate
Phoneme	98.37	96.87	86.55
Page-blocks	95.58	94.26	89.27
Texture	99.56	98.67	86.59
Optdigits	98.79	96.56	86.49
Mushroom	95.78	94.62	87.97
Coil2000	91.37	92.56	88.35
Nursery	99.82	97.45	93.89
Letter	99.33	96.69	89.30
Shuttle	99.68	95.45	86.71
Connect4	97.90	96.43	97.96

16.5 Conclusion

Dual data reduction of bulky datasets involves a very large search space and unacceptably long run-time. Thus, there is a need of an advanced parallel genetic algorithm that brings more reduction by removing redundant instances and noisy features. In this paper, we have put forward a proposal for implementation of parallel GA for dual data reduction. The proposed PGA uses migration in-between the individuals in each subpopulation and joins the chromosomes at last. This algorithm will decrease the size of dataset without sacrificing the classification performance. The main disadvantage of this method is that it cannot work appropriately while dealing with multiple objective simultaneously.

In future, we will cope up with the problems related to parallel genetic algorithm for dual selection. A multi-objective parallel genetic algorithm (MOPGA) would be a good choice to handle multiple objectives simultaneously in fairly small runtimes. We also intend to experiment with multi-objective and parallel CHC genetic algorithm for dual selection.

References

1. Liu H, Yu L (2005) Toward integrating feature selection algorithms for classification and clustering. IEEE Trans Knowl Data Eng 17:491–502
2. Silva J, Aguiar A, Silva F. (2015) A parallel computing hybrid approach for feature selection. In: 2015 IEEE 18th international conference on computational science and engineering, pp 97–104
3. García-Pedrajas N, de Haro-García A, Pérez-Rodríguez J (2013) A scalable approach to simultaneous evolutionary instance and feature selection. Inf Sci 228:150–174
4. García-Pedrajas N, de Haro-García A, Pérez-Rodríguez J (2014) A scalable memetic algorithm for simultaneous instance and feature selection. Evol Comput 22:1–45
5. Goldberg DE (1989) Genetic algorithms in search, optimization and machine learning, 1st edn. Addison-Wesley Publishing Company, Boston, MA, USA
6. Lee H, Oh B, Yang J, Kim S (2009) Distributed genetic algorithm using automated adaptive migration. In: Proceedings of the eleventh conference on congress on evolutionary computation. IEEE Press, Piscataway, NJ, USA, pp 1835–1840
7. Kuncheva LI, Jain LC (1999) Nearest neighbor classifier: simultaneous editing and feature selection. Pattern Recogn Lett 20:1149–1156
8. Ros F, Guillaume S, Pintore M, Chretien J (2007) Hybrid genetic algorithm for dual selection. Pattern Anal Appl 11:179–198
9. García-Pedrajas N, del Castillo JAR, Ortiz-Boyer D (2009) A cooperative coevolutionary algorithm for instance selection for instance-based learning. Mach Learn 78:381–420
10. Derrac J, Triguero I, Garcia S, Herrera F (2012) Integrating instance selection, instance weighting, and feature weighting for nearest neighbor classifiers by coevolutionary algorithms. IEEE Trans Syst Man Cybern B Cybern 42:1383–1397
11. Zhang L, Chen C, Bu J, He X (2012) A unified feature and instance selection framework using optimum experimental design. IEEE Trans Image Process 21:2379–2388
12. Ratnoo S, Rathee S, Ahuja J (2018) A clustering based hybrid approach for dual data reduction. Int J Intell Eng Inform 6:468–490
13. Al-ramadin TA (2011) Reduction technique for instance-based learning using distributed genetic algorithms. Int J Grid Distrib Comput 4:47–60
14. Soufan O, Kleftogiannis D, Kalnis P, Bajic V (2015) DWFS: a wrapper feature selection tool based on a parallel genetic algorithm. PLoS One 10(2):1–23
15. Chen Z, Lin T, Tang N, Xia X (2016) A parallel genetic algorithm based feature selection and parameter optimization for support vector machine. Sci Program 1–11
16. Adi SI, Aldasht M (2018) Parallel evolutionary algorithms for feature selection in high dimensional datasets. Am J Comput Sci Eng Surv 6:013–021

Chapter 17
Exploring Alternative Financial Route in the Form of Private Investor Capital: A Case Study of Gramiksha

Bhawna Mehta and Archana Singh

17.1 Introduction

India ranks 131st out of 188 countries in the Human Development Index as per the 2018 HDI report. Education and per capita income are 2 of the 3 key parameters taken under consideration by the HDI. It is the world's second most populous countries just about to overtake China in terms of population. India's current population, 1.3 billion, has quadrupled since 1947. WHO projects it will reach 1.7 billion by 2050 [1].

17.1.1 Proliferation of NGOs in India

Non-Governmental Organizations with the objectives of social welfare, literacy and relief rather than profits have been present in India for a long time, as early as medieval era, flowing from the principles of Vedas and Puranas like Daana and Seva. In the nineteenth century, organizations like Prathana Samaj, Satya Shodhan Samaj, Arya Samaj, the National Council for Women in India, and the Indian National Conference were established; the success of which paved the way for nationwide recognition and reception of NGOs as registered entities. One of the first such acts was Societies Registration Act of 1860 [2] (Table 17.1).

There are over 1.5 million NGOs operating in India at local, national and international level. NGOs can be registered under Societies, Trusts or Charitable Companies. They are mostly volunteer dependent. Traditionally, NGOs in India have relied on some key sources of funding for their projects like Government grants, Donations,

B. Mehta (✉) · A. Singh
Delhi Technological University, Delhi 110042, India
e-mail: bhawnamehta6@gmail.com

A. Singh
e-mail: archanasingh205@gmail.com

P. K. Kapur et al. (eds.), *Decision Analytics Applications in Industry*, Asset Analytics,
https://doi.org/10.1007/978-981-15-3643-4_17

Table 17.1 Constitution of different acts

Societies Registration Act	1860
The Central Social Welfare Board	1953
M.P. Society Act	1973

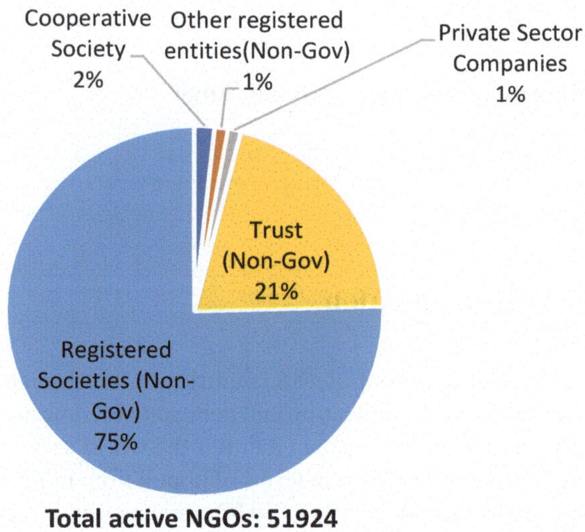

Fig. 17.1 Sector wise distribution of NGOs in India [4] *Source* http://ngodarpan.gov.in/index.php/search/

Gifts in-kind, and Sponsorships from Corporates, Institutions and Brands as part of CSR mandate of organizations like these. Only 2% of the 1.5 million NGOs are registered with the Foreign Contribution Regulation Act [3] (Fig. 17.1).

17.1.2 Child Education: A Sector Deserving Critical Attention

The Indian education system emphasizes on learning by rote. All the studies by different organizations, the Government and agencies unanimously argue that students' level of education does not match with the expected level of learning in their grade.

According to NGO Darpan, an initiative of NITI AAYOG and NIC, there are a total of 3005 registered Non-Governmental Organizations working in the education and literacy sector in India (Fig. 17.2).

The atmosphere is such that it prohibits the enhancement of creativity and inquisitiveness among children, "Urgent Needs of NGOs in the Education Sector," a report by CII, Deloitte and Give India says. Their study surveyed 17 NGOs associated with Give India. Fundraising emerged as the second most prominent area of need among the NGOs surveyed (Fig. 17.3).

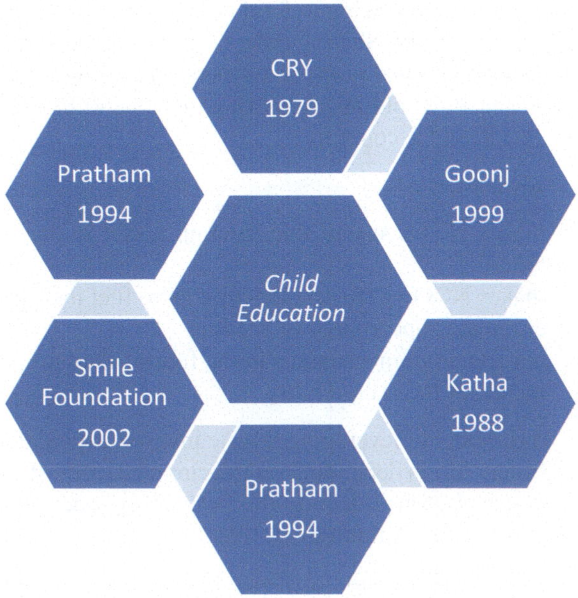

Fig. 17.2 Top NGOs in India working for child education and welfare

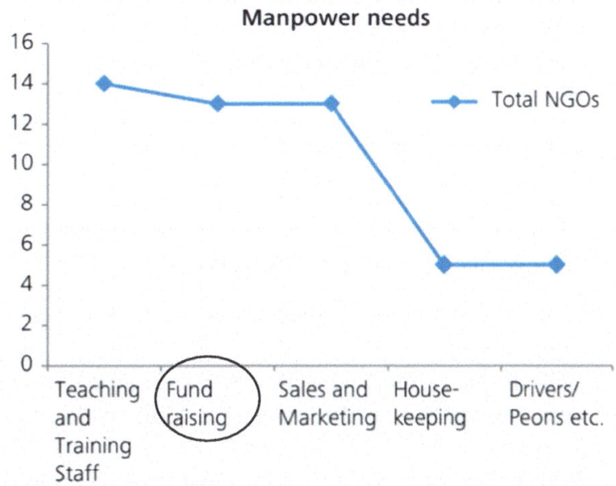

Fig. 17.3 Major manpower needs of NGOs in present world. *Source* www2.deloitte.com/content/dam/Deloitte/in/Documents/IMO/in-imo-ngo%27s-in-the-education-sector-noexp.pdf

17.1.3 The Role of Private Investor Capital

The fundraising or business model is linear with the funds flowing from the donors to the NGO who uses the funds and delivers the impact to the beneficiary. The prevalent model is fraught with challenges. A Report by PSI and Zurich International Ltd., "Using private investor capital to increase NGO impact: framework and key considerations to facilitate engagement" [5] classifies these as

(1) Timing: Borrowing to fund immediate project needs working capital
(2) Accessibility: Sharing risks of performance-based projects (PPP)
(3) Shortage: Dual purpose assets->> recycling, reinvestment
(4) Capacity Building: Acquiring the capabilities for business planning, management, and accounting.

Advantages

1. makes NGOs sustainable through social impact partnerships and low interest loans
2. makes NGOs self-sufficient or able to fuel their projects on their own
3. eliminates uncertainty of funding
4. attracts more investments in the future when properly realized
5. increases NGO credibility.

This model of leveraging private investor capital to finance the NGO projects can take many forms depending on the circumstances of the NGO and the specific stage of project. Before selecting the perfect approach, there are certain questions that NGOs need to ask themselves like

1. How much mission drift-perceived or real do the profit-generating or revenue accumulation intensive activities create?
2. There will always be a tension between profit and impact when joining hands with a private investor on the investor's part and operating income and impact on our part. How much harm can it cause?
3. How much prepared are we to manage financial distress in terms of
   (a) more expensive financing
   (b) opportunity cost of products
4. How can internal assets be monetised to generate income outside the project to be leveraged to pay back the returns?

33% Percentage of all investment deals in impact enterprises since 2000 which are at Seed stage.

30% Impact enterprises that have received at least two rounds of follow-on investments validating the investor's faith in the NGO's scalability and financial viability.

50% Percentage of Impact enterprises tracked by Intellecap that have taken off post-2010.

17.2 Definitions

There are many different versions of achieving impact through alternative ways by repaying the principle to the investors and the required return or by becoming a self-sufficient organization, building productive partnerships, scaling up the existing systems and building capacity. All the related terms are differentiated on some aspects.

Impact investments are "investments made into companies, organizations, and funds with the intention to generate social and environmental impact alongside a financial return" [6].

Social enterprises "Businesses that trade to intentionally tackle social problems, improve communities, provide people access to employment and training, or help the environment" [7].

NGOs can derive greater benefit by capitalizing on internal assets.

Dual purpose assets: "These are internal assets NGOs have developed in the course of implementing donor-funded projects that could be monetized to generate income outside the project. These include physical assets such as buildings or equipment or intangible assets, such as brand, knowledge, and data that can be used to generate impact and income. Income can be generated while generating impact, could be through the provision of services to beneficiaries at a basic fee or use of the assets to deliver services at a fee for non-impact related activities while they would otherwise be idle" ([5], p. 56).

Capacity building: National Council of Non-Profits, US [8] defines capacity building as "whatever is needed to bring a non-profit to the next level of operational, programmatic, financial, or organizational maturity, so it may more effectively and efficiently advance its mission into the future. Capacity building is not a one-time effort to improve short-term effectiveness, but a continuous improvement strategy toward the creation of a sustainable and effective organization."

17.3 Literature Review

Deloitte and Give India's report, "Urgent Needs of NGOs in the Education Sector," [9] puts forward critical recommendations in the face of challenges surrounding the education sector in India. National Teacher Training Program, Amending RTE Act to allow recognition of special schools run by NGOs, and Modifying Sarva Shiksha Abhiyan intervention model are some of the key recommendations on the list.

Report by PSI and Zurich International Ltd., 2014, "Using private investor capital to increase NGO impact framework and key considerations to facilitate engagement," [5] identifies the challenges that NGOs operating throughout the globe face and explains how these challenges can be wiped out using alternative model of financing by using private investor capital. The study also develops models applicable in different contexts.

Nisha Dutt and Usha Ganesh, The Evolution of Impact Investment in India [10], shed light in their report on the status of Impact investing in India, what attracts the impact investors in today's economy and outlines what challenges lie ahead.

Robert S. Kaplan and Allen S. Grossman's HBR article: The Emerging Capital Market for Non-profits [11], highlights issues such as lack of good information and the general lack of accountability of NGOs as being obvious hurdles to impact investing. We find an example of The Robinhood Foundation in New York which works with 240 organizations who measures their outcomes which are then translated into key metric by Robinhood and compared across organizations. The data is used to collect information which is sold to foundations and the raised funds are allocated to NGOs. In the case study, Gramiksha suffers from uncertainty of funding sources. The lack of financial data to show income streams and impact measurement at Gramiksha is a severe obstacle in making it attractive to investors. Adopting a sustainable approach first requires them to bridge this gap. We analyze the needs and deliverables stemming from the objectives of programs like Udaan and explore the best models available to attract private investors.

17.4 Methodology

In this study, a qualitative research methodology was used. To support the hypothesis and arrive at the conclusion, relevant existing literature was consulted and secondary data was extracted from studies conducted by research organizations in the sector.

Structured and unstructured interviews with the NGO (Gramiksha) representatives (National finance manager, operations manager and others associated in different capacities) were conducted through email and conference call for understanding the environment and challenges. This report is the result of all those interactions and investigations.

Further, during the course of this study an event of special mention occurred in India and was part of media articles and news, which substantiated a version of the proposed financing solution for NGOs. This was also recorded as an illustration.

17.5 Case Study

17.5.1 An Introduction of Gramiksha

Gramiksha is a Not-for-profit organization working for the escalation of child education standards for 7 years since 2011. It is registered under the M.P. Society Act, 1973. It currently operates in six major cities of India: Bhopal, Delhi, Indore, Gwalior, Jaipur, and Laxmangarh.

Udaan: By using “activity-based learning models,” Udaan achieves to integrate theoretical learning with the real world and the challenges it manifests.

Dream School Project: A resource center in Gwalior in association with Tribe of Lambs for 300+ children. One of the short-term objectives is to create more of its kind.

Across all its projects, Gramiksha tackles the issues of

1. Getting assimilated in the school environment and at the same time transforming it with respect to preset objectives
2. Seeking permissions from education functionaries and other authorities.

17.5.1.1 Existing Systems of Financing and Fundraising Channels

Presently, Gramiksha relies on 4 sources of funds and revenue generation. These are

17.5.1.2 Fundraisers

Fundraisers are events conducted by Gramiksha members and volunteers in different regions across its operations in the six cities. The event primarily aims to raise funds from institutions, other organizations and even general public. One such example is a City Level Quiz Competition in Jaipur targeting Private Schools and students thereof across a city. The secondary objective is to create awareness among the masses about the impact the organization is creating and to add recognition and acceptance.

17.5.1.3 Registration Fees from Volunteers

Every year Gramiksha adds volunteers for a definite term to join the different teams and assist in their projects. The volunteers are required to pay a registration fee.

17.5.1.4 Donations

A network of 16 organizations supports the projects and operations of Gramiksha in the form of donations. In 2016, Gramiksha teamed with one of its supporters, Tribe of Lambs, to make happen their biggest project till date in terms of investment. The Dream School Project, with an investment of $3000 raised by Tribe of Lambs through their product revenues (jewelry), was implemented to create a resource center out of an old building in Gwalior which is their largest project till date.

17.5.1.5 Sponsorships

Since auditing of financial data to arrive at statement is under progress Gramiksha is not eligible for CSR donations yet.

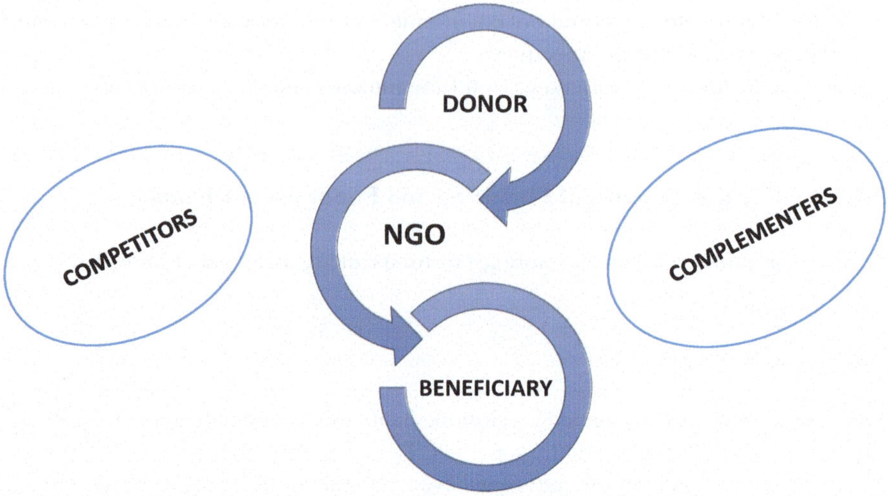

Fig. 17.4 Conventional model of fundraising by NGO

17.5.1.6 Challenges Inherent in the Existing System

1. Uncertainty in the availability of funds, insufficient funds
2. Access to capital
3. Reputational risks
4. Poor system of securing capital for immediate requirement of funds
5. Inability to adopt the best practices
6. No flexibility
7. High transition rate of volunteers
8. Presence across multiple fundraising channels is lagging.

17.6 Proposal

17.6.1 Illustrating Fundraising Through Schematic Diagram

17.6.1.1 The Traditional Model

See Fig. 17.4.

17.6.1.2 An Alternative Approach: Redefining the Role of Stakeholders

See Fig. 17.5.

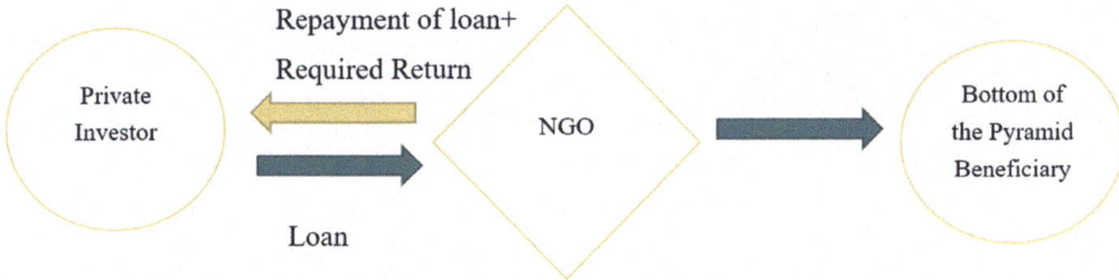

Fig. 17.5 Alternative approach to fundraising by NGO

17.6.2 The Required Return

Internal assets of the NGO can serve as perfect source of repayment to the investor. These assets are called Dual purpose assets.

PSI and Zurich International Ltd.'s report outlines 3 approaches:

(a) Using donations to repay the loan
(b) Creating income source out of Dual purpose assets
(c) Generating income by becoming a social enterprise.

Outlining the first two approaches which are best suited to be explored by Gramiksha owing to their objectives at this stage

1. Using donations to repay the loans
(a) Borrowing for immediate needs (Fig. 17.6)
(b) Sharing risk of performance (Fig. 17.7)
2. Creating income out of "Dual purpose assets"
(a) Private investor capital used to develop assets (Fig. 17.8)
(b) Loans repaid using assets (Fig. 17.9).

Insights are derived from "Report by PSI and Zurich International Ltd., 2014, Using private investor capital to increase NGO impact framework and key considerations to facilitate engagement" [5].

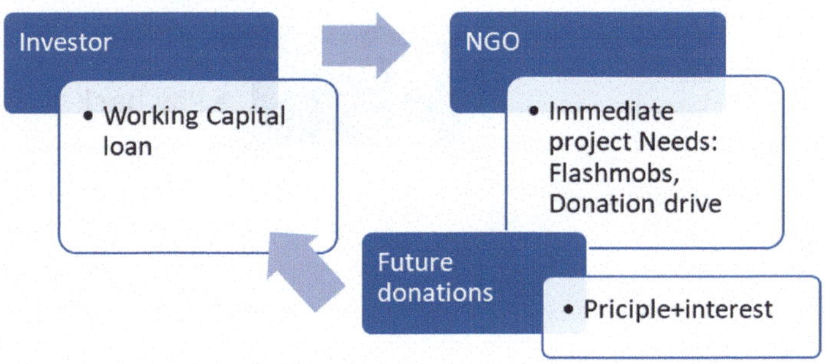

Fig. 17.6 Borrowing for immediate needs

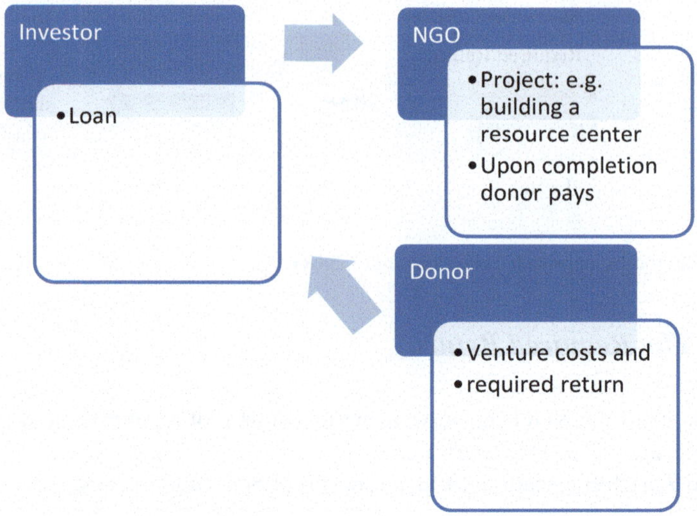

Fig. 17.7 Sharing risk of performance

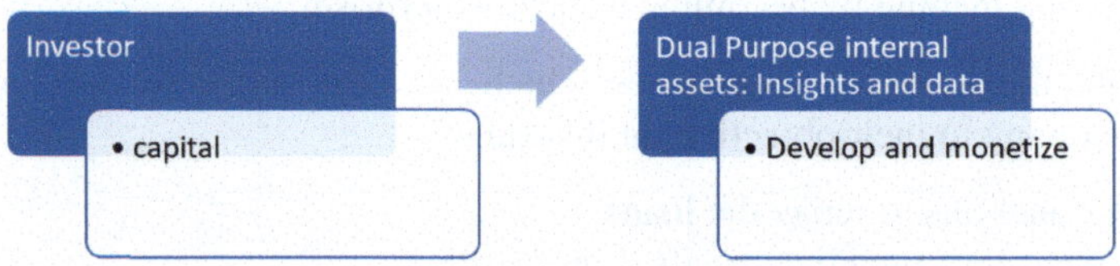

Fig. 17.8 Using private investor capital to develop assets

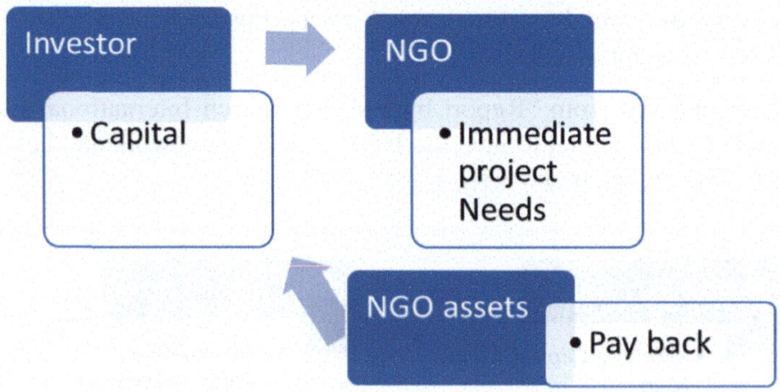

Fig. 17.9 Loans repaid using assets

17.6.2.1 Information at Its Best: Monetizing Research Base Toward Research Institute Programs

One example is insights and data about the sector of their operation. Collecting data in their operating sector and cities related to the number of children in the school-going age group, proportion of children not having access to particular facilities, pupil–teacher ratio in the government schools of the area, number of schools at particular stages with respect to improvements in education and infrastructure and other such attributes. The information collected in this way can be used to solicit funds from different organizations seeking the information for authentic purposes. These funds can then be allocated to different projects.

17.6.2.2 Impact Metrics: The Investor's Perspective

An investor would look for two things: a measurable impact and more importantly the income streams. To attract a good class of investors, Gramiksha needs to show the social return of its activities.

What they (Gramiksha) do?

(1) baseline and **end line** assessment
(2) continuous assessment on parameters.

Suggestions: The study suggests defining metrics to calculate this return on impact.

The Global Impact Investing Network (GIIN) in its IRIS defines metrics to assess the social, environmental, and financial impact and performance of an investment. We can derive the relevant metrics for our sector and corresponding to our objectives.

This requires

1. Creation and maintenance of centralized database for all the cities of operations
2. Timely update of database
3. Definition of performance metrics as per the projects
4. Assessment and review

Illustration 1: A project of duration say, 4 years is implemented in the city of Gwalior in 2013 covering 10 schools of the area, aimed at enhancement of learning experience and higher student engagement. Impact review for the year 2013 (Table 17.2).

A similar evaluation for all the 4 periods is prepared to compare the results of the reporting periods and amount of resulting impact.

Illustration 2: A resource center for children of the nearby area (Table 17.3).

Table 17.2 Assessment of classroom enhancement program across metrics

Performance metrics/rating	1	2	3	4	5
Student to classroom ratio			*		
Student to teacher ratio	*				
Student engagement	*				
Student attendance rate		*			
School enrollment	*				
Extra-curricular programs offered	*				
Student transition rate					*
Value of the new education material				*	

Table 17.3 Assessment of resource center program across metrics

Performance metrics	1	2	3	4	5
Number of books per class of students	*				
Availability of copies	*				
Reading facilities		*			
Renewal of stock	*				

17.7 Conclusion

The potential number of children that can be impacted is huge and hence warrants the presence of Gramiksha in the sector. Gramiksha enjoys a support network of over 16 organizations but it is still in the nascent stage with respect to the fundraising channels. The integration of conventional domestic donors with private investor capital can boost the fundraising approach provided Gramiksha capitalizes on the opportunities by developing their internal assets to achieve scale to offer lucrative returns in the form of insights to the profit-seeking investors while maximizing their impact.

17.8 Future Scope

By using the appropriate parameters for measurement of social impact created by NGOs like Gramiksha to value social and monetary returns generated, NGOs can effectively tap Social Impact oriented (partially or fully) private sector investors. This system can become a vehicle for making profitable investments in NGOs by the private sector thereby increasing NGOs overall performance and providing sustainability.

References

1. Academic claims (2017) India is world's most populous nation with 1.32 bn people. http://theguardian.com/
2. Asian Development Bank Report, June 2009, Overview of Civil Society Organizations India, Civil Society Briefs
3. http://indpaedia.com/ (2014)
4. http://ngodarpan.gov.in/index.php/home/sectorwise_ngo/30048/11/3005
5. PSI and Zurich International Ltd. (2014) Using private investor capital to increase NGO impact framework and key considerations to facilitate engagement
6. GIIN. http://thegiin.org/impact-investing/
7. http://socialtraders.com.au/
8. National Council of Non-Profits, US, What is Capacity Building? www.councilofnonprofits.org/tools-resources/what-capacity-building
9. CII, Deloitte and Give India, Urgent Needs of NGOs in the Education Sector a Report. www2.deloitte.com/content/dam/Deloitte/in/Documents/IMO/in-imo-ngo%27s-in-the-education-sector-noexp.pdf
10. Dutt N, Ganesh U, The evolution of impact investment in India. https://ssir.org/articles/entry/the_evolution_of_impact_investment_in_india
11. Kaplan RS, Grossman AS (2010) The emerging capital market for nonprofits. https://hbr.org/2010/10/the-emerging-capital-market-for-nonprofits

Chapter 18
Performance Analysis of Optimal Placement of Multiple DGs Using PSO

Harsh Maurya, Nidhi Singh Pal, and Vikas Singh Bhadoria

18.1 Introduction

When power is transmitted over long distances then some power losses are caused by Joule effect in power lines. Some losses are also occurring in transformers and other transmission line equipments. Total loss from generation to the reaching of consumer is in the range of 8–15%. These losses can be reduced if power has to travel less distance. This can be done by use of DG. DG is production of electricity near and on-site to the load center which is connected to the grid in bidirectional fashion. DGs are small in size and are generally renewable in nature. DGs are the future of the smart grid system and making the grid efficient in terms of losses, reliability, and stability. DG is far better than traditional power generation methods because it is closer to the end user more than the main generations, which reduces the transmission losses as well as cost. It is very capable in energy loss reduction.

By installing DG, a consumer can reduce the energy consumption from main supply which results in low electricity bill. DGs are mostly renewable resources; hence these are better for the environment. Also, the use of DG is beneficial to government because of low maintenance cost of distribution and transmission [1]. Providing energy near the end user will also improve the voltage profile. This is why Optimal DG Allocation (ODGA) is a heeded topic for researchers all over the world. It is discussed by the researchers with different objectives depending upon

H. Maurya · N. S. Pal
Department of Electrical Engineering, Gautam Buddha University, Greater Noida, UP, India
e-mail: harshmaurya38@gmail.com

N. S. Pal
e-mail: nidhi@gbu.ac.in

V. S. Bhadoria (✉)
Department of Electrical and Electronics Engineering, ABES Engineering College, Ghaziabad, UP, India
e-mail: vikasbhadoria@gmail.com

P. K. Kapur et al. (eds.), *Decision Analytics Applications in Industry*, Asset Analytics,
https://doi.org/10.1007/978-981-15-3643-4_18

the ODGA problems such as improvement in the voltage profile, system stability, reliability, efficiency, and lowering of the system losses [2]. Different types of DG's and their impacts on the distribution system are discussed in detail in [3]. Optimal DG placement can have single objective or multiple objectives [4]. But DG must be installed at a proper bus and of proper size because system losses depend upon both location and size of DG. Hence the location and size of the DG must be optimized by analyzing the distribution system data. This is a part of power system planning.

For a good approach, proper system data is taken and location of the DG is calculated mathematically. By placing a DG in the system, the overall system gets changed. As per the size of DG, losses may increase or decrease depending on its value. So the ending results are the location and size of the DG. The wrong location of the DG results in increase of loss in distribution system. Optimally placed DG also improves the reliability of the system [5].

There are many optimization algorithms and techniques available to solve ODGA problem. In general, these optimization techniques can be classified as Analytical, Numerical, and Heuristic. Analytical technique is found useful in optimal placement of DG, which provides reactive power support only [6]. D. B. Prakasha have analyzed different type of DG injection with both real and reactive power [7]. A hybrid DG optimization technique of Genetic Algorithm (GA) and Artificial Bee Colony (ABC) is used for the placement of DG and capacitor in [8]. Arulraj R proposed Competitive Swarm Optimizer (CSO) Algorithm [9]. M. Mosbah used NSGA-II and Fuzzy Logic Combination in his paper [10]. A new mathematical expression power Voltage Stability Constant (PVSC) is used in [11]. Witoon used an Adaptive Weight PSO for the placement of DG [12]. A novel Hybrid PSO Algorithm with Quasi-Newton method [13] is used for optimal placement of DG in IEEE-33 and IEEE-69 bus system. Several other advanced optimization techniques like Loss Sensitivity Method [14], Artificial Immune Based PSO [15], Genetic Algorithm (GA) and DEA [16], Cuckoo Search Algorithm (CSA) [17], Improved PSO [18], Discrete PSO and Optimal Power Flow [19], Artificial Bee Colony [20], Firefly Algorithm [21], and Shuffled Frog Leaping Algorithm (SFLA) [22] are also found in the literature. PSO is also found useful in optimal placement of UPFC [23]. Heuristic approach, PSO is used in this paper as it is faster, more efficient, and accurate than other approaches and algorithms.

In this paper, Sect. 18.1 discusses the introduction. Problem formulation is given in Sect. 18.2 and optimization technique used is given in Sect. 18.3. Results and conclusion are given in Sects. 18.4 and 18.5, respectively.

18.2 Literature Survey

Any optimization problem may have single objective or a multi-objective function can be formed by combining several objectives. Objective of the optimization problem may be minimization of losses, maximization of profit, improvement in voltage profile, improvement in reliability, etc. Objective of this optimization is to decrease the overall system losses by positioning DG at optimum location with proper size. Hence objective function of the optimization problem can be given by Eq. (18.1)

$$\text{Objective} = \text{Minimize}\,(P_{\text{loss}}) \tag{18.1}$$

where P_{loss} denotes overall system losses. These losses are calculated by the standard equations of load flow analysis. Here, author entered the different parameters of the power system such as bus number, bus code, voltage, angle, load parameters (P and Q), and generator parameters in the form of a matrix. Other system parameters such as resistance and reactance are entered which are mandatory for Ybus calculations. Additional data, i.e., bus codes of sending and receiving ends along with half susceptance and transformer taps are also entered thereby forming a matrix.

Ybus is a square matrix of size $n \times n$ where 'n' denotes number of buses in the system. Formation of Y_{bus} is done by addition of two different matrices, one of which consists of diagonal elements while other consists of off-diagonal elements.

The values of P and Q are calculated by Eqs. (18.2) and (18.3), respectively.

$$P_i = \sum_{\substack{k=1 \\ k \neq i}}^{n} |Y_{ik} V_i V_k| \cos(\theta_{ik} + \delta_k - \delta_i) \tag{18.2}$$

$$Q_i = \sum_{\substack{k=1 \\ k \neq i}}^{n} |Y_{ik} V_i V_k| \sin(\theta_{ik} + \delta_k - \delta_i) \tag{18.3}$$

Equations (18.2) and (18.3) are used to calculate the total losses of the system, which can be expressed by Eq. (18.4).

$$P_{\text{loss}} = \sqrt{P_i^2 + Q_i^2} \tag{18.4}$$

Since each optimization problem is having some equality and inequality constraints. The constraints of this optimization problem are:

(i) Voltage of the buses should be within permissible limit

(ii) Active power generation should be equal to the active demand and real losses

(iii) Reactive power generation should be equal to the reactive demand and reactive losses

(iv) Current flow in the lines should be within limits

18.3 Particle Swarm Optimization

Optimization is a technique to find the maximum or minimum value for well-defined function with some constraints in a defined domain. A group or set of particles is taken to form a model with a defined purpose. It is taken from observing living creatures with group movement in nature, e.g., bee swarm, bird flock, fruit flies, and fish school and they have a purpose of searching food. It is important to note that they serve a common purpose or activity. Initially, PSO is proposed by Keneddy in 1995 [24]. A search space is initialized with some particles, and i^{th} particle of the swarm is represented by a D-dimensional vector $X_i = (x_{i1}, x_{i2} \ldots x_{iD})$. For this optimization problem the number of particles is selected as 40. Among the swarm, there is particle with the least value of the function which is denoted by the local best; for each iteration and of all the iterations we have a particle which is found to be the least, and is known as global best. The best previous position (i.e., the position corresponding to the best function value) of the ith particle is recorded and represented as $P_i = (p_{i1}, p_{i2} \ldots p_{i\text{D}})$ and the position change (velocity) of the ith particle is $V_i = (v_{i1}, v_{i2} \ldots v_{i\text{D}})$. The particles are finagled according to Eqs. (18.5) and (18.6). In these equations superscripts denote the iteration.

$$V_i^{k+1} = wV_i^k + c_1 r_{i1}^k (P_i^k - X_i^k) + c_2 r_{i2}^k (P_g^k - X_i^k) \tag{18.5}$$

$$X_i^{k+1} = X_i^k + V_i^{k+1} \tag{18.6}$$

where $i = 1, 2, \ldots N$ and N is the size of the population, w is the inertia weight; c_1 and c_2 are positive constants, called the cognitive and social parameter, respectively; r_{i1} and r_{i2} are random numbers uniformly distributed within the range [0, 1]. Velocity equation is used to determine the ith particle's new velocity, in each iteration, while position equation provides the new position of the ith particle, adding its new velocity, to its current position. Application of the PSO in optimum placement of DG can be explained with the flowchart shown in Fig. 18.1.

In PSO, inertial weight plays an important role. It controls the convergence response of the PSO. If the weight is high then particles explore new places or values suitable for function whereas if the weight is kept low it traverses to the nearer values of the global best particle. There are many ways one can vary the weight of PSO. Some of the weight techniques are global local best inertia weight, time-varying inertia weight, constant inertia weight, linear decreasing, etc. [25]. Oscillating Inertia Weight (OIW) is better than constant inertia weight and it varies with every iteration. OIW is used in this paper because in each iteration the value of inertia weight of the particle facilitates exploitation rather than exploring the new values of particle weight [26].

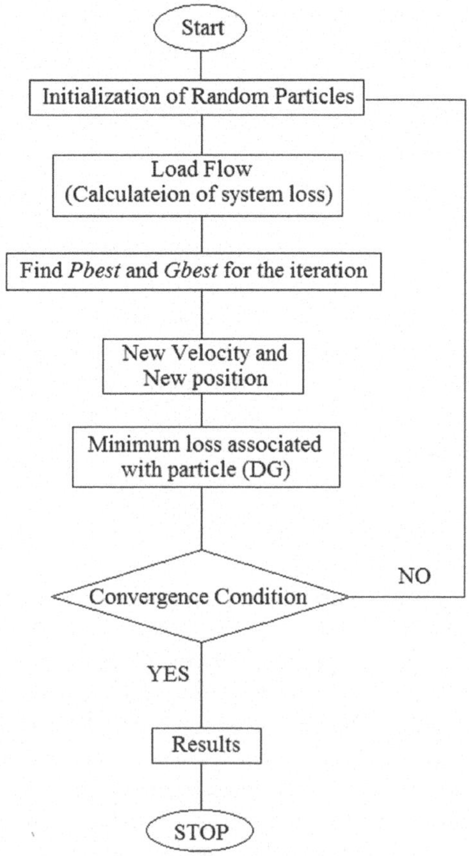

Fig. 18.1 Flow chart of PSO for optimal placement of DG

18.4 Result and Analysis

IEEE 33 bus standard radial distribution system is considered for analyzing the effect of DG allocation at different nodes. Data for the calculation purpose is given in [27]. One line diagram of the test system is shown in Fig. 18.2. First of all system losses without any DG placement comes out to be 231.0619 kW. Results of the PSO are

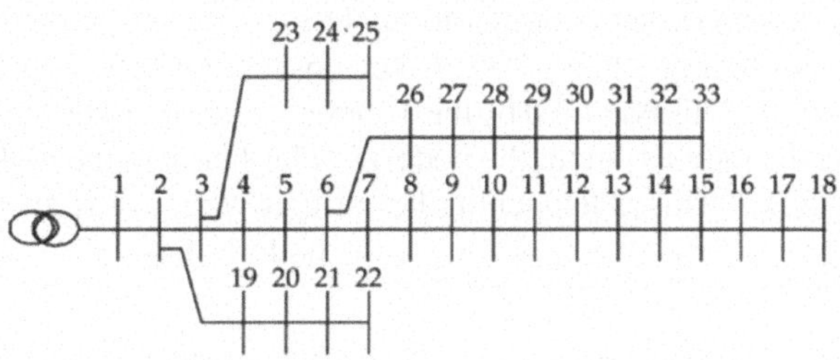

Fig. 18.2 One line diagram of IEEE 33-bus radial distribution system

Table 18.1 Results and comparison of optimal placement of DG

Base	Size (MW)	Location	Losses (kW)	% Loss reduction (%)
Loss sensitivity method [14]	Single DG—2.7019	6	69.6668	69.84
	Two DGs—1.9024, 0.5649	6, 15	89.85	55.66
CSA [17]	Single DG—2.469	6	182.1	21.18
	Two DGs—1.4, 0.8	6, 31	144	37.67
FA [21]	Single DG—1.1904	30	116.7	49.49
	Two DGs—0.6128, 1.0131	14, 30	96.9	58.06
SFLA [22]	Single DG—1.199	30	118.2	48.84
	Two DGs—0.6022, 1.1246	14, 30	105.4	54.38
PSO	Two DGs—2.7, 0.39082	6, 30	53.280	76.94

compared with other existing methods like Loss Sensitivity Method [14], CSA [17], FA [21], SFLA [22]. These results are shown in Table 18.1. From results it is clear that the size proposed by the CSA is very less but at the same time loss reduction in that case is very low. Size and loss reduction in FA and SFLA are approximately similar. If the size proposed by the PSO is installed in the IEEE 33-bus distribution system at bus number 6 and bus number 30 then this will give the maximum loss reduction. After placement of the first DG the losses drop to 69.6668 kW which means by placing a DG of size 2702.2 kW, at node 6, the system losses are reduced by 161.3951 kW. Second DG of size 390.828 kW is placed at node 30 and the system losses are dropped to 53.28 kW, hence the total reduction in system losses is 177.7819 kW. Thus results of PSO are found better among compared techniques. In case of PSO, loss reduction in single DG Placement is 69.84% and for two DGs the placement is 76.94%. This loss reduction is highest among the compared methods.

As the DGs are placed the losses of the system decreases and voltage profile of the system is improved depending upon the node locations. By placing two DGs at locations 6 and 30 of the corresponding sizes the voltage profile is also improved. This voltage profile is shown in Fig. 18.3. From the figure it is clear that the node closer to the DG will show maximum improvement in voltage profile as compared to the node that is farther to the DG.

From Fig. 18.3, it can also be observed that there is a shoot between the nodes 18 and 19. This is caused by the different locations of the nodes in the distribution system. In the radial system, the node 18 is farther than the node 19 from the generation and the DG. This is happening because as we move farther from the source the voltage the profile gets poorer. Hence the node 19 has more improved voltage than the node 18.

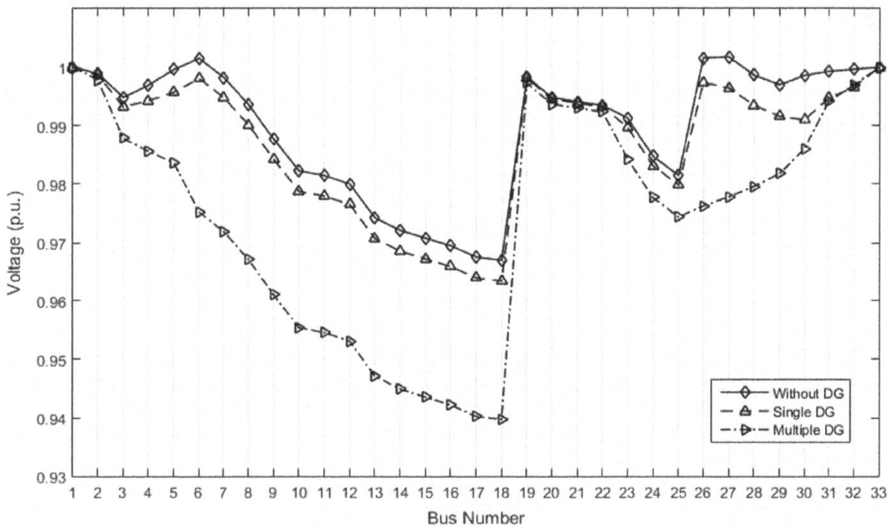

Fig. 18.3 Voltage profile of the IEEE 33-bus system with and without DG

As seen in the results that placement of the first DG shows large amount of change in the voltage profile but the impact of second DG is not too much on the whole system losses. The impact of second DG is observed by the neighboring nodes as their node voltage reaches near to the unity. By placing more DG in the system the losses get reduced by a small fraction hence there are only two DGs that are placed in this paper.

18.5 Conclusion

By allocating DG of appropriate size the losses of the whole system can be reduced. The system losses will increase or decrease depending upon the size and location of the allocated DG. In this paper, two DGs are placed at different locations, i.e., 6 and 30 which results in the 76.96% decrease in the whole system losses which makes the radial distribution system more efficient and reliable. Thus, PSO is found better than the compared techniques. By allocation of the DG, the voltage profile is also affected. Voltage profile of the radial distribution system is improved for the nodes which are farther from the generation.

Hence a DG plays an important role to improvise the whole distribution system by improving the voltage profile as well as decreasing the system losses. It is beneficial for the Government because it reduces the distribution cost of the supplier. Also, the DG is renewable so it is a better futuristic option as it reduces the main generation which results in low pollution, because most of the main generation plants are thermal-based.

References

1. Ramavat SR, Jaiswal SP, Goel N, Shrivastava V (2018) Battery energy storage technology integrated for power system reliability improvement. Lect Notes Electr Eng 508:131–140
2. Adefarati T, Bansal RC (2016) Integration of renewable distributed generators into the distribution system: a review. IET Renew Power Gener 10(7):873–884
3. Bhadoria VS, Pal NS, Shrivastava V (2013) A review on distributed generation definitions and DG impacts on distribution system. In: International conference on advanced computing and communication technologies (ICACCT^{TM}-2013)
4. Bhadoria VS, Pal NS, Shrivastava V (2014) Installation of DG for optimal demand compensation. In: Proceedings of the 2014 international conference on issues and challenges in intelligent computing techniques, ICICT 2014
5. Bhadoria VS, Pal NS, Shrivastava V, Jaiswal SP (2017) Reliability improvement of distribution system by optimal sitting and sizing of disperse generation. Int J Reliab Qual Saf, Eng
6. Bhadoria VS, Pal NS, Shrivastava V, Jaiswal SP (2018) Optimal sitting and sizing of capacitor using iterative search method for enhancement of reliability of distribution system. Lect Notes Electr Eng 508:123–129
7. Prakash DB, Lakshminarayana C (2016) Multiple DG placements in distribution system for power loss reduction using PSO algorithm. Procedia Technol 25:785–792
8. Jegadeesan M, Babulal CK, Subathra V (2017) Simultaneous placement of multi-DG and capacitor in distribution system using hybrid optimization method. In: IEEE international conference on innovations in green energy and healthcare technologies - 2017, pp 1–7
9. Kumarappan N, Arulraj R (2016) Optimal installation of multiple DG units using competitive swarm optimizer (CSO) algorithm. In: 2016 IEEE congress on evolutionary computation, CEC 2016, pp 3955–3960
10. Mosbah M, Arif S, Mohammedi RD (2017) Multi-objective optimization for optimal multi DG placement and sizes in distribution network based on NSGA-II and fuzzy logic combination. In: 5th international conference on electrical engineering - Boumerdes, ICEE-B 2017, pp 1–6
11. Nawaz S, Imran M, Singh S (2018) A novel approach for multiple DG allocation in distribution system of Jaipur City. In: IEEE international conference on information, communication, instrumentation and control, pp 1–6
12. Prommee W, Ongsakul W (2008) Optimal multi-distributed generation placement by adaptive weight particle swarm. In: International conference on control, automation and systems, ICCAS 2008, pp 1663–1668
13. Tolba MA, Tulsky VN, Diab AAZ (2017) Optimal allocation and sizing of multiple distributed generators in distribution networks using a novel hybrid particle swarm optimization algorithm. In: Proceedings of the 2017 IEEE Russia section young researchers in electrical and electronic engineering conference, ElConRus 2017, pp 1606–1612
14. Kumar A, Bhimarasetti RT (2018) Multiple distribution generation location in reconfigured radial distribution system distributed generation in distribution system multiple distribution generation location in reconfigured radial distribution system distributed generation in distribution. IOP Conf Ser Earth Environ Sci 1–9
15. Bhadoria VS, Pal NS, Shrivastava V (2019) Artificial immune system based approach for size and location optimization of distributed generation in distribution system. Int J Syst Assur Eng Manag 10(3):339–349
16. Kumar M, Mukherjee V, Kumar V (2016) Electrical power and energy systems greenfield distribution network expansion strategy with hierarchical GA and MCDEA under uncertainty. Int J Electr Power Energy Syst 79:245–252
17. Priyadarshini R, Kori S (2017) Cuckoo search algorithm based multiple dg placement and voltage profile improvement in a radial distribution system. In: Second international conference on electrical, computer and communication technologies (ICECCT), Coimbatore, pp 1–6
18. Prommee W, Ongsakul W (2011) Optimal multiple distributed generation placement in microgrid system by improved reinitialized social structures particle swarm optimization. Eur Trans Electr Power 21(1):489–504

19. Gomez-gonzalez M, López A, Jurado F (2012) Optimization of distributed generation systems using a new discrete PSO and OPF. Electr Power Syst Res 84(1):174–180
20. Abu-Mouti FS, El-Hawary ME (2011) Optimal distributed generation allocation and sizing in distribution systems via artificial bee colony algorithm. IEEE Trans Power Deliv 26(4):2090–2101
21. Nadhir K, Chabane D, Tarek B (2013) Distributed generation location and size determination to reduce power losses of a distribution feeder by firefly algorithm. Int J Adv Sci Technol 56:61–72
22. Afzalan E, Taghikhani MA, Sedighizadeh M (2012) Optimal placement and sizing of DG in radial distribution networks using SFLA. Int J Energy Eng 2(3):73–77
23. Jaiswal SP, Shrivastava V (2018) A PSO based search for optimal tuning and fixing of UPFC to improve usefulness of distribution system. J Intell Fuzzy Syst 1–9
24. Kennedy J, Eberhart R (1995) Particle swarm optimization. In: Proceedings of ICNN'95 - international conference on neural networks, pp 1942–1948
25. Bansal JC, Singh PK, Saraswat M, Verma A, Jadon SS, Abraham A (2011) Inertia weight strategies in particle swarm. In: Third world congress on nature and biologically inspired computing, pp 633–640
26. Maurya H, Bohra VK, Pal NS, Bhadoria VS (2019) Effect of inertia weight of PSO on optimal placement of DG. IOP Conf Ser Mater Sci Eng
27. Baran ME, Wu FF (1989) Network reconfiguration in distribution systems for loss reduction and load balancing. IEEE Trans Power Deliv 4(2):1401–1407

Chapter 19
Cognitive Activities Help Delay Dementia in 60 Plus Population

Z. Rahman, S. Bansal, S. Malik, B. Mitra, D. Das, A. Mohan, J. Dhingra, A. Suneja, S. Gulati, A. Gupta, S. C. Dabas, and R. Jain

19.1 Introduction

Dementia is an age-related neurodegenerative and terminal clinical syndrome/disorder, characterized by progressive deterioration and ultimately loss of memory and other cognitive abilities. Some epidemiologic investigations and intervention studies indicate that engaging in physical exercise and mental activities markedly reduces the risk of Alzheimer's disease and ageing-induced decline of mental faculties [1].

19.1.1 Significance of Study

At the global level, an estimated 50 million elderly persons were living with dementia in the year 2018. This number is expected to touch 82 million by the year 2030 and surpass the 150 million mark by 2050. Currently 60% of dementia patients inhabit developing countries. This figure is expected to rise further and touch 68% by the year 2050 [2]. China, India and countries of South Asia and the Western Pacific neighbouring these nations are set to witness the most explosive growth in the elderly population.

As India prepares to step up to the challenge of possibly becoming home to the world's largest population of senior citizens, mental health of the elderly is receiving

Z. Rahman · S. Bansal · S. Malik · B. Mitra · D. Das · A. Mohan · J. Dhingra · A. Suneja · S. Gulati · A. Gupta · R. Jain (✉)
Department of Statistics, Ram Lal Anand College, University of Delhi, South Campus, Benito Juarez Road Dhaula Kuan, New Delhi 110021, India
e-mail: ritajain313@gmail.com

S. C. Dabas
Department of Hindi, Ram Lal Anand College, University of Delhi, South Campus, Benito Juarez Road Dhaula Kuan, New Delhi 110021, India

P. K. Kapur et al. (eds.), *Decision Analytics Applications in Industry*, Asset Analytics,
https://doi.org/10.1007/978-981-15-3643-4_19

greater attention. The number of Indian citizens over 60 years of age is expected to exceed 324 million by 2050. While the percentage of geriatric population (60+) in rural areas has gone up from 5.3 to 5.7%, the corresponding figures for urban areas have risen from 6.0 to 8.0% [3, 4]. In India, the prevalence rate of dementia has ranged between 0.9 and 7.5% in the last two decades. With an imminent rise in the population of senior citizens, India can expect to see a commensurate increase in the number of persons with MCI and dementia.

Keeping in view the above facts, this project studies whether performing different cognitive interventions in everyday life helps in delaying dementia in a population aged 60 and over. This survey done on 96 older adults staying in various old age homes in Delhi, studies the effect of cognitive activities on their working memory. A formal permission was taken from authorities to conduct the workshops. All residents consented to attend the workshop. In all old age homes, there was a limited capacity. Generally there are two occupants in a room, very few have a private room. Volunteers visited the rooms of occupants with restricted movement. Only few who were above the age of ninety were excluded due to limited understanding. The aim of these workshops was to study the effect of cognitive intervention and to manage and halt the progressive decline into dementia without the side effects of medication.

19.1.2 Methodology

All the participants were of age **60–93** and a total of 96 participated in the programme. On the first day of the workshop, a questionnaire was filled containing questions regarding their age, gender, morbidities, and physical and mental health. Standard MMSE test (Mini-mental-state-exam) and Mini-Cog test were administered at the beginning and at the culmination of the workshop to access and evaluate any changes brought about by the programme (25 were assigned as a control group who attended the workshops quite irregularly).

Each participant in experimental groups received 21 ninety-minute sessions (five sessions per week). Structured intervention dealt with the cognitive functions of attention and orientation, memory, language, vasoconstrictive ability, executive functions and visual–manual coordination. The group that received structured intervention, solely scored the highest in nearly all cognitive tests. Computerized cognitive training was proposed as well as it boosted memory/attention but it was dropped due to the residents' advanced years/age.

Each session of the workshop was divided into two parts. The first consisted of stimulating games and activities to encourage the participation and regular presence of the participants at the workshops as well as to increase their ease with the investigators. This helped them to come out of their isolation and interact with their fellow residents. Sessions began with theatrical performances for the participants, followed by group activities like carom, cards, identifying movie actors through dumb charades, cup stacking, BINGO, etc. Participants were also encouraged to share their experiences, sing, etc.

The second part of the workshop focused on evaluating the effect of cognitive activities on the memory skills of the participants. It consisted of tests based on verbal episodic memory and inductive reasoning.

During the course of the programme, the worksheets were explained to each individual properly and solved under the guidance of the investigators. Tests included word search, crossword and maze, finding the hidden objects in pictures, identifying colours and symbols, basic arithmetic, wordlist recall, wordlist delayed recall, digits backwards, letter–number sequencing, etc. Pictorial questions such as spotting the difference, clock drawing, identification of objects such as famous personalities, symbols and alphabet series.

19.2 Outcome Measures

In this study, the MMSE and COG test were the primary outcome measures, in both the intervention and control groups, the tests were administered at the beginning of the workshop and 21 days after the intervention had taken place. Another MMSE was done after a gap of six months when no intervention was taking place. In addition, throughout the 21 days intervention, scores of each day are assessed and the summary score was analyzed.

19.3 Results

96 elderly people formed the study population of which 69 were females and 27 were males. 33 females were residing in Arya Mahila old age home; therefore, more females participated in the programme. 7 males and 21 females were more than 80 years of age (29.2%) and 7.26% had no formal education. There were 3 residents who were Ph. D. qualified and 48.95% were graduates. Out of the 96 screened for dementia using MMSE, 21 scored 24 and below, a prevalence rate of 21.87% which is far higher than the prevailing rate in the country (5–8%). This again may point out the importance of social and family environment.

Higher level of dementia was observed as age advances (Table 19.1). Dementia is more prevalent among females (Table 19.2). Higher prevalence of dementia was found in females (27.53%), aged above 80 years (26.3%) and those with more than 10 years of education (30.2%) (Table 19.3). The mean age among dementia patients was 76 years.

At the beginning of the workshop, all subjects took the MMSE test and we found that there no difference in the cognition level of both the groups (2-sample t-test, $t = 0.137$ with p-value > 0.01) But at the culmination of the workshop, those who attended the interventions showed remarkable improvement after training (Fig. 19.1).

On completing the training for 21 days, another MMSE test was conducted to see the impact of this training and we found that scores achieved by members of the

Table 19.1 Prevalence of dementia in subjects according to age

Age (Years)	Dementia		Total (n)	Prevalence (%)
	Absent	Present		
60–64	14	2	16	12.5
65–69	18	6	24	25.0
70–74	16	3	19	15.8
75–79	13	5	18	27.8
>80	14	5	19	26.3

Table 19.2 Gender-wise prevalence of dementia

Gender	Dementia		Total (n)	Prevalence (%)
	Absent	Present		
Male	25	2	27	7.4
Female	50	19	69	27.53
$X^2 = 4.601$ d.f. = 1 p-value = 0.032				

Table 19.3 Education- wise prevalence of dementia

Education	Dementia		Total (n)	Prevalence (%)
	Absent	Present		
<10	31	2	33	6.1
≥10	44	19	63	30.2
$X^2 = 7.359$ d.f. = 1 p-value = 0.007				

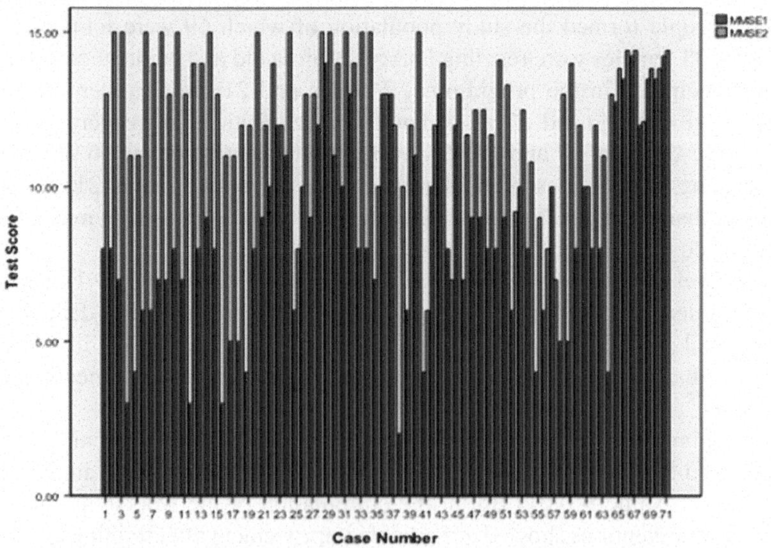

Fig. 19.1 Score difference before and after training

intervention group were better than those of the control group. The MMSE scores pre- and post-cognitive intervention are shown in Table 19.4 and the results of paired t-test before and after training MMSE scores in all the participants of the intervention group were statistically improved, whereas there was a decline in the MMSE score of the control group.

Effect of cognitive intervention is more visible in males than females (refer to Fig. 19.2). There is 73.8% change in MMSE score of males as compared to 40.8% change in the scores of females. Another glaring fact was that the education or active memory reserve in earlier years was not making any visible impact on cognitive

Table 19.4 Cognitive function score before and after workshop in both groups

MMSE	Intervention		Control		p-value (Intervention vs control)
	Mean	±SD	Mean	±SD	
pre/post	8.46/12.61	±3.15/±2.17	9.89/10.52	±2.55/±2.22	0.209/0.474
% change	49.05%		6.37%		
p-value (pre vs post)	0.00		0.137		

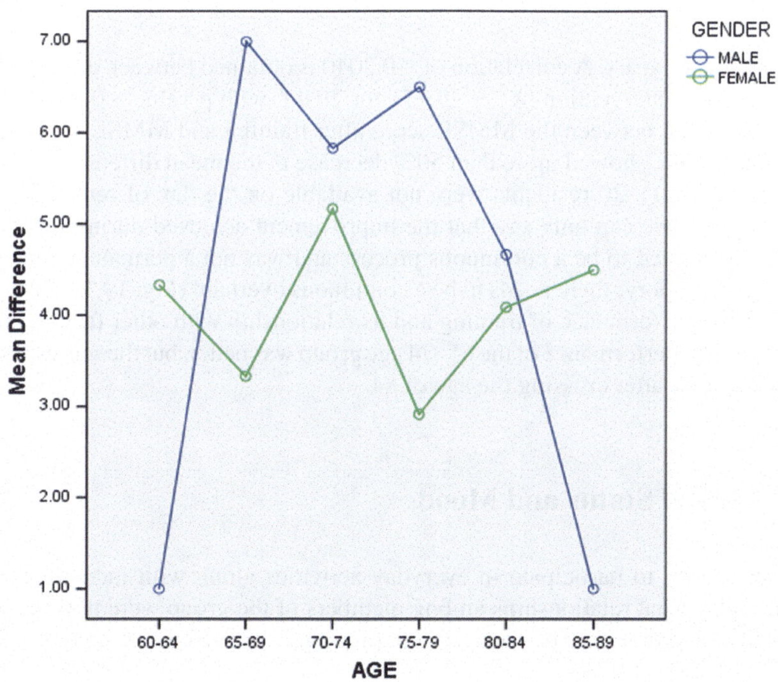

Fig. 19.2 Gender difference

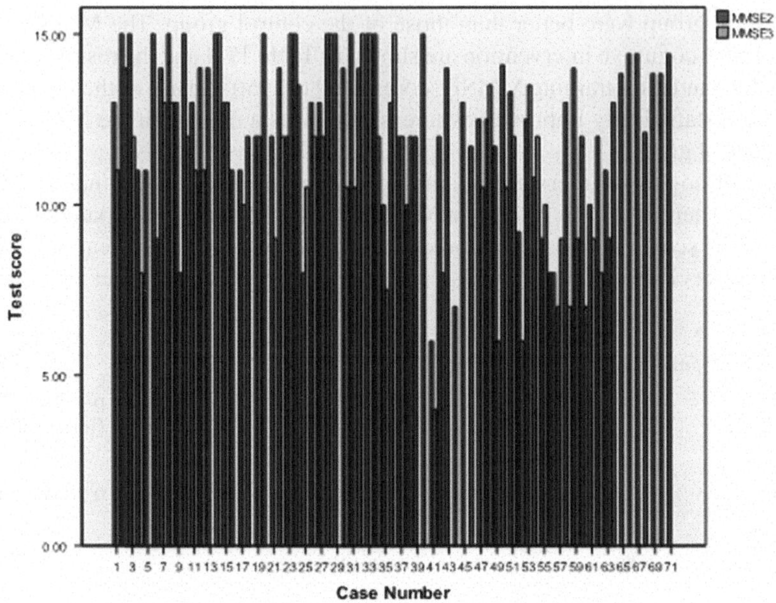

Fig. 19.3 Score difference after a gap of one month

functions at later years. A correlation of −0.2040 is obtained between education and test scores after intervention ($X^2 = 29.171$ for 70 d.f. with p-value > 0.05).

A comparison between the MMSE score after training and MMSE score after a gap of one month, showed more than 50% decrease in the mean difference in scores (1.6096 ± 2.3061). 20 residents were not available on the day of retest. Through p-value < 0.05, we can only say that the improvement achieved during the 21-day workshop appeared to be a continuous process and was not a permanent change in the working memory; there needs to be a continuous overhaul (Fig. 19.3). Table 19.5 illustrates the performance of training and its relationship with other factors in both the groups. The performance of the 65–74 age group was better, but the scores showed a steep downfall after crossing the age of 84.

19.4 Mental Status and Mood

Improved ability to participate in everyday activities along with increase in willingness and cordial relationships among members of the group were noticed. Their mood and motivation were boosted as well. Investigators also observed greater social and emotional adaptation.

Table 19.5 Performance of training and relationship to other potential factors

Variable	Level	Intervention Group		Control Group	
		Sample Size	Difference between MMSE1 and 2	Sample Size	Difference between MMSE1 and 2
Center	Sandhya	20	−7.20	6	−2.167
	Aradhna	9	−3.81	4	−1.25
	Guru nanak Sukh Shala	10	−4.10	7	−0.86
	Arya Mahila	25	−2.88	8	0.25
	Ashirvaad	7	−0.45	–	–
Age	60–64	6	−3.02	5	2.0
	65–69	9	−4.78	1	0
	70–74	19	−4.79	3	4.67
	75–79	18	−3.99	7	−1.0
	80–84	15	−4.0	4	1.5
	85–89	4	−2.59	4	0
	90+	–	–	1	−1.0
Sex	Male	21	−5.26	6	−2.33
	Female	50	−3.68	19	−0.42
Education	<10 years	23	−4.32	15	−0.73
	≥ 10 years	48	−4.06	10	−1.1
Total		71	−4.15 (p-value = $0.00 < 0.05$)	25	−0.88 (p-value = $0.137 > 0.05$)

19.4.1 Case Examples

Case 3046: An 89-year-old woman refused to participate at the commencement of the workshop, but came around after persuasion and joined the workshop. She started attending the workshops on a regular basis and began socializing along with sharing various episodes of her life, such as despite being barred from receiving an education she took matters into her own hands and learnt English. Though, she did not solve all the worksheets seriously and so her performance kept on fluctuating.

Case 1015: A 76-year-old man with a MSME score 8 initially refused to participate in the workshop due to inhibitions. Thereafter, seeing other inmates participate actively in the workshop, he joined as well. He began verbalizing his opinions with greater ease and frequency and also participated in the other entertainment-based activities during the workshop. His worksheet score also showed improvement.

Case 4063: An 83-year-old woman with a MSME score 4 suffering from an eye ailment had to be motivated to participate in the workshop. She enjoyed verbalizing her life experiences and also shared the problems she faced in her day-to-day life. Her performance improved significantly and her MSME score reflected the same. Her post-intervention MSME score was 13. She was spiritually inclined and sang in the same line.

Case 2034: A 70-year-old woman refused to attend the workshops at commencement. Observing other inmates enjoy the workshop she joined in, thereafter, she diligently completed all the worksheets. There was a remarkable improvement in her MSME score, from 8 at the beginning and to 15 at the end of the workshop. There was a considerable increase in her social interactions and willingness to share her life experiences and opinions.

19.5 Discussion

Dementia is one of the global health epidemics effecting disability and dependency among the ageing population. It is a disease which gradually progresses with deteriorating cognitive functions noted more than from the age factor in the ageing population. It affects the thinking capacity of the individual. Gradually it also affects individual's sense of orientation and comprehension. There is deterioration in the emotional control and social behaviour due to the impairment in the cognitive function. The cause of dementia is damage to or loss of nerve cells and their connections in the brain. The symptoms experienced by a person may be unique and depend on the location and extent of neural damage in his own brain. Therefore, clinical manifestations of dementia may widely vary in a population.

Carrion et al. established the importance of cognitive intervention for treating geriatric persons experiencing dementia by demonstrating the positive impact of skills training trials and mixed trials on decline of mental faculties [5]. Konda et al. reported a low cognitive impairment prevalence in the urban elderly population in our society as compared to many published reports from Western countries. The prevalence among women was comparable with the Western rates. Also higher obesity rates were found to be associated with cognitive impairment among the women. The higher mortality rates along with the geriatric problems may therefore contribute to increase in cognitive impairment in the future [6]. Similarly, Kyunghwa Jo et al. too have advocated the benefits of cognitive intervention in senior citizens with or without dementia, at least in the short term and reported a comparatively greater improvement in mental faculties in the dementia group [7].

In the present study, the research was designed with the objective to investigate the impact of cognitive and social intervention in the population aged 60 and above. The hypothesis of the study was to test the cognitive and social intervention among the ageing population; the symptoms of dementia may be detected in the early stages of the disease; and its progression may be reduced with the correct care and timely treatment. The population residing in the old age homes in different parts of Delhi were selected as the target respondents for the study. The study showed statistically significant improvement in MMSE score before and after 21 days of training. There was also a general upliftment of mood among the residents. This result has proven to be consistent with the prior studies conducted in other countries [8, 9].

Mental exercises, especially memory training, stimulating leisure and social activities are considered vital to a healthy brain. In order to sustain the brain, active exercises through building reserves of healthy brain cells and connections between them are important. Simple exercises like reading aloud and attempting elementary arithmetic problems involve multiple cognitive processes such as recognition of visually presented words or numbers, inferring appropriate meanings of words, and hand–eye coordination, to name a few. Performing these tasks leads towards the strengthening of executive functions, episodic memory and attention. Therefore, these cognitive interventions helped in improving the working memory. Similarly, positive and frequent social interactions benefitted the residents, as they looked forward to these interactions every day and shared their experiences and stories gladly. It was also found that in the smaller groups, the interactions became highly personalized.

Future Plans: We intend to expand the current research by taking a much wider sample and eliminating all the shortcomings we faced as much as possible. Along with that, we wish to compare the cognitive abilities of elderly individuals living with their friends and family, in a more active and wholesome life with those living at old age homes. An analysis and inquiry of the contributing differences might lead to slowing down of dementia in the overall geriatric population of India. Along with that, we will take the research further by evaluating and conducting workshops in rural India as well, and not just urban cities like Delhi. This would throw further light on the factors contributing to or preventing the onset of dementia.

19.6 Conclusion

The study demonstrated that cognitive intervention with individual attention was effective, not only in improving working memory but also in improving socialization. With positive encouragement and considering individual capability, the members of the intervention group when measured by standardized test showed improvement in their scores, indicating a better mental and temporal constitution. The group exercises also encouraged involvement and helped uplift their overall state of mind. As of now, an older individual is not provided adequate health care against decline into dementia unless they are identified as a sufferer of the disease, and have an almost zero chance of regaining their mental health. Therefore, this study hopes to draw attention to dementia and to create awareness about it, thus enabling us to take actions towards prevention and delay of the disease by intervention and management.

However, this study has a few limitations. The first, cognitive intervention could take place for small number of days. The workshops conducted in old age homes leave many things uncontrolled. Second, the design of a control group was not planned initially. A carefully designed randomized controlled study with a longer duration is suggested to obtain a more definitive conclusion.

Acknowledgments The authors wish to thank Dr. Prerna Diwan for her invaluable help. The authors greatly acknowledge University of Delhi, for financial assistance provided under Innovation Project Scheme RLA-303.

References

1. Bloom DE, Luca DL (2016) The global demography of aging: facts, explanations, future. Handb Econ Popul Aging 1:3–56
2. World Health Organization: Global Action Plan on the Public Health Response to Dementia 2017–2025. Geneva, World Health Organization, 2017. Dement Geriatr Cogn Disord 46:1–26. https://doi.org/10.1159/000490851
3. Government of India. Census of India 2011. Population composition, chapter 2. 2011
4. Sumana M, Sreelatha CY, Sreeranga A, Arpitha B, Akshatha SP, Anand HD (2016) Prevalence of dementia and other psychiatric morbidities among geriatric population of Salagame primary health centre in Hassan district, Karnataka, India. Int J Community Med Public Health 3(5):1315–1317. http://www.ijcmph.com, pISSN 2394-6032 | eISSN 2394-6040 Research Article
5. Carrion C, Folkvord F, Anastasiadou D, Aymerich M (2018) Cognitive therapy for dementia patients: a systematic review. Dement Geriatr Cogn Disord 46(1–2):1–26. https://doi.org/10.1159/000490851
6. Konda P, Sharma PK, Gandhi AR, Ganguly E (2018) Correlates of cognitive impairment among Indian Urban Elders. J Gerontol Geriatr Res 7(6): 489. https://doi.org/10.4172/2167-7182.1000489
7. Jo K, Jhoo JH, Mun Y-J, Kim YM, Kim SK, Kim S, Lee S-H, Jang J-W (2018) The effect of cognitive intervention on cognitive improvement in patients with dementia. Dement Neurocogn Disord 17(1):23–31
8. Alloway TP, Alloway RG (2008) The efficacy of working memory training in improving crystallized intelligence Jungle memory training program. Memosyne Ltd, UK
9. George D, Rebok W, Ball K, Guey LT et al (2014) Ten-year effects of the active cognitive training trial on cognition and everyday functioning in older adults. J Am Geriatr Soc 62(1):16–24. https://doi.org/10.1111/jgs.12607

Chapter 20
Predicting the Fix Time of a Reported Bug using Radoop: A Big Data Approach

Madhu Kumari, Meera Sharma, Sameer Anand, and V. B. Singh

20.1 Introduction

When it comes to big data, it is clear that the data is produced at an astronomical rate. In fact, 90% of the world's data were created in the last two years. The term "Big Data" can be defined as data that becomes too large to be processed using traditional methods. The volume of data that can be considered as large data is constantly changing, and new tools are continually being developed to deal with this massive data. It completely changes our world and shows no signs of transit that will disappear at any time in the near future. To understand this vast amount of data, it is often defined by five V's: Velocity, Volume, Value, Variety, and Veracity.

Velocity is the speed at which large amount of data is generated, collected, and analyzed. Every day, emails, twitter messages, photos, and videos are increasing around the world at lightning speed. Data is increased every second of a day. Big data technology now allows us to analyze data as it is generated without having to put it into the database. A large number of bug reports are reported on different bug tracking systems at a higher speed. From different geographical locations, researchers collected and analyzed this big data of bug reports.

M. Kumari · V. B. Singh (✉)
Delhi College of Arts & Commerce, University of Delhi, Delhi, India
e-mail: vbsingh@dcac.du.ac.in

M. Kumari
e-mail: mesra.madhu@gmail.com

M. Sharma
Swami Shraddhanand College, University of Delhi, Delhi, India
e-mail: meerakaushik@gmail.com

S. Anand
SS College of Business Studies, University of Delhi, Delhi, India
e-mail: sananddu@gmail.com

P. K. Kapur et al. (eds.), *Decision Analytics Applications in Industry*, Asset Analytics,
https://doi.org/10.1007/978-981-15-3643-4_20

Volume refers to the large amount of data that is generated each second from social media, mobile phones, cars, credit cards, photos, videos, etc. In fact, such data have become so big that we cannot store and analyze it using traditional database technologies. Facebook has 10 billion messages, clicked the "Like" button for 4.5 billion times, and uploaded more than 350 million new images every day. It is clear that the collection and analysis of such data is a huge engineering challenge. A large number of bug reports are reported on different bug tracking systems every day.

Value refers to the worth of the data being extracted. Having endless amounts of data is one thing, but it is useless unless it can be converted into value. Although there is a clear correlation between data and ideas, this does not always mean having value in big data. The most important part of the development of the big data plan is to understand the costs and benefits of data collection and analysis to ensure that the resulting data can be monetized. The bug reports data has the value of being extracted and analyzed for different bug attributes prediction [9–14] and bug fix time prediction [1–4, 7, 8].

Variety is defined as the different types of data we can use. Today's data is very different from previous data. We no longer only have structured data (names, phone numbers, addresses, financial statements, etc.), which is great for spreadsheets. Today's data are not organized. In fact, 80% of the world's data fall into this category, including images, video sequences, social-network updates, and more. Big data technology now allows the use of structured and unstructured data that is collected, stored, and used simultaneously. Bug reports on bug tracking systems consist of a collection of attributes. Some textual attributes such as patch, summary, and long description are unstructured attributes. Some attributes such as bug-id, platform, cc-list, assignee, operating system, hardware, component, reporter, resolution, product, status, severity, priority [23] are structured attributes. It means that bug reports deal with both types of unstructured and structured attributes.

Veracity denotes uncertainty and distortion in data. The data which is mined should be meaningful for the problem being analyzed. Veracity is the biggest challenge when comparing with volume and velocity. Various bug attributes are filed by the reporter of bugs during bug reporting on bug repository. Based on these attributes different prediction models have been proposed to improve the software quality. During bug reporting high irregular pattern has been observed. Bug repository size increases at a high rate with irregularities and uncertainty.

For bug-related analysis like software quality measurement [16] and development effort coordination in bug triaging [17], bug fix time plays an important role. It assists in software quality improvement. It also assists in resource allocation and release time management. "If bugs in a file take a relatively long time to be fixed, the file may have some structural problems that make it difficult to make changes" [16].

In this paper, we have applied an extension called Radoop to enable the integration of RapidMiner with Hadoop to handle big bug report datasets. We have used Hive-based Naïve Bayes (NB) and Spark-based Decision Tree (DT) machine learning techniques in RapidMiner open source software [6] for bug fix time prediction of a reported bug. The experimental analysis is validated on 1,23,849 bug reports of Eclipse and 67,178 bug reports of Mozilla projects. The results show that Decision

Tree performs better than Naïve Bayes for both the projects in terms of accuracy. The proposed approach shows improved accuracy in comparison with the work mentioned in [20] for fix time prediction.

The rest of the paper has been divided into four sections: Sect. 20.2 discusses description of datasets, bug attributes, and model building. Section 20.3 describes the results and discusses the results. Section 20.4 presents related work. Conclusion and future research directions have been given in Sect. 20.5.

20.2 Data Collection and Model Building

In this section, we have described the data collection and model building.

20.2.1 Data Collection

We have considered seven independent bug attributes, namely, product, component, number of comments, operating system, priority, severity, and hardware. Product, component, operating system, priority, severity, and hardware are nominal attributes, whereas the number of comments is a continuous attribute. The bug attributes have been described in Table 20.1 [15].

We have calculated the time to fix a bug by subtracting bug opened date from last resolved date.

Bug fix time = Last_resolved date − Bug_open date.

"The bugs which are reported by different users are assigned to different developers. The bugs take a reasonable amount of time in fixing" [22]. These bugs are having an open status till they get fixed.

To validate our proposed approach, we have considered datasets of Eclipse [21] and Mozilla [5] open source projects. "We have considered bug reports of resolution 'fixed', 'works for me' and status 'verified', 'resolved' and 'closed'. Only these bug reports contain meaningful and static information" [22]. In this paper, we have taken different independent attributes such as product, component, number of comments, operating system, priority, severity, and hardware to predict bug fix time.

"We observed a large variation in bug fix time, which can affect the results. We have drawn a distribution graph shown in Fig. 20.1 for number of bugs versus fix time and found that the maximum number of bugs is having the fix time of 0-99 days across both the datasets" [22].

We classified the bug fix time in three ranges, i.e., 0–32 days (Bug fix time1), 33–65 days (Bug fix time2), and 66–99 days (Bug fix time3) for our study.

Table 20.2 shows the bug reports of Eclipse and Mozilla projects for the observed time periods.

Table 20.1 Description of different bug attributes

Bug attribute	Short description
Bug-Id	This represents a unique numeric id for the bug
Resolution	This describes what happened to this bug, e.g., fixed, Worksforme, etc.
Status	It determines the current state (New, Verified, Resolved, etc.) of bug
Severity	Severity of a bug gives its impact on the software or its components. It is divided into seven levels: Blocker (1) to Enhancement (7)
Priority	Bug priority determines the importance of a bug in the presence of others. It ranges from P1 (most important) to P5 (least important)
Number of comments	Number of different comments on a bug that have been given by different users
CC count	When a change is done in a bug during fixing, e-mail is sent to different concerned people. The number of such people indicates the value of the CC Count field
Summary	A brief text about the bug
Component	It refers to the subdivision of the product in which bug lies
OS	It refers to different operating systems in relation to which bug was filed
Product	Bugs are categorized into Products and Components, where a Product has one or more Components in it
Hardware	The computing environment in which the bug has been detected

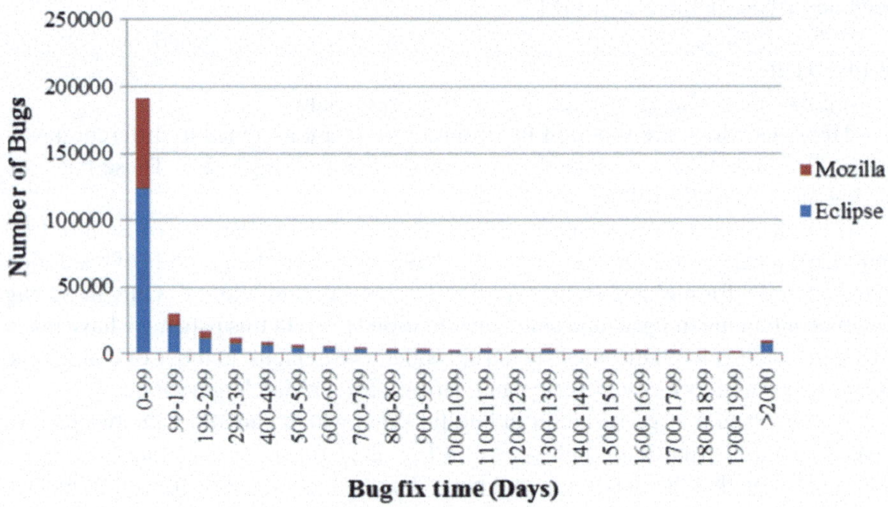

Fig. 20.1 Distribution of bugs in different fix time range

Table 20.2 Number of bug reports of Eclipse and Mozilla projects

Projects	Number of bugs	Observation period
Eclipse	1,23,849	Oct. 2001–May 2015
Mozilla	67,178	July 1998–Nov. 2017

20.2.2 Model Building

RapidMiner Radoop is a user-friendly graphical interface and client software in the Hadoop cluster for handling and analyzing large amounts of data. It can be installed on a RapidMiner Studio and/or RapidMiner server, and provides a platform for storing data and running computations in a Hadoop cluster. RapidMiner Radoop runs on any Java-enabled platform. It is a code-free environment for advanced analytics processes in which the computations come down to Hadoop cluster. It works directly in Hadoop so that the value of data is unlocked in a variety of machine learning applications [6] (Fig. 20.2).

In this paper, we have applied an extension called Radoop: RapidMiner with Hadoop to handle big bug report datasets. This extension provides an additional operator for RapidMiner and communicates with the Hadoop cluster to run the job. The Radoop process starts by adding the Radoop Nest meta-operator. It contains general group settings (such as Hadoop's master node IP address), and all other Radoop operators can only be used within this meta-operators. We have used some of the data analysis features of Hive and Mahout because they are highly optimized [18]. We have used Hive-based Naïve Bayes and Spark-based Decision Tree machine learning techniques to predict bug fix time.

We have downloaded the Cloudera Quickstart VM (version 5.13) from the Cloudera website [19] and connected it with RapidMiner Radoop. Figure 20.3 shows the connection of Cloudera Hadoop with RapidMiner Radoop.

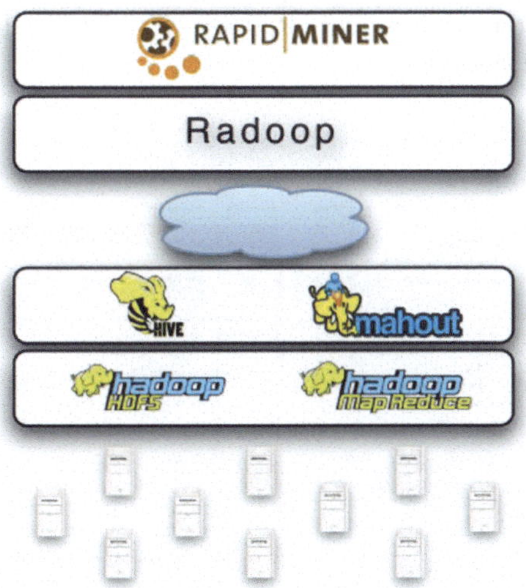

Fig. 20.2 An architecture of the RapidMiner with Hadoop integration (Radoop) [18]

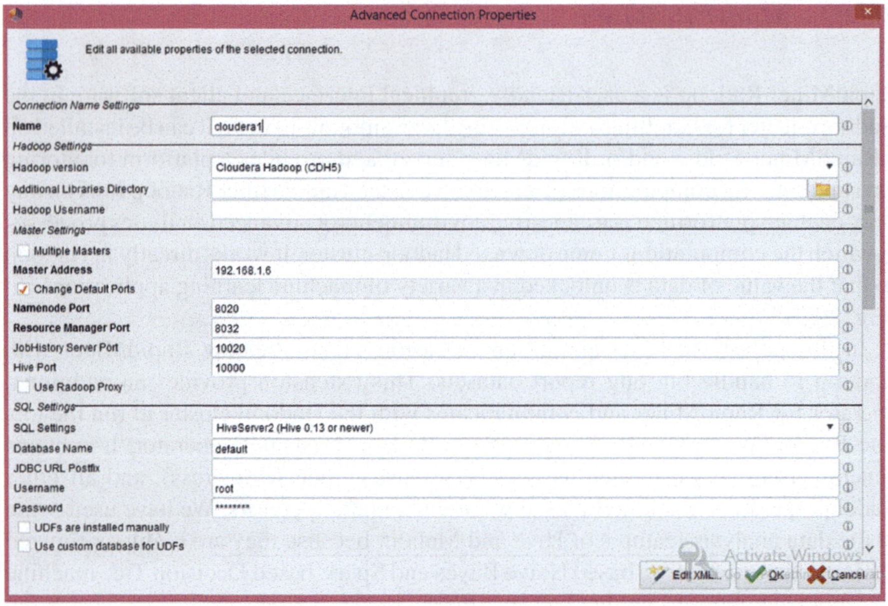

Fig. 20.3 Connection of RapidMiner Radoop with Cloudera Hadoop

We have extracted the data of Eclipse project and saved it into CVS format. After that we uploaded that data into Hive database as shown in Fig. 20.4.

Figure 20.5 shows the main process Radoop Nest. This is the main operator for running processes on Hadoop.

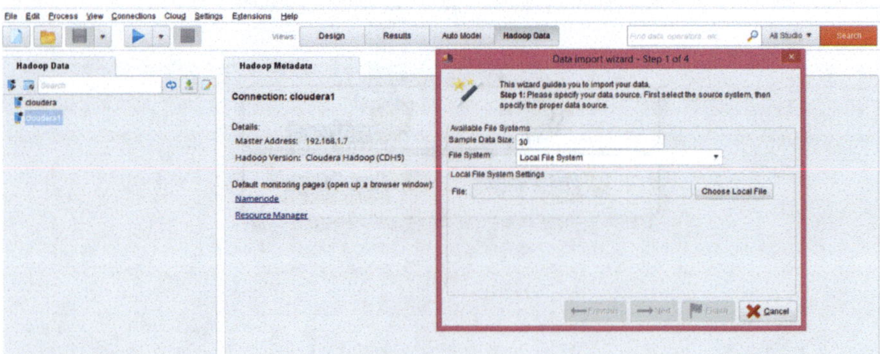

Fig. 20.4 Data upload into Hive database

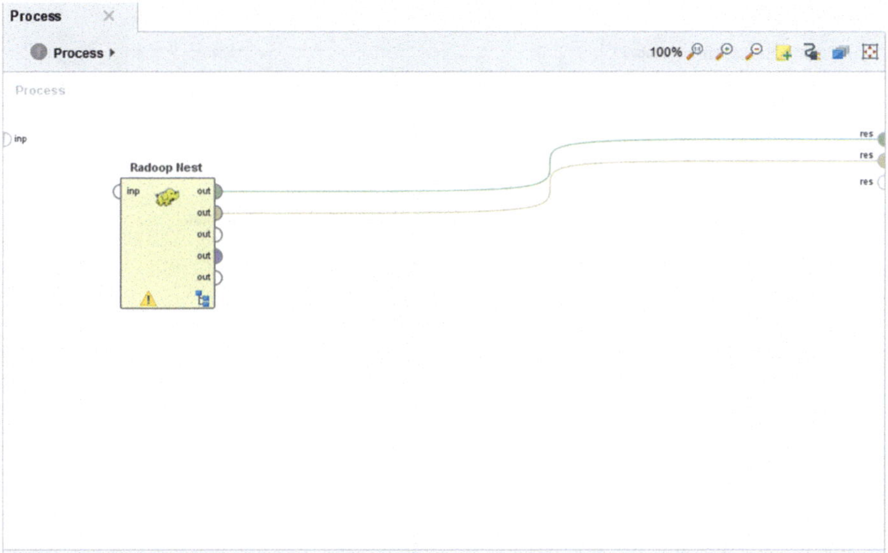

Fig. 20.5 Radoop nest process in Radoop

Figure 20.6 shows the subprocess within Radoop Nest, which consists of Retrieve, Set Role, and Validation operators. Hive-based Retrieve operator retrieves a Hive table for analysis. Set Role operator can be used to change the attribute role. The

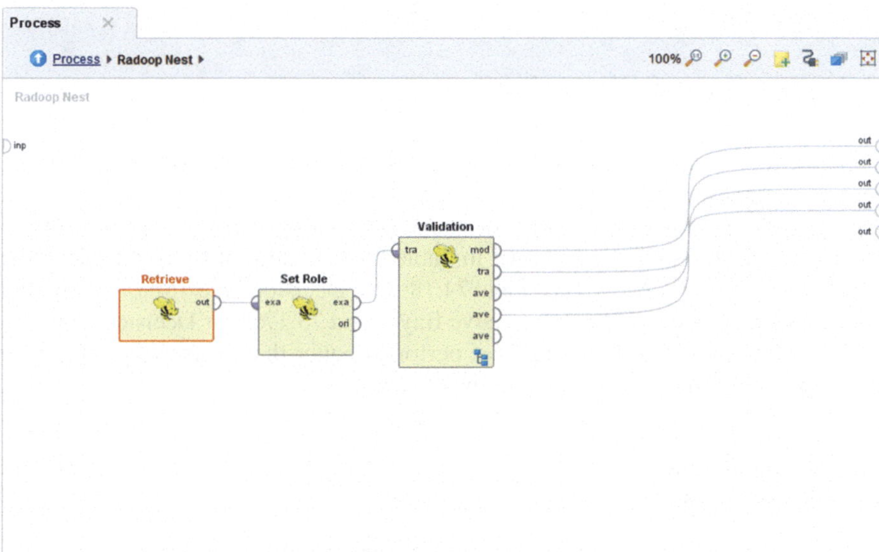

Fig. 20.6 Subprocess of Radoop nest in Radoop

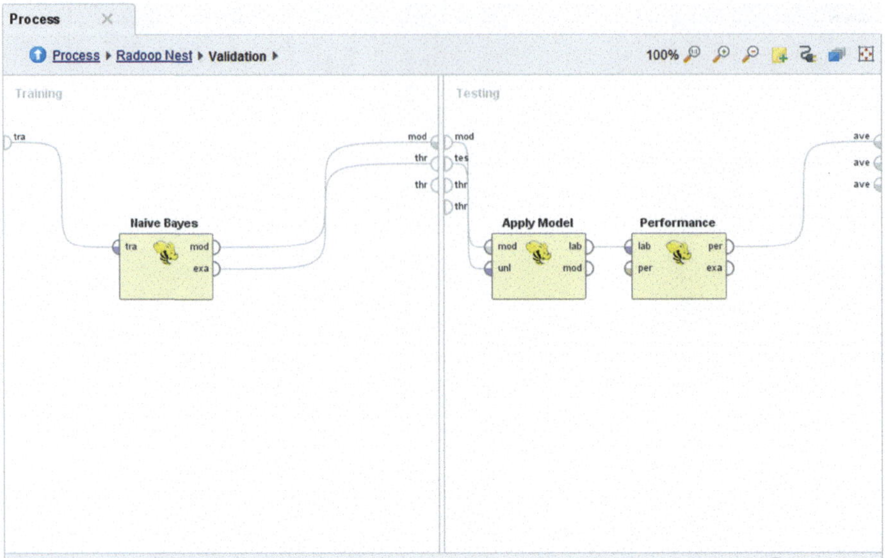

Fig. 20.7 Subprocess of validation operator in Radoop

example set has been randomly split into training and testing sets by using Validation operator.

Figure 20.7 shows the subprocess within Validation operator which builds the Naïve Bayes classification model.

20.3 Results and Discussion

In this section, we have presented and discussed the results in terms of performance measure accuracy.

Bug fix time prediction accuracy of Naïve Bayes and Decision Tree for Eclipse and Mozilla projects has been shown in Table 20.3. In case of Eclipse project, the accuracy is 72.57% for Naïve Bayes and 74.18% for Decision Tree. In case of Mozilla project, the accuracy is 71.95% for Naïve Bayes and 73.17% for Decision Tree.

The results show that Decision Tree performs better than Naïve Bayes for both the Eclipse and Mozilla projects in terms of accuracy.

Table 20.3 Accuracy for Bug fix time prediction

Projects	Accuracy (%)	
	NB	DT
Eclipse	72.57	74.18
Mozilla	71.95	73.17

We have also compared our proposed approach with the state-of-the-art approach as mentioned in Panjer [20]. The author has used 0-R, 1-R, Naive Bayesian Networks (NB), C4.5 Decision Trees (C4.5 DT), and Logistic Regression (LR) to perform data mining and analysis of the constructed datasets for predicting bug lifetimes. The author achieved prediction accuracy 29.10%, 31.00%, 31.90%, 32.50%, and 34.90% for 0-R, 1-R, NB, C4.5 DT, and LR algorithm, respectively. Our proposed approach achieved bug fix time prediction accuracy 72.57% for Naïve Bayes and 74.18% for Decision Tree. The comparisons of the proposed approach with the Panjer [20] have been shown in Figs. 20.8 and 20.9. We observed that the proposed approach improves the accuracy significantly.

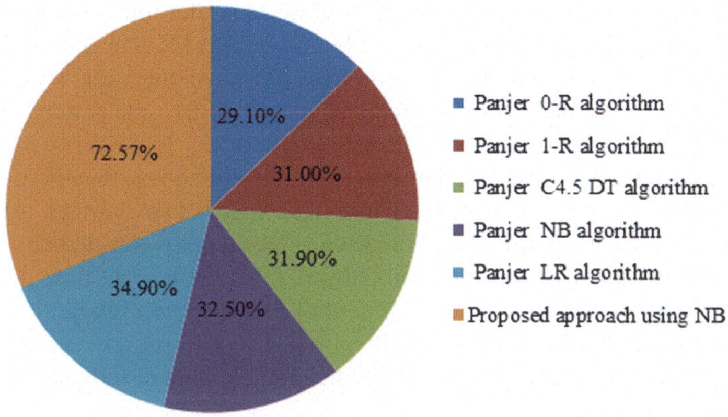

Fig. 20.8 Eclipse project Radoop (NB) accuracy comparison with Panjer [20]

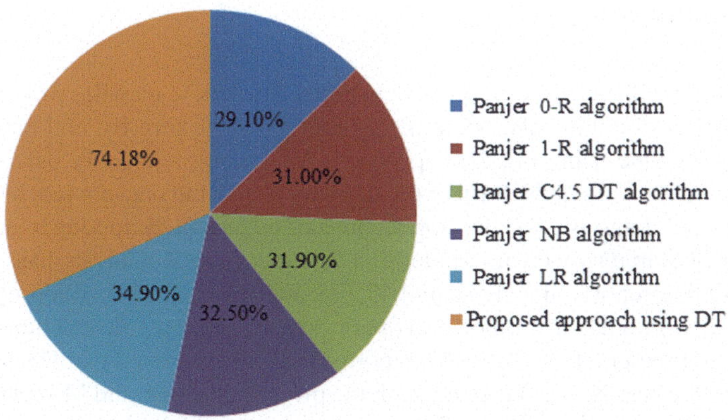

Fig. 20.9 Eclipse project Radoop (DT) accuracy comparison with Panjer [20]

20.4 Related Work

Researchers have contributed a significant contribution in the development of bug fix time prediction models. An attempt has been done for 72,482 bug reports of Linux software [1]. The authors observed that the people, who have participated in the range from 1 to 8 users, corrected 95% bug reports. The study shows that during bug fixing 92% developers have a linear relationship with bug fix time. The proposed model improved the result in terms of R^2. In [2], the authors test the prediction performance of previously used models by using regression models. The previously used models' predictive performance lies in the range of 30–49%. No correlation was found between bug-fix likelihood, the reputation of the developer who opens the bug, and bug fix time. A model has been proposed to determine how much time a bug will take to get fixed by using different bug attributes [3]. The authors observed that the performance measure accuracy improved if developers and comments are included. In [4], the authors study the tendencies of bug fix time pattern in Mozilla and Apache datasets [4]. Result shows that bugs of priority levels 5 and 4 take more than 100 days to get fixed. Bugs of the priority level 2 take less than 80 days to get fixed and bugs of the priority level 1 or 3 are fixed in less than 30 days. An attempt has been made to focus on the delays incurred by developers during bug fixing [7]. In [8], the authors identify and filter the outliers from the bug fix time distribution [8]. Filtering outliers resulted in improvement of prediction accuracy.

To handle the increasing volume of software bug repositories, we need to use big data approach. We have used Radoop to predict bug fix time of Eclipse and Mozilla projects bug reports.

20.5 Conclusion

In this paper, we have taken seven independent attributes, namely, product, component, number of comments, operating system, priority, severity, and hardware to predict bug fix time using big data approach. As the bug repository data keeps on increasing and growing in the form of big data, we need a big data approach to handle big bug reports' datasets. We have applied an extension called Radoop to enable the integration of RapidMiner with Hadoop. The experimental analysis has validated on 1,23,849 bug reports of Eclipse and 67,178 bug reports of Mozilla projects. We have used Hive-based Naïve Bayes and Spark-based Decision Tree machine learning techniques to predict bug fix time of a reported bug. We observed that Decision Tree performs better than Naïve Bayes for both the projects, Eclipse and Mozilla in terms of accuracy. We have also compared our proposed approach with the state-of-the-art work proposed by Panjer [20] for bug fix time prediction. The proposed approach shows improved accuracy. In future, the study can be extended on more open source and closed source projects.

References

1. Anbalagan P, Vouk M (2009) On predicting the time taken to correct bug reports in open source projects. International conference on software management (Edmonton, AB). IEEE, pp 523–526
2. Bhattacharya P, Neamtiu I (2012) Bug-fix time prediction models: can we do better?. In: 8th working conf. mining software repositories. ACM, New York, NY, USA, pp 207–210
3. Giger E, Pinzger M, Gall H (2010) Predicting the fix time of bugs. In: International workshop recommendation systems on software engineering. ACM, New York, NY, USA, pp 52–56
4. Mockus A, Fielding RT, Herbsleb JD (2002) Two case studies of open source software development: apache and mozilla. ACM Trans Softw Eng 11:3
5. http://bugzilla.mozilla.org
6. https://rapidminer.com
7. Zhang F, Khomh F, Zou Y, Hassan AE (2012) An empirical study on factors impacting bug fixing time. In: 19th working conference on reverse engineering (WCRE), pp 225–234
8. AbdelMoez W, Kholief M, Elsalmy FM (2013) Improving bug fixtime prediction model by filtering out outliers. In: International conference on technological advances in electrical, electronics and computer engineering (TAEECE), pp 359–364
9. Tian Y, Lo D, Sun C (2013) DRONE: predicting priority of reported bugs by multi-factor analysis. In: IEEE International conference on software maintenance, pp 200–209
10. Kanwal J, Maqbool O (2012) Bug prioritization to facilitate bug report triage. J Comput Sci Technol 27(2):397–412
11. Menzies T, Marcus A (2008) Automated severity assessment of software defect reports. In: IEEE international conference on software maintenance, ICSM 2008. IEEE, pp 346–355
12. Lamkanfi A, Demeyer S, Giger E, Goethals B (2010) Predicting the severity of a reported bug. In: 2010 7th IEEE working conference on mining software repositories (MSR), pp 1–10
13. Lamkanfi A, Demeyer S, Soetens QD, Verdonck T (2011) Comparing mining algorithms for predicting the severity of a reported bug. In: 2011 15th European conference on software maintenance and reengineering (CSMR), pp 249–258
14. Chaturvedi KK, Singh VB (2012) Determining bug severity using machine learning techniques. In: 2012 CSI sixth international conference on software engineering (CONSEG), pp 1–6
15. Sharma M, Kumari M, Singh VB (2013) Understanding the meaning of bug attributes and prediction models. In: 5th IBM collaborative academia research exchange workshop, I-CARE, Article No. 15. ACM
16. Kim S, Whitehead JE (2006) How long did it take to fix bugs?. In: International workshop mining software repositories. ACM, New York, NY, USA, pp 173–174
17. Hooimeijer P, Weimer W (2007) Modeling bug report quality. ASE
18. Prekopcsak Z, Makrai G, Henk T, Gaspar-Papanek C (2011) Radoop: analyzing big data with rapidminer and hadoop. In: Proceedings of the 2nd RapidMiner community meeting and conference, pp 1–12
19. https://www.cloudera.com/downloads/quickstart_vms/5-13.html
20. Panjer LD (2007) Predicting eclipse bug lifetimes. In: Proceedings of the fourth international workshop on mining software repositories, pp 1–4
21. http://bugs.eclipse.org/bugs/
22. Sharma M, Kumari M, Singh VB (2019) Multi-attribute dependent bug severity and fix time prediction modeling. Int J Syst Assur Eng Manag 10(5):1328–1352
23. Kumari M, Singh VB (2018) An improved classifier based on entropy and deep learning for bug priority prediction. In: International conference on intelligent systems design and applications. Springer, Cham, pp 571–580

Chapter 21
Measurement and Governance of Health of Information Technology Projects Through Use of Discriminant Analysis Technique

Vinayak Oak and Kavita Laghate

21.1 Introduction

In their paper [1] authors have discussed financial impact on IT industry due to failure of the projects. If the projects are not controlled properly and proper mitigation are not placed in, then organizations faces financial losses. These losses can go into multimillion dollars is what authors have shown. Many times the investment in an IT project can be small, but if that project fails then it can lead to very high losses to organizations [2].

There are many factors that contribute to success or failure of the project. This includes "Soft Skills" of project manager [3] [4], how scope gets impacted due to evolvement of requirements over time [5] to failure of projects in spite of use of appropriate governance of the projects [6]. In their paper, [1] authors, through literature survey, studied the challenges in governance of Information technology projects and it was found that there are six major aspects which are responsible for failure of the IT projects. Also, authors checked if Discriminant analysis can be used to statistically separate projects from one another in terms of failure. Discriminant analysis has been used in various fields like construction projects [7], Psychology [8], Finance [9] for the classification of given object into various groups.

With the above background, the current study has been carried out based on the pilot survey. Project data for 100 projects have been collected from 50 practicing project managers during this pilot study. An attempt has been made to establish a discriminant analysis model. SPSS tool was used to complete the statistical analysis.

Why this study is critical: Six important aspects that impact health of the project was established based on the earlier study of research papers and subsequent research

V. Oak (✉) · K. Laghate
Jamnalal Bajaj Institute of Management Studies, Mumbai University, Mumbai, India
e-mail: oakvinayak@yahoo.com

K. Laghate
e-mail: kavitalaghate@jbims.edu

P. K. Kapur et al. (eds.), *Decision Analytics Applications in Industry*, Asset Analytics,
https://doi.org/10.1007/978-981-15-3643-4_21

papers published by authors. During the further research and use of discriminant analysis a model has been established. This model is useful and project managers can use this model to check the possibility of project success at any stage of the project. Governance of health of the project can be achieved through the use of this model. Based on the output of the model appropriate actions can be planned to bring back the given project on track. Since this is a critical action which will lead to saving the project from failure and hence this research work is very crucial for IT industry.

The high-level mission statement: With issues in execution of IT projects, financial burden comes on organization and extent of this burden is very high. The aim of this research is, using discriminant analysis, establish a model and test the hypothesis that the aspects identified can predict success or failure of the project. Project Managers can check the health of the project and take corrective actions by making use of this model.

21.2 Research Methodology

Figure 21.1 shows, the overall design framework based on the research methodology framework [10]. The secondary data has been collected through the questionnaire and the corresponding survey. Using these variables, a relation between these variables was established. The questionnaire is used to measure scores for each independent variable and used as data input in the SPSS tool.

Various parameters which were considered to determine acceptance or rejection of the null hypothesis were: (a) Wilk's Lambda (b) Eigenvalues (c) the relative importance of independent variables (Standardize Canonical Discriminant Function Coefficient) and (d) the group centroids. These parameters and tests are based on

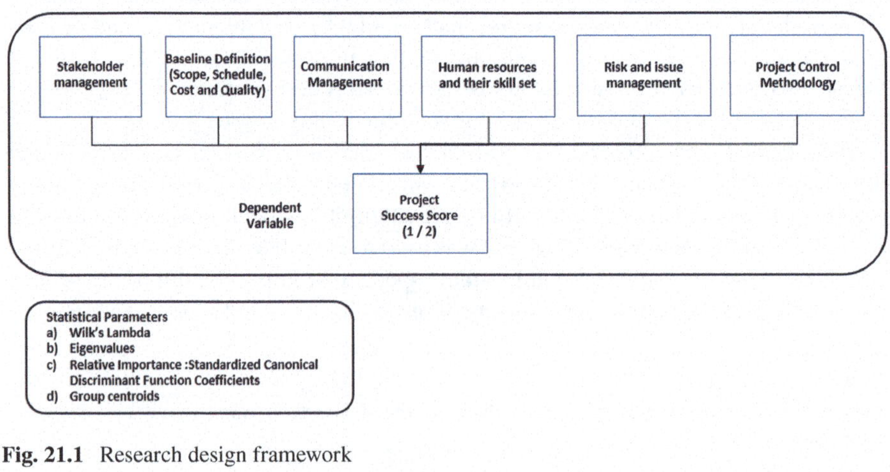

Fig. 21.1 Research design framework

discussions in various papers on discriminant analysis [11] and discussions in books on statistics [12].

As shown in Tables 21.1 and 21.2, below are the dimensions of the independent variables identified as established by authors [1]. These dimensions were used for questionnaire development. It was ensured that the questions will cover these dimensions.

Table 21.1 Dimensions Set 1

Stakeholder management	Baseline definition	Communication management
Types of project stakeholders	Clarity about scope and deliverables	Communication with stakeholders
Communication with stakeholders	Accuracy of the estimation	Communication system to connect to virtual teams
Stakeholder management	Scope creep	Unavailability of documentation
Perception of stakeholders about the final objective of the project	Definition of scope, schedule, cost boundaries	Effective communication/lack of communication
Stakeholder mapping	Scope boundaries and concent of stakeholders—contract clarity about final deliverables	

Table 21.2 Dimensions Set 2

People and their skills	Management of issues/risks	Methods for project governance
Resource management	Management of risks and mitigation plans	Project governance methodology
Adequate resource allocation/resource shortage	Risk allocation	Software and tools for governance of projects
Contractor's inadequate experience		Missing skills in methods for project control
Skills of people and the availability of people		Earned value analysis
		Project evaluation or maturity models

21.2.1 The Hypothesis Formation

Based on the variables and their types (viz. Dependent, Independent) hypothesis was defined. The corresponding variables were: Management of stakeholders, Definition of Scope/Schedule/Cost/Quality boundaries, Communication Management, Skills of people working, Issue/Risk management, Methods for project governance. The Authors [1] established that projects can get into troubled or can be completed successfully due to these six aspects.

Hypothesis statements:
Null Hypothesis H_0: Management of stakeholders, Definition of Scope/Schedule/Cost/Quality boundaries, Communication Management, Skills of people working, Issue/Risk management and Methods for project governance have no impact on the project success, i.e., discriminating power in these variables is not significant.

Alternate Hypothesis H_1: These six aspects have an impact on the project failure, i.e., significant discriminating power exists.

21.2.2 Data Collection Method

Questionnaire feedback from project managers from IT industry was the key input. The feedback from Project Managers was collected through the structured questionnaire. This questionnaire was shared with the participating project managers through an online URL using Google forms. For each independent variable following were the number of questions asked (as shown in Table 21.3). The questionnaire was designed in such a manner that it will determine whether the project will be successful or will be in troubled status. Following were the number of questions asked for each independent variable.

Table 21.3 Variables: independent

Independent variable	No of questions
Methods to govern projects	6
Managing issues/risks	7
People and their skills	6
Management of communication	5
Cost/Schedule/Scope/Quality boundaries	12
Managing stakeholders	8
Total questions	44

21.2.3 Sample Size for the Pilot Study

As per report from PMI-EY survey [13] it is projected that India will have approximately 400,000 Project Managers by 2022. Hence it can be inferred that there will be minimum 400,000 IT projects that organization in India would be executing assuming 1 project per Project Manager. Hence the sample size needed for broader study would be 384 projects [10]. As a part of the broader survey this pilot study was conducted for 50 project managers. Each project managers have shared data for two projects, i.e., the paper is based on data of 100 projects.

21.2.4 Reliability and Validity of the Questionnaire

After the questionnaire was developed, it was required to get the questions validated. Two methods were used to validate the questions (a) Review by Experts and (b) sample survey by sending this questionnaire to project managers from IT organizations—The questionnaire was sent to total 10 reviewers. It was studied if the project managers were able to understand the questions and were able to answer those. Feedback from Experts and observations of project Managers chosen for this pilot survey were suitably incorporated in the questionnaire.

21.3 Analysis and Findings

21.3.1 SPSS Output and Analysis of the Results'

21.3.1.1 Group Statistics Table

The group statistics table shows a good variation between mean values indicates that the selected independent variables (six) may be good discriminators as the difference in mean is large. (20–40%) [14] (Fig. 21.2).

21.3.1.2 Equality of the Group Means Test

Based on the Wilks' Lambda value shown in below table, it can be concluded that Variable "Baseline Definition" and "Stakeholder Management" could be the most important discriminating factor [11] (Fig. 21.3).

Group Statistics

S		Mean	Std. Deviation	Valid N (listwise) Unweighted	Weighted
1	S_BD	38.58	6.440	50	50.000
	S_CM	17.10	2.178	50	50.000
	S_HR	18.76	3.450	50	50.000
	S_PC	19.10	5.027	50	50.000
	S_RM	23.50	3.177	50	50.000
	S_SM	26.84	3.930	50	50.000
2	S_BD	22.30	8.379	50	50.000
	S_CM	12.12	4.650	50	50.000
	S_HR	13.78	4.325	50	50.000
	S_PC	13.42	6.643	50	50.000
	S_RM	14.98	6.968	50	50.000
	S_SM	16.04	6.537	50	50.000
Total	S_BD	30.44	11.055	100	100.000
	S_CM	14.61	4.394	100	100.000
	S_HR	16.27	4.627	100	100.000
	S_PC	16.26	6.519	100	100.000
	S_RM	19.24	6.882	100	100.000
	S_SM	21.44	7.632	100	100.000

Fig. 21.2 Group statistics

Tests of Equality of Group Means

	Wilks' Lambda	F	df1	df2	Sig.
S_BD	.452	118.652	1	98	.000
S_CM	.676	47.037	1	98	.000
S_HR	.708	40.515	1	98	.000
S_PC	.808	23.243	1	98	.000
S_RM	.613	61.892	1	98	.000
S_SM	.494	100.247	1	98	.000

Fig. 21.3 Equality of group means test

21.3.1.3 Eigenvalues and Wilks' Lambda

The explanation of proportion of variance can be elaborated by Eigen Value. Higher the value indicates that the function is very strong. Correlation value 1 means the

Eigenvalues

Function	Eigenvalue	% of Variance	Cumulative %	Canonical Correlation
1	1.972[a]	100.0	100.0	.815

a. First 1 canonical discriminant functions were used in the analysis.

Wilks' Lambda

Test of Function(s)	Wilks' Lambda	Chi-square	df	Sig.
1	.336	103.479	6	.000

Fig. 21.4 Eigenvalues and Wilks' Lambda

function is excellent. The value in this study was found to be 1.972 which indicates a very good correlation (Fig. 21.4).

Below are the values of various parameters which was calculated using SPSS

(A) Wilk's Lambda: 0.336.
Chi-Square: 103.479.
DOF: 6.

Since Wilk's Lambda is much lower than 1 it indicates that the variability within group is very small with respect to total variability. It can be concluded that group mean differs from each other.

(B) At Level of confidence of 95% value of a = 0.05.
P value which we have is 0.000.

Since 0.05 > 0.000, null hypothesis should be rejected and accept alternate hypothesis. Which means that six aspects have significant power for discrimination.

21.3.1.4 Test for Independent Variable's Relative importance

In the below table, SPSS output of Standardized Canonical Discriminant Function coefficient has been mentioned. Higher the power indicates higher standardize discriminant coefficient.

Highest discriminating coefficient value is observed for "Cost/Schedule/Scope/Quality boundaries": 0.615. Other KPIs can be ranked based on the decreasing value of coefficient and the sequence of importance is "Management of Stakeholders", "Management of Issues/risks", "People and Skills", "Communication Management" and "Methods to control projects". This indicates that "Cost/Schedule/Scope/Quality boundaries" has a best predictor of whether the project will get into a troubled status or not (Fig. 21.5).

Fig. 21.5 Standerdized canonical discriminant coefficient

Standardized Canonical Discriminant Function Coefficients

	Function
	1
S_BD	.615
S_CM	-.093
S_HR	.227
S_PC	-.329
S_RM	.284
S_SM	.574

Fig. 21.6 Functions at group centroids

-1.390 0 +1.390

Functions at Group Centroids

S	Function 1
1	1.390
2	-1.390

Unstandardized canonical discriminant functions evaluated at group means

21.3.1.5 Functions at the Group Centroids

Below table shows the functions at the group centroids. Since the number of cases for success and failures are same, we can take average of these extreme points to come to center which is "0". So, if the value of *B* for any project lies between 0 and -1.390 then given project has high possibility of failure (Fig. 21.6).

21.3.1.6 Graphical Representation of the Discriminant Function

Below are the graphs which SPSS has plotted, and graphs clearly indicates that the six aspects have significant power to differentiate (Fig. 21.7).

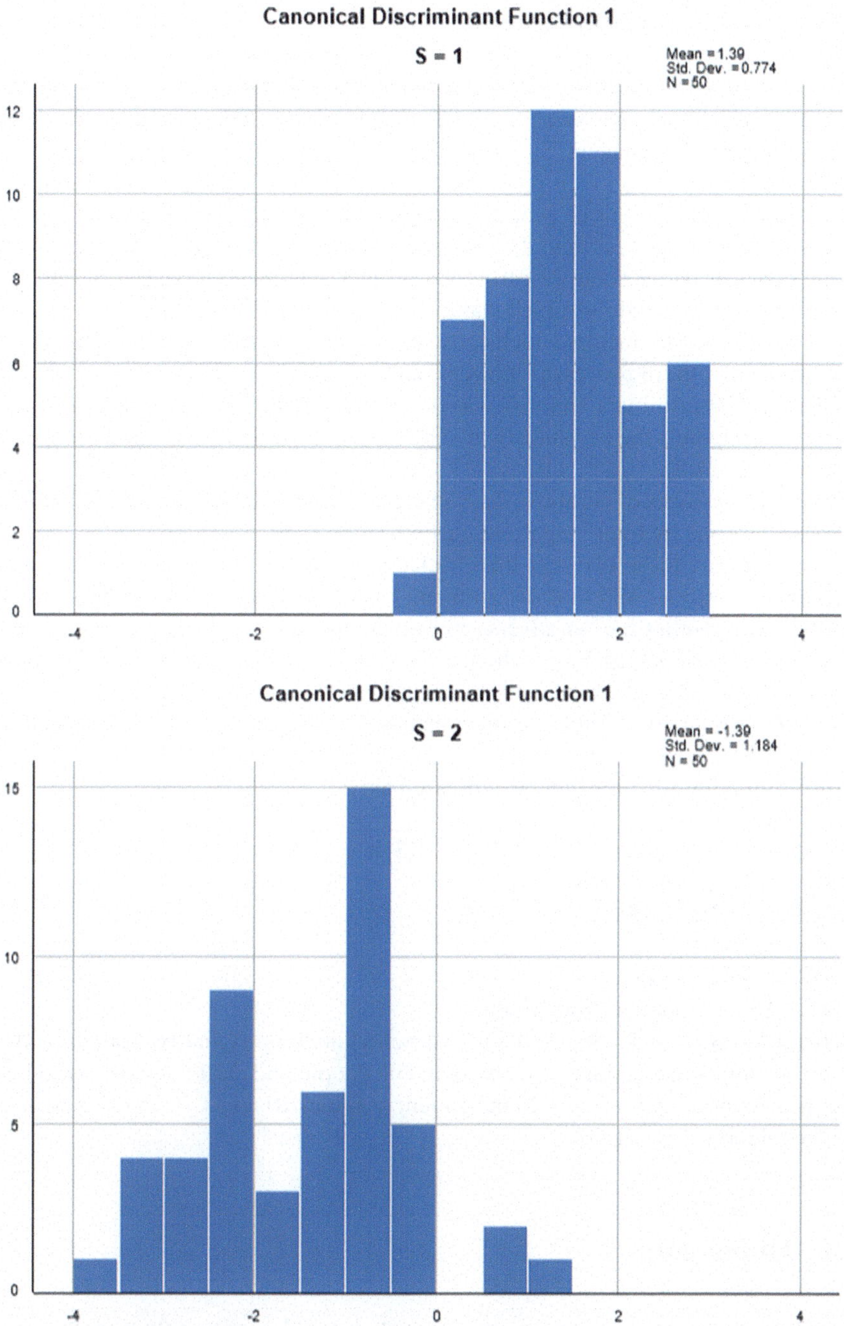

Fig. 21.7 Graphical representation of discriminant function

21.3.1.7 Outliers in Data Collected

In this pilot study, data from 50 project managers was collected for 100 projects. Marks were given to each answer ranging from four to zero depending upon the option selected by Project Managers. Sum of marks were calculated for each parameter and then the sum total for a given project was calculated. As shown in the Table 21.4 it was observed that for three project managers (six projects) the total score for the successful project was less than the total score for troubled projects. In other words, despite equal or better control of troubled project, compared to successful project, the said project went into troubled status. Which also means that for these projects there were some other factors which caused these projects to get into troubled status. So, it seems that these four projects were outliers.

When the Discriminant analysis was done on the complete set of data including above six data points the key statistical values Wilks' Lambda and Eigenvalues were as below (Fig. 21.8).

Whereas the values of the above 2 parameters were as below when the discriminant analysis was done by removing these outliers (Fig. 21.9).

If we compare these two sets then it clearly shows the impact of outlier on the discriminant analysis and that discriminant analysis is very sensitive for outliers. Pathology like outlier has an adverse impact on the accuracy and interpretation of discriminant analysis [15]. From which it can be inferred that if from the discriminate equation if it is wrongly predicted that project will be successful then the cost of wrong prediction will be high as project manager might get misguided. Hence, these four data points were removed from the sample data collected [11].

Discriminating function can be expressed as

$$B = a + a1A1 + a2A2 + a3A3 + a4A4 + a5A5 + a6A6.$$

B = Dependent Variable
'a' = constant
'$a1 \ldots a6$' are coefficients
'$A1 \ldots A6$' are independent variables.

Therefore $B = -5.458 + 0.82$(Cost/Schedule/Scope/Quality boundaries) $-$ 0.026(Communication Management) + 0.058(People and skills) $-$ 0.056(Methods of project governance) + 0.052(Management of Issues/Risks) + 0.106(Management of stakeholders) (Fig. 21.10).

21.4 Discussion

With more and more digitization and connected world coming into play there is a constant need of organizations to take up IT projects to meet their end goal. Organizations are investing in these IT projects in a big way. On the other hand, financial losses for IT projects are very high due to failure of the projects. Hence, there is need

Table 21.4 An Outliers

Project manager	Total score	Baseline definition	Communication management	Human resources and skills	Project control methodology	Risk and issue management	Stakeholder management	Total score
A	Successful project	35	13	19	13	22	27	129
	Troubled project	39	13	19	20	22	21	134
B	Successful project	43	17	19	14	23	31	147
	Troubled project	41	20	20	14	23	31	149
C	Successful Project	18	7	8	10	17	25	83
	Troubled project	42	16	23	24	27	29	161

Eigenvalues

Function	Eigenvalue	% of Variance	Cumulative %	Canonical Correlation
1	1.353[a]	100.0	100.0	.758

a. First 1 canonical discriminant functions were used in the analysis.

Wilks' Lambda

Test of Function(s)	Wilks' Lambda	Chi-square	df	Sig.
1	.425	86.418	6	.000

Fig. 21.8 Wilks' Lamda and Eigenvalues before removing outliers

Eigenvalues

Function	Eigenvalue	% of Variance	Cumulative %	Canonical Correlation
1	1.972[a]	100.0	100.0	.815

a. First 1 canonical discriminant functions were used in the analysis.

Wilks' Lambda

Test of Function(s)	Wilks' Lambda	Chi-square	df	Sig.
1	.336	103.479	6	.000

Fig. 21.9 Revised Eigenvalues and Wilks' Lambda

Fig. 21.10 Revised canonical discriminant function coefficient

Canonical Discriminant Function Coefficients

	Function
	1
S_BD	.082
S_CM	-.026
S_HR	.058
S_PC	-.056
S_RM	.052
S_SM	.106
(Constant)	-5.458

Unstandardized coefficients

to check and govern what are the possible reason for failure of IT projects so that the risk of failure of an IT project can be minimized or a mitigation plan can be designed by Project Managers and management team.

With that background in phase 1 of research undertaken by the authors [1], it was established through the literature survey that there six main parameters (independent variables) on which success or failure of a project depends. These parameters are Cost/Schedule/Scope/Quality boundaries, Communication Management, People and their skills, Methods of project governance, Management of Issues/Risks, Management of stakeholders.

A sample of 50 project managers and their 100 projects were taken for this pilot study to establish the model to measure the impact of these factors using discriminant analysis. In the above discussion, the model has been established. It was observed that Cost/Schedule/Scope/Quality boundary has maximum influence on the success of the project.

By using this model project managers can measure the possible outcome for their project based on the current situation on these six factors. If the model predicts that project might get into troubled status and may fail, then accordingly project manager can take appropriate measures to avoid failure of the project. Project Manager can use this model at any point in time during the execution of the project which can help project manager to keep appropriate control over the project.

21.5 Conclusion

In this study an analysis of the data collected from the survey of 50 project managers (and their 100 projects) was carried out. The corresponding questionnaire was developed to collect the data of the six independent variables which influence the success or failure of a project. Using discriminant analysis, it was established and the Null Hypothesis that variables do not have discriminating power was rejected. SPSS software was used to analysis the data collected. A model which was derived out of the analysis is $B = -5.458 + 0.82$(Cost/schedule/scope/Quality boundaries) $-$ 0.026(Communication Management) $+$ 0.058(People and skills) $-$ 0.056(Methods of project governance) $+$ 0.052(Management of Issues/Risks) $+$ 0.106(Management of stakeholders).

And if the value of the dependent variable B lies between 0 and -1.390 then that given project has a high possibility of failure and vice versa.

It has been established through this pilot study that the six parameters do influence the success or failure of the project and that further study can be undertaken. Moreover, this model can be used to calculate discriminant score to categories a project into a successful or troubled project. Project managers can then take corrective actions.

Future scope: Authors will be expanding the survey and will be taking more samples to further enhance the study and to finalize the model. Moreover, authors are also working on deriving a proposed suggestion through the use of this discriminant analysis by helping project managers to pin point the area where improvements are required to bring the project on track. There few areas for further research were identified like why the quality of subcontractor resources in India is not up to mark and what organization can do for the same.

References

1. Oak VJ, Laghate K (2016) Analysis of project management issues in information technology industry: an overview of literature. Int J Syst Assur Eng Manag 7(4):418–426. https://doi.org/10.1007/s13198-016-0469-4
2. Flyvbjerg B, Budzier A (2011) Why your IT project may be riskier than you think. Harward Bus Rev. https://hbr.org/2011/09/why-your-it-project-may-be-riskier-than-you-think
3. Creasy T, Anantatmula VS (2013) From every direction—how personality traits and dimensions of project managers can conceptually affect project success. Proj Manag J 44(6):36–51. https://doi.org/10.1002/pmj.21372
4. PMI (2013) Organizational agility the high cost of low performance: the essential role of communications. https://www.pmi.org/-/media/pmi/documents/public/pdf/learning/thought-leadership/pulse/the-essential-role-of-communications.pdf
5. Schneider S, Wollersheim J, Krcmar H, Sunyaev A (2018) How do requirements evolve over time? A case study investigating the role of context and experiences in the evolution of enterprise software requirements. J Inf Technol 33:151–170. https://doi.org/10.1057/s41265
6. Terlizzi MA, de Souza Meirelles F, de Moraes HROC (2016) Barriers to the use of an IT Project Management Methodology in a large financial institution. Int J Proj Manag 34(3):467–479. https://doi.org/10.1016/j.ijproman.2015.12.005
7. Kim DY, Han SH, Kim H (2008) Discriminant analysis for predicting ranges of cost variance in international construction projects. J Constr Eng Manag 134. https://doi.org/10.1061/(ASCE)0733-9364(2008)134:6(398)
8. Betz NE (1987) Use of discriminant analysis in counseling psychology research. J Couns Psychol 34(4):393–403. https://doi.org/10.1037/0022-0167.34.4.393
9. Fiskovi C (2013) Discriminant analysis of bank profitability levels. Rozga Ante Faculty of Economics, University of Split Kundid Ana, vol 4. Hrvatsko društvo za operacijska istraživanja. https://hrcak.srce.hr/index.php?show=clanak&id_clanak_jezik=143352
10. Sekaran B (2013) Research methods for business: a skill-building approach. http://as.wiley.com/WileyCDA/WileyTitle/productCd-111994225X.html#
11. Uddin N, Meah MS (2013) Discriminant analysis as an aid to human resource selection and human resource turnover minimization decisions. Int J Bus Manag 8(17):153–169. https://doi.org/10.5539/ijbm.v8n17p153
12. Johnson R, Dean W (2008) Applied multivariate statistical analysis, 6th edn. Pearson (Prentice Hall). https://www.scribd.com/document/339648080/Applied-Multivariate-Statistical-Analysis-by-Johnson-and-Wichern-pdf
13. PMI-EY (2014) PMI-EY study on perspectives of Indian organizations on project management skills 2 content. https://www.pmi.org.in/downloads/PMI-EY-Study.pdf
14. Ayogu C (2014) Application of discriminant function analysis in agricultural extension research, pp 117–146
15. Klecka WR (1980) Discriminant analysis. SAGE. https://books.google.co.in/books/about/Discriminant_Analysis.html?id=9vrgGJy7XOQC&printsec=frontcover&source=kp_read_button&redir_esc=y#v=onepage&q&f=false

Chapter 22
An S-Shaped Fault Detection and Correction SRGM Subject to Gamma-Distributed Random Field Environment and Release Time Optimization

Vishal Pradhan, Joydip Dhar, Ajay Kumar, and Ashish Bhargava

22.1 Introduction

The software is a set of instructions or programs in the form of a line of codes. Software control and coordinate the functionalities of the systems. In the twenty-first century, every organization or even a single person depends on software systems either directly or indirectly. Software becomes an essential component of our modern society and organizations like defense, nuclear reactor, industrial process, patient health monitoring, banking, telecommunications, transportation, home appliances, personal entertainment, and so on. Today, almost everyone is linked with computer systems in one or another way. Therefore dependency on software systems increases, and system failures may have disastrous results. Software faults have resulted in catastrophic failures, some with dreadful consequences. There are some examples of catastrophic failures, only because of software failures. In 1986, Boeing727 airliner of Mexican Airlines collapsed within a hill area because of the software system negotiation problem [1]. Recently, many banks and digital currency organizations suffered from software failures because they either lost the information or information are mishmash.

V. Pradhan (✉) · J. Dhar · A. Kumar · A. Bhargava
ABV-Indian Institute of Information Technology and Management, Gwalior 474015, MP, India
e-mail: vishal.iiitmg@gmail.com

J. Dhar
e-mail: jdhar.iiitmg@gmail.com

A. Kumar
e-mail: ajay1dma@gmail.com

A. Bhargava
e-mail: ashishbhargava39@gmail.com

P. K. Kapur et al. (eds.), *Decision Analytics Applications in Industry*, Asset Analytics,
https://doi.org/10.1007/978-981-15-3643-4_22

In this digital world, healthy competition in software development and maintenance is must, so there is a need for reliable software development in any organization to maintain its reputation and to survive in the market. Definition of reliable software is the system success probability or the possibility to perform the intended function within the given environmental conditions during a specified period under specified design constraints [2]. In software development, the prime reason for software failure is the complexity in software design and coding. Over the past four decades, numerous SRGMs are developed for the reliability growth estimation of software [3–8]. Since these software errors can be detected and corrected, reliability growth is possible. The most competent software developers cannot make a software product completely error free. Because it is not about the only skills of software professionals or methods they have used, instead it is the conceptual and logical framework complexity behind the software development. Mainly two kinds of software reliability model exist, deterministic and probabilistic. The probabilistic growth model describes the failure experiences and the fault removals as a probabilistic event.

These SRGMs can be classified into failure rate, curve fitting, error seeding, Markov structure, time-series, and NHPP. In general, the NHPP-based SRGMs are accessible and attractive for a quantitatively assessing software reliability. Almost every NHPP-based models are convenient to use, and most widely applied models in industries. Since the 1970s, many NHPP-based models are proposed with different assumptions. These assumptions help to build a more realistic model [9]. These proposed models also help to estimate the failure rate, software reliability, total, and the remaining amount of errors in a software [10–14]. All SRGMs work either in imperfect debugging or perfect debugging. Perfect debugging is the certainty of removing a fault in the case of a failure and remaining faults decreases with respect to debugging time. In an imperfect debugging environment, there is no certainty of removing a fault, and the remaining faults may increase or decrease with respect to debugging time due to new fault introduction [15–18]. In reliability growth modeling some authors used perfect debugging assumption and few use imperfect debugging assumption.

One regular assumption in most of the previous literature was perfect debugging of faults identification. But it is impracticable, due to the probability of new faults introduction during the period of debugging or testing, and there is a notable delay between correction and detection process [19, 20]. Recently, some researcher describes the effect of random operating environments on software failure and its reliability [21–28]. Hoang Pham presents, a generalized fault detection SRGM considering the impact of RFE in fault detection rate [21]. But it was deemed to be perfect debugging, which is impracticable in general. In 2014 Pachauri et al., develop an SRGM with dynamic faults and release time optimization using multi-attribute utility theory and genetic algorithm, and discussed the optimal release policy [29]. Many SRGMs are proposed till date, but most of them are based on fault detection process only [21–28]. In literature very few researchers work on both fault detection and correction process but they not considered RFE [15, 30–32]. In all the above literature, we found that the model with RFE is not considering both fault correction and detection process; they only study the fault detection process. Also, some models are not considering imperfect debugging environment.

The most important use of the SRGMs, for every software industry managers, is to determine the optimal point of deliverance [33, 34]. The excellent quality of a software system depends on, which testing techniques are in use and how much time allocated on testing. More time given on testing directs to more reliable software, but testing costs also increase. There is a venture to deliver a software system by low quality or high price. Therefore, many researchers introduce a model for operation release time, and these cost models give trade-off between reliability and cost [35]. Over the last two decades, many researchers have been discussed for optimal deliverance strategy based on cost and reliability measures [36, 37]. The point of software deliverance is calculated in different ways, and we give priority to one of them either cost or reliability [38–40]. One basic approach implies to determine the release time of the software system for the pre-fixed required reliability. Therefore the overall price throughout the rest phases of the software life-cycle is reduced. In our model, the debugging cost before and after the testing phase, and per unit testing cost are considered.

In practical users use the software in a random field environment. Therefore in this article, we incorporate random field environment (RFE) in the proposed model. In previous literature, we found the detection model with RFE is available. But to the best of my knowledge we did not find the reliability growth model with both detection and correction process in imperfect debugging RFE. So in this article, we consider both fault correction and detection process in imperfect debugging with RFE. This paper is organized as follows. In Sect. 22.2, we discuss some related work and their mean value function. Section 22.3 describes an integrated fault correction and detection model in an imperfect debugging environment with RFE. Section 22.4 presents the numerical experiment and data analysis. In Sect. 22.5, the optimal release policy is discussed. Section 22.6 presents the parameter's sensitivity analysis. Section 22.7 and 22.8 explain discussion and conclusions, respectively.

22.2 Related Work

In 2000, Yang and Xie consider that the testing reliability and operational reliability are not same [41]. After that Teng and Pham incorporated RFE in SRGM and used gamma distribution to represent the RFE [28]. A testing coverage SRGM with uncertainty of operating environments is studied by Chang et al. [26]. Pham et al. introduced two parameter-generalized probability density function for random operating environment [21]. A SRGM with three parameter fault detection rate and exponential RFE is studied by Song et al. [25]. Here, in Table 22.1 we consider some existing SRGMs and their mean value functions for parameter estimation on actual software failure data-set.

In next section, we introduce an integrated fault correction and detection SRGM with gamma distributed FRE under a set of standard assumptions.

Table 22.1 Some existing models with mean value function

Model	Mean value function (MVF) ($m(t)$)
G-O model [42]	$a\left(1-e^{-bt}\right)$
Delayed S-shaped model [43]	$a\left(1-(1+bt)e^{-bt}\right)$
PNZ model [44]	$\frac{a\left(1-e^{-bt}\right)\left(1-\frac{\alpha}{b}\right)+a\alpha t}{1+\beta e^{-bt}}$
Vtub-shaped model [45]	$\mu(1-\left(\frac{\beta}{\beta+a^{t^b}-1}\right)^{\alpha})$
Yang et al. single release model with exponential time delay [32]	$a\left(1-(1+\beta t)e^{-\beta t}\right) \quad \alpha=\beta$ $a\left(1-\frac{\alpha}{\alpha-\beta}e^{-\beta t}+\frac{\beta}{\alpha-\beta}e^{-\alpha t}\right) \quad \alpha\neq\beta$
Yang et al. single release model with gamma time delay [32]	$a\Gamma\left(t,\alpha,\frac{1}{\beta}\right)-\frac{ae^{-\gamma t}}{\Gamma(\alpha)\beta^{\alpha}}\times$ $\int_0^t e^{-(\beta-\gamma)(t-x)(t-x)^{\alpha-1}}dx$

22.3 A Unified Fault Detection and Correction Reliability Growth Model in an Imperfect Debugging with RFE

In this work, we introduced an SRGM for both fault detection and correction process with the impact of RFE in an imperfect debugging environment. The following assumptions addressed for forming the fault detection and correction models:

1. The number of software failures follow an NHPP as counting process, where $m_d(t)$ and $m_c(t)$ are mean value functions of fault detection and correction at time t.
2. All the identified faults in software are independent.
3. The expected number of detected faults $m_d(t)$ in the time interval $(t, t+\Delta t)$ is proportional to the number of remaining unidentified faults at that time. Moreover, the proportionality is time-dependent inflection S-shaped function.
4. The uncertainty of the operating environment is depicted by multiplying the fault detection rate $b(t)$ and a random variable ζ.
5. Faults can be introduced during the debugging process with constant fault introduction rate and the complete fault content function is linear time-dependent.
6. The expected number of corrected faults $m_c(t)$ in the time interval $(t, t+\Delta t)$ is proportional to the mean number of identified but not yet removed faults in the system. Here the proportionality is assumed to be constant.

Keeping in view of the above assumptions, we derive both fault detection and fault correction process. Here, $m_d(t)$ and $m_c(t)$ denote the expected number of detected and corrected faults in the time interval (0, t), respectively. The proposed SRGM can be written as

$$\frac{dm_d(t)}{dt}=\zeta b(t)[a(t)-m_d(t)] \tag{22.1}$$

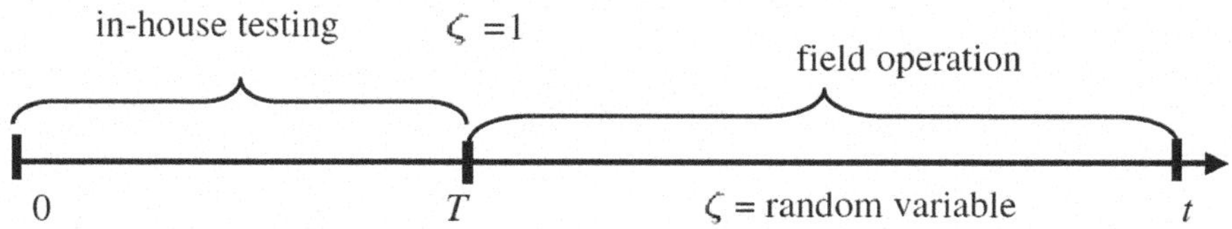

Fig. 22.1 Random field environment

$$\frac{dm_c(t)}{dt} = \gamma[a(t) - m_c(t)]. \tag{22.2}$$

where,

$$\frac{da(t)}{dt} = \alpha \frac{dm_d(t)}{dt}, \tag{22.3}$$

$$b(t) = \frac{b}{1 + \beta e^{-bt}}, \quad b > 0, 0 < \beta < 1. \tag{22.4}$$

The marginal conditions are $m_d(0) = 0$, $m_c(t) = 0$, and $a(0) = a$. Where $a(t)$ is fault content function at any time t. Here, $b(t)$ represent time-dependent inflection S-shaped fault detection rate and γ denote constant fault correction rate at any time t. In $b(t)$ b represent the asymptotic unit software failure detection rate and β denote the shape of the learn curve. In environment of imperfect debugging fault introduction rate is represented by α. The RFE (ζ) is shown in Fig. 22.1, where T is the software deliverable time. Where RFE ζ is nonnegative random variable (r.v.) with probability density function (pdf) $f(\zeta)$, i.e.,

$$\begin{aligned} \zeta &= 1, & t < T \\ &= \text{r.v. with pdf } f(\zeta), & t > T \end{aligned}$$

Now, for testing and operational phases, we solve the above differential equations for both the fault correction and detection models.

22.3.1 Fault Detection Model

Now solving the differential Eq. (22.1) for testing and operational phases are given as follows:

Case 1: For testing phase ($t < T$)

$$m_d(t) = \frac{a}{1 - \alpha}\left[1 - e^{-(1-\alpha)\int_0^t b(v)dv}\right]. \tag{22.5}$$

Case 2: For operation phase ($t > T$)

$$m_d(t) = \frac{a}{1-\alpha}\left[1 - e^{-(1-\alpha)\int_0^T b(\upsilon)d\upsilon} F^*\left((1-\alpha)\int_T^t b(\upsilon)d\upsilon\right)\right]. \tag{22.6}$$

where $F^*(x)$ is the laplace-transform of gamma distribution function of ζ. After simplifying Eqs. (22.5) and (22.6), the MVF ($m_d(t)$) for the detection model in testing and operational phases are given as follows:

Case 1: For testing phase ($t < T$)

$$m_d(t) = \frac{a}{1-\alpha}\left[1 - \left(\frac{1+\beta}{\beta+e^{bt}}\right)^{1-\alpha}\right]. \tag{22.7}$$

Case 2: For operation phase ($t > T$)

$$m_d(t) = \frac{a}{1-\alpha}\left[1 - \left(\frac{1+\beta}{\beta+e^{bT}}\right)^{1-\alpha}\left(\frac{\lambda}{\lambda+(1-\alpha)\log\left(\frac{\beta+e^{bt}}{\beta+e^{bT}}\right)}\right)^{\theta}\right]. \tag{22.8}$$

where λ and θ are scale and shape parameters of gamma distribution function of RFE (ζ).

22.3.2 *Fault Correction Model*

Again corresponding MVF ($m_c(t)$ for the correction model (22.2), are as follows:

Case 1: For testing phase ($t < T$)

$$m_c(t) = \frac{a(1-e^{-\gamma t})}{1-\alpha}\left[1 - \left(\frac{1+\beta}{\beta+e^{bt}}\right)^{1-\alpha}\right]. \tag{22.9}$$

Case 2: For operational phase ($t > T$)

$$m_c(t) = \frac{a(1-e^{-\gamma t})}{1-\alpha}\left[1 - \left(\frac{1+\beta}{\beta+e^{bT}}\right)^{1-\alpha}\left(\frac{\lambda}{\lambda+(1-\alpha)\log\left(\frac{\beta+e^{bt}}{\beta+e^{bT}}\right)}\right)^{\theta}\right]. \tag{22.10}$$

where γ is a constant fault correction rate. In next section we discuss about the numerical experiment and comparison criteria of proposed model with existing models.

Table 22.2 Estimated value of model parameters for fault detection data

Models/Parameters	a	b	α	β	γ	μ/λ^*
G-O model [42]	59.64	0.0745	–	–	–	–
Delayed S-shaped model [43]	54.76	0.1998	–	–	–	–
PNZ model [44]	23.47	0.2529	0.0355	0.0447	–	–
Vtub-shaped model [45]	1.11	0.3076	0.1576	0.5023	–	73.01
Yang et al. with exponential delay [32]	57.28	–	0.0490	0.0862	–	–
Yang et al. with gamma delay [32]	60.60	–	0.2511	0.0063	0.071	–
Proposed testing phase model	40.36	0.1749	0.1078	0.0001	–	–
Proposed operational phase model	24.31	0.1204	0.7565	0.7923	–	4.59

λ^* for operational phase model

22.4 Numerical Experiment and Data Analysis

Now, we set some evaluation criteria for their performance analysis. For that, we first need to estimate the model parameters from few benchmark data [32]. Therefore, in this study we estimate the model parameters using nonlinear least square estimation (LSE). Few researchers also used maximum likelihood estimation (MSE) for parameter estimation.

22.4.1 Parameter Estimation and Evaluation Criteria

After obtaining expression for $m_d(t)$ and $m_c(t)$ next important part is to get the fittest value of parameters for the model. In this paper, we have used LSE method for parameter estimation and obtained parameters are shown in table. In Table 22.2 the models parameters are estimated for fault detection data and in Table 22.3 models parameters are estimated for fault correction data. For analysis and comparison of our proposed SRGM with previous growth models the following measures are used: Mean Squared Error (MSE), coefficient of determination (R^2), and adjusted R^2. In Tables 22.4 and 22.5 the comparison is made for fault detection and correction models. All analysis are done on real software failure data.

22.4.2 Performance Analysis and Results

We have used the secondary dataset from existing article [32] and this is the fault tracking data of version Firefox 3.0 from Bugzilla. This is 53 weeks data, so we consider first half data for testing phase and other half data for operational phase. Here, we consider that the release time is $T = 27$ for the given data-set. We take the

Table 22.3 Estimated value of model parameters for fault correction data

Models/Parameters	a	b	α	β	γ	μ/λ^*
G-O model [42]	57.56	0.0279	–	–	–	–
Delayed S-shaped model [43]	42.75	0.101	–	–	–	–
PNZ model [44]	12.63	0.4333	0.0533	4.07	–	–
Vtub-shaped model [45]	6.508	0.2044	0.6209	0.105	–	46.11
Yang et al. with exponential delay [32]	52.03	–	0.0244	0.0629	–	–
Yang et al. with gamma delay [32]	55.24	–	0.3034	0.0027	0.051	–
Proposed testing phase model	27.78	0.1692	0.1078	0.0001	0.245	–
Proposed operational phase model	19.41	0.1204	0.7565	0.7923	0.459	6.104

λ^* for operational phase model

Table 22.4 Comparison of fault detection function with proposed and existing models

Models/Comparison criteria	MSE	R^2	adj-R^2
G-O model [42]	16.098	0.9123	0.9105
Delayed S-shaped model [43]	40.647	0.7784	0.7741
PNZ model [44]	13.919	0.9212	0.9163
Vtub-shaped model [45]	7.7981	0.9558	0.9521
Yang et al. single release model with exponential time delay [32]	17.912	0.9527	0.9484
Yang et al. single release model with gamma time delay [32]	16.168	0.9661	0.9640
Proposed testing phase model	2.9839	0.9756	0.9736
Proposed operational phase model	1.1600	0.9623	0.9571

Table 22.5 Comparison of fault correction function with proposed and existing models

Models/Comparison criteria	MSE	R^2	adj-R^2
G-O model [42]	4.4528	0.9691	0.9685
Delayed S-shaped model [43]	13.5698	0.9060	0.9041
PNZ model [44]	2.4933	0.9911	0.9902
Vtub-shaped model [45]	7.3585	0.9490	0.9448
Yang et al. single release model with exponential time delay [32]	22.5427	0.9786	0.9777
Yang et al. single release model with gamma time delay [32]	6.0826	0.9807	0.9790
Proposed testing phase model	0.4425	0.9929	0.9916
Proposed operational phase model	1.4676	0.9549	0.9463

value $\theta = 1$ in this analysis because we observe from different value of θ, the smaller value of θ is good in experimental analysis. Estimated values of model parameters for fault detection data are shown in Table 22.2. In Table 22.3 the estimated values of model parameters for fault correction data are given. Different performance measures and comparison with existing models for detection and correction model are shown in Tables 22.4 and 22.5 respectively. Figures 22.2 and 22.4 shows predicted number of detected and corrected faults curve respectively. Figures 22.3 and 22.5 shows the testing and operational phases predicted number of detected and corrected faults respectively, with respect to time.

In Fig. 22.2, we see that the fault detection curve of G-O model, Delayed S-shaped model, PNZ model and V-tub S-shaped model over actual data. In next Fig. 22.3, we see the proposed model detection curve for testing and operational phase over actual data. We found that the Fig. 22.3, detection curve is more close to actual data and we also see by the Table 22.4, where MSE is less than the other comparison models.

Similarly, in Fig. 22.5, we see that the fault correction curve of G-O model, Delayed S-shaped model, PNZ model and V-tub S-shaped model over actual data. In next Fig. 22.5, we see the proposed model correction curve for testing and operational phase over actual data. We found that the Fig. 22.5, correction curve is more close to actual data and we also see by the Table 22.5, where MSE is less than the other comparison models.

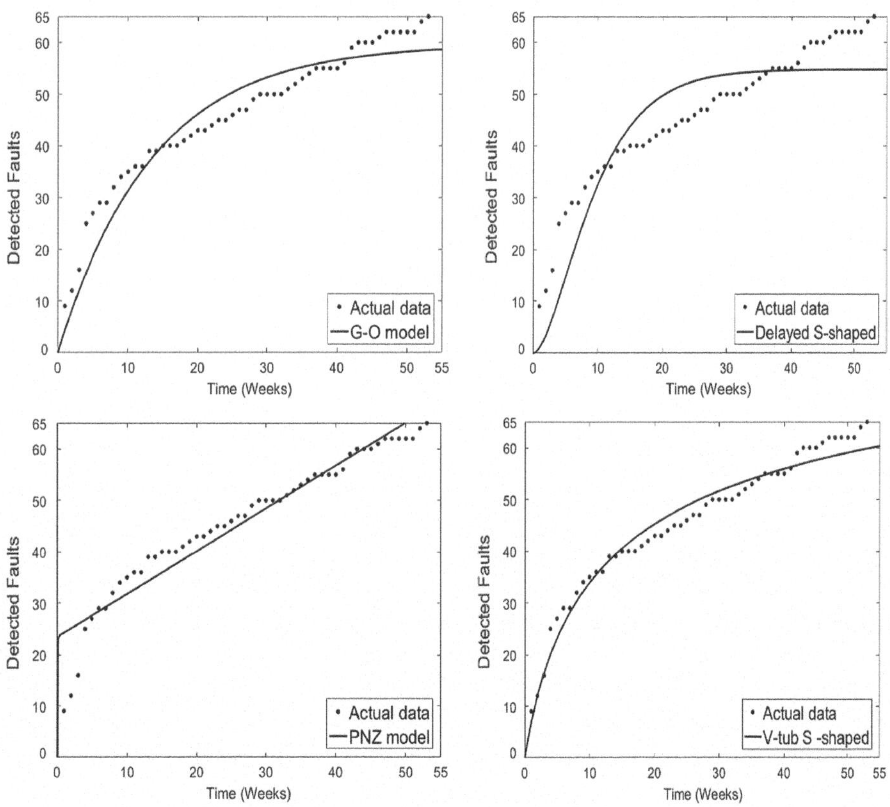

Fig. 22.2 Fault detection curves for SRGMs on fault detection data

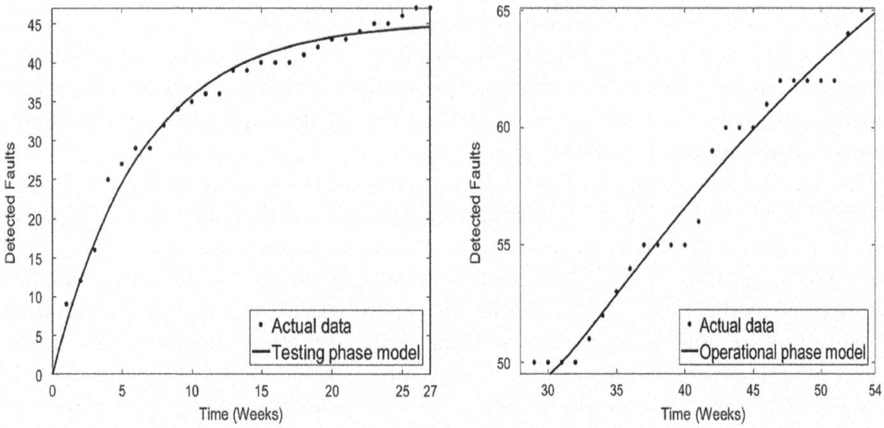

Fig. 22.3 Fault detection curves for testing and operational phase model on fault detection data

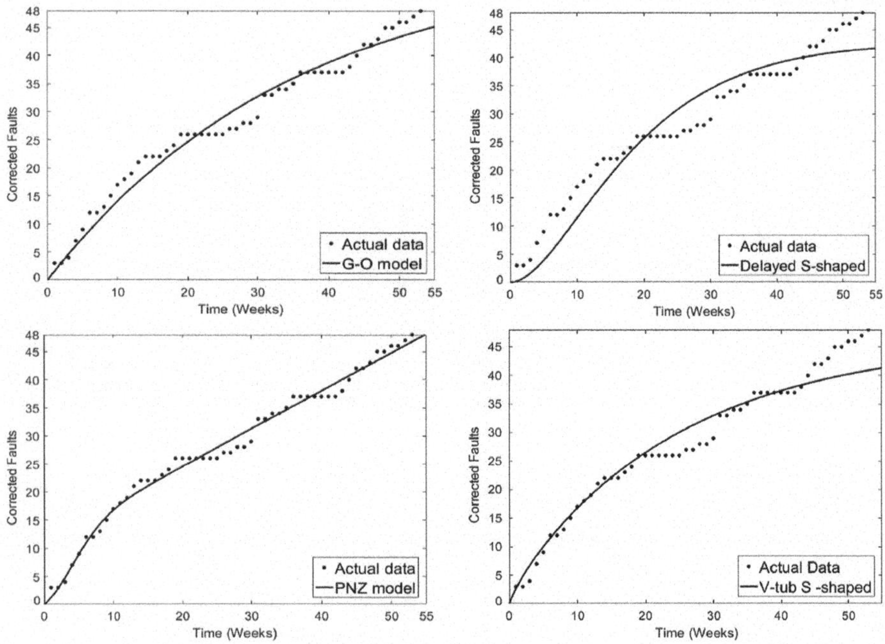

Fig. 22.4 Fault correction curves for SRGMs on fault correction data

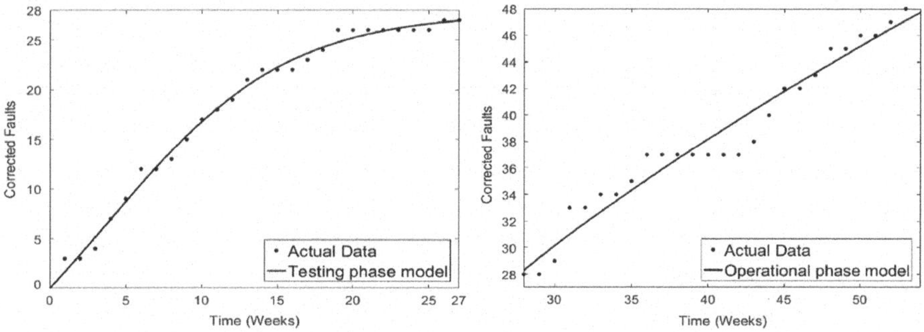

Fig. 22.5 Fault correction curves for testing and operational phase model on fault correction data

22.5 Optimal Release Policy

Optimal release policy is a trade-off between software cost and reliability with respect to the software release time [38–40]. There are two criteria for optimal release time, either we give priority to cost or reliability. But here, we make a tread-off between them. In this section, we study the effectiveness of the proposed detection model to calculate the software optimal release time.

22.5.1 *Software Optimal Release Time*

In this release time model, the cost is minimized subject to reliability.

The conditional reliability at a given time t is described as

$$R(x/T) = e^{-[m_d(T+x)-m_d(T)]}, \tag{22.11}$$

The software development cost is calculated using given formula [40]:

$$\text{Cost}(T) = Q_1(T) + Q_2 m_d(T)y + Q_3[1 - R(x/T)], \tag{22.12}$$

where Q_1T is cost to perform testing and Q_1 is software test cost per unit time. $Q_2m_d(T)y$ the expected cost to correct all error by time T and Q_2 is the cost of removing each error per unit time during testing. Here, y is the expected time to remove an error during testing the phase. Where $Q_3[1 - R(x/T)]$ is the risk cost due to software failure after software release and $R(x/T)$ is the conditional reliability by time T for a mission time x. Hence, the total expected software cost is calculated using above mentioned equation. Given Q_1, Q_2, Q_3, x and y are assumed value. The assumed values in this equation are taken from article [40]. Where in case 1 values are $Q_1 = 25$, $Q_2 = 200$, $Q_3 = 7000$, $x = 0.05$, and $y = 0.1$. In case 2 $Q_3 = 10{,}000$ is assumed. Table 22.6 shows total cost and reliability of the software at different time

Table 22.6 Total cost and reliability with testing time

Testing time	Total cost (for $Q_3 = 7000$)	Reliability (for $Q_3 = 7000$)	Total cost (for $Q_3 = 10{,}000$)	Reliability (for $Q_3 = 10{,}000$)
2	2847.29211	0.598922203	4050.525502	0.598922203
4	2282.487139	0.687144091	3221.054867	0.687144091
6	1825.535552	0.759853121	2545.976189	0.759853121
8	1470.648159	0.817904521	2016.934595	0.817904521
10	1204.748403	0.863187017	1615.187353	0.863187017
12	1012.462008	0.897912047	1318.725867	0.897912047
14	878.9500756	0.924210517	1106.318526	0.924210517
16	791.2248939	0.943945911	959.3871605	0.943945911
18	738.5547691	0.958657257	862.5829986	0.958657257
20	712.3927436	0.969569886	803.6830857	0.969569886
22	706.0978428	0.977635685	773.1907884	0.977635685
24	714.5971884	0.983581674	763.8521657	0.983581674
26	734.0618341	0.987956548	770.1921907	0.987956548
28	761.6262329	0.991170914	788.1134896	0.991170914
30	795.158888	0.993530185	814.5683329	0.993530185

instances for both cases. According to reliability based release policy the software should be delivered in the market between 27th and 28th week for both the cases. Because better reliability is closed to 1. But according to cost based release policy the software should be delivered in the market between 22th and 23th for case 1, and between 24th and 25th for case 2.

22.6 Sensitivity Analysis

We have used normalized forward sensitivity index to analyze the sensitive parameters of MVF ($m(t)$) in the operational phase [46]. We derived analytical expressions with respect to each parameter, for sensitivity index of detection model ($m_d(t)$). The general formula to calculate the normalized forward sensitivity index of a variable, v that depends on a parameter, q, is given as

$$ {}^{v}_{q}\Upsilon = \frac{\partial v}{\partial q} \times \frac{q}{v}. $$

Obtained results of sensitivity analysis are categorized and shown in Table 22.7.

Table 22.7 Sensitive parameters

Most sensitive	Moderate sensitive	Least sensitive
α	a, b, T	β, λ

22.7 Discussion

Table 22.4 compares proposed detection model with some previous models and proposed model has the lowest MSE, i.e., 2.9839 in testing phase and 1.1600 in operational phase. The value for R^2 is 0.9756 in testing and 0.9623 in operation phase, which is closest to 1. Adjusted R^2 are also calculated and compared to previous models. So our model is giving better performance results compared to previous models on given dataset. Similarly from Table 22.5, the proposed correction model is also giving better results than previous model for both fault correction and detection.

For case 1, we can see that initially, cost was too high, and reliability was also too low but after that reliability increases and cost also decreases and reaches the lowest value on 22th week and after that again starts increases. So from cost-based policy software release time should be in between 22th and 23th week, we can see that by Fig. 22.6. For case 2, we can see that initially, cost was too high, and reliability was also too low but after that reliability increases and cost also decreases and reaches the lowest value on 24th week and after that again starts increases. So from cost-based policy software release time should be in between 24th and 25th week, we can see that by Fig. 22.7. We have used sensitivity analysis for $m_d(t)$ and as seen from Table 22.7, α is the most while λ and β are the least sensitive parameters.

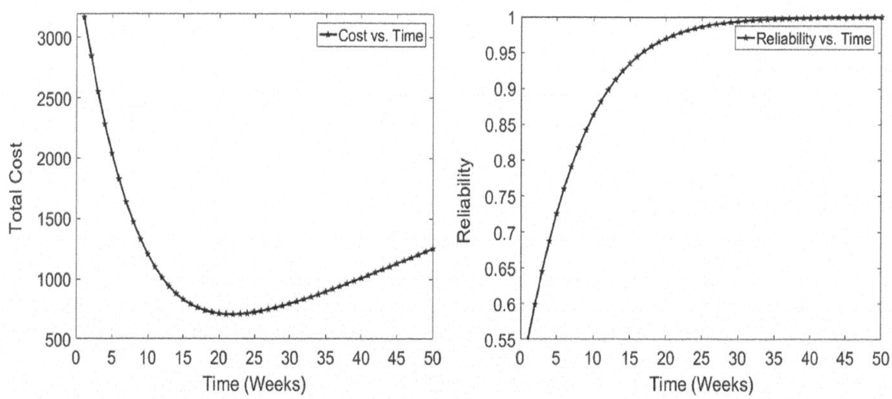

Fig. 22.6 For $Q_3 = 7000$. Total cost versus Time and Reliability versus Time curve

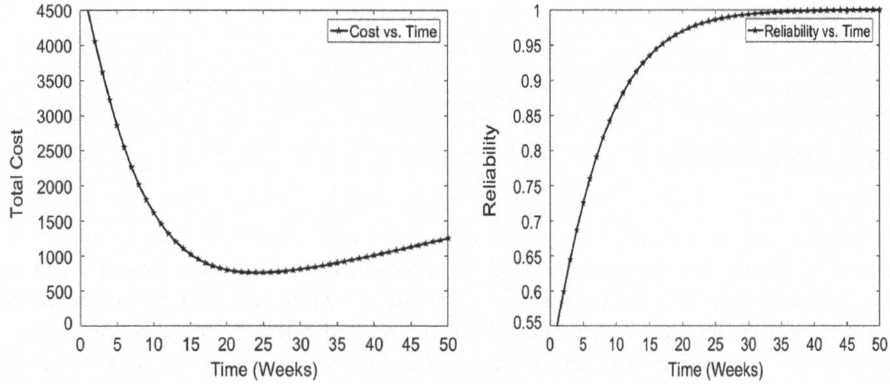

Fig. 22.7 For $Q_3 = 10{,}000$. Total cost versus Time and Reliability versus Time curve

22.8 Conclusions and Future Work

In this work, an SRGM for both fault correction and detection processes in imperfect debugging with RFE is developed. Some of the important observations of this work are: here detection and correction processes are considered separately, inflection s-shaped fault detection function and constant fault correction rate is considered. Analysis of proposed model is done, and it is found that model with smaller value of θ is good. According to sensitivity analysis α is the most sensitive parameter. From results, it is clear that our proposed model is giving better performance than previous models.

Here single release policy is given, multiple release policy can be calculated on this model. We have considered constant fault correction rate; time-dependent fault correction could also be considered in future work.

References

1. Pham H (2007) System software reliability. Springer Science & Business Media
2. Lyu MR (1996) Handbook of software reliability engineering. McGraw Hill, New York
3. Pachauri B, Kumar A, Dhar J (2019) Reliability analysis of open source software systems considering the effect of previously released version. Int J Comput Appl 41:31–38
4. Mishra G, Kapur P, Shrivastava A (2017) Multi release cost model a new perspective. Int J Reliab Qual Saf Eng 24(06):1740007
5. Li Q, Pham H (2017) Nhpp software reliability model considering the uncertainty of operating environments with imperfect debugging and testing coverage. Appl Math Model 51:68–85
6. Pachauri B, Dhar J, Kumar A (2015) Incorporating inflection s-shaped fault reduction factor to enhance software reliability growth. Appl Math Model 39(5–6):1463–1469
7. Kapur P, Anand S, Inoue S, Yamada S (2010) A unified approach for developing software reliability growth model using infinite server queuing model. Int J Reliab Qual Saf Eng 17(05):401–424

8. Kapur P, Goswami D, Bardhan A, Singh O (2008) Flexible software reliability growth model with testing effort dependent learning process. Appl Math Model 32(7):1298–1307
9. Lai R, Garg M (2012) A detailed study of NHPP software reliability models. J Softw 7(6):1296–1306
10. Gandhi N, Sharma H, Aggarwal AG, Tandon A (2019) Reliability growth modeling for oss: a method combining the bass model and imperfect debugging. In: Smart innovations in communication and computational sciences. Springer, Berlin, pp 23–34
11. Majumdar R, Kapur P, Khatri SK, Shrivastava A (2018) Evolutionary algorithm based faults optimization of multi-modular software. In: Smart computing and informatics. Springer, Berlin, pp 281–291
12. Kapur P, Shrivastava A, Singh O (2017) When to release and stop testing of a software. J Indian Soc Probab Stat 18(1):19–37
13. Kumar R, Kumar S, Tiwari SK, A study of software reliability on big data open source software. Int J Syst Assur Eng Manag 1–9
14. Dhar J, Ingle S, Sheshker Y et al (2014) Software reliability growth model with logistic exponential tef in imperfect debugging environment. In: Recent advances and innovations in engineering (ICRAIE). IEEE, pp 1–4
15. Pachauri B, Kumar A, Raja S (2019) Imperfect software reliability growth model using delay in fault correction. In: Performance prediction and analytics of fuzzy, reliability and queuing models. Springer, pp 119–126
16. Wang J, Wu Z (2016) Study of the nonlinear imperfect software debugging model. Reliab Eng Syst Saf 153:180–192
17. Kapur P, Pham H, Anand S, Yadav K (2011) A unified approach for developing software reliability growth models in the presence of imperfect debugging and error generation. IEEE Trans Reliab 60(1):331–340
18. Kapur P, Singh O, Shrivastava A, Singh JN (2015) A software up-gradation model with testing effort and two types of imperfect debugging. In: 2015 international conference on futuristic trends on computational analysis and knowledge management (ABLAZE). IEEE, pp 613–618
19. Pham H, Zhang X (2003) Nhpp software reliability and cost models with testing coverage. Eur J Oper Res 145(2):443–454
20. Ohba M (1984) Software reliability analysis models. IBM J Res Dev 28(4):428–443
21. Pham H (2016) A generalized fault-detection software reliability model subject to random operating environments. Vietnam J Comput Sci 3:145–150
22. Song KY, Chang IH, Pham H (2017) A software reliability model with a weibull fault detection rate function subject to operating environments. Appl Sci 7(10):983
23. Zhu M, Pham H (2018) A software reliability model incorporating martingale process with gamma-distributed environmental factors. Ann Oper Res 1–22
24. Li Q, Pham H (2017) NHPP software reliability model considering the uncertainty of operating environments with imperfect debugging and testing coverage. Appl Math Model 51:68–85
25. Song KY, Chang IH, Pham H (2017) A three-parameter fault-detection software reliability model with the uncertainty of operating environments. J Syst Sci Syst Eng 26(1):121–132
26. Chang IH, Pham H, Lee SW, Song KY (2014) A testing-coverage software reliability model with the uncertainty of operating environments. Int J Syst Sci: Oper Logist 1(4):220–227
27. Mao C, Li Q (2016) A testing-coverage software reliability growth model considering the randomness of the field environment. In: 2016 IEEE international conference on software quality, reliability and security companion (QRS-C), pp 402–403
28. Teng X, Pham H (2006) A new methodology for predicting software reliability in the random field environments. IEEE Trans Reliab 55(3):458–468
29. Pachauri B, Kumar A, Dhar J (2014) Software reliability growth modeling with dynamic faults and release time optimization using GA and MAUT. Appl Math Comput 242:500–509
30. Lo J-H, Huang C-Y (2006) An integration of fault detection and correction processes in software reliability analysis. J Syst Softw 79(9):1312–1323
31. Zhu M, Pham H (2016) A software reliability model with time-dependent fault detection and fault removal. Vietnam J Comput Sci 3(2):71–79

32. Yang J, Liu Y, Xie M, Zhao M (2016) Modeling and analysis of reliability of multi-release open source software incorporating both fault detection and correction processes. J Syst Softw 115:102–110
33. Chatterjee S, Singh J (2014) A NHPP based software reliability model and optimal release policy with logistic–exponential test coverage under imperfect debugging. Int J Syst Assur Eng Manag 5(3):399–406
34. Majumdar R, Shrivastava A, Kapur P, Khatri SK (2017) Release and testing stop time of a software using multi-attribute utility theory. Life Cycle Reliab Saf Eng 6(1):47–55
35. Rafi SM, Akthar S (2010) Software reliability growth model with logistic-exponential testing effort function and analysis of software release policy. In: Proceedings of international conference on advances in computer science
36. Pachauri B, Kumar A, Dhar J (2013) Modeling optimal release policy under fuzzy paradigm in imperfect debugging environment. Inf Softw Technol 55(11):1974–1980
37. Kapur P, Pham H, Aggarwal AG, Kaur G (2012) Two dimensional multi-release software reliability modeling and optimal release planning. IEEE Trans Reliab 61(3):758–768
38. Koch HS, Kubat P (1983) Optimal release time of computer software. IEEE Trans Softw Eng 3:323–327
39. Yamada S, Osaki S (1985) Cost-reliability optimal release policies for software systems. IEEE Trans Reliab 34(5):422–424
40. Zhang X, Pham H (1998) A software cost model with error removal times and risk costs. Int J Syst Sci 29(4):435–442
41. Yang B, Xie M (2000) A study of operational and testing reliability in software reliability analysis. Reliab Eng Syst Saf 70(3):323–329
42. Goel AL, Okumoto K (1979) Time-dependent error-detection rate model for software reliability and other performance measures. IEEE Trans Reliab 28(3):206–211
43. Yamada S, Ohba M, Osaki S (1984) S-shaped software reliability growth models and their applications. IEEE Trans Reliab 33(4):289–292
44. Pham H, Nordmann L, Zhang Z (1999) A general imperfect-software-debugging model with S-shaped fault-detection rate. IEEE Trans Reliab 48(2):169–175
45. Pham H (2014) A new software reliability model with V-tub-shaped fault-detection rate and the uncertainty of operating environments. Optimization 63(10):1481–1490
46. Chitnis N, Hyman JM, Cushing JM (2008) Determining important parameters in the spread of malaria through the sensitivity analysis of a mathematical model. Bull Math Biol 70(5):1272

Chapter 23
Autoregressive Model for Multivariate Crime Prediction

Romika Yadav and Savita Kumari Sheoran

23.1 Introduction

Crime is a legal offence possessing a great challenge before modern civilized societies. It has deeply penetrated to the society as an unproductive attitude. The severity of situation can be adjudged from the fact that the National Judicial Report in USA reveals that every out of 100 people every 12 are victims of criminal events. The severity of issue is so deep that United State had to issue advisory in many cities for public safety. A report submitted by Federal Bureau of Investigation (FBI) in 2018 emphasizes that in many areas the heinous crimes like homicide and criminal breach of confidence have achieved peak in past few years. It is undesired social behavior where prevention is considered always better then cures. In order to mitigate the crimes, the law enforcement and security agencies always focus getting advance information of crime to deploy forces and issue public advisories as preventive measure.

The prediction of crime in advance is always helpful in preventing the crime to happen. Crime prediction includes predicting offenders, offender's crime patterns and predicting victim of crime along with crime prone site. With the societal advancement leverages by the technological developments the crimes have many folds in volume, varieties as well as velocity (dynamism). This leap and bound hike in crime events forced the stakeholders to be more proactive and pre-judgmental on crime events. Recently the nature of crime data has appeared as big challenge in the path of forecasting the crime events. The present day crime data set is propelled by

R. Yadav (✉) · S. K. Sheoran
Department of Computer Science & Engineering, Indira Gandhi University Meerpur, Rewari, India
e-mail: romikasim@gmail.com

S. K. Sheoran
e-mail: savita.sheoran@igu.ac.in

P. K. Kapur et al. (eds.), *Decision Analytics Applications in Industry*, Asset Analytics,
https://doi.org/10.1007/978-981-15-3643-4_23

technological advancement completely indifferent from routine datasets and hence traditional data analysis techniques failed drastically.

Data analytics refers to quantitative and qualitative techniques used to enhanced data activity. Data is categorized and extracted on behavioral patterns fit to organizational requirements. The main reason behind poor performance of data analytics in effective forecasting of the crime event is the complex nature of timeline data. The available crime dataset is either semi-structure or unstructured with high volume, velocity, variety and veracity. The dataset with these characteristics fall under recent category of Big Data, which is suffice to deal with such novel data regime.

It is the process of analyzing, collecting and organizing large datasets called Big Data analytics to infer meaningful patterns from the information. Big Data analytics provide better information of crime dataset and reveals the complex and hidden pattern to map the future predictable events. Crime increased day by day due to this advancements and new techniques is required to accommodate and for prevention for these crimes. Computer technologies in today's environment give fine systems and techniques, i.e., Big Data analytics. Also fuzzy rule base inference has recently been popularized in such complex scenario where decision-making require a lot of iterations. Therefore, predictive analytics involving Big Data, fuzzy logic and multivariate functions is quite a significant tools in analyzing the crime events which can infer evocative results from the existing crime dataset. As we concerned with the cyber crime it is increased in various forms which include financial fraud, hacking and e-mail spoofing are major crimes that need to be analysed properly in crime prediction. Intrusion detection method can be applied to prevent cyber crime and Big Data techniques also provide helpful way to overcome the cyber crime.

The research studies conducted across the globe reveals that using Big Data blended with fuzzy logic is a powerful tool to make future prediction on the basis of past data. Few researches in China focus on measurement of trajectories for security reasons [1]. Similarly, a database approach is applied in a research conducted in Chicago city for spatio-temporal forecasting with experimental valuation statistics [2]. These studies suffer of their intricate issues like accuracy, precession etc.

Also many of studies are based upon Univariate dataset; however, in real situation many situations are there which involve multivariate data. We have explored multivariate crime data over blended techniques of Big Data and fuzzy membership functions, where later is used as analytic platform and former as function. Such experimentation is expected to generate real-time solution for Crime Prediction.

The further organization of this paper comprises as follows: Sect. 23.2 includes literature analysis of various research works. Section 23.3 devoted to define the problem of hybrid approach work. Section 23.4 discussed the methodology of the proposed work and Sect. 23.5 provides results ad discussions of proposed work.

23.2 Literature Survey

The regime of crime prediction is naïve area and a lot of research studies have been carried out by research community. But main issue is that this area is highly dynamic in nature and hence change at every instance of time. The research work by Krause Josua et al. discover a technique technique which is capable to describe a method that combine spatial likelihood modeling and logistic regression modeling for mapping the hidden patterns. Based upon experimental work they have developed the fusion method which can improves the scalability of forecasting methods subsequently available in spatial domain [1]. The author of these researches explored the role of mathematical modeling in criminology. Criminality is an interdisciplinary subject and confront at various levels of study. The research by Krause Josua et al. reveals that the crime events have varied dynamic and depends on situation and locations. The research outcomes envisages that population in developed countries in urban areas is around 85%. Therefore in these areas the crime events mostly happen in urban locality while in the country like India majority of population lies in rural areas and hence these geographical localities are more prone to crime.

The research work in various papers extensively improves the prediction over a given geographic area but they drastically in a varied situation [2, 3]. The research work by Orsogna Maria and Catlett Charlie et al. involve a predictive modeling of multi-dimensional and diverse data with semantic features. The work done by few researchers show that non-informative features can be best estimated by applying the feature selection algorithm. The work of Brown et al. is confined to rank the predictive features based upon diagonal selection algorithm through validation and classifiers [4]. Further, forecasting of societal occurrences like elections, epidemic outbreaks, domestic violences and political crises etc. were modeled to cross verify the research outputs. The work of Samuel et al. focuses on design consideration, system architecture and hardware performance [5].

The research of Ramakrishnan N et al. mainly focus was on theoretical modeling of crime preventive measures adopted in real-time situation [6]. They have devised naïve Bayes classifier with feature reduction for intrusion detection. In order to categorize important features a vitality-based reduction method is proposed [7]. Few researchers have These researches have highlighted the complex systems which are used interdisciplinary areas [8]. Based on the work reported in the literature review this piece of research work formulated and addressed a more real-time issue [9].

23.3 Problem Definition

Crime is a worldwide public concern and expected to be severe in the time to come. The main research obstacle in this regime is that situation is highly complex, uncertain and dynamic one while the dataset is non-traditional. In past many geographical techniques were successfully deployed for univariate analysis of data but in real situation many times we require multivariate data set. Our work is to accommodate new data functionalities and multivariate analysis of data. We develop a technique to add the results of different input variables and generate a useful prediction for law enforcement agencies and security departments.

23.4 Methodology

The Autoregressive models carry out prediction in behavioural trend using the past pattern. The prediction find correlation between time-series values and succeed with mapping of past with future based on certain rules auto developed through hidden pattern.

The web portal of 'National Crime Records Bureau (NCRB)' is authenticated and reliable sources of crime data in India. For the purpose this study we use data extracted from NCCRB portal. We are only analyzing the dataset of union territory Delhi and state Haryana of country India. Further illustrate Multivariate analysis of crime data.

$$Y_t = \mu + \emptyset_1 Y_{t-1} + \emptyset_2 Y_{t-2} \ldots \emptyset_p Y_{t-p} \ldots e_t$$

where Y_t is time series component in the equation; μ_t, is intercept value of time series; $\emptyset_1$, $\emptyset_2$ and $\emptyset_p$ are stationary time series while e_t is noise occurring at time t.

The series data obtained from NCRB portal are analysed through vector Autoregression models with multivariate data. The variable in the dataset is a complex function of past lag which is used to formulate future lags of other variable in the same domain.

Further enhance to make hybrid approach of vector autoregressive model with fuzzy set to identify the membership degree of crime for the Delhi and Haryana.

Representation of fuzzy set as defined with the crime variables. A fuzzy set '***F***' can be defined using the expression

$$F = \sum \mu F\,(c_i)/c_i$$

The translation from c to $\mu F\,(c)$ is known as fuzzification of the membership, μF (c_i) is a membership degree, and $\mu F\,(c_i)/c_i$—refers to a fuzzy set containing exactly one element c with a membership degree. Here 0—less Crime, 0.5—Medium crime and 1—High Crime.

23.5 Results and Discussions

In this section, results for the hybrid approach is shown in which vector autoregressive model of multivariate used for the crime prediction. Then fuzzy approach used to identify which crime high membership degree means high occurring crime in the respective states Haryana and Delhi.

Figure 23.1 show two variable crime forecasting in favor of Delhi for the murder and rape in which data from the year 2001 to 2016 are taken as time series data and predicted values for the year 2016, 2017 and 2018.

Subsequently Fig. 23.1 shows a multivariate forecasting of crime which comprise arson, theft, assault, vehicle theft, robbery and kidnapping of Delhi. The forecasting of crime by vector autoregressive model it takes input as a multiple data set and correspond to the output in one graph.

Figure 23.2 shows a multivariate forecasting of crime which include assault, arson and burglary of Haryana. The prediction of crime by vector autoregressive model it take input as a multiple data set and correspond to the output in one graph. Consequently take the fuzzy approach in which the membership function is categorized according to their values are associated with it. In the membership degree the values

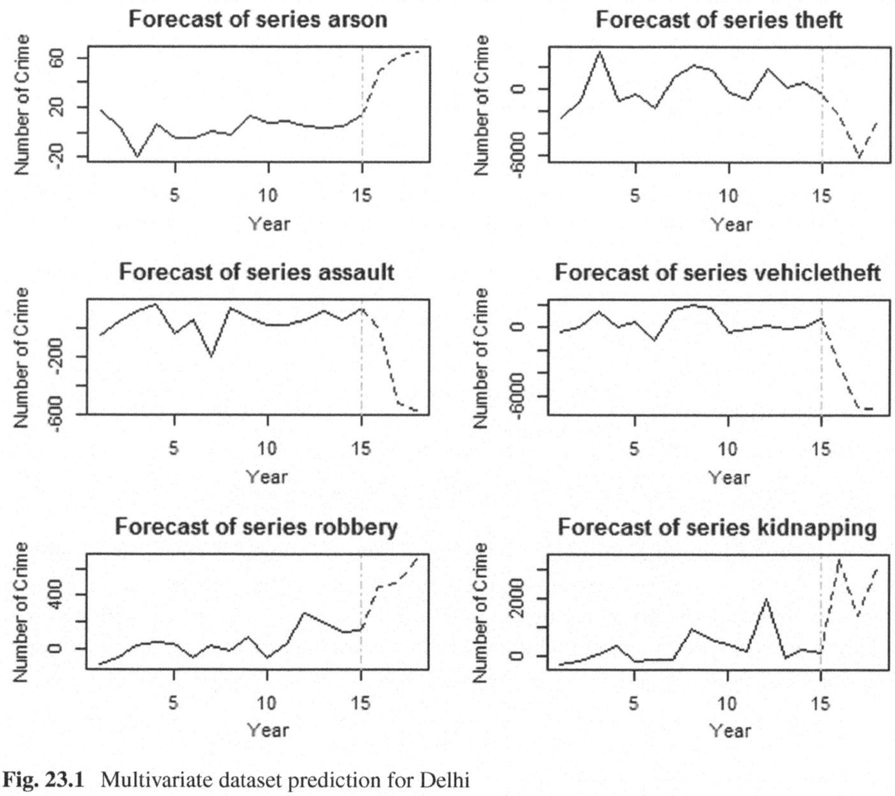

Fig. 23.1 Multivariate dataset prediction for Delhi

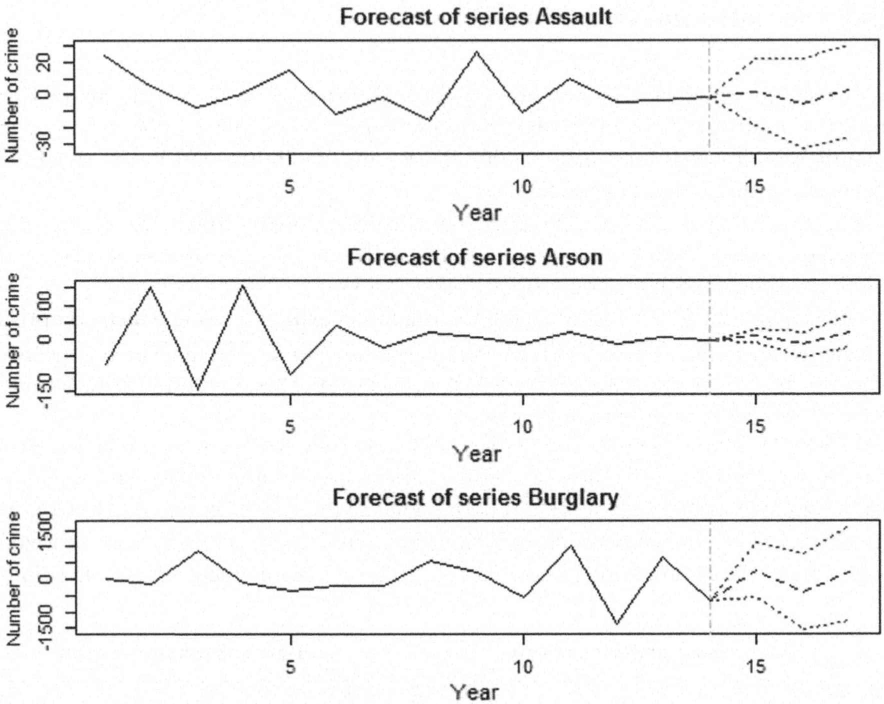

Fig. 23.2 Multivariate dataset prediction for Haryana

are ranging from the 0 to 1. This is also called as a crisp set in those values only belong the range from 0 to 1. The degree of membership firstly calculated for the Delhi.

$$\text{Less crime} - (0/\text{Arson},\ 0/\text{criminal breach of trust},\ 0/\text{Murder})$$

$$\text{Medium Crime} - \begin{pmatrix} 0.5/\text{Assault},\ 0.5/\text{Burglary},\ 0.5/\text{Robbery}\ ,\ 0.5/\text{Rape}, \\ 0.5/\text{Kidnapping} \end{pmatrix}$$

$$\text{High Crime} - (1/\text{Theft},\ 1/\text{Vehicle Theft})$$

The degree of membership was firstly calculated for the state Haryana. In which data would be considered from the predicted values of all crime types of state Haryana. In which the predicted are added with the actual one and then which crime is more and highly happened in which state is identified using the fuzzy membership function.

$$\text{Less crime} - (0/\text{Arson},\ 0/\text{Assault},\ 0/\text{Robbery})$$

$$\text{Medium Crime} - \begin{pmatrix} 0.5/\text{criminal breach of trust},\ 0.5/\text{Rape},\ 0.5/\text{Murder}, \\ 0.5/\text{Kidnapping} \end{pmatrix}$$

$$\text{High Crime} - (1/\text{Burglary},\ 1/\text{Theft},\ 1/\text{Vehicle Theft})$$

From the above fuzzy membership function clearly stated that the both the states have less Arson crime and high Vehicle Theft crime.

23.6 Conclusions

This research paper had applied the Autoregressive model on time series data for crime trend prediction and subsequently to forecast the crime sites. The results presented in Sect. 23.5, signify that the multivariate analysis using vector auto regressive model consequently applied with fuzzy membership functions is capable to segregate the less spreader and high spreader crime. As the application of fuzzy membership function simplify multivariate crime prediction. It signifies that soft computing techniques like fuzzy neural networks have a better scope in this regime. Therefore, vector auto regressive with neural network and AI may be investigated to further explore this issue.

References

1. Josua K, Adam P, Enrico B (2014) INFUSE: interactive feature selection for predictive modeling of high dimensional data. IEEE Trans Visual Comput Graph 20(12):1614–1623
2. Catlett C, Cesario E; Domenico T, Andrea V (2018) A data-driven approach for spatio-temporal crime predictions in smart cities. In: IEEE international conference on smart computing, pp 17–24
3. D'Orsogna MR, Perc M (2015) Physics for better human societies reply to comments on "statistical physics of crime: a review. Phys Life Rev 12:40–43. Elsevier
4. Donald B, Jason D, Heidi H (2004) Spatial forecast methods for terrorist events in urban environments. Notes in computer science, vol 3073. Springer, Berlin, pp 426–435
5. Huddleston SH, Brown DE (2009) A statistical threat assessment. IEEE Trans Syst Man Cybern – Part - A: Syst Hum 39(6):1307
6. Ramakrishnan Naren L, Chang-Tien MM, Achla M, Anil V, Stephen E, Scotland L, Michael R, Brownstein JS, Kristen S, Lise G, Aravind S, Tanzeem C, Dipak G, David M (2015) Model-based forecasting of significant societal events. IEEE Intell Syst 30(5):86–90
7. Wang L, Yu Z, Guo B, Yi F, Ku T (2017) Moving destination prediction using sparse dataset: a mobility gradient descent approach. ACM Trans Knowl Discovery Data 11. https://doi.org/10.1145/3051128
8. Mukherjee S, Sharma N (2012) Intrusion detection using naive bayes classifier with feature reduction 4:119–128. Elsevier
9. Primicerio M (2015) The role of mathematical modelling in modern criminology comment on "statistical physics of crime: a review In: D'Orsogna MR, Perc M (eds) Physics of life reviews. Elsevier, pp 34–35

Chapter 24
Reliability and Profit Analysis of a Power Generating System with Effect of Ambient Temperature and Priority for Repair to the Gas Turbine over Steam Turbine on System Failure

Rajesh, Gulshan Taneja, and Jagdish Prasad

24.1 Introduction

Systems consisting one or two or more similar/dissimilar units have been studied by various researchers including [1–3] in the area of reliability. In the studies, based on two dissimilar units, most of them have taken one unit as operative and the other one as standby. However, there may be situations where being two dissimilar units nature of their output is the same. Such a situation was discussed by Singh and Taneja [4, 5] for a G. T. power plant but they made the study without considering the parameter 'Ambient Temperature' (A. T.). However, consideration of A. T. is very important as it affects the production capacity of the G. T. power plant. Effect of A. T. was taken into consideration by [6–8] but they did not find the reliability and performance measures like Mean Time to System Failure (MTSF), Availability in different Cycles, etc. Rajesh et al. [9, 10] discussed a reliability model on a G. T. power generating system consisting 2 G. T. and 1 S. T. with effect of A. T. on the production of the plant. In their study they followed FCFS repair pattern throughout. However, if the repair of S. T. is continued on the failure of all the three turbines also, it may lead to increase in the down time of the system as even on completion of the repair of the

Rajesh (✉) · J. Prasad
Department of Statistics, University of Rajasthan, Jaipur, India
e-mail: rajesh.sos@ignou.ac.in

J. Prasad
e-mail: jagdish55_singh@yahoo.co.in

G. Taneja
Department of Mathematics, Maharshi Dayanand University, Rohtak, India
e-mail: drgtaneja@gmail.com

J. Prasad
Presently Amity School of Applied Sciences, Amity University Rajasthan, Jaipur, India

P. K. Kapur et al. (eds.), *Decision Analytics Applications in Industry*, Asset Analytics,
https://doi.org/10.1007/978-981-15-3643-4_24

S. T.; it cannot be made operative without the G. T. Hence, there is need to examine that as to whether the repair of the S. T. should be continued or may be kept in abeyance by undertaking the failed G. T. for repair. To examine this, there is need to develop a reliability model for such a system by giving priority for repair to the G. T. over the S. T. on complete system failure.

Keeping all this in view, the present paper, discuss cost analysis of a G. T. system with change in season (winter and summer). Working of the system in two different seasons, i.e., summer and the winter have been taken into consideration while developing the Model. The power generating system under consideration comprises 2 G. T. and 1 S. T. Initially, both the Gas Turbines and the S. T. are in operating mode and this period of working is called working in Full Capacity (F. C.). When the system works with failure of at least one of the turbines, its working is termed as working under reduced capacity. However, in the case of failure of both the Gas Turbines, the S. T. cannot work and hence the system is put to down mode. When all the three turbines get failed, the priority for repair is given to the G. T. over the S. T., otherwise the repair pattern is considered as FCFS. The analysis of the system is done using semi-Markov processes and Regenerative Point Technique. MTSF, availability in different capacities, etc. have been obtained. The availability analysis is done in case of Full as well as in reduced Capacities and also for different seasons. Computational work along with graphical study has been done using R by making use of the contributions made by [11, 12]. Using the estimated values (on the basis of one year collected data/information from a G. T. power plant) of different functions and parameters involved in the present study various interesting conclusions have been drawn including the minimum electricity price to be charged from the buyer of the power generated in Type I and Type III for different values of probabilities of paying higher price of electricity by the buyer.

24.2 Notation, Description of the Model and State Transition Probabilities

24.2.1 Notation

Symbol	Meaning
G. T./S. T./A. T.	Gas Turbine/Steam Turbine/Ambient Temperature
PDF/CDF	Probability Density Function/Cumulative Distribution Function
RS/FS/FPT	Regenerative State/Failed State/First Passage Time
$q_{ij}(t)/Q_{ij}(t)$	PDF/CDF of FPT from RS i to a RS j or a FS j without visiting to any other RS in time (0, t]
$q_{ij}^{(k)}(t)/Q_{ij}^{(k)}(t)$	PDF/CDF of FPT from RS i to a RS j or a FS j visiting state k only once in time (0, t]
$\phi_i(t)$	CDF of the FPT from RS i to FS

(continued)

(continued)

Symbol	Meaning
$O_{gt}^{wi},\ O_{st}^{wi}/O_{gt}^{su},\ O_{st}^{su}$	G. T., S. T. operative in winter/summer
$U_{rgt}^{wi},\ U_{rst}^{wi}/U_{rgt}^{su},\ U_{rst}^{su}$	G. T., S. T. under repair in winter/summer
$d_{gt}^{wi},\ d_{st}^{wi}/d_{gt}^{su},\ d_{st}^{su}$	G. T., S. T. put to down mode in winter/summer
$W_{rgt}^{wi},\ W_{rst}^{wi}/W_{rgt}^{su},\ W_{rst}^{su}$	G. T., S. T. waiting for repair in winter/summer
$U_{Rst}^{wi}/U_{Rst}^{su}$	Repair of S. T. continuing from previous state in winter/summer
$h_1(t)/h_2(t)$	PDF of time of changing the season from winter to summer/(summer to winter)
$\lambda_1, \alpha_1/\lambda_2, \alpha_2$	Failure rate of G. T., S. T. in winter/summer
$g_1(t)/g_2(t)$	PDF of repair time of gas/steam turbine
p_{11}/p_{21}	Probability that there is demand for higher rate but this demand is not greater than production of one gas turbine in winter/summer
p_{12}/p_{22}	Probability that there is demand for higher rate and this demand is greater than production of one gas turbine in winter/summer
$p_1 = p_{11} + p_{12}/$ $p_2 = p_{21} + p_{22}$	Probability that there is demand for higher rate and this demand is either greater than or not greater than production of one gas turbine in winter/summer
$q_1 = 1 - p_1/q_2 = 1 - p_2$	Probability that no customer is agree to pay higher price in winter/summer
$AW_i^f(t),\ AW_i^I(t),$ $AW_i^{II}(t),\ AW_i^{III}(t)/$ $AS_i^f(t),\ AS_i^I(t),$ $AS_i^{II}(t),\ AS_i^{III}(t)$	Probability that system is up in F. C., Type I, Type II, Type III during winter/summer at instant t given that it entered state i at t = 0
$q^*(s)/Q^{**}(s)$	Laplace Transform (LT) of q(t)/Laplace Stieltjes Transform (LST) of Q(t)
©/Ⓢ	Laplace Convolution/(Laplace Stieltjes Convolution)

24.2.2 *Description of the Model*

State transition diagram for the model can be seen in Fig. 24.1. States 0 to 5, 9, 12, 13, 14, 15, 16, 20 and 23 are regeneration points and thus 0, 1, 2, 3, 4, 5, 9, 12, 13, 14, 15, 16, 20 and 23 are RS. States 9, 11, 20 and 22 are FS. States 0 and 1 are states where the system works in F. C. (both gas turbines and S. T. are operating) in winter and summer respectively. States 4 and 15 are the states where the system works at reduced capacity when only two gas turbines are operative [Type I Capacity, say] in winter and summer respectively. Similarly, states 5 and 16 are the states corresponding to when one G. T. and one S. T. are operating [Type II Capacity, say] in winter and summer respectively. States 3, 6, 8 and 12/(14, 17, 19 and 23) are the states corresponding to when only one G. T. is operating [Type III Capacity, say] in winter/(summer). States

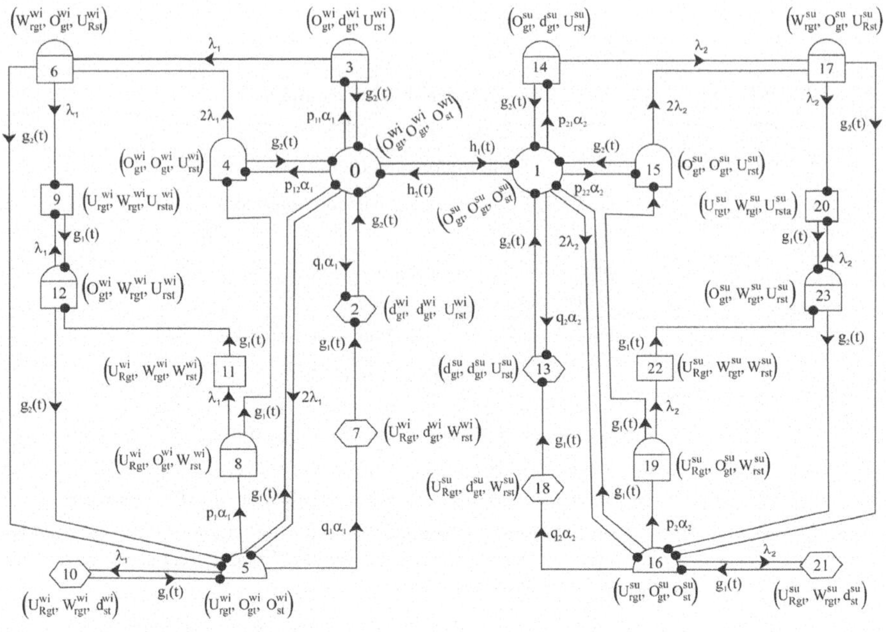

Fig. 24.1 State transition diagram

10 and 21 are the states where the S. T. is compulsorily put to down mode on failure of both gas turbines as the former cannot work without the operability of at least one G. T. in winter and summer respectively. States 7 and 2/(18 and 13) are the states the gas turbine(s) is/are put to down mode when the S. T. is failed and there is no buyer of electricity at higher price so as to make the system to work at reduced capacities in winter/(summer).

24.2.3 Transition Probabilities

The values of $q_{ij}(t)$, PDF of FPT from RS i to RS j or FS j are obtained and the same are given as follows:

$q_{01}(t) = e^{-(2\lambda_1+\alpha_1)t}.h_1(t)$		$q_{02}(t) = q_1\alpha_1 \cdot e^{-(2\lambda_1+\alpha_1)t}.\overline{H}_1(t)$	
$q_{03}(t) = p_{11}\alpha_1 \cdot e^{-(2\lambda_1+\alpha_1)t} \cdot \overline{H}_1(t)$		$q_{04}(t) = p_{12}\alpha_1 \cdot e^{-(2\lambda_1+\alpha_1)t} \cdot \overline{H}_1(t)$	
$q_{05}(t) = 2\lambda_1 \cdot e^{-(2\lambda_1+\alpha_1)t} \cdot \overline{H}_1(t)$		$q_{10}(t) = e^{-(2\lambda_2+\alpha_2)t} \cdot h_2(t)$	
$q_{1,13}(t) = q_2\alpha_2 \cdot e^{-(2\lambda_2+\alpha_2)t} \cdot \overline{H}_2(t)$		$q_{1,14}(t) = p_{21}\alpha_2 \cdot e^{-(2\lambda_2+\alpha_2)t} \cdot \overline{H}_2(t)$	
$q_{1,15}(t) = p_{22}\alpha_2 \cdot e^{-(2\lambda_2+\alpha_2)t} \cdot \overline{H}_2(t)$		$q_{1,16}(t) = 2\lambda_2 \cdot e^{-(2\lambda_2+\alpha_2)t} \cdot \overline{H}_2(t)$	
$q_{20}(t) = g_2(t)$	$q_{30}(t) = g_2(t) \cdot e^{-\lambda_1 t}$	$q_{35}^{(6)}(t) = \lambda_1 t e^{-\lambda_1 t} g_2(t)$	$q_{39}^{(6)}(t) = \lambda_1^2 t e^{-\lambda_1 t}\overline{G}_2(t)$
$q_{40}(t) = g_2(t) \cdot e^{-2\lambda_1 t}$		$q_{45}^{(6)}(t) = 2[e^{-\lambda_1 t} - e^{-2\lambda_1 t}]g_2(t)$	

(continued)

(continued)

$q_{49}^{(6)}(t) = 2\lambda_1[e^{-\lambda_1 t} - e^{-2\lambda_1 t}]\overline{G}_2(t)$	$q_{50}(t) = g_1(t) \cdot e^{-(\alpha_1+\lambda_1)t}$		
$q_{52}^{(7)}(t) = \frac{q_1\alpha_1}{\alpha_1+\lambda_1}[1 - e^{-(\alpha_1+\lambda_1)t}]g_1(t)$	$q_{54}^{(8)}(t) = p_1[e^{-\lambda_1 t} - e^{-(\alpha_1+\lambda_1)t}]g_1(t)$		
$q_{55}^{(10)}(t) = \frac{\lambda_1}{\alpha_1+\lambda_1}[1 - e^{-(\alpha_1+\lambda_1)t}]g_1(t)$	$q_{5,11}^{(8)}(t) = p_1\lambda_1[e^{-\lambda_1 t} - e^{-(\alpha_1+\lambda_1)t}]\overline{G}_1(t)$		
$q_{5,12}^{(8,11)}(t) = \frac{p_1}{\alpha_1+\lambda_1}[\alpha_1(1 - e^{-\lambda_1 t}) - \lambda_1(e^{-\lambda_1 t} - e^{-(\alpha_1+\lambda_1)t})]g_1(t)$			
$q_{9,12}(t) = g_1(t)$	$q_{12,5}(t) = e^{-\lambda_1 t}g_2(t)$	$q_{12,9}(t) = \lambda_1 e^{-\lambda_1 t}\overline{G}_2(t)$	$q_{13,1}(t) = g_2(t)$
$q_{14,1}(t) = g_2(t) \cdot e^{-\lambda_2 t}$	$q_{14,16}^{(17)}(t) = \lambda_2 t e^{-\lambda_2 t}g_2(t)$	$q_{14,20}^{(17)}(t) = \lambda_2^2 t e^{-\lambda_2 t}\overline{G}_2(t)$	
$q_{15,1}(t) = g_2(t) \cdot e^{-2\lambda_2 t}$	$q_{15,16}^{(17)}(t) = 2[e^{-\lambda_2 t} - e^{-2\lambda_2 t}]g_2(t)$	$q_{15,20}^{(17)}(t) = 2\lambda_2[e^{-\lambda_2 t} - e^{-2\lambda_2 t}]\overline{G}_2(t)$	
$q_{16,1}(t) = g_1(t) \cdot e^{-(\alpha_2+\lambda_2)t}$	$q_{16,13}^{(18)}(t) = \frac{q_2\alpha_2}{\alpha_2+\lambda_2}[1 - e^{-(\alpha_2+\lambda_2)t}]g_1(t)$		
$q_{16,15}^{(19)}(t) = p_2[e^{-\lambda_2 t} - e^{-(\alpha_2+\lambda_2)t}]g_1(t)$	$q_{16,16}^{(21)}(t) = \frac{\lambda_2}{\alpha_2+\lambda_2}[1 - e^{-(\alpha_2+\lambda_2)t}]g_1(t)$		
$q_{16,22}^{(19)}(t) = p_2\lambda_2[e^{-\lambda_2 t} - e^{-(\alpha_2+\lambda_2)t}]\overline{G}_1(t)$			
$q_{16,23}^{(19,22)}(t) = \frac{p_2}{\alpha_2+\lambda_2}[\alpha_2(1 - e^{-\lambda_2 t}) - \lambda_2(e^{-\lambda_2 t} - e^{-(\alpha_2+\lambda_2)t})]g_1(t)$			
$q_{20,23}(t) = g_1(t)$	$q_{23,16}(t) = e^{-\lambda_2 t}g_2(t)$	$q_{23,20}(t) = \lambda_2 e^{-\lambda_2 t}\overline{G}_2(t)$	

The transition probabilities p_{ij} can be obtained using the relation $p_{ij} = \lim_{s\to 0} q_{ij}^*(s)$.

Mean Sojourn Time (μ_i) gives expected time of stay of the system in state i. The obtained values of (μ_i) are given as follows:

$\mu_0 = \frac{1 - h_1^*(\alpha_1 + 2\lambda_1)}{\alpha_1 + 2\lambda_1}$	$\mu_1 = \frac{1 - h_2^*(\alpha_2 + 2\lambda_2)}{\alpha_2 + 2\lambda_2}$	$\mu_2 = \int_0^\infty \overline{G}_2(t)dt$	$\mu_3 = \frac{1 - g_2^*(\lambda_1)}{\lambda_1}$
$\mu_4 = \frac{1 - g_2^*(2\lambda_1)}{2\lambda_1}$	$\mu_5 = \frac{1 - g_1^*(\alpha_1 + \lambda_1)}{\alpha_1 + \lambda_1}$	$\mu_9 = \int_0^\infty \overline{G}_1(t)dt$	$\mu_{12} = \frac{1 - g_2^*(\lambda_1)}{\lambda_1}$
$\mu_{13} = \mu_2$	$\mu_{14} = \frac{1 - g_2^*(\lambda_2)}{\lambda_2}$	$\mu_{15} = \frac{1 - g_2^*(2\lambda_2)}{2\lambda_2}$	
$\mu_{16} = \frac{1 - g_1^*(\alpha_2 + \lambda_2)}{\alpha_2 + \lambda_2}$	$\mu_{20} = \mu_9$	$\mu_{23} = \frac{1 - g_2^*(\lambda_2)}{\lambda_2}$	

24.3 MTSF

To obtain MTSF of the system, we need recursive relations for $\phi_i(t)$, CDF of the FPT from RS i to FS and the same are given as follows:

$$\phi_0(t) = Q_{01}(t)Ⓢ\phi_1(t) + Q_{02}(t)Ⓢ\phi_2(t) + Q_{03}(t)Ⓢ\phi_3(t) + Q_{04}(t)Ⓢ\phi_4(t) + Q_{05}(t)Ⓢ\phi_5(t)$$
$$\phi_1(t) = Q_{10}(t)Ⓢ\phi_0(t) + Q_{1,13}(t)Ⓢ\phi_{13}(t) + Q_{1,14}(t)Ⓢ\phi_{14}(t) + Q_{1,15}(t)Ⓢ\phi_{15}(t) + Q_{1,16}(t)Ⓢ\phi_{16}(t)$$
$$\phi_2(t) = Q_{20}(t)Ⓢ\phi_0(t)$$
$$\phi_3(t) = Q_{30}(t)Ⓢ\phi_0(t) + Q_{35}^{(6)}(t)Ⓢ\phi_5(t) + Q_{39}^{(6)}(t)$$
$$\phi_4(t) = Q_{40}(t)Ⓢ\phi_0(t) + Q_{45}^{(6)}(t)Ⓢ\phi_5(t) + Q_{49}^{(6)}(t)$$
$$\phi_5(t) = Q_{50}(t)Ⓢ\phi_0(t) + Q_{52}^{(7)}(t)Ⓢ\phi_2(t) + Q_{54}^{(8)}(t)Ⓢ\phi_4(t) + Q_{55}^{(10)}(t)Ⓢ\phi_5(t) + Q_{5,11}^{(8)}(t)$$
$$\phi_{13}(t) = Q_{13,1}(t)Ⓢ\phi_1(t)$$
$$\phi_{14}(t) = Q_{14,1}(t)Ⓢ\phi_1(t) + Q_{14,16}^{(17)}(t)Ⓢ\phi_{16}(t) + Q_{14,20}^{(17)}(t)$$
$$\phi_{15}(t) = Q_{15,1}(t)Ⓢ\phi_1(t) + Q_{15,16}^{(17)}(t)Ⓢ\phi_{16}(t) + Q_{15,20}^{(17)}(t)$$
$$\phi_{16}(t) = Q_{16,1}(t)Ⓢ\phi_1(t) + Q_{16,13}^{(18)}(t)Ⓢ\phi_{13}(t) + Q_{16,15}^{(19)}(t)Ⓢ\phi_{15}(t) + Q_{16,16}^{(21)}(t)Ⓢ\phi_{16}(t) + Q_{16,22}^{(19)}(t)$$

On taking LST and solving resulting equations for $\phi_0^{**}(s)$, we have
$\phi_0^{**}(s) = \dfrac{N(s)}{D(s)}$, where

$$D(s) = -q_{01}^{*}q_{10}^{*}\left\{1 - q_{16,17}^{(18)*}q_{17,16}^{(20)*} - q_{17,17}^{(22)*}\right\}\left[1 - q_{45}^{(6)*}q_{54}^{(8)*} - q_{55}^{(10)*}\right]$$
$$+\begin{bmatrix} 1 - q_{45}^{(6)*}q_{54}^{(8)*} - q_{55}^{(10)*} - q_{02}^{*}q_{20}^{*} + q_{02}^{*}q_{20}^{*}q_{45}^{(6)*}q_{54}^{(8)*} \\ +q_{02}^{*}q_{20}^{*}q_{55}^{(10)*} - q_{03}^{*}q_{30}^{*} + q_{03}^{*}q_{30}^{*}q_{45}^{(6)*}q_{54}^{(8)*} \\ +q_{03}^{*}q_{30}^{*}q_{55}^{(10)*} - q_{20}^{*}q_{52}^{(7)*}q_{03}^{*}q_{35}^{(6)*} - q_{20}^{*}q_{52}^{(7)*}q_{04}^{*}q_{45}^{(6)*} \\ -q_{20}^{*}q_{52}^{(7)*}q_{05}^{*} - q_{50}^{*}q_{03}^{*}q_{35}^{(6)*} - q_{50}^{*}q_{04}^{*}q_{45}^{(6)*} - q_{50}^{*}q_{05}^{*} \\ -q_{40}^{*}q_{04}^{*} - q_{40}^{*}q_{03}^{*}q_{35}^{(6)*}q_{54}^{(8)*} - q_{40}^{*}q_{05}^{*}q_{54}^{(8)*} + q_{40}^{*}q_{04}^{*}q_{55}^{(10)*} \end{bmatrix}$$
$$\begin{bmatrix} \left(1 - q_{17,17}^{(22)*}\right)\left(1 - q_{1,14}^{*}q_{14,1}^{*} - q_{1,15}^{*}q_{15,1}^{*} - q_{1,16}^{*}q_{16,1}^{*}\right) \\ -q_{16,17}^{(18)*}q_{17,16}^{(20)*}\left(1 - q_{1,14}^{*}q_{14,1}^{*} - q_{1,15}^{*}q_{15,1}^{*}\right) \\ -\left(q_{1,17}^{*} + q_{1,15}^{*}q_{15,17}^{(18)*}\right)\left(\begin{matrix} q_{17,1}^{*} + q_{16,1}^{*}q_{17,16}^{(20)*} + \\ q_{14,1}^{*}q_{17,14}^{(19)*} \end{matrix}\right) \\ -q_{1,16}^{*}q_{16,17}^{(18)*}\left(q_{17,1}^{*} + q_{14,1}^{*}q_{17,14}^{(19)*}\right) \end{bmatrix}$$

$$N(s) = -q_{01}^{*}\left\{\begin{array}{r} q_{1,15}^{*}q_{15,21}^{(18)*}q_{16,17}^{(18)*}q_{17,16}^{(20)*} - q_{1,17}^{*}q_{16,21}^{(18)*}q_{17,16}^{(20)*} - q_{1,15}^{*}q_{15,17}^{(18)*}q_{16,21}^{(18)*}q_{17,16}^{(20)*} \\ -q_{1,17}^{*}q_{17,23}^{(20)*} - q_{1,15}^{*}q_{15,17}^{(18)*}q_{17,23}^{(20)*} - q_{1,16}^{*}q_{16,17}^{(18)*}q_{17,23}^{(20)*} + q_{1,15}^{*}q_{15,21}^{(18)*}q_{17,17}^{(22)*} \\ +q_{1,16}^{*}q_{16,21}^{(18)*}q_{17,17}^{(22)*} - q_{1,16}^{*}q_{16,21}^{(18)*} - q_{1,15}^{*}q_{15,21}^{(18)*} \end{array}\right\}$$

$$\left[1 - q_{45}^{(6)*}q_{54}^{(8)*} - q_{55}^{(10)*}\right]$$

$$+\left[\begin{array}{r} \left\{1 - q_{17,17}^{(22)*}\right\}\left\{1 - q_{1,14}^{*}q_{14,1}^{*} - q_{1,15}^{*}q_{15,1}^{*} - q_{1,16}^{*}q_{16,1}^{*}\right\} \\ -q_{1,17}^{*}q_{17,1}^{*} - q_{1,15}^{*}q_{17,1}^{*}q_{15,17}^{(18)*} - q_{1,16}^{*}q_{17,1}^{*}q_{16,17}^{(18)*} - q_{1,17}^{*}q_{14,1}^{*}q_{17,14}^{(19)*} \\ -q_{1,15}^{*}q_{14,1}^{*}q_{15,17}^{(18)*}q_{17,14}^{(19)*} - q_{1,16}^{*}q_{14,1}^{*}q_{16,17}^{(18)*}q_{17,14}^{(19)*} - q_{1,17}^{*}q_{16,1}^{*}q_{17,16}^{(20)*} \\ -q_{1,15}^{*}q_{16,1}^{*}q_{15,17}^{(18)*}q_{17,16}^{(20)*} - q_{16,17}^{(18)*}q_{17,16}^{(20)*} + q_{1,14}^{*}q_{14,1}^{*}q_{16,17}^{(18)*}q_{17,16}^{(20)*} \\ +q_{1,15}^{*}q_{15,1}^{*}q_{16,17}^{(18)*}q_{17,16}^{(20)*} \end{array}\right]$$

$$\left[\begin{array}{l} q_{03}^{*}q_{39}^{(6)*}\left\{1 - q_{45}^{(6)*}q_{54}^{(8)*} - q_{55}^{(10)*}\right\} \\ +q_{5,11}^{(8)*}\left\{q_{03}^{*}q_{35}^{(6)*} + q_{04}^{*}q_{45}^{(6)*} + q_{05}^{*}\right\} \\ -q_{40}^{*}q_{49}^{(6)*}\left\{\begin{array}{l} q_{04}^{*} + q_{03}^{*}q_{35}^{(6)*}q_{54}^{(8)*} \\ +q_{05}^{*}q_{54}^{(8)*} - q_{04}^{*}q_{55}^{(10)*} \end{array}\right\} \end{array}\right]$$

If T_0 denotes value of MTSF when the system starts from the state '0', then $T_0 = \lim_{s \to 0} \frac{1 - \varphi_0^{**}(s)}{s} = \frac{N}{D}$, where $N = \lim_{s \to 0}\left(D'(s) - N'(s)\right)$, $D = \lim_{s \to 0} D(s)$

24.4 Availability in Full Capacity During Winter

Recursive relations for $AW_i^f(t)$ (refer notations) satisfy:

$$AW_0^f(t) = MW_0^f(t) + q_{01}(t)©AW_1^f(t) + q_{02}(t)©AW_2^f(t) + q_{03}(t)©AW_3^f(t) + q_{04}(t)©AW_4^f(t) + q_{05}(t)©AW_5^f(t)$$

$$AW_1^f(t) = q_{10}(t)©AW_0^f(t) + q_{1,13}(t)©AW_{13}^f(t) + q_{1,14}(t)©AW_{14}^f(t) + q_{1,15}(t)©AW_{15}^f(t) + q_{1,16}(t)©AW_{16}^f(t)$$

$$AW_2^f(t) = q_{20}(t)©AW_0^f(t)$$

$$AW_3^f(t) = q_{30}(t)©AW_0^f(t) + q_{35}^{(6)}(t)©AW_5^f(t) + q_{39}^{(6)}(t)©AW_9^f(t)$$

$$AW_4^f(t) = q_{40}(t)©AW_0^f(t) + q_{45}^{(6)}(t)©AW_5^f(t) + q_{49}^{(6)}(t)©AW_9^f(t)$$

$$AW_5^f(t) = q_{50}(t)©AW_0^f(t) + q_{52}^{(7)}(t)©AW_2^f(t) + q_{54}^{(8)}(t)©AW_4^f(t)$$

$$+ q_{55}^{(10)}(t)©AW_5^f(t) + q_{5,12}^{(8,11)}(t)©AW_{12}^f(t)$$

$$AW_9^f(t) = q_{9,12}(t)©AW_{12}^f(t)$$

$$AW_{12}^f(t) = q_{12,5}(t)©AW_5^f(t) + q_{12,9}(t)©AW_9^f(t)$$

$$AW_{13}^f(t) = q_{13,1}(t)©AW_1^f(t)$$

$$AW_{14}^f(t) = q_{14,1}(t)©AW_1^f(t) + q_{14,16}^{(17)}(t)©AW_{16}^f(t) + q_{14,20}^{(17)}(t)©AW_{20}^f(t)$$

$$AW_{15}^f(t) = q_{15,1}(t)©AW_1^f(t) + q_{15,16}^{(17)}(t)©AW_{16}^f(t) + q_{15,20}^{(17)}(t)©AW_{20}^f(t)$$

$$AW_{16}^f(t) = q_{16,1}(t)©AW_1^f(t) + q_{16,13}^{(18)}(t)©AW_{13}^f(t) + q_{16,15}^{(19)}(t)©AW_{15}^f(t) + q_{16,16}^{(21)}(t)©AW_{16}^f(t) + q_{16,23}^{(19,22)}(t)©AW_{23}^f(t)$$

$$AW_{20}^f(t) = q_{20,23}(t)©AW_{23}^f(t)$$

$$AW_{23}^f(t) = q_{23,16}(t)©AW_{16}^f(t) + q_{23,20}(t)©AW_{20}^f(t)$$

where $MW_0^f(t) = e^{-(2\lambda_1+\alpha_1)t}\overline{H}_1(t)$.

On taking LT and solving resulting equations for $AW_0^{f*}(s)$, we have

$$AW_0^{f*}(s) = \frac{N_1(s)}{D_1(s)},$$

where

$$D_1(s) = \begin{vmatrix}
1 & -q_{01}^*(s) & -q_{02}^*(s) & -q_{03}^*(s) & -q_{04}^*(s) & -q_{05}^*(s) & 0 & 0 & 0 & 0 & 0 & 0 & 0 & 0 \\
-q_{10}^*(s) & 1 & 0 & 0 & 0 & 0 & 0 & 0 & -q_{1,13}^*(s) & -q_{1,14}^*(s) & -q_{1,15}^*(s) & -q_{1,16}^*(s) & 0 & 0 \\
-q_{20}^*(s) & 0 & 1 & 0 & 0 & 0 & 0 & 0 & 0 & 0 & 0 & 0 & 0 & 0 \\
-q_{30}^*(s) & 0 & 0 & 1 & 0 & -q_{35}^{(6)*}(s) & -q_{39}^{(6)*}(s) & 0 & 0 & 0 & 0 & 0 & 0 & 0 \\
-q_{40}^*(s) & 0 & 0 & 0 & 1 & -q_{45}^{(6)*}(s) & -q_{49}^{(6)*}(s) & 0 & 0 & 0 & 0 & 0 & 0 & 0 \\
-q_{50}^*(s) & 0 & -q_{52}^{(7)*}(s) & 0 & -q_{54}^{(8)*}(s) & 1-q_{55}^{(10)*}(s) & 0 & -q_{5,12}^{(8,11)*}(s) & 0 & 0 & 0 & 0 & 0 & 0 \\
0 & 0 & 0 & 0 & 0 & 0 & 1 & -q_{9,12}^*(s) & 0 & 0 & 0 & 0 & 0 & 0 \\
0 & 0 & 0 & 0 & 0 & -q_{12,5}^*(s) & -q_{12,9}^*(s) & 1 & 0 & 0 & 0 & 0 & 0 & 0 \\
0 & -q_{13,1}^*(s) & 0 & 0 & 0 & 0 & 0 & 0 & 1 & 0 & 0 & 0 & 0 & 0 \\
0 & -q_{14,1}^*(s) & 0 & 0 & 0 & 0 & 0 & 0 & 0 & 1 & 0 & -q_{14,16}^{(17)*}(s) & -q_{14,20}^{(17)*}(s) & 0 \\
0 & -q_{15,1}^*(s) & 0 & 0 & 0 & 0 & 0 & 0 & 0 & 0 & 1 & -q_{15,16}^{(17)*}(s) & -q_{15,20}^{(17)*}(s) & 0 \\
0 & -q_{16,1}^*(s) & 0 & 0 & 0 & 0 & 0 & 0 & -q_{16,13}^{(18)*}(s) & 0 & -q_{16,15}^{(19)*}(s) & 1-q_{16,16}^{(21)*}(s) & 0 & -q_{16,23}^{(19,22)*}(s) \\
0 & 0 & 0 & 0 & 0 & 0 & 0 & 0 & 0 & 0 & 0 & 0 & 1 & -q_{20,23}^*(s) \\
0 & 0 & 0 & 0 & 0 & 0 & 0 & 0 & 0 & 0 & 0 & -q_{23,16}^*(s) & -q_{23,20}^*(s) & 1
\end{vmatrix}$$

$$\text{or } D_1(s) = \left| C_1\ C_2\ C_3\ C_4\ C_5\ C_6\ C_7\ C_8\ C_9\ C_{10}\ C_{11}\ C_{12}\ C_{13}\ C_{14} \right|$$

where C_i, $(1 \le i \le 14)$ denotes ith column of the determinant $D_1(s)$.

$$N_1(s) = \left| C_1^{n_1}\ C_2\ C_3\ C_4\ C_5\ C_6\ C_7\ C_8\ C_9\ C_{10}\ C_{11}\ C_{12}\ C_{13}\ C_{14} \right|$$

where $C_1^{n_1} = [MW_0^{f*}(s), 0, 0, 0, 0, 0, 0, 0, 0, 0, 0, 0, 0, 0] =$ First column of $N_1(s)$.

In steady-state, the availability in Full Capacity during winter of the system is given by

$$AW_0^f = \lim_{s\to 0} sAW_0^{f*}(s) = \frac{N_1}{D_1}, \text{ where } N_1 = \lim_{s\to 0} N_1(s), \quad D_1 = \lim_{s\to 0} D_1'(s)$$

24.5 Other Performance Measures of the System Effectiveness

Similarly, in steady-state, the availability in Type I, Type II, and Type III respectively during winter of the system are given by

$$AW_0^I = \lim_{s\to 0} sAW_0^{I*}(s) = \frac{N_2}{D_1} \quad AW_0^{II} = \lim_{s\to 0} sAW_0^{II*}(s) = \frac{N_3}{D_1},$$

$$AW_0^{III} = \lim_{s\to 0} sAW_0^{III*}(s) = \frac{N_4}{D_1},$$

In steady-state, the availability in F. C., Type I, Type II, and Type III during summer of the system are given by

$$AS_0^f = \lim_{s\to 0} sAS_0^{f*}(s) = \frac{N_5}{D_1}, \quad AS_0^I = \lim_{s\to 0} sAS_0^{I*}(s) = \frac{N_6}{D_1}$$

$$AS_0^{II} = \lim_{s\to 0} sAS_0^{II*}(s) = \frac{N_7}{D_1}, \quad AS_0^{III} = \lim_{s\to 0} sAS_0^{III*}(s) = \frac{N_8}{D_1},$$

In steady-state, the total fraction of time for which the system is down during winter and summer respectively excluding failed state are given by

$$DW_0 = \lim_{s\to 0} sDW_0^*(s) = \frac{N_9}{D_1}, \quad DS_0 = \lim_{s\to 0} sDS_0^*(s) = \frac{N_{10}}{D_1},$$

In steady-state, the total fraction of time for which the system is under repair during winter and summer respectively are given by

$$BW_{11} = \lim_{s\to 0} sBW_0^*(s) = \frac{N_{11}}{D_1}, \quad BS_{12} = \lim_{s\to 0} sBS_0^*(s) = \frac{N_{12}}{D_1},$$

In steady-state, the number of visits per unit time by the repairman during winter and summer respectively are given by

$$VW_0 = \lim_{s\to 0} sVW_0^*(s) = \frac{N_{13}}{D_1}, \qquad VS_0 = \lim_{s\to 0} sVS_0^*(s) = \frac{N_{14}}{D_1},$$

where $D_1(s)$ already specified

$$N_2(s) = \left| C_1^{n_2}\ C_2\ C_3\ C_4\ C_5\ C_6\ C_7\ C_8\ C_9\ C_{10}\ C_{11}\ C_{12}\ C_{13}\ C_{14} \right|$$

$$C_1^{n_2} = [0, 0, 0, 0, MW_4^{I*}(s), 0, 0, 0, 0, 0, 0, 0, 0, 0] = \text{First column of } N_2(s)$$

where C_i, $(2 \le i \le 14)$ denotes ith column of the determinant $D_1(s)$.

Similarly, replacing first column of $D_1(s)$ by the columns $C_1^{n_3}, C_1^{n_4}, \ldots, C_1^{n_{14}}$, we can get values of the determinants $N_3(s)$, $N_4(s)$, $N_5(s)$, …, $N_{14}(s)$, respectively where

$$\begin{aligned}
C_1^{n_3} &= [0, 0, 0, 0, 0, MW_5^{II*}(s), 0, 0, 0, 0, 0, 0, 0, 0]\\
C_1^{n_4} &= [0, 0, 0, MW_3^{III*}(s), 0, 0, 0, MW_{12}^{III*}(s), 0, 0, 0, 0, 0, 0]\\
C_1^{n_5} &= [0, MS_1^{f*}(s), 0, 0, 0, 0, 0, 0, 0, 0, 0, 0, 0, 0]\\
C_1^{n_6} &= [0, 0, 0, 0, 0, 0, 0, 0, 0, 0, MS_{15}^{I*}(s), 0, 0, 0]\\
C_1^{n_7} &= [0, 0, 0, 0, 0, 0, 0, 0, 0, 0, 0, MS_{16}^{II*}(s), 0, 0]\\
C_1^{n_8} &= [0, 0, 0, 0, 0, 0, 0, 0, 0, MS_{14}^{III*}(s), 0, 0, 0, MS_{23}^{III*}(s)]\\
C_1^{n_9} &= [0, 0, NW_2^*(s), NW_3^*(s), 0, 0, 0, 0, 0, 0, 0, 0, 0, 0]\\
C_1^{n_{10}} &= [0, 0, 0, 0, 0, 0, 0, 0, NS_{13}^*(s), NS_{14}^*(s), 0, 0, 0, 0]\\
C_1^{n_{11}} &= [0, 0, RW_2^*(s), RW_3^*(s), RW_4^*(s), RW_5^*(s), RW_9^*(s), RW_{12}^*(s), 0, 0, 0, 0, 0, 0]\\
C_1^{n_{12}} &= [0, 0, 0, 0, 0, 0, 0, 0, RS_{13}^*(s), RS_{14}^*(s), RS_{15}^*(s), RS_{16}^*(s), RS_{20}^*(s), RS_{23}^*(s)]\\
C_1^{n_{13}} &= [Q_{02}^*(s) + Q_{03}^*(s) + Q_{04}^*(s) + Q_{05}^*(s), 0, 0, 0, 0, 0, 0, 0, 0, 0, 0, 0, 0, 0]\\
C_1^{n_{14}} &= [0, Q_{1,13}^*(s) + Q_{1,14}^*(s) + Q_{1,15}^*(s) + Q_{1,16}^*(s), 0, 0, 0, 0, 0, 0, 0, 0, 0, 0, 0, 0]
\end{aligned}$$

where D_1 already specified and

$$N_2 = \lim_{s\to 0} N_2(s),\ N_3 = \lim_{s\to 0} N_3(s),\ N_4 = \lim_{s\to 0} N_4(s),\ N_5 = \lim_{s\to 0} N_5(s),\ N_6 = \lim_{s\to 0} N_6(s),$$
$$N_7 = \lim_{s\to 0} N_7(s),\ N_8 = \lim_{s\to 0} N_8(s),\ N_9 = \lim_{s\to 0} N_9(s),\ N_{10} = \lim_{s\to 0} N_{10}(s),$$
$$N_{11} = \lim_{s\to 0} N_{11}(s),\ N_{12} = \lim_{s\to 0} N_{12}(s),\ N_{13} = \lim_{s\to 0} N_{13}(s),\ N_{14} = \lim_{s\to 0} N_{14}(s)$$

24.6 Cost-Benefit Analysis

Expression for expected Profit for the system in steady state can be expressed as:

$$\begin{aligned}
\text{Profit} = {} & CW_0^f \times AW_0^f + CS_0^f \times AS_0^f + CW_1^I \times AW_0^I + CW_1^{II} \times AW_0^{II} + CW_1^{III} \times AW_0^{III} + CS_1^I \times AS_0^I\\
& + CS_1^{II} \times AS_0^{II} + CS_1^{III} \times AS_0^{III} - CW_2 \times BW_0 - CS_2 \times BS_0 - CW_3 \times VW_0 - CS_3 \times VS_0
\end{aligned}$$

where

$CW_0^f, CW_1^I, CW_1^{II}, CW_1^{III}/$ $CS_0^f, CS_1^I, CS_1^{II}, CS_1^{III}$:	Revenue per unit uptime when the system is working in F.C., Type I, Type II, Type III during winter/summer
CW_2/CS_2:	Cost per unit time for which the repairman remains busy in doing repair in winter/summer
CW_3/CS_3:	Per visit cost of the repairman in winter/summer

24.7 Results and Discussion

Using R programming language linear models are fitted between Ambient Temperature (predictor variable) and production in different Capacities (response variable) and the obtained results are given as follows:

$$\begin{aligned} &GT1 = 103.6116 - 0.4442TEMP \\ &GT1\_ST = 174.2383 - 0.7237TEMP \\ &GT1\_GT2 = 196.9855 - 0.5614TEMP \\ &GT1\_GT2\_ST = 310.7873 - 0.5815TEMP \end{aligned}$$

where GT1 denotes production of gas turbine 1, similarly, GT1_ST denotes production of gas turbine 1 and steam turbine together, etc. TEMP denotes predictor variable, Ambient Temperature. Negative coefficient of the predictor variable (TEMP) in each case shows that production of electricity decreases as Ambient Temperature increases.

Using fitted linear models production (in MW) and revenue (in INR) of electricity per unit time in different Capacities during winter and summer are given as follows:

Different capacities	Production in megawatts		Revenue in Indian Rupees	
	Winter	Summer	Winter	Summer
Full capacity:	299.3257	291.6628	13,46,965.65	13,12,483
Reduced capacity of Type I:	185.9201	178.5221	1,85,920.1P_3	1,78,522.1P_5
Reduced capacity of Type II:	159.9739	150.4371	7,19,882.55	6,76,966.95
Reduced capacity of Type III:	94.8562	89.0027	94,856.2P_4	89,002.7P_6

where normal price of electricity = INR 4.5 per unit when steam turbine is working, and

- P_3 Price of electricity per unit to be charged from the buyer when production is under Reduced Capacity of Type I during winter
- P_4 Price of electricity per unit to be charged from the buyer when production is under Type III during winter
- P_5 Price of electricity per unit to be charged from the buyer when production is under Type I during summer

P_6 Price of electricity per unit to be charged from the buyer when production is under Type III during summer

Using the estimates of the parameters involved in the present study obtained from collected data, we have studied the behaviour of various performance measures explained as follows:

Figure 24.2 visualizes behaviour of MTSF with respect to the four failure rates α_1, α_2, λ_1, λ_2. It clearly shows that MTSF decreases as any one of the four failure rates increase.

Behaviour of the availabilities in Full Capacity in winter as well as summer with respect to each of the four types of failure rates α_1, α_2, λ_1, λ_2 has been shown in Fig. 24.3. Following is observed:

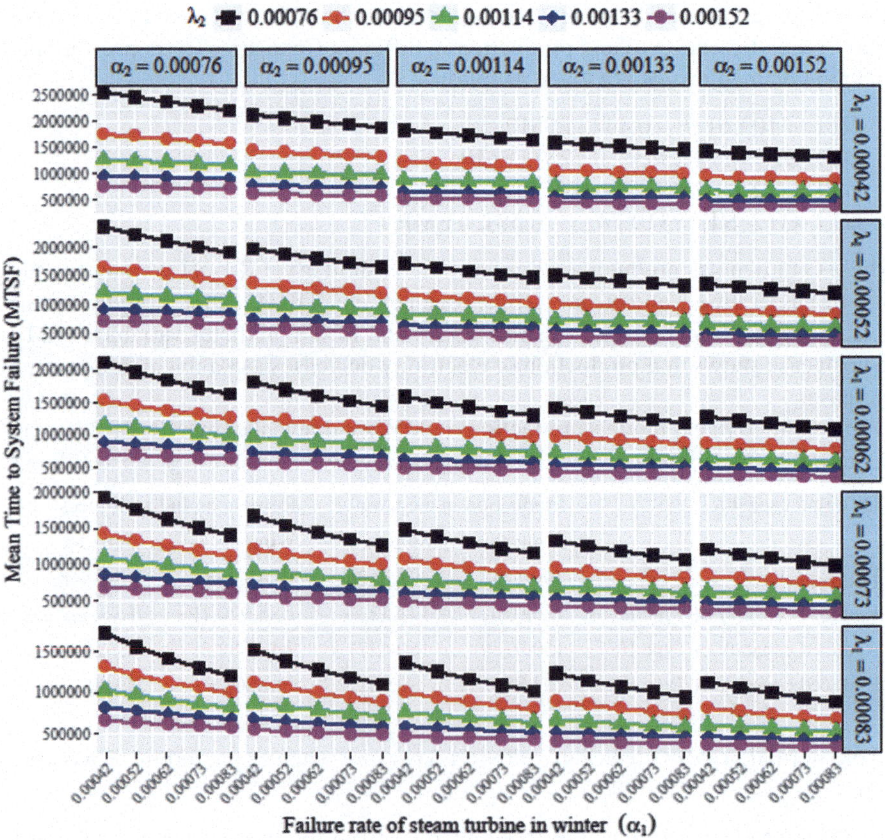

Fig. 24.2 MTSF with respect to failure rates of steam turbines (α_1/α_2) and gas turbines (λ_1/λ_2) during winter/summer

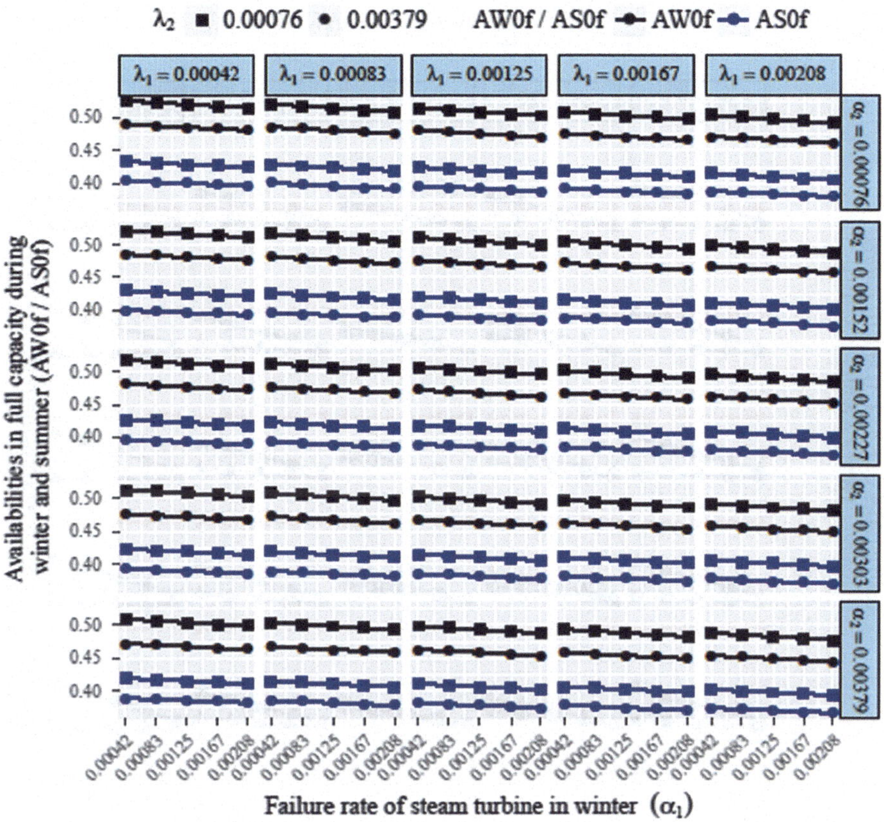

Fig. 24.3 Availability in Full Capacity during winter/summer (AW0f/AS0f) simultaneously with respect to failure rates of steam turbines (α_1/α_2) and gas turbine (λ_1/λ_2) during winter/summer

- Both the Full Capacity availabilities during winter (AW0) and summer (AS0) decrease as any one among the four failure rates increase; and
- Full Capacity availability during winter (AW0) remains greater than the Full Capacity availability during summer (AS0) throughout the range of variation of the four failure rates.

Behaviour of three types (Type I, Type II and Type III) of availabilities under reduced capacity during winter with respect to α_1, λ_1, and p_{12} is depicted in Fig. 24.4. Following may be observed for the three types of availability under reduced capacity:

- AW0I increases as failure rate of S. T. (α_1) during winter and p_{12} increase.
- AW0II increases as failure rate of G. T. (λ_1) during winter increases, but it decreases as α_1 increases.

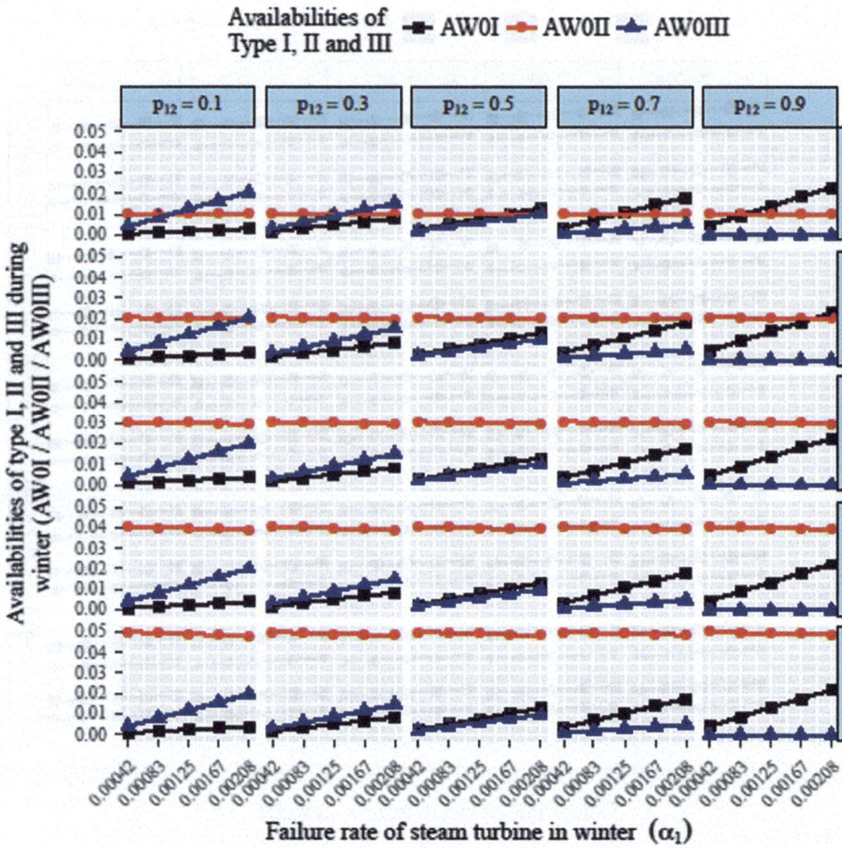

Fig. 24.4 Availabilities of Type I, II and III during winter (AW0I/AW0II/AW0III) simultaneously with respect to failure rates of steam and gas turbines (α_1/λ_1) during winter as well as for different probabilities of paying higher price of electricity in reduced capacities

- AW0III increases as the failure rate of S. T. (α_1) during winter and probability p_{11} increases (or p_{12} decreases, since $p_{11} + p_{12} = 1$).

Behaviour of three types (Type I, Type II and Type III) of availabilities under reduced capacity during summer is depicted in Fig. 24.5. Following may be observed for the three types of availabilities under reduced capacity:

- AS0I increases as failure rate of S. T. during summer and probability p_{22} increases.
- AS0II increases as failure rate of G. T. during summer increases and is almost independent of the demand probability and p_{22}.
- AS0III increases as the failure rate of S. T. during summer and probability p_{21} increases (or p_{22} decrease, since $p_{21} + p_{22} = 1$).

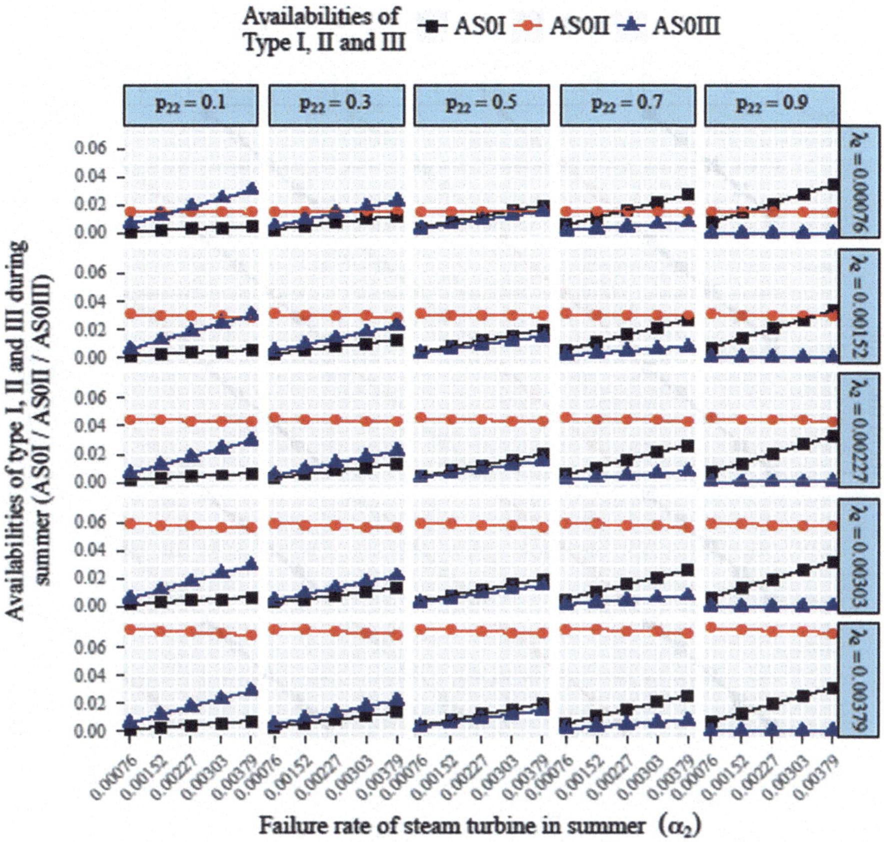

Fig. 24.5 Availabilities of Type I, II and III during summer (AS0I/AS0II/AS0III) simultaneously with respect to failure rates of steam and gas turbines (α_2/λ_2) during summer as well as for different probabilities of paying higher price of electricity in reduced capacities

Figure 24.6 visualizes behaviour of Profit function with respect to plant operating expenses (K_0) per unit time and varying values of per unit price of electricity in Type I in both the seasons. Following may be concluded/obtained by observation/calculation:

- Profit of the plant decreases with increase in plant operating expenses while it increases as the price of electricity to be charged in Type I during winter increases.
- The Profit gets increased on increasing the price of electricity in Type I during summer also.
- The values of K_0 for positive Profit are obtained at different price of electricity in Type I during winter (P_3) simultaneously with different price of electricity in Type I during summer (P_5) and the same are shown in Table 24.1, called cut-off points.

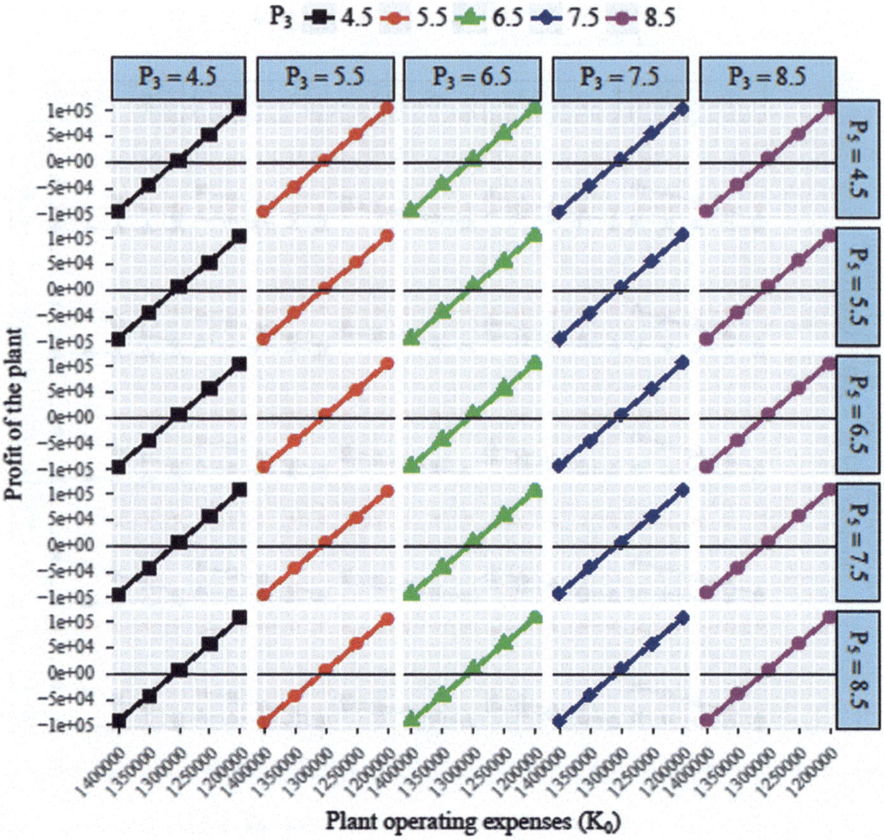

Fig. 24.6 Profit with respect to variation in plant operating expenses and price of electricity to be charged in Type I during winter/summer

Table 24.1 Cut-off Values of K_0 for different values of P_3 and P_5

$P_3 = 4.5$	$P_3 = 5.5$	$P_3 = 6.5$	$P_3 = 7.5$	$P_3 = 8.5$	
13,04,751.398	13,05,247.582	13,05,743.767	13,06,239.951	13,06,736.136	$P_5 = 4.5$
13,05,474.084	13,05,970.268	13,06,466.453	13,06,962.637	13,07,458.822	$P_5 = 5.5$
13,06,196.77	13,06,692.954	13,07,189.139	13,07,685.323	13,08,181.507	$P_5 = 6.5$
13,06,919.456	13,07,415.64	13,07,911.825	13,08,408.009	13,08,904.193	$P_5 = 7.5$
13,07,642.142	13,08,138.326	13,08,634.511	13,09,130.695	13,09,626.879	$P_5 = 8.5$

- For profitability of the system, the value of K_0 should be less than the value corresponding to cut-off point.

For the system to be beneficial, the value of K_0 should not be more than the value corresponding to the cut-off point.

Behaviour of the Profit function with respect to price of electricity per unit to be charged from the buyer in Type I during winter, price of electricity in Type I during summer and probability that there is demand for higher rate and this demand is greater than production of one gas turbine in winter is shown in Figs. 24.7 and 24.8. It reveals that

- The Profit increases as price of electricity (P_3/P_5) per unit in Type I during winter/summer increases, and it also increases as p_{12} increases.
- P_3 should be increased with increase in K_0 so as to avoid the decrease in the Profit.
- Cut-off points for P_3 have been obtained to avoid negative Profit at different values of P_5 and p_{12} and are shown in Table 24.2.
- Cut-off points for P_5 have been obtained to avoid negative Profit for different values of P_3 and p_{22} and are shown in Table 24.3.

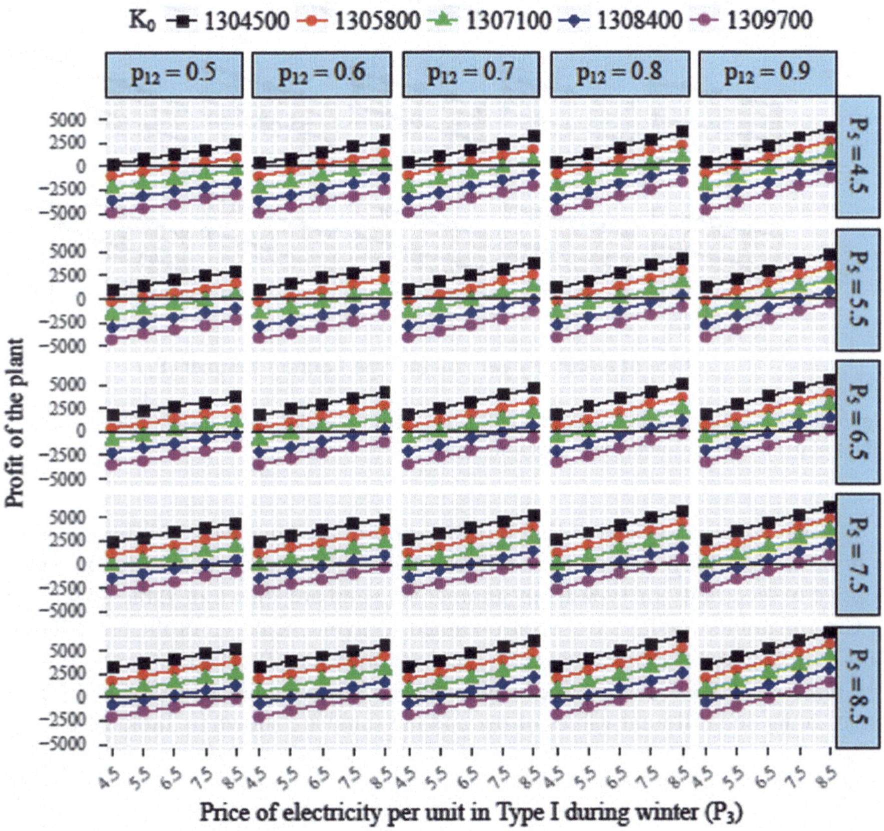

Fig. 24.7 Profit versus price of electricity to be charged in Type I during winter/summer for different values of p_{12} and variation in plant operating expenses

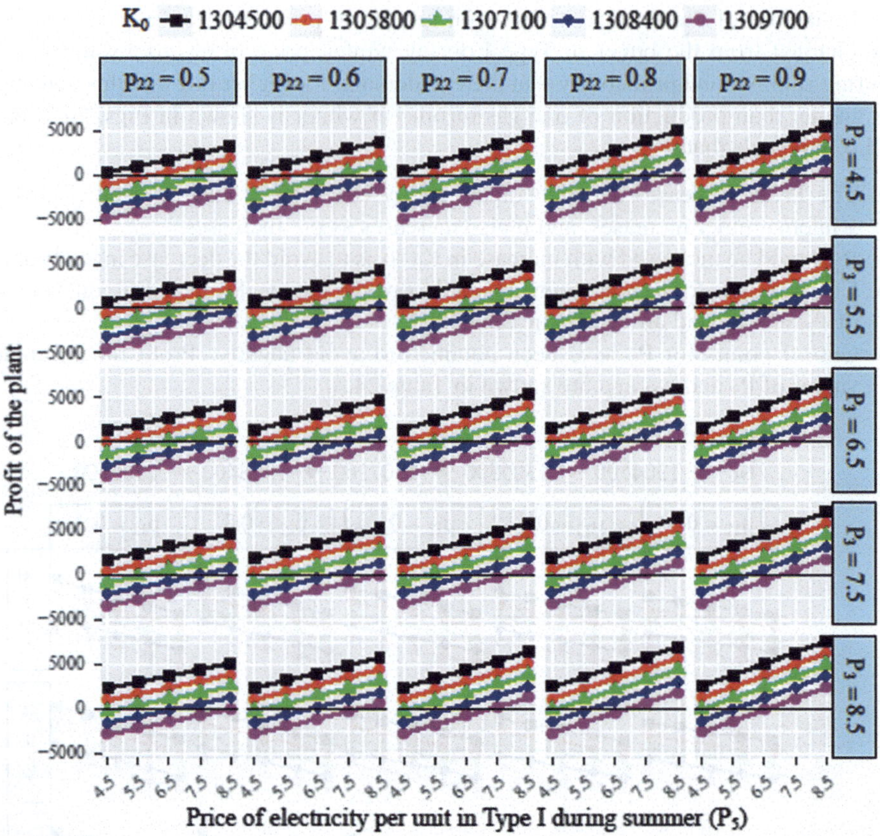

Fig. 24.8 Profit versus price of electricity to be charged in Type I during summer/winter for different values of p_{22} and variation in plant operating expenses

Similarly, Cut-off values of K_0 for positive Profit are obtained for different values of price of electricity in Type III in winter (P_4) simultaneously with different values of price of electricity in Type III in summer (P_6) and are shown in Table 24.4.

Cut-off points for P_4 have been obtained to avoid negative Profit for different values of P_6 and different values of p_{11} and are shown in Table 24.5.

Cut-off points for P_6 have been obtained to avoid negative Profit for different values of P_4 and different values of p_{21} and the same are shown in Table 24.6.

Table 24.2 Minimum values of price of electricity to be charged in Type I during winter for different values of P_5, K_0, and p_{12}

	$p_{12} = 0.5$	$p_{12} = 0.6$	$p_{12} = 0.7$	$p_{12} = 0.8$	$p_{12} = 0.9$	
$K_0 =$ 1305800	$P_3 \geq 6.62$	$P_3 \geq 6.19$	$P_3 \geq 5.89$	$P_3 \geq 5.65$	$P_3 \geq 5.47$	$P_5 = 4.5$
	$P_3 \geq 5.16$	$P_3 \geq 4.97$	$P_3 \geq 4.83$	$P_3 \geq 4.73$	$P_3 \geq 4.65$	$P_5 = 5.5$
	$P_3 \geq 3.71$	$P_3 \geq 3.75$	$P_3 \geq 3.78$	$P_3 \geq 3.81$	$P_3 \geq 3.83$	$P_5 = 6.5$
	$P_3 \geq 2.25$	$P_3 \geq 2.53$	$P_3 \geq 2.73$	$P_3 \geq 2.89$	$P_3 \geq 3.01$	$P_5 = 7.5$
	$P_3 \geq 0.79$	$P_3 \geq 1.31$	$P_3 \geq 1.68$	$P_3 \geq 1.96$	$P_3 \geq 2.18$	$P_5 = 8.5$
$K_0 =$ 1307100	$P_3 \geq 9.24$	$P_3 \geq 8.39$	$P_3 \geq 7.78$	$P_3 \geq 7.31$	$P_3 \geq 6.95$	$P_5 = 4.5$
	$P_3 \geq 7.78$	$P_3 \geq 7.17$	$P_3 \geq 6.72$	$P_3 \geq 6.39$	$P_3 \geq 6.13$	$P_5 = 5.5$
	$P_3 \geq 6.33$	$P_3 \geq 5.95$	$P_3 \geq 5.67$	$P_3 \geq 5.47$	$P_3 \geq 5.31$	$P_5 = 6.5$
	$P_3 \geq 4.87$	$P_3 \geq 4.73$	$P_3 \geq 4.62$	$P_3 \geq 4.55$	$P_3 \geq 4.48$	$P_5 = 7.5$
	$P_3 \geq 3.41$	$P_3 \geq 3.51$	$P_3 \geq 3.57$	$P_3 \geq 3.62$	$P_3 \geq 3.66$	$P_5 = 8.5$
$K_0 =$ 1308400	$P_3 \geq 11.85$	$P_3 \geq 10.58$	$P_3 \geq 9.66$	$P_3 \geq 8.97$	$P_3 \geq 8.43$	$P_5 = 4.5$
	$P_3 \geq 10.40$	$P_3 \geq 9.36$	$P_3 \geq 8.61$	$P_3 \geq 8.05$	$P_3 \geq 7.61$	$P_5 = 5.5$
	$P_3 \geq 8.95$	$P_3 \geq 8.14$	$P_3 \geq 7.56$	$P_3 \geq 7.13$	$P_3 \geq 6.78$	$P_5 = 6.5$
	$P_3 \geq 7.49$	$P_3 \geq 6.92$	$P_3 \geq 6.51$	$P_3 \geq 6.20$	$P_3 \geq 5.96$	$P_5 = 7.5$
	$P_3 \geq 6.03$	$P_3 \geq 5.70$	$P_3 \geq 5.46$	$P_3 \geq 5.28$	$P_3 \geq 5.14$	$P_5 = 8.5$

Table 24.3 Minimum values of price of electricity to be charged in Type I during summer for different values of P_3, K_0, and p_{22}

	$p_{22} = 0.6$	$p_{22} = 0.7$	$p_{22} = 0.8$	$p_{22} = 0.9$	$p_{22} = 1$	
$K_0 =$ 1305800	$P_5 \geq 5.96$	$P_5 \geq 5.65$	$P_5 \geq 5.42$	$P_5 \geq 5.25$	$P_5 \geq 5.12$	$P_3 = 4.5$
	$P_5 \geq 5.27$	$P_5 \geq 5.07$	$P_5 \geq 4.92$	$P_5 \geq 4.81$	$P_5 \geq 4.73$	$P_3 = 5.5$
	$P_5 \geq 4.58$	$P_5 \geq 4.49$	$P_5 \geq 4.42$	$P_5 \geq 4.37$	$P_5 \geq 4.33$	$P_3 = 6.5$
	$P_5 \geq 3.90$	$P_5 \geq 3.91$	$P_5 \geq 3.93$	$P_5 \geq 3.93$	$P_5 \geq 3.94$	$P_3 = 7.5$
	$P_5 \geq 3.21$	$P_5 \geq 3.34$	$P_5 \geq 3.43$	$P_5 \geq 3.50$	$P_5 \geq 3.55$	$P_3 = 8.5$
$K_0 =$ 1307100	$P_5 \geq 7.75$	$P_5 \geq 7.16$	$P_5 \geq 6.73$	$P_5 \geq 6.40$	$P_5 \geq 6.14$	$P_3 = 4.5$
	$P_5 \geq 7.07$	$P_5 \geq 6.58$	$P_5 \geq 6.23$	$P_5 \geq 5.96$	$P_5 \geq 5.75$	$P_3 = 5.5$
	$P_5 \geq 6.38$	$P_5 \geq 6.01$	$P_5 \geq 5.73$	$P_5 \geq 5.52$	$P_5 \geq 5.36$	$P_3 = 6.5$
	$P_5 \geq 5.70$	$P_5 \geq 5.43$	$P_5 \geq 5.23$	$P_5 \geq 5.08$	$P_5 \geq 4.97$	$P_3 = 7.5$
	$P_5 \geq 5.01$	$P_5 \geq 4.85$	$P_5 \geq 4.73$	$P_5 \geq 4.65$	$P_5 \geq 4.58$	$P_3 = 8.5$
$K_0 =$ 1308400	$P_5 \geq 9.55$	$P_5 \geq 8.68$	$P_5 \geq 8.04$	$P_5 \geq 7.55$	$P_5 \geq 7.17$	$P_3 = 4.5$
	$P_5 \geq 8.87$	$P_5 \geq 8.10$	$P_5 \geq 7.54$	$P_5 \geq 7.11$	$P_5 \geq 6.78$	$P_3 = 5.5$
	$P_5 \geq 8.18$	$P_5 \geq 7.52$	$P_5 \geq 7.04$	$P_5 \geq 6.67$	$P_5 \geq 6.39$	$P_3 = 6.5$
	$P_5 \geq 7.49$	$P_5 \geq 6.94$	$P_5 \geq 6.54$	$P_5 \geq 6.23$	$P_5 \geq 6.00$	$P_3 = 7.5$
	$P_5 \geq 6.81$	$P_5 \geq 6.36$	$P_5 \geq 6.04$	$P_5 \geq 5.80$	$P_5 \geq 5.60$	$P_3 = 8.5$

Table 24.4 Cut-off Values of K_0 for different values of P_4 and P_6

$P_4 = 4.5$	$P_4 = 5.5$	$P_4 = 6.5$	$P_4 = 7.5$	$P_4 = 8.5$	
13,06,683.674	13,06,883.295	13,07,082.916	13,07,282.537	13,07,482.158	$P_6 = 4.5$
13,06,965.075	13,07,164.696	13,07,364.317	13,07,563.938	13,07,763.559	$P_6 = 5.5$
13,07,246.475	13,07,446.097	13,07,645.718	13,07,845.339	13,08,044.96	$P_6 = 6.5$
13,07,527.876	13,07,727.498	13,07,927.119	13,08,126.74	13,08,326.361	$P_6 = 7.5$
13,07,809.277	13,08,008.898	13,08,208.52	13,08,408.141	13,08,607.762	$P_6 = 8.5$

Table 24.5 Minimum price of electricity (P_4) to be charged in Type III during winter for different values of P_6, K_0, and p_{11}

	$p_{11} = 0.6$	$p_{11} = 0.7$	$p_{11} = 0.8$	$p_{11} = 0.9$	$p_{11} = 1.0$	
$K_0 = 1307150$	$P_4 \geq 6.84$	$P_4 \geq 8.24$	$P_4 \geq 9.18$	$P_4 \geq 9.85$	$P_4 \geq 10.35$	$P_6 = 4.5$
	$P_4 \geq 5.43$	$P_4 \geq 7.12$	$P_4 \geq 8.24$	$P_4 \geq 9.04$	$P_4 \geq 9.65$	$P_6 = 5.5$
	$P_4 \geq 4.02$	$P_4 \geq 5.99$	$P_4 \geq 7.30$	$P_4 \geq 8.24$	$P_4 \geq 8.94$	$P_6 = 6.5$
	$P_4 \geq 2.61$	$P_4 \geq 4.86$	$P_4 \geq 6.36$	$P_4 \geq 7.43$	$P_4 \geq 8.24$	$P_6 = 7.5$
	$P_4 \geq 1.20$	$P_4 \geq 3.73$	$P_4 \geq 5.42$	$P_4 \geq 6.63$	$P_4 \geq 7.53$	$P_6 = 8.5$
$K_0 = 1307650$	$P_4 \geq 9.35$	$P_4 \geq 10.25$	$P_4 \geq 10.85$	$P_4 \geq 11.28$	$P_4 \geq 11.60$	$P_6 = 4.5$
	$P_4 \geq 7.94$	$P_4 \geq 9.12$	$P_4 \geq 9.91$	$P_4 \geq 10.48$	$P_4 \geq 10.90$	$P_6 = 5.5$
	$P_4 \geq 6.53$	$P_4 \geq 7.99$	$P_4 \geq 8.97$	$P_4 \geq 9.67$	$P_4 \geq 10.19$	$P_6 = 6.5$
	$P_4 \geq 5.12$	$P_4 \geq 6.87$	$P_4 \geq 8.03$	$P_4 \geq 8.86$	$P_4 \geq 9.49$	$P_6 = 7.5$
	$P_4 \geq 3.71$	$P_4 \geq 5.74$	$P_4 \geq 7.09$	$P_4 \geq 8.06$	$P_4 \geq 8.78$	$P_6 = 8.5$
$K_0 = 1308150$	$P_4 \geq 11.85$	$P_4 \geq 12.25$	$P_4 \geq 12.52$	$P_4 \geq 12.71$	$P_4 \geq 12.86$	$P_6 = 4.5$
	$P_4 \geq 10.44$	$P_4 \geq 11.13$	$P_4 \geq 11.58$	$P_4 \geq 11.91$	$P_4 \geq 12.15$	$P_6 = 5.5$
	$P_4 \geq 9.03$	$P_4 \geq 10.00$	$P_4 \geq 10.64$	$P_4 \geq 11.10$	$P_4 \geq 11.45$	$P_6 = 6.5$
	$P_4 \geq 7.62$	$P_4 \geq 8.87$	$P_4 \geq 9.70$	$P_4 \geq 10.30$	$P_4 \geq 10.74$	$P_6 = 7.5$
	$P_4 \geq 6.21$	$P_4 \geq 7.74$	$P_4 \geq 8.76$	$P_4 \geq 9.49$	$P_4 \geq 10.04$	$P_6 = 8.5$

Table 24.6 Minimum values ofā price electricity (P_6) to be charged in Type III during summer for different values of P_4, K_0, and p_{21}

	$p_{21} = 0.6$	$p_{21} = 0.7$	$p_{21} = 0.8$	$p_{21} = 0.9$	$p_{21} = 1.0$	
$K_0 = 1307150$	$P_6 \geq 6.16$	$P_6 \geq 7.75$	$P_6 \geq 8.81$	$P_6 \geq 9.56$	$P_6 \geq 10.13$	$P_4 = 4.5$
	$P_6 \geq 5.45$	$P_6 \geq 7.18$	$P_6 \geq 8.33$	$P_6 \geq 9.16$	$P_6 \geq 9.77$	$P_4 = 5.5$
	$P_6 \geq 4.74$	$P_6 \geq 6.61$	$P_6 \geq 7.86$	$P_6 \geq 8.75$	$P_6 \geq 9.42$	$P_4 = 6.5$
	$P_6 \geq 4.03$	$P_6 \geq 6.04$	$P_6 \geq 7.39$	$P_6 \geq 8.34$	$P_6 \geq 9.06$	$P_4 = 7.5$
	$P_6 \geq 3.32$	$P_6 \geq 5.48$	$P_6 \geq 6.91$	$P_6 \geq 7.94$	$P_6 \geq 8.71$	$P_4 = 8.5$
$K_0 = 1307650$	$P_6 \geq 7.94$	$P_6 \geq 9.17$	$P_6 \geq 9.99$	$P_6 \geq 10.58$	$P_6 \geq 11.02$	$P_4 = 4.5$
	$P_6 \geq 7.23$	$P_6 \geq 8.60$	$P_6 \geq 9.52$	$P_6 \geq 10.17$	$P_6 \geq 10.66$	$P_4 = 5.5$
	$P_6 \geq 6.52$	$P_6 \geq 8.03$	$P_6 \geq 9.04$	$P_6 \geq 9.77$	$P_6 \geq 10.31$	$P_4 = 6.5$
	$P_6 \geq 5.81$	$P_6 \geq 7.47$	$P_6 \geq 8.57$	$P_6 \geq 9.36$	$P_6 \geq 9.95$	$P_4 = 7.5$
	$P_6 \geq 5.10$	$P_6 \geq 6.90$	$P_6 \geq 8.10$	$P_6 \geq 8.96$	$P_6 \geq 9.60$	$P_4 = 8.5$

(continued)

Table 24.6 (continued)

	$p_{21} = 0.6$	$p_{21} = 0.7$	$p_{21} = 0.8$	$p_{21} = 0.9$	$p_{21} = 1.0$	
$K_0 = 1308150$	$P_6 \geq 9.72$	$P_6 \geq 10.59$	$P_6 \geq 11.18$	$P_6 \geq 11.60$	$P_6 \geq 11.91$	$P_4 = 4.5$
	$P_6 \geq 9.01$	$P_6 \geq 10.02$	$P_6 \geq 10.70$	$P_6 \geq 11.19$	$P_6 \geq 11.55$	$P_4 = 5.5$
	$P_6 \geq 8.30$	$P_6 \geq 9.46$	$P_6 \geq 10.23$	$P_6 \geq 10.78$	$P_6 \geq 11.20$	$P_4 = 6.5$
	$P_6 \geq 7.59$	$P_6 \geq 8.89$	$P_6 \geq 9.76$	$P_6 \geq 10.38$	$P_6 \geq 10.84$	$P_4 = 7.5$
	$P_6 \geq 6.88$	$P_6 \geq 8.32$	$P_6 \geq 9.28$	$P_6 \geq 9.97$	$P_6 \geq 10.49$	$P_4 = 8.5$

24.8 Conclusion

For a gas turbine power station comprising two gas turbines and one S. T., various performance measures of system effectiveness are obtained using R programming language for winter and summer given as follows.

Performance measure	Winter	Summer
Availability in full capacity	0.5257605	0.4337524
Availability in Type I	0.0026688	0.0040481
Availability in Type II	0.0104854	0.0156835
Availability in Type III	0.0021045	0.0031617
Down time	0.0026184	0.0039154
Busy period	0.0160583	0.0244289
Number of visits	0.0006572	0.0009853
MTSF	2430372	
Profit	INR 1851.466 per hour	

Cut-off points for plant operating expenses have been obtained for different values of P_3 and P_5 shown in Table 24.1 and for P_4 and P_6 shown in Table 24.4. Minimum price of electricity for different values of plant operating expenses have been obtained for Type I and Type III during winter as well as summer shown in Tables 24.2, 24.3, 24.5, 24.6. This will be helpful for users of such systems to take various important decisions regarding quality of the turbines and use of gas turbine during the failure time of steam turbine. Profit of the gas turbine power plant is Indian Rupees 1851.466 per hour whereas profit of the same system is Indian rupees 1698.552 per hour if throughout FCFS repair pattern is used. Therefore, it is recommended that FCFS repair pattern is used till all the three turbines do not fail. On failure of all the three turbines priority should be given to the repair of gas turbine over steam turbine. This will generate INR 152.914 more profit per hour.

Acknowledgments The authors are thankful to the 'Pragati Power Station, Gas Power Station, IP Estate, Ring Road, New Delhi', India for providing yearly data regarding maintenance of gas turbine power generating system for different values of Ambient Temperature. First author is thankful to the Indira Gandhi National Open University, New Delhi for granting study leave to pursue his Ph.D. work.

References

1. Baohe Su (1997) On a two-dissimilar-unit system with three modes and random check. Microelectron Reliab 37:1233–1238
2. Chow D.K. (1973) Reliability of some redundant systems with repair. IEEE Transactions on Reliability R-22(4): 223–228
3. Tuteja RK, Arora RT, Taneja G (1991) Analysis of a two-unit system with partial failures and three types of repairs. Reliability Engineering and System Safety 33:199–214
4. Singh D, Taneja G (2012) Reliability analysis of a power generating system through gas and steam turbines with scheduled inspection. Aryabhatta Journal of Mathematics & Informatics 5(2):373–380
5. Singh D, Taneja G (2014) Reliability and cost-benefit analysis of a power plant comprising two gas and one steam turbines with scheduled inspection. International Journal of Soft Computing and Engineering 4(3):30–37
6. Abigail GD, Agustin MAC, Maria OGD, Angel MA, Mathieu L, Jose MGS (2017) Effect of the ambient conditions on gas turbine combined cycle power plants with post-combustion CO_2 capture. Elsevier Ltd. 134:221–233
7. Farouk N, Sheng L (2013) Effect of ambient temperature on the performance of gas turbines power plant. International Journal of Computer Science 10(3):439–442
8. Saleh S, Baakeem JO, Hany A (2015) Performance of a typical simple gas turbine unit under Saudi weather conditions. International Journal of Energy and Power Engineering 1(3):59–71
9. Rajesh Taneja G, Prasad J (2018) Reliability of a gas turbine system with change in weather and optimisation of electricity price when working in single cycle. International Journal of Agricultural and Statistical Sciences 14(1):119–128
10. Rajesh Taneja G, Prasad J (2018) Reliability and availability analysis for a three-unit gas turbine power generating system with seasonal effect and FCFS repair pattern. International Journal of Applied Engineering Research 13(12):10948–10964
11. H. Wickham. ggplot2 (2009) Elegant Graphics for Data Analysis. Springer-Varlag New York, (2009)
12. R Core Team R (2018) A language and environment for statistical computing. R Foundation for Statistical Computing, Vienna, Austria URL https://www.R-project.org/

Chapter 25
Forces Driving a Trading Company Towards CSR: A Case Study of GCF

Nandita Singh and Nitin Arora

25.1 Introduction

In an economy, sustaining in the cutthroat competition is the main goal. The border less economy and environment makes it more difficult for both the company and the environment to sustain itself. Nowadays, importance of CSR is increasing in the Indian economy as organisations have realized that trust building and maintaining relationship with the society is also important. The CSR has become a compulsion for every organization as it helps in maintaining a balance between expectations and needs without effecting the resources for the next generations.

The significance of CSR rose considerably in the last decade. As per the compliance [2], the activities to be covered under CSR should create a difference in the surroundings. Over a period of time there is a shift of companies thought from profit making to creating a social impact. The companies have been transparent about the funding and expenditure done under CSR and making it accessible to the public. CSR is an way in for incorporating various public issues present and including them in a company policy and strategy. However, there is a worldwide and major view on protecting the interest of both stakeholders and environment. Emerging economies like India have also witnessed a number of firms proactively engaging in CSR activities, and adhering to the guidelines and framework established by the Ministry of Corporate Affairs [7–10] for firms. As per the mandate 2% of companies net profit should be allocated for CSR. The survey done by KPMG [6] shows various companies have been spending more than the prescribed limit towards CSR. It was found that the Energy and power sector spent the highest amount of Rs. 2525 crore followed by BFSI(Banking and finance) which spent Rs. 1281 crore and IT&software

N. Singh · N. Arora (✉)
Amity International Business School, Amity University Campus, Sector 125, Noida, India
e-mail: narora4@amity.edu

N. Singh
e-mail: nanditasingh86@gmail.com

P. K. Kapur et al. (eds.), *Decision Analytics Applications in Industry*, Asset Analytics,
https://doi.org/10.1007/978-981-15-3643-4_25

which spent Rs. 959 crore. Globe Capital Market Ltd. [13] is one such part of finance industry which has started its CSR in the year 2015.

25.1.1 Globe Capital Market Ltd [13]

Founded in 1994, Globe is well-liked and trusted financial service provider, operating in Indian capital market for more than 23 years offering diverse range of services devoted to help its clients achieve their goals. Its varied range of services and a broad client base is an result of proficiency in guiding through the financial markets.

With more than 3000 offices across 400 more locations Globe's rising network has spread in India and overseas offices through a subsidiary in DUBAI. It offer varied services including Equities, Commodities and Currencies Markets Intermediation, Portfolio Management Schemes, Depository Services, Mutual fund and IPO distribution, Qualified Depository Participant services, Securities Lending and Borrowing Services, Corporate Advisory and International Broking, etc. Globe Capital today accounts for a significant share of NSE clearing volumes in the Equity and Currency derivative segments. It is also one of the leading Depository Participants with NSDL and CDSL. Globe Capital Market Limited is a continuous contributor towards India's economic ongoing journey and remains committed to investing and innovating for India. Accounting for a significant share of NSE, BSE, MCX and others clearing volumes in the equity and currency derivative segments makes it one of the leading clearing house in the industry. A vision of running a marathon, not a 100 m race with strong yet simple thoughts Globe Capital Market Limited, is the biggest clearing company accounting for 25% of the exchange clearing volumes across segments and across exchanges. Globe since past 20 years has been contributing to India's economy and yet to continue to do so and help its client grow higher. For achieving its CSR objectives, Globe Capital Market Limited after discussion with the board members will keep aside the amount of its net profits as a part of the CSR budget for the specific financial year. The allocation shall be done in accordance with the Companies Act, 2013 and rules formed and amended accordingly. Any surplus amount will be added to the budget of next financial year which can be used in future activities. Any amount which is surplus amounting out of CSR cannot be considered as a part of companies profit whatsoever.

25.1.2 Globe Capital Foundation

Globe Capital Foundation [14], the philanthropic arm of the Globe Capital Market Ltd., was established in 2015. Since the day it came to light, it has acted a bridge between the users and providers of capital and knowledge by bringing the skills, resources and talents of the profit focussed world to the non-profit focus area. The initiatives planned include our employees where each employee contributes its capabilities and expertise which help the non-profits in the projects and financial help. Globe Capital Foundation aims to benefit the society irrespective of creed, class, community and for general public. The organisation strongly believes that a true business not only caters to the needs and interests of the company but goes beyond to serve the society from which it has created the wealth.

The company always measured the progress of a community by the degree which youth has achieved. Globe Capital believes that an empowered youth holds a massive power to create lasting and positive change in the community. Not losing its sight, Globe Capital Market working on the three basic principles Simplicity, Foresightedness and Humility has always contributed for the society as it believes while climbing the ladder for great business, one should also feel for the society. This has in turn motivated a financial service organisation to aggressively invest in CSR.

The aim of GCF:

i. Promotion of healthcare and sanitation, eradicating hunger, poverty and malnutrition.
ii. Imparting education to the underprivileged members of the society. Spreading awareness about special needs learning and work opportunety. Providing vocational skills training among children, women, elderly and the specially abled.
iii. Women empowerment and gender equality to be promoted. Establishing old-age homes, daycare centers and such other facilities for senior citizens and homes for creating equality in socially and economically backward groups.
iv. Retention and management of art and culture, national heritage which includes restoration of structure and location of historical relevance and work of art, setting up public libraries, promotion and development of traditional arts and handicrafts.
v. Providing help to the victims effected by natural calamity like floods, earthquake, droughts and other similar situations in the country by supporting directly or indirectly or contributing towards the government recognized relief centres or funds.

Being a financial services organisation, the forces that drive the trading company towards CSR:

- Economic responsibility: The main reason is to increase the stakeholder welfare [4], ensure growth and profitability by financial originality. Due to regular change in various financial interest for individuals or corporate, new opportunities are being created by financial institutions for effective mediation and minimizing the risk. This can be done by developing new products and services redesigning the existing and forming the new products and channels. Involving the stakeholders and board members is crucial for developing these new channels.
- Reconnecting with stakeholders: When focus is on financial goals and profits, CSR brings purpose to the organisation. To reach satisfaction among the organisation, positive psychology is the key. The emotional quotient helps bring together an organisation and people involved with it.

- Ethical Responsibility [5]: Individual conscience and the expectation of the external factors and stakeholders can forma part of ethical norms. It brings out the transparent, ethical, fair conduct approach in the market.

25.2 Findings

25.2.1 CSR Initiatives by GCF [12]

25.2.1.1 Health Initiatives for Fit India

Bridging the health gap for healthier and fit India, Globe Capital Foundation has taken various initiatives to provide a better and improved health infrastructure.

a. **Vision for All**

Globe Capital foundation strives to work for the society by addressing the most prevailing causes in the society in contributing to achieve the World Health Organization's resolution on Vision 20/20 in collaboration with Shroff's Eye Care centre. Cataract and diabetic has been documented to be the most significant reason of bilateral blindness in India. Approx 50–80% of bilaterally blindness reposted is due to cataract In India.

GCF under its CSR mandate, has expressed a desire to hold eye camps in association with SCEH to bring quality eye care services to the poor and the underserved in Delhi NCR and neighboring states. It has decided in this regard to establish a **Mobile Eye Care program** and hold camps in underserved areas to screen, identify and treat adults and children suffering from Refractive error, Cataract, Glaucoma, Retina and others. It has decided to sponsor free spectacles and free eye surgeries to the poor patients identified through the camps for **3 years** from the starting date of the project. A mobile medical bus will be used to conduct eye camps for both adult and pediatric. The mobile bus will also act as a back bone for the old age people, who are not able to reach the center for the eye care treatment. The initiative for providing eye care service at the doorstep by the mobile bus, will be taking a step further in eradicating avoidable blindness from the community. The other facilities which can be provided by mobile bus include providing transportation facility to the patients from the camp site to SCEH base hospital for the surgery to strengthen activities in a more systematic way for eradicating avoidable blindness. The bus would provide to and fro facility to the poor and needy patients who are not able to afford the cost of the surgery but they are also not able to bear the travel cost incurred during the treatment (Table 25.1).

b. Since Feb 2016, every year Super Specialty Health Camp and Talk with Medanta-The Medicity at Haryana Bhawan which included free blood test, ECG, mammography, eye check-up, euro check-up, etc. is held. In this camp, approx. 500 people are provided free check and free medicine is also distributed to the needy.

Table 25.1 Free eye checkup camps data—FY16-till now

Camp type	No of camps	OPD	Cataract surgery	Specialty surgery done	Pediatric surgery	Specs distributed
General	57	10,363	740	31	NA	1426
Pediatric	17	1201	1	20	26	75
Total	74	11,564	741	51	26	1501

c. GCF made contribution to Apna Ghar sponsoring residents for healthier life.
d. Blood Donation Camp-Human blood is an essential element of human life with no substitute. There is a lack of 3 million blood units in a year faced in India. This shortage can be easily eliminated if only an additional 2% of India's youth donated blood, just once in a year. Globe Capital Foundation in association with Indian Red Cross Society held a blood donation camp at Ansal Bhavan office site in June 2018.
e. Blood Test Camp-Globe Capital Foundation in association with SRL Diagnostic Lab organised a free blood test camp. Complete Blood Count Test (CBC) was done which can be used to evaluate overall health and detect a wide range of disorders, including anemia, infection and leukemia of an individual.

25.2.1.2 Education Initiative

a. **Education for Success**
Education is one of the foundation stones of any nation; at Globe Capital Foundation the aim is to improve academic standards in economically backward schools that prepare students to become equally competitive for future events. Launched in March 2017, by Globe Capital Foundation, **Educate a Child** program aims to identify and expand opportunities to reach the marginalised and underprivileged children and families and help them access and complete their education. GCF received total of 175 applications. GCF sponsored fees, books and uniforms of the children.
b. Bharat Lok Shikhsha Parishad was identified and listed on 3 March, 2000. Its vision is to bring an overall development of the remote tribal and rural villages of India. It operates in over 53,000 schools and educates more than 1.5 million children. GCF sponsored 100 EKAL schools in Pilibhit and 100 EKAL schools in Saket, UP respectively. These schools work towards the education of underprivileged children.
c. GCF sponsored JMC fest-Com'Acumen by dept of commerce included commerce students from all over colleges in Delhi. GCF also hosted an event Trade smart wherein the students will be participating in online test and will take part in virtual trading and cash prize will be distributed.

25.2.1.3 Human Rights Initiative

a. **New Year Celebrations**—GCF planned to celebrate New Year eve with Bal Sahyog, NGO which works for the welfare of orphan children. These kids are homeless or separated from parents or referred by government agencies end up being here. GCF decided to renovate their facility as a token of appreciation at Bal Sahyog. This was a small step towards a big change and a bright tomorrow. In Lucknow, GCF distributed school supplies and other daily necessities to the underprivileged kids and needy in Kakori.100 underprivileged students studying in Prathmik Vidyalaya, Kakori were provided stationery set and blanket.
b. **Christmas for a Cause**—Globe Capital Foundation tied up with DCCW running PALNA—"A Cradle of Hope" Programme for abandoned, physically and mentally challenged children. The employees contribute food items, sanitary products, clothes, bedsheets, bedcovers, blankets, baby food. GCF contributed desk and chairs for kids aged 2–10 years. Cupboards for kids room were also contributed.
c. **Contributing towards Kerala Floods**—Kerala faced one of the worst floods in the past decade. It will take months to recover and rebuild itself. GCF made its contribution by providing cloths, bedsheets, utensils and other daily use items.

25.2.1.4 Environment Initiative

a. Green Initiative for Healthier India
Working towards a better tomorrow, At Globe Capital Foundation the crucial point of CSR initiatives for environment has been in encouraging for green initiatives for clean, green and healthy India. Thirty trees were planted under project "Green Roads Green Rajasthan with Aashadeep (NGO).

25.2.1.5 Reporting—For GCF Funded Activities/Camps

Pre-event

- "Project Implementation Schedule "detailing particulars of the programme will be organized in advance by GCF.
- As a part of employee engagement, GCF also observes/deputes volunteers in the programmes to oversee the activities.

Post-event

- The data maintained by GCF after every programme:
    a. Name and address of participants;
    b. Photographs of venue;
    c. List of beneficiaries covered under the event.

25.3 Conclusion

The study concluded that GCF is implementing different CSR activities in field of health, education and environment. Globe Capital Foundation is serving society since its launch in 2015 and is adhering to the guidelines setup by the government. It highlights the effective ways companies can benefit the society and be a part of its upliftment and improvement. The initiatives undertaken are helping large strata of society. Being a trading company and a part of financial sector, GCF is fulfilling its role and responsibility by covering focused areas as a way of giving back to the society. It is also helping in creating a balance with the user and provider so that the upcoming generation can benefit from it. The work done under CSR has not been the same every year. The various activity vary every year [3]. It has been observed that startups or new companies to be started should be of social nature to the society and business nature to the company like hospitals and educational institutes. This will be helpful in fulfilling the social and commercial objectives.

References

1. Basu K, Palazzo G (2008) Corporate social responsibility: a process model of sensemaking. Acad Manag Rev 33(1):122–136
2. Davies R (2002) Corporate citizenship and socially responsible investment in Asia. In: Unpublished paper delivered at the conference of the association for sustainable and responsible investment in Asia, Hong King, 10 June 2002
3. Friedman M (1970) The social responsibility of business is to increase its profits. The New York Times Magazine. The New York Times Company, 13 September 1970
4. Gangopadhyay S (2012) Profiting from CSR: mandatory spending on corporate social responsibility does not fit in a market-driven society, Business Standard, 28 January 2012
5. Goenka H (2012) Why it is a good idea to mandate corporate social responsibility, Economic Times. http://articles.economictimes.indiatimes.com/2012-09-01/news/33535456_1_net-profit-companies-bill-crore-last-year. Accessed 1 Sept 2012
6. KPMG-UNEP (2010) Carrots and sticks-promoting transparency and sustainability: an update on trends in voluntary and mandatory approaches to sustainability reporting. KPMG Advisory NV, UNEP GRI
7. MCA (2009) Corporate social responsibility voluntary guidelines in 2009 by the Ministry of Corporate Affairs, Government of India

8. MCA (2011) The national voluntary guidelines on socio economic and environmental responsibilities of business. Ministry of Corporate Affairs, Government of India, New Delhi
9. MCA (2013) Companies Act, 2013. Ministry of Corporate Affairs, Government of India, New Delhi. http://www.mca.gov.in/Ministry/pdf/CompaniesAct2013.pdf
10. MCA (2014) CSR Rules, 2014. Ministry of Corporate Affairs, Government of India, New Delhi. http://www.mca.gov.in/Ministry/pdf/CompaniesActNotification2_2014.pdf
11. PWC (2013) Handbook on corporate social responsibility in India. Price Waterhouse Coopers Private Limited. http://www.pwc.in/assets/pdfs/publications/2013/handbook-on-corporate-social-responsibility-in-india.pdf
12. https://csrbox.org/India_Company_Globe-Capital-Market-Limited-Delhi_6152
13. http://www.globecapital.com
14. http://www.globecapital.com/about-us/milestones

Chapter 26
Automated Censoring of Cigarettes in Videos Using Deep Learning Techniques

Swapnil Dhanwal, Vishnu Bhaskar, and Tanya Agarwal

26.1 Introduction

Videos are recordings of moving visual images that convey some information or may just provide entertainment. However, they may contain age-inappropriate or harmful content that needs to be censored. Censoring is the act of examining a book, a video, a film, et cetera and suppressing objectionable or harmful content from it. Video censoring, therefore, is a process that consists of watching an entire video, detecting and locating the object to be censored (referred to as the target object) in each frame and finally censoring the target object. The entire process can become cumbersome, tedious and drawn out. The purpose of automating a process is to improve efficiency and reduce human effort—the only human effort required in such a process lies in its development and maintenance. Hence, automation of video censoring can possibly lead to a more efficient and accurate outcome. This involves extracting frames from the video, calculating and processing their segments and running them through an object classifier and localiser or object detector to locate the target object. Studies [15] have shown that exposure to media containing cigarettes can have profound effects on people, especially the young, and can lead to the development of smoking habits in the future. Smoking at young ages can have dire consequences in adulthood such as respiratory and cardiovascular diseases. This habit is injurious to not only the smoker, but also to the people in the vicinity. Motivated to remediate this, we have tried to ameliorate the process of censoring cigarettes in videos, by automating an otherwise manual effort. This poses many challenges. Firstly, there are challenges associated

S. Dhanwal (✉) · V. Bhaskar · T. Agarwal
Netaji Subhas University of Technology, Delhi, India
e-mail: swapnild.co@nsit.net.in

V. Bhaskar
e-mail: vishnub.co@nsit.net.in

T. Agarwal
e-mail: tanyaa.co@nsit.net.in

P. K. Kapur et al. (eds.), *Decision Analytics Applications in Industry*, Asset Analytics,
https://doi.org/10.1007/978-981-15-3643-4_26

with object detection itself—such as deformation and variation in lighting, aspect-ratio and occlusion. These challenges are compounded by having a cigarette as the target object. Unlike detection of objects such as cars and animals, cigarettes can be present in different sizes, orientations and aspect-ratios in an image. Differences in illumination can confuse even human beings as to the identity of a cigarette. Consider an unlit cigarette placed on a table. It is difficult for a human being to differentiate between it and a piece of chalk, let alone for a deep neural network to do so. The second challenge is the lack of availability of a cigarette image dataset. The ImageNet dataset does contain images of cigarette butts, about 1,300 of them, but since we want to detect all parts of cigarettes, it is insufficient for our purposes. Thus, data need to be collected, cleaned and processed manually. This lack of data also necessitates the use of a pre-trained model. Another challenging aspect is performance—deep neural networks require considerable resources to train and run. Training and predicting can be made faster by using a GPU but sequential tasks such as segmentation are CPU bound. It is thus necessary to intelligently extract portions of the image which are likely to contain cigarettes, prior to censoring. In the following sections, we explain in detail, works related to our research, the deep learning technique used, our methodology, our experimentation, the results of our efforts and the future scope of improvement. Through the approach proposed by us in Sect. 26.4, we hope to arrive at a suitable model for detecting and censoring the target object, cigarettes.

26.2 Related Work

Several researchers have worked in the field of object classification, localisation, image segmentation and development of fast techniques for object detection. There have also been works that aim to detect the presence of cigarettes or smoking events in a scene—for instance, images or video-feed from a CCTV camera. In our research, however, we found that the technique of Deep Learning has not been exploited for the same. Several works attempt the detection of smoking events in a scene. For instance, Harikrishnan et al. [5] proposed a method to detect smoking events by using a combination of sensors to detect smoke and facial detection to detect the smoker. Here, facial detection was performed by using Discrete Wavelet Transform [10]. mPuff, a hardware approach proposed by Ali et al. [1], classified respiratory recordings as smoking-puffs or regular breathing. Iwamoto et al. [9] investigated smoke detection from captured image sequences. Their system proposes to address estimation of candidate areas of smoke and detection of smoke in the scene. Kavitha et al. [8] proposed a technique in which, initially, moving objects are detected in the video frame. Then, template matching is used to detect cigarettes and the Haar classifier is used to detect smokers' faces. The Inception family of deep CNNs proposed by Ioffe and Szegedy [13] and Szegedy et al. [3] also possesses the ability to detect cigarette butts—a class of the ImageNet dataset containing 1,300 images. A big challenge in object detection is localisation. This entails segmenting the image into smaller parts and passing these segments through the CNN to get predictions. A naive method would use a sliding window approach to extract image segments, however this would be very inefficient. Several researchers have worked on increasing the performance of object detection by using specialised segmentation techniques. Chen et al. [2] described R-CNN, or Region-based CNN, which aims to simplify the problem of selecting regions in conventional CNNs, by pre-defining 2000 regions to make predictions from. To make these region-proposals, it uses Selective Search [6]—an algorithm which creates larger regions by combining perceptually similar, smaller regions in a bottom up manner. Ren et al. [14], in Faster R-CNN, introduced the concept of Region Proposal Networks (RPNs), which try to eliminate the cost of calculating region-proposals by sharing features with the main CNN. A modified version of Faster R-CNN was employed to speed up the detection of small objects in remote sensing applications by Ren et al. [17]. One of the most revolutionary papers in the field of fast object detection is You Only Look Once (YOLO) by Redmon et al. [7]. In this method, instead of using a sliding window approach or region-proposals, the dataset images were divided into grids, with the target objects annotated by bounding boxes. The deep neural network would now predict two features—bounding boxes and the class of the detected object. These models are aptly called single-shot detectors because they perform classification and localisation simultaneously. Liu et al. [16] proposed a faster single-shot detector compared to YOLO. It works by fixing boxes of various sizes and aspect ratios and then, at test time, scores these boxes for object presence. In this paper, we have attempted to apply the collective knowledge and inspiration gained from the above works to create a model that can detect and censor cigarettes in videos.

26.3 Deep Learning Techniques

Deep learning models consist of greater than two layers, are better suited for end-to-end learning and thrive on large datasets containing millions of samples. However, they are also excellent for feature extraction. One such model is Inception V3.

The Inception V3 model is a deep Convolutional Neural Network trained on the ImageNet dataset [11]. The ImageNet dataset contains 14 million images belonging to 1000 classes ranging from 'cat' and 'dog' to 'dishwasher' and 'plane'. This model was chosen for three reasons. Firstly, the lack of a public dataset of cigarette images meant we had little data to work with; hence, some form of transfer learning was necessary. The second reason was its performance in the ImageNet Large Scale Visual Recognition Challenge (ILSVRC) (2015), of which it was the first runner up. Thirdly, the ImageNet dataset contains approximately 1300 images of cigarette butts, which would give our model some intuition about the shape, colour and texture of cigarettes from the get-go. In later sections, we show results of using a deep neural network instead of conventional machine learning approaches.

26.4 Proposed Approach

We now describe the steps undertaken by us in creating our model, which is divided into two parts—the classifier and the censor. As shown in Fig. 26.1, initially, data collection and cleaning were performed by downloading images from various online sources. Then, to balance the skewed dataset, data augmentation techniques were applied and synthesised images were incorporated. The next step was to perform transfer learning by building a model on top of Inception V3. This involved using the Inception V3 model for feature extraction and using the extracted features to train a fully connected neural network—forming the basis of the classifier. The second part of our model, the censor, used Selective Search to calculate region-proposals. These region-proposals were fed into the classifier and tagged according to the generated predictions. Finally, the tags were used to censor the video. This process is described in detail in the following subsections.

26.4.1 Data Collection

Generally, with an increase in the size of the dataset, performance of neural networks increases as opposed to traditional learning algorithms. However, in the cases where transfer learning is used, smaller datasets can suffice. To account for the lack of images of cigarettes in the ImageNet dataset, we created our dataset using images downloaded from Google Images, Bing Images, Getty Images and DuckDuckGo.

After downloading, the images were segmented to into the following sizes:

1. 200 × 200
2. 300 × 300
3. 400 × 400

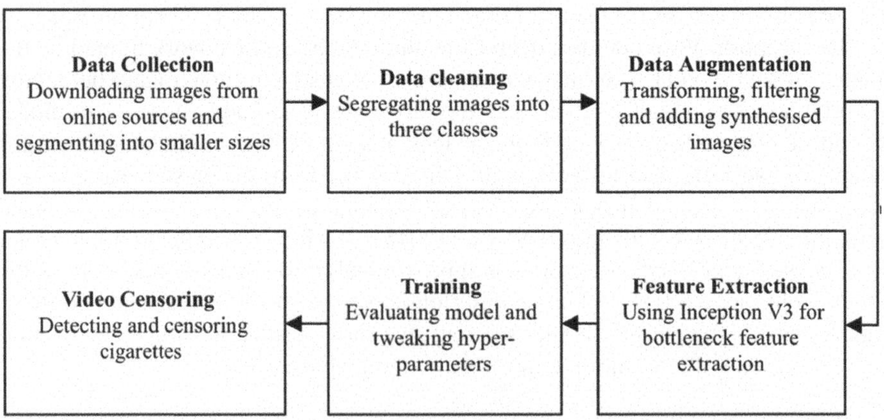

Fig. 26.1 Methodology

4. 500 × 500
5. 600 × 600
6. 700 × 700
7. 800 × 800
8. 900 × 900.

With the following overlaps:

1. 1–3: 50% Overlap
2. 4–6: 25% Overlap
3. 7–8: No Overlap.

Applying the above transformations yielded a coarse dataset of 105,000 images, the size of which reduced to about 70,000 after cleaning. These were used to create the training, development and the positive class of the test dataset. The negative class of the test set was populated by randomly sampling the ImageNet dataset.

26.4.2 Data Cleaning

Data cleaning can be defined as the process of detecting and correcting corrupt or inaccurate records from database, i.e. identifying and replacing dirty or coarse data. It may be performed interactively with data wrangling tools or as batch processing through scripting. We manually segregated the collected images into three categories. Here, we do not use the word category and class interchangeably. The word category refers to the descriptions assigned to the data by us; and the word class refers to the labels used for training the model.

The collected data conform to the following categories:

1. Cigarette with coloured background
2. Cigarette with white background
3. Not a cigarette.

The first two categories were used for data augmentation purposes as explained in the next subsection. A positive and negative class was used for training. However, since the dataset was severely skewed in favour of the negative-class samples, data augmentation techniques, as described below, were applied.

26.4.3 Data Augmentation

As explained previously, the training set was skewed in the favour of images belonging to the negative class. To remedy this problem, more images belonging to the positive class needed to be obtained. Figure 26.2 shows some samples from our augmented dataset.

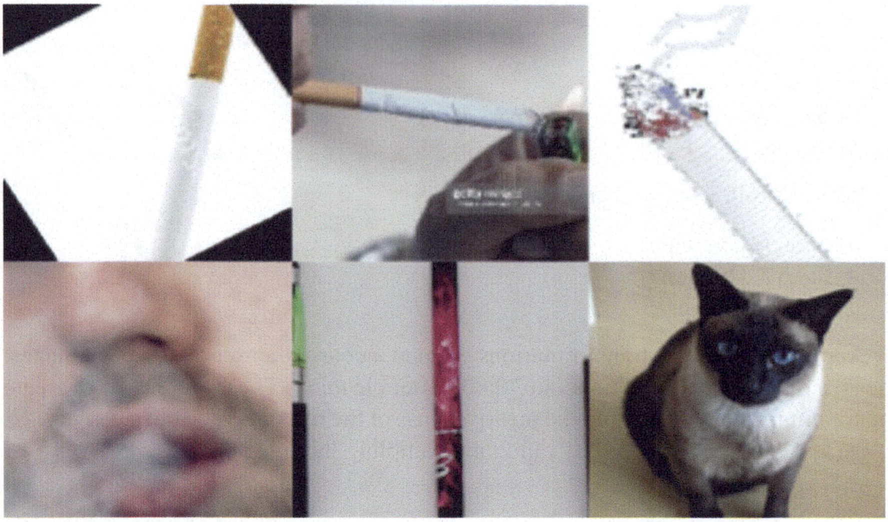

Fig. 26.2 The top and bottom rows show positive and negative samples respectively

Rotation. Images belonging to the positive classes were rotated about their centre by 30 degrees successively to obtain a total of 12 images.

Colour filtering. A histogram of colour values of the images belonging to the 'cigarette with white background' category—to isolate the colours of cigarettes—was generated. Then, pixels with colours having less than a predefined threshold frequency were set to white. However, this approach produced mixed results—some of the resulting images were noisy and lossy. In some cases, cigarettes were isolated perfectly so that they resembled images from the category 'cigarette with white background'. In others, the loss of information and generated noise was too large for the image to be classified as positive. Different values of the threshold frequency were experimented with and finally, about 12,000 images were successfully filtered and moved to the 'cigarette with white background' category.

Synthesising images. To further remedy the problem of a skewed dataset, a small set of images—synthesised by superimposing cigarettes on different textures—was generated.

26.4.4 Classifier Architecture

Our proposed model consists of all the layers of the Inception V3 model except the fully connected layers which are replaced by a combination of dense and dropout layers used to generate the final prediction, as shown in Fig. 26.3. Images scaled to 150×150 pixels were input into Inception V3 convolutional layers and the extracted features were used to train the fully connected layers.

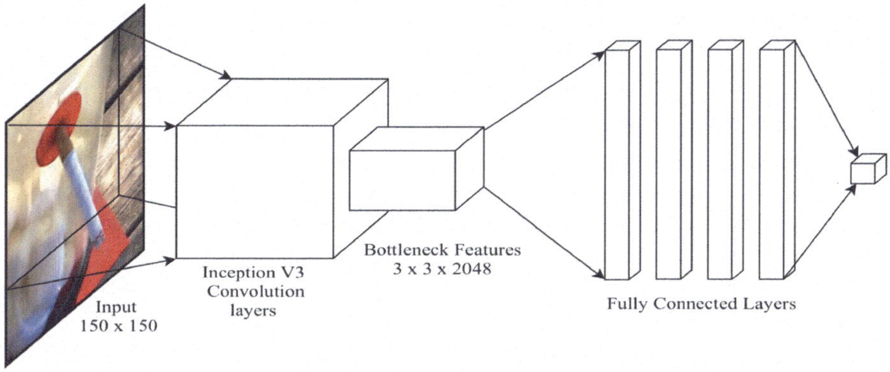

Fig. 26.3 Classifier architecture

26.4.5 Video Censoring

This process was composed of region-proposal generation, prediction, tagging and finally, video-frame censoring. Region-proposals can be defined as the perceptually important portions of an image i.e. they are more likely to represent objects. To optimise performance, it made sense to run the model on a subset of all possible image segments (i.e. region-proposals) instead of segments generated through a sliding-window approach. The algorithm used for generating region-proposals in this paper is Selective Search. Selective Search starts by over-segmenting the image using the graph-based segmentation algorithm described by Felzenszwalb and Huttenlocher [12]. Many of these fundamental segments can collectively represent individual objects. Selective Search is able to combine these smaller segments into larger ones based on similarity, in a bottom-up manner. The size, scale and number of segments can be controlled using the corresponding parameters to the algorithm. The output is a set of principal segments called region-proposals. The region-proposals thus obtained were tagged with the frame-number and frame-location before being input to the model for prediction. Non-max suppression was applied on the positively predicted regions to reduce the number of overlapping bounding boxes. Figure 26.4 showcases the video frame censoring process.

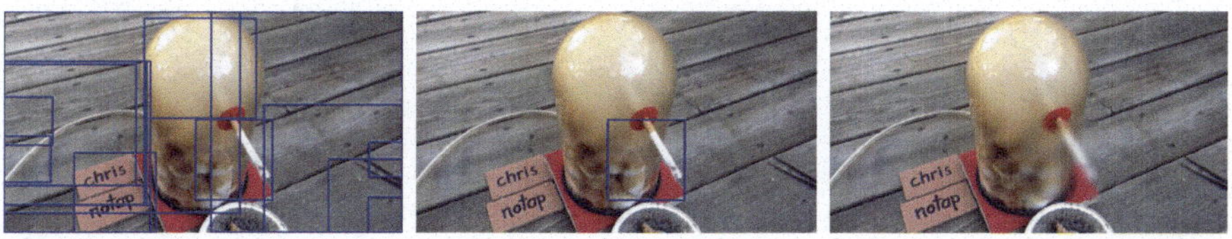

Fig. 26.4 From left to right: region proposal generation using Selective Search, prediction, censoring

26.5 Experimentation and Results

Using Google Colaboratory as our development environment, we created and experimented with different model architectures (referred to as the 'precursors') and hyperparameter settings. To create the models, we used Keras and Tensorflow (Inception V3 and Dense layers) and Scikit-learn (for classical machine learning models).

Since selective search is not optimised for the GPU, we used a laptop with a Core i7 9820HQ processor and 16 GB RAM for the censoring process.

We refer to the final model used for censoring as the 'conclusive model'.

Precursors to the conclusive model had the following shortcomings:

1. Misclassified objects with similar colour and shape as cigarettes
2. Misclassified text, smoke, water, skin and wood textures as cigarettes
3. Misclassified small objects as cigarettes—likely due to their similarity with cigarette butts.

Corrective measures included adding images of smoke and water textures to the negative class. With these measures, the conclusive model showed test-set accuracy of 89.99%. We took a unique approach to calculating test-set accuracy wherein the test set was created by taking a random sample of ImageNet (for the negative samples) and interspersing it with positive samples from our own data. Images from the cigarette butt class of ImageNet were also added to the positive class.

The fully connected layers of the conclusive model had 2 dense and 2 dropout layers in an alternating manner, an ADAM [4] optimiser, binary-cross-entropy as the loss function and was trained for 100 epochs. The conclusive model's performance metrics and confusion matrix are shown in the Tables 26.1 and 26.2, respectively.

Table 26.1 Characteristics of the conclusive model

Parameters	Value
Training samples	20,000
Validation samples	10,000
Testing samples	38,000
Training accuracy	95.74%
Testing accuracy	89.99%
Precision	97.03%
Recall	84.12%
F1 score	90.01%

Table 26.2 Confusion matrix for test-set

	Predicted: true	Predicted: false
Actual: true	True positives: 18,559	False negatives: 3503
Actual: false	False positives: 568	True negatives: 15,624

Table 26.3 Comparison of different classifiers

Model	Testing accuracy (%)
Logistic regression	73.06
Decision tree	68.43
SVM	71.11
Random forest	71.76
Conclusive model	89.99

Further, Table 26.2 shows a comparison of the conclusive model with classical models—Logistic Regression, Random Forest, Support Vector Machine (SVM) and Decision Tree.

The Decision Tree was the lowest performing model followed by the SVM and Random Forest classifiers. Logistic Regression came second at 73% but was sufficiently outperformed by the conclusive mode.

Content with the above results, we chose the conclusive model to perform video-censoring. With our current approach, censoring a one-minute video took 4 h—a number we hope to decrease in the future. Our findings translated well to images taken from dissimilar distributions. For instance, in videos taken from YouTube, we observed high precision in detecting cigarettes, however, there were still some misclassifications (as evident from the slightly lower recall metric). Although every cigarette was correctly censored, we observed that our model would seldom misclassify small, cylindrical objects as cigarettes. There were also some false-negatives, however, these were not missed by the censor considering that the smaller regions they were comprised of had been detected earlier.

26.6 Conclusion and Future Work

Exposure to media depicting cigarette consumption can have harmful effects on young viewers. Hence, censoring of such content is of paramount importance. Video censoring, a tedious process in which the entire video needs to be examined manually can be replaced by an automated process using Deep Learning. This is what we aimed for, and we have achieved 89.99% accuracy with our proposed model. Our model is able to censor videos containing cigarettes which are visually distinguishable from other objects, with satisfactory performance. Future work lies in making cigarette censoring a real-time process. This involves creating an annotated dataset—complete with bounding boxes to feed single-shot detectors like YOLO. Our model also has other limitations. It misclassifies some objects of similar shape, colour and texture as cigarettes and its performance precludes it from being classified as a real-time censor. Further, it fails in situations when the cigarette is so small as to be indistinguishable from tiny objects. In the future, we aim to remove these limitations by gathering more data, performing multi-label classification and using a single-shot detector. With this in mind, in conclusion, we believe that we have taken a step in the right direction and hope that that our work can inspire other researchers to make media safer for younger generations.

References

1. Ali AA, Hossain SM, Hovsepian K, Mahbubur Rahman Md, Plarre K, Kumar S (2012) mPuff: automated detection of cigarette smoking puffs from respiration measurements
2. Chen C, Liu M-Y, Tuzel CO, Xiao J (2016) R-CNN for small object detection
3. Felzenszwalb PF, Huttenlocher DP (2003) Efficient graph-based image segmentation
4. Harikrishnan K, Akhilkrishna P, Varghese B, Shylesh S, Chandrasekhar V, Mathai MB (2015) Smoke detection captured from image features
5. Heatherton TF, Sargent JD (2010) Does watching smoking in movies promote teenage smoking
6. Ioffe S, Szegedy C (2015) Batch normalization: accelerating deep network training by reducing internal covariate shift
7. Iwamoto K, Inoue H, Matsubara T, Tanaka T (2010) Cigarette smoke detection from captured image sequences
8. Kavitha CS, Gayathri Devi M, Pradeep Kumar, Reshma (2015) Automated system for smoking detection
9. Kingma DP, Ba J (2014) Adam: a method for stochastic optimization
10. Liu W, Anguelov D, Erhan D, Szegedy C, Reed S, Fu C-Y, Berg AC (2015) SSD: single shot MultiBox detector
11. Redmon J, Divvala S, Girshick R, Farhadi A (2016) You only look once: unified, real-time object detection
12. Ren S, He K, Girshick R, Sun J (2016) Faster R-CNN: towards real-time object detection with region proposal networks
13. Ren Y, Zhu C, Xiao S (2018) Small object detection in optical remote sensing images via modified faster R-CNN
14. Russakovsky O, Deng J, Hao S, Krause J, Satheesh S, Ma S, Huang Z, Karpathy A, Khosla A, Bernstein M, Berg AC, Li F-F (2015) ImageNet large scale visual recognition challenge
15. Shensa MJ (1992) The discrete wavelet transform: wedding the a trous and Mallat algorithms
16. Szegedy C, Vanhoucke V, Ioffe S, Shlens J, Wojna Z (2016) Rethinking the inception architecture for computer vision
17. Uijlings JRR, van de Sande KEA, Gevers T, Smeulders AWM (2012) Selective search for object recognition

Chapter 27
Qualimetry in the Consumer Quality Assessment of Cars Class C

Nellie V. Klassen

27.1 Introduction

The automotive industry is one of the fundamental industries in Russia, which employs human and production facilities that stimulate the development of other sectors of the economy, for example, metallurgy, electrical, engineering and others. All these industries are suppliers and partners to create high-quality automotive products [1, 2].

The requirements for a modern car are increasing every day: consumers want to see a reliable, safe, eco-friendly car with minimal operating costs.

The main management and quality assurance tool is the quality system. Increasing consumer requirements for the quality of goods leads to an increase in the requirements for the quality systems of the supplying enterprises and requires their improvement [1, 2].

Since one of the basic principles of quality management, according to ISO 9001: 2015, is consumer orientation, it is also necessary to take into account the opinion of consumers in the quality assessment. Therefore, the purpose and objectives of this study is to find out which indicators of the quality and competitiveness of a car are the most important for consumers; to form control and expert groups of consumers, analyze the results obtained and use them in the future activities of the organization in the field of quality [3–6].

The first step in assessing the quality level is to determine the groups of quality indicators that most fully characterize the quality level of this product. These are indicators of reliability and maintainability (costs for repairs, labor intensity of repairs, costs of operating (consumable) materials), safety and environmental friendliness, purpose and ergonomics, aesthetic indicators [7–10].

N. V. Klassen (✉)
Peter the Great St. Petersburg Polytechnic University, Polytechnicheskaya, 29, 195251 St. Petersburg, Russia
e-mail: nellieklassen@gmail.com

P. K. Kapur et al. (eds.), *Decision Analytics Applications in Industry*, Asset Analytics,
https://doi.org/10.1007/978-981-15-3643-4_27

27.2 Research Methods

Methods for determining the numerical values of indicators of product quality are divided into two groups:

(1) on the methods of obtaining information (instrumental, calculated, organoleptic, registration);
(2) on the sources of information (traditional, expert, sociological).

The first of these methods—instrumental—is based on the use of technical measuring instruments (reagents, etc.). The measuring method determines the majority of quality indicators, the numerical indicators of which we can take directly from the technical means, without resorting to calculations and formulas. This is an objective and accurate method [11]. And the results are easily reproducible repeatedly, which means that they are not difficult to verify. However, for using this method it can be too long or difficult to fulfill, requiring highly qualified measuring specialists.

The next of these methods—organoleptic—is based on the analysis of sensations and perceptions obtained with the help of five human senses. The method is quite simple to use, therefore it is often used first, takes less time and does not require a large number of resources, unlike the measurement one. Taste and smell can be estimated only by the presented method, for their assessment there are some varieties, such as sensory, tasting and other methods [12].

The calculation method is based on the use of theoretical and empirical dependencies to determine the numerical indicators of product quality. Usually used to determine reliability, performance, labor intensity, etc.

The last method from the first subgroup—registration—is based on the analysis and calculation of certain events. So calculate the indicators of durability, reliability, reliability and shelf life of products. The difficulty in applying this method lies in the need for lengthy observation to obtain a representative sample [13, 14].

The second subgroup of methods. The traditional method involves obtaining information about the quantitative assessment of quality indicators from traditional sources of information in the enterprise (organization): laboratories, TCI [15].

The expert method is based on the opinions of specialists. The method is used in cases of lack of information, the need to develop special technical means, etc. It is desirable that the expert group was formed not for one examination, but as a constantly functioning body with a fairly stable composition of experts. [16].

The last method from the second subgroup—sociological—is based on the collection and analysis of consumer opinions. Opinions can be collected in different ways, for example, conducting an oral survey, distributing questionnaires, organizing sales exhibitions, conferences or auctions. [17]. The sociological method is widely used at the stage of marketing research, in the study of demand, to determine quality indicators, quality assessment [18, 19].

All these indicators are formed by an expert and sociological method.

27.3 Key Research Findings

To assess the prospective quality level, the quality indicators of the best samples of the world's leading firms are analyzed. For exemplary (ideal), you can take a product of modern production, which includes all the best performance achieved by various enterprises and firms in cars of this type. This sample must be of the same class; in this case, class C cars are considered, the type and purpose in fixed consumption conditions as the car being analyzed [20, 21].

When assessing the quality of a complex method, it was necessary to establish the weight of the properties of the product, their importance from the point of view of the degree of customer satisfaction [22]. The weight, the significance of individual properties, the quality characteristics of the product were established using weighting factors (g) [23].

Quality assessment conducted on a three-point scale

- "excellent"—$g = 2$ (in the case of an important parameter);
- "good"—$g = 1$ (in the case when the expert finds it difficult to assess the importance of the parameter);
- "satisfactory"—$g = 0$ (in the case of an unimportant parameter).

Each expert differently evaluates the importance of a particular parameter. Therefore, it is necessary to determine the average value of the weight coefficient (G) for each of the parameters

$$G = \frac{1}{M} \sum_{m=1}^{M} g_{im}$$

where

M number of expert;
i number of quality indicators

The results of the evaluation of the averaged value of the weight coefficient are shown in Table 27.1.

Also determined the integrated indicator of the quality of a real and perfect product. Each expert in all parameters forms the numerical value of the complex parameter of quality K:

$$K = \sum_{i=1}^{N} G_i \times q_{im}$$

For an ideal car, when all parameters correspond to the $q_{\text{im}} = 3$ level, the complex K_{ideal} indicator is defined:

Table 27.1 The average value of the weighting factors of quality indicators

№ quality indicator	Expert weight coefficients				Average weight coefficient
	g_1	g_2	g_3	g_4	
1	1	1	2	2	1.5
2	2	2	2	1	1.75
3	0	1	0	1	0.5
4	2	0	0	1	0.75
5	2	2	2	2	2
6	1	1	1	2	1.25
7	2	1	2	1	1.5
8	2	2	2	2	2
9	0	2	0	1	0.75
10	2	1	2	1	1.5
11	2	2	2	2	2
12	2	1	2	1	1
13	2	2	2	2	2
14	2	2	2	1	1
15	2	2	2	2	2
16	2	1	0	1	1
17	2	1	2	1	1.5
18	1	0	0	1	0.5
19	1	1	0	1	0.75
20	1	1	1	2	1.25
21	1	2	1	1	1.25
22	2	2	2	2	2
23	2	1	2	1	1.5
24	2	2	2	2	2
25	2	1	2	2	1.75
26	1	2	2	2	1.75
27	1	1	1	1	1
25	2	1	1	2	1.5
29	2	2	1	1	1.5
30	0	0	1	1	0.5
31	2	2	2	2	2
32	0	2	0	1	0.75
33	2	1	2	1	1.5
34	2	2	2	2	2
35	2	1	2	1	1

(continued)

Table 27.1 (continued)

№ quality indicator	Expert weight coefficients				Average weight coefficient
	g_1	g_2	g_3	g_4	
36	2	2	2	2	2
37	2	2	2	1	1
38	2	2	2	2	2

$$K_{ideal} = 3 \times \sum_{i=1}^{N} G_i = 3 \times 38 = 114$$

In this case, we accept $K_{\text{ideal}} = 114$ as the maximum attainable level of quality, at which all consumer needs are fully satisfied. This is the highest indicator with which further comparisons will be made of the complex indicator of an actual product.

The results of the evaluation and comparison of the collective assessment of quality indicators are shown in Table 27.2.

Subsequently, the obtained complex indicator of the quality of the investigated vehicle K was compared with the complex indicator of the quality of the ideal K_{ideal} product.

Qualimetric assessment means expert assessment, i.e. assessment of the quality of the object and work based on or engaging the experience of experts.

A qualified expert group will conduct a qualifying assessment.

The main condition for expert evaluation (assignment of weight coefficients) is the ability to reapply, from a slightly different point of view, assessments of the same directions. For this, the procedure of partial pairwise comparisons is used. It allows you to compare objects with each other, and then several ways to calculate the weights, which provides the control operation [24].

This operation is a control because it can be repeated several times for different groups of experts in order to collect the most complete picture of opinions, which is a different assessment direction.

Paired comparisons are performed as follows.

Take a group of objects, in our case, the area or areas of research to be evaluated. Build a table of dimension $(n-1) \times (n-1)$, where n is the number of areas (directions) in the group. On the left and upper sides of the table enter the names or numbers of objects. The first is to record a more significant object and then—in descending order.

Using the individual filled matrix of pairwise comparisons, shown in Table 27.3, we calculate the weight ratio in Table 27.4. As you can see, an unacceptable discrepancy (more than 0.2) between the pairs is not observed, therefore, the internal estimates are consistent. Then we calculate individual average scores (Table 27.4, individual average scores are highlighted).

Table 27.2 Evaluation and comparison of collective assessment of quality indicators

№ quality indicator	G_i	Qualification q_m				
		m = 1	m = 2	m = 3	m = 4	—
1	1.5	0	1	2	1	—
2	1.75	1	2	1	0	—
3	0.5	2	2	2	2	—
4	0.75	0	0	2	1	—
5	2	2	2	2	1	—
6	1.25	2	2	2	2	—
7	1.5	1	1	0	1	—
8	2	2	2	2	2	—
9	0.75	2	2	2	2	—
10	2	2	2	2	2	—
11	2	2	1	2	2	—
12	2	2	2	2	2	—
13	2	0	1	1	2	—
14	1.75	2	2	2	2	—
15	2	0	1	0	0	—
16	1	0	0	0	1	—
17	1.5	0	1	0	1	—
18	0.5	1	1	2	0	—
19	0.75	1	1	2	0	—
20	1.25	2	2	2	2	—

(continued)

Table 27.2 (continued)

№ quality indicator	G_i	Qualification q_m				
		m = 1	m = 2	m = 3	m = 4	—
21	1.25	1	1	2	0	—
22	2	2	2	2	2	—
23	1.5	2	2	2	2	—
24	2	0	1	1	0	—
25	1.75	0	1	0	1	—
26	1.75	1	1	1	2	—
27	1	0	2	2	1	—
25	1.5	0	2	1	1	—
29	2	2	2	2	2	—
30	0.5	2	1	2	2	—
31	2	1	1	2	2	—
32	0.75	1	2	2	2	—
33	1.5	2	1	1	1	—
34	2	2	2	2	2	—
35	1	1	1	1	1	—
36	2	1	1	2	2	—
37	1	1	2	2	2	—
38	2	2	2	2	2	—
The value of a complex indicator		—	—	—	—	Collective assessment
Real product		84.75	89.75	88.50	83.50	86.625
Perfect product		114.00	114.00	114.00	114.00	114.00

Table 27.3 Individual completed matrix of pairwise comparisons

Indicators	(22)	(23)	(21)	(19)	(20)
(17)	9	7	5	3	1
(22)		9	7	5	3
(23)			9	7	5
(21)				9	7
(19)					9

Table 27.4 Individual average ratings

(20):(19)	(19):(21)	(21):(23)	(23):(22)	(22):(17)
0.33	0.6	0.71	0.87	0.90
0.6	0.71	0.87	0.90	
0.71	0.87	0.90		
0.87	0.90			
0.90				
0.68	**0.77**	**0.83**	**0.89**	**0.90**

Denote by X the weight of the most insignificant, twentieth, indicator. Then the weight of the twenty-fourth indicator will be, the weight of the nineteenth indicator will be $\frac{X}{0.55*0.83}$ etc. Since the sum of the weights of all indicators is equal to 1, we obtain the equation:

$$x + \frac{X}{0.68} + \frac{X}{0.68*0.77} + \frac{X}{0.68*0.77*0.83} + \frac{X}{0.68*0.77*0.83*0.89} + \\ + \frac{\lambda}{0.68*0.77*0.83*0.89*0.90} = 1$$

From where X = $g_{20} = 0.08$, $g_{19} = 0.12$, $g_{21} = 0.16$, $g_{23} = 0.19$, $g_{22} = 0.21$, $g_{17} = 0.24$. Since the weight of the indicator $g_{20} < 0.1$, we exclude the last from the list of indicators and recalculate the weights:

$$X + \frac{X}{0.77} + \frac{X}{0.83*0.77} + \frac{X}{0.83*0.77*0.89} + \frac{X}{0.83*0.77*0.89*0.90} = 1$$

From where X = $g_{19} = 0.13$, $g_{21} = 0.17$, $g_{23} = 0.21$, $g_{22} = 0.23$, $g_{17} = 0.26$. As a result, we obtain the equation of the generalized index:

$$Q = 0.13g_{19} + 0.17g_{21} + 0.21g_{23} + 0.23g_{22} + 0.26g_{17}.$$

27.4 Conclusions

According to the results of the comparison in Table 27.2, one can draw conclusions about the quality level of the investigated product (the quality of a real car is 24% lower than the quality of an ideal sample), about the quality parameters that should be given the most attention in the manufacture of a car according to consumers, thereby increasing its competitiveness on the market.

The use of qualimetry in consumer assessment of the quality of class C cars will allow manufacturers to understand what a potential buyer expects from their product, to which parameters he pays the most attention and what range of quality indicators he identifies.

And also groups of indicators by which quality assessment was carried out were selected, matrices of paired comparisons were compiled. Compiled by the equation of the generalized index, taking into account the influence of the highest rates on the quality of the car.

Acknowledgements The paper is based on the research which is carried out with the financial support of the Russian Science Foundation grant (Project No 14-38-00009). Peter the Great St. Petersburg Polytechnic University.

References

1. Chekmarev AN (2010) Qualimetry and quality management. Part 1: qualimetry: studies manual. Aerospace, University, Samar Publishing House, Samara, 172 pp
2. Chekmarev AN (2010) Qualimetry and quality management. Part 2: quality management: studies manual. Aerospace, University, Samar Publishing House, Samara, 138 pp
3. Klochkov Y, Gazizulina A, Golovin N (2016) Assessment of organization development speed based on the analysis of standards efficiency. Paper presented at the proceedings—2nd international symposium on stochastic models in reliability engineering, life science, and operations management, SMRLO 2016, 530–532. https://doi.org/10.1109/smrlo.2016.93 Accessed www.scopus.com
4. Klochkov Y, Klochkova E, Krasyuk I, Krymov S, Gasyuk D, Akobiya N (2018) An approach to decrease the risk of losing customers. Paper presented at the 2017 6th international conference on reliability, infocom technologies and optimization: trends and future directions, ICRITO 2017, 133–142 Jan 2018. https://doi.org/10.1109/icrito.2017.8342414 Accessed www.scopus.com
5. Klochkov Y, Klochkova E, Volgina A, Dementiev S (2016) Human factor in quality function deployment. Paper presented at the proceedings—2nd international symposium on stochastic models in reliability engineering, life science, and operations management, SMRLO 2016, 466–468. https://doi.org/10.1109/smrlo.2016.81 Accessed www.scopus.com
6. Utkin LV, Podolskaja AV, Zaborovsky VS (2018) A robust interval autoencoder. Paper presented at the proceedings—2017 international conference on control, artificial intelligence, robotics and optimization, ICCAIRO 2017, 115–120 Jan 2018. https://doi.org/10.1109/iccairo.2017.32

7. Petrovski SV, Kozlovski VN, Petrovski AV, Skripnuk DF, Schepinin VE, Telitsyna E (2018) Intelligent diagnostic complex of electromagnetic compatibility for automobile ignition systems. Paper presented at the 2017 6th international conference on reliability, infocom technologies and optimization: Trends and future directions, ICRITO 2017. 282–288. https://doi.org/10.1109/icrito.2017.8342439
8. Kozlovskiy V, Aydarov D (2017) Analytical models of mass media as a method of quality management in the automotive industry. Qual-Access Success 18(160):83–87
9. Kozlovskiy V, Aydarov D (2017) System of customer satisfaction monitoring by new cars in view of perceived quality. Qual-Access Success 18(161):54–58. Accessed www.scopus.com
10. Kozlovskiy VN, Antipov DV, Zayatrov AV (2016) Methodology for analysis and forecasting of cars in operation quality. Actual Probl Econ 186(12):387–398. Accessed www.scopus.com
11. Didenko NI, Skripnuk DF, Mirolyubova OV (2017) Big data and the global economy. Paper presented at the proceedings of 2017 10th international conference management of large-scale system development, MLSD 2017. https://doi.org/10.1109/mlsd.2017.8109611
12. Gazizulina AY, Mirolyubova OV, Konakhina NA, Kikkas KN, Grigorieva AA, Danilova SY (2018) Problems of forming requirements to training of specialists for industrial and economic complex. Paper presented at the proceedings of 2017 IEEE 6th forum Strategic Partnership of Universities and Enterprises of Hi-Tech Branches (Science. Education. Innovations), SPUE 2017, 196–198 Jan 2018. https://doi.org/10.1109/ivforum.2017.8246090
13. Ziniakov VY, Gorodetskiy AE, Tarasova IL (2016) Control of vitality and reliability analysis. Stud Syst, Decis Control 49:193–204. https://doi.org/10.1007/978-3-319-27547-5_18
14. Ziniakov VY, Gorodetskiy A. E, Tarasova IL (2016) System failure probability modelling. Stud Syst, Decis Control 49:205–215. https://doi.org/10.1007/978-3-319-27547-5_19
15. Alasas BM, Ishkildina S, Zvonov S, Antipov D, Ushanova N, Demidenko E (2017) Evaluation of the quality of the process by measuring the index of opportunities. Int J Qual Res 11(2):457–468. https://doi.org/10.18421/ijqr11.02-13
16. Leoro J, Krutitskiy S, Tarasov A, Borovkov A, Aleshin M, Klyavin O (2017) Vehicle dynamics prediction module. Mater Phys Mech 34(1):82–89. https://doi.org/10.18720/MPM.3412017_10
17. Kurtc VV, Anufriev IE (2017) Car-following model with explicit reaction-time delay: linear stability analysis of a uniform solution on a ring. Math Model Comput SimulS 9(6):679–687. https://doi.org/10.1134/s2070048217060096
18. Voskresenskaya E, Snetkov V, Tebryaev A, Askarov Z (2017) Atypical real estate objects: legal regime and control system. Paper presented at the MATEC web of conferences, 106. https://doi.org/10.1051/matecconf/201710608055
19. Fradkov A, Gorlatov D, Tomchina O, Tomchin D (2016) Control of oscillations in vibration machines: start up and passage through resonance. Chaos 26(11). https://doi.org/10.1063/1.4966632
20. Kasatkin I, Egorov M, Kotov E, Zakhlebaev E (2018) Cooling of a battery pack of a car, working on renewable energy. Paper presented at the MATEC web of conferences, 245. https://doi.org/10.1051/matecconf/201824515003
21. Efanov D, Sapozhnikov V, Sapozhnikov V, Plotnikov D (2018) The evaluation of error detection probability at the outputs of combinational circuits under concurrent error detection on the basis of summation codes. Paper presented at the proceedings of 2018 IEEE east-west design and test symposium, EWDTS 2018. https://doi.org/10.1109/ewdts.2018.8524735
22. Kovalenko II, Sokolitsyn AS, Sokolitsyna NA (2018) The enterprise's automated management stability system taking into account its life cycle stage. Paper presented at the proceedings of the 2018 international conference "quality management, transport and information security, information technologies", IT and QM and IS 2018, 357–360. https://doi.org/10.1109/itmqis.2018.8524966
23. Platovskikh MJ, Vetyukov MM (2017) Self-oscillations of machines and mechanisms. Lect Notes Mech Eng 2017:87–103. https://doi.org/10.1007/978-3-319-53363-6_10
24. Gazizulina A, Eskina E, Vasilieva I, Valeeva O (2017) The reasons for the increase in self-organization in companies. Int J Reliab, Qual Saf Eng 24(6). https://doi.org/10.1142/s0218539317400022

Chapter 28
Algorithms for Managing the Sustainability of the Development of Production Systems in the Context of the Technological Level of Industries

Evgenii A. Konnikov and Olga A. Konnikova

28.1 Introduction

Despite the academic dominance of the "post-industrial" economy concept, industry provides the genesis of economic growth in the economy of a lot of countries. Production system is the central category defining the essence of an industrial enterprise. The sustainability of an industrial enterprise means its ability to maintain the current state in the presence of external influences and it is functionally dependent on the sustainability of the enterprise production system. Therefore, ensuring economic growth at the micro and macro level is possible as a consequence of ensuring the sustainability of the industrial enterprise production system [10, 11].

The production system is dynamical category, and its condition changes over time. Its condition can be characterized by a set of indicators that determine the key characteristics of the production system. The sustainability of the development of production system is a dynamic characteristic and is determined by production system ability to withstand the negative effects of the environment in the process of transition from condition №0 to condition №1. Consequently, the problem of ensuring the sustainability of the development of production system is complex, and its solution lies both in the effective assessment and forecasting of key characteristics of the production system, and in the forming management advice [9, 12].

Thus, the relevance of this study is due to the need to update the directions of ensuring the sustainability of the development of the production system.

E. A. Konnikov (✉)
Peter the Great St. Petersburg Polytechnic University (SPbPU), St. Petersburg, Russia
e-mail: konnikov.evgeniy@gmail.com

O. A. Konnikova
St. Petersburg State University of Economics, St. Petersburg, Russia
e-mail: olga.a.konnikova@gmail.com

P. K. Kapur et al. (eds.), *Decision Analytics Applications in Industry*, Asset Analytics,
https://doi.org/10.1007/978-981-15-3643-4_28

28.2 Literature Survey

The concept of "sustainability of development" is complex. In the framework of this study, it is necessary to consider the elements of this concept. Main approaches of the defining of the concepts "development" and "growth" (as applied to production systems) were formulated in the research of J. Schumpeter, M. Torado, S. Smith, V. Lewis, T. Schulz, A. Sen, J. Stiglitz, D. A. Gaynanova. Authors agree with the approach of D. A. Gaynanova and R. L. Akoff, according to which the "development" of the production system is a consequence of economic growth, shown in the systems qualities changing [15, 16].

The essence of the "sustainability" concept was investigated in the works of A. Poincaré, A. M. Lyapunov, Le Chatelier, A. A. Bogdanov and many others. Authors share the point of view of A. M. Lyapunov, that sustainability is the ability of a system (production system) to return to a condition of equilibrium under the influence of negative effects of the environment [3, 5].

Thus the concept of "Sustainability of development" (as applied to production systems) treated as the systems ability to adapt, providing potential development, under the destabilizing environmental factors influence. This concept interpretation is broad, but the display of the sustainability of development, as a property of the production system, is largely differentiated for the facilitys under study. The facility of this study is the production system of an industrial enterprise. The concept of "sustainable development of the production system" was revealed in the works of O. A. Singer, A. V. Ilyasova, T. S. Merzlikina, I. I. Yermolenko, E. B. Shevrina, V. A. Medvedev and many others. Authors proposes to specify this concept to production system of an industrial enterprises. In this way "the sustainability of the development of production systems" means the ability of industrial enterprise to implement management measures aimed at a quantitative change in key economic indicators of the production system, the purpose of which is to increase the enterprise's potential while maintaining the influence of negative effects of the environment.

28.3 Research Methodology

The instruments of assessing the sustainability of the development of the production system is primarily based on instruments of assessing the financial condition of the enterprise. In modern practice, there is a large number of these instruments, the vast majority of which are methods of diagnosing the probability of bankruptcy [8]. At the same time, the main source of development, which means existence of fluctuations in the enterprise financial condition, is a certain qualitative change, which can be achieved either through the process of integrating new technologies into production, inte-grating new materials, or the process of creating new consumer values. These three ways can be implemented in any combination, which allows to formulate 8 possible scenarios [2, 7].

However, the possibility of using one or another scenario is differentiated depending on the level of technological level of the industrial enterprise. Low-tech industries are recognized, the proportion of spending on research and development in which less than 1.5% [13]. The consequence of this is a limited opportunity of implementation of qualitative changes due to the extreme insufficiency of the budget. In order to confirm or refute this hypothesis, as well as to determine effective algorithms to ensure the sustainability of the development of production system, it is necessary to study the influence of internal factors on the sustainability of the development of production system, depending on technological level of the industrial enterprise.

The nature of the sustainability of the development of production systems may vary depending on the technological level of industries. Authors have formed a sample of 9 manufacturing sectors clustered according to the Ward method. The indicator of clustering is the share of the cost of technological innovations in the total volume of goods shipped (for the period 2010–2016). A sample of 9 manufacturing industries was divided into 4 clusters [17].

- «low-tech» (cluster 1: food production, wood processing, pulp production);
- «medium-low-tech» (cluster 2: textile production and production of petroleum products);
- «medium-high-tech» (cluster 3: production of machinery and equipment, manufacture of electrical machinery and electrical equipment);
- «high-tech» (cluster 4: chemical production, aircraft and spacecraft production).

Representatives of each of this industries are characterized by number of their own internal factors, however, the smallest share of research and development expenses in the low-tech cluster implies a relative limitation in the implementation of qualitative changes. The key factors of the internal environment are the characteristics of the financial stability, such as dependence on borrowed capital, solvency, etc. In the case of representatives of the high-tech cluster, the situation is the opposite. The multiplication of development potential of this enterprises is based on the generation of innovations and their integration into the production process. The key factor of the internal environment is the intensity of innovation, expressed in both financial and non-financial indicators. For the purposes of this study, the financial independence ratio (indicator X1) was chosen as indicators of the financial sustainability, and the share of expenses on technological innovations in the total volume of goods shipped (indicator X2) was used as an indicator of the intensity of innovation activity. The result indicator in this case is the number of initiated insolvency (bankruptcy) cases (Y) in each of the industries.

28.4 Research Results

Correlation coefficient was chosen as the measure of definition. Correlation coefficients have a negative value, which confirms the nature of their influence on the sustainability of the development of the production system. Figure 281.1 shows the

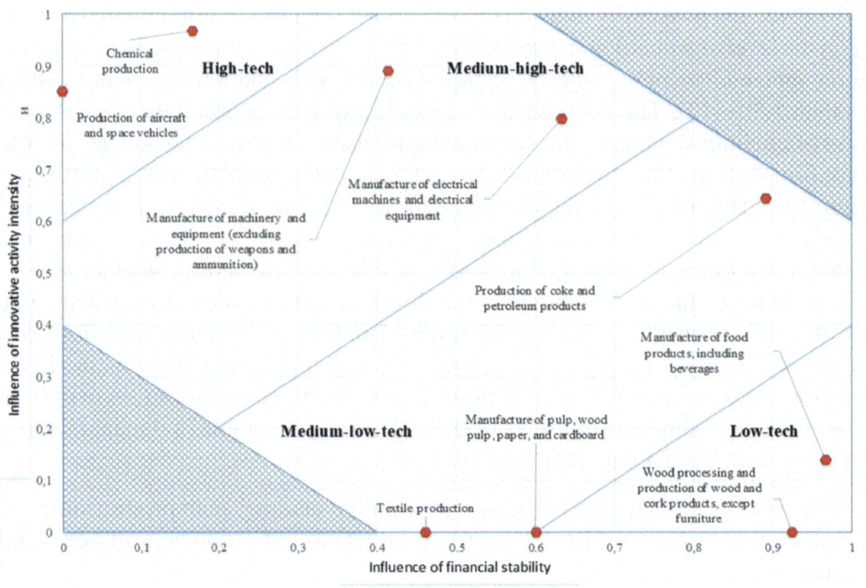

Fig. 28.1 Assignment of the industries in accordance with the correlation of Y with indicators X1 and X2 (rx1y and rx2y)

assignment of the industries in accordance with the obtained indicators.

This figure confirms the hypothesis regarding the nature of the enterprise internal factors that influence the sustainability of the development of production systems of various technological level. The industries can be divided into 4 groups, repeating previously identified clusters from the point of view of their composition. As you can see, 3 industries (food processing, wood processing and pulp production) can be attributed to low-tech industries, while chemical manufacturing and aircraft and spacecraft manufacturing are high-tech industries. Other industries are related to medium-low-tech and medium-high-tech industries. None of the industries showed both a high correlation with X1 and X2, and a low correlation too. The figure clearly demonstrates a decrease in the impact of financial stability (rx1y), with a simultaneous increase in the influence of the intensity of innovation activity (rx2y), with an increase of technological level of the industry. In this case, the nature of the distribution of this influence is of particular interest. The dependence of the distribution of industries is described by the function of a second-degree polynomial (the coefficient of determination is 0.65). However, exclusion from the studied sample of low-tech and high-tech industries, allows to achieve the maximum value of the coefficient of determination:

$$r_{x^2_{M.T.} y_{M.T.}} = 2{,}935 \times r^2_{x^1_{M.T.} y_{M.T.}} + 2{,}557 \times r_{x^1_{M.T.} y_{M.T.}} - 0{,}461 \qquad (1.1)$$

where: rx2M.TyM.T.—the impact of the intensity of innovation activity on the dynamics of initiated insolvency (bankruptcy) cases in medium-tech industries; rx1M.TyM.T.—the impact of financial stability on the dynamics of initiated insolvency (bankruptcy) cases in medium-tech industries.

Thus, the differentiation of the nature of the sustainability of the development of production systems depending on the technological level of industrial enterprises is clearly demonstrated. Therefore, the instrument for managing sustainability of the development of production systems should also be vary.

28.5 Discussion

In the case of production systems of low-tech industrial enterprises, the enterprise's ability to predict possible adverse changes in the external environment and to react to them in a timely manner plays an exceptional role. Consequently, the key instruments of ensuring sustainability of development will be instruments of forecasting, which includes instruments of assessing the consequences of the influence of the external environment and instruments of identifying areas of influence and forming recommendations for forming management advice regarding the necessary changes [6].

In the case of the high-tech industrial enterprises production systems, it should be noted that the integration of innovations inevitably involves a lot of risks. Consequently, one of the most relevant instruments for ensuring sustainability of the development of its production systems in this case will be an instrument of assessing the level of prospectiveness of possible innovative solutions.

At the same time, both approachers are relevant for medium-tech industries. The degree of significance of the results of the application of this approachers can be distributed in accordance with the model (1). Thus, for each of the selected clusters, we can offer an own methodology for ensuring the sustainability of development of the production system in the form of a closed loop, consistently combining the use of specialized instruments. These algorithms (in simplified form) are presented in Figs. 28.2, 28.3 and 28.4.

Each of the steps of the simplified algorithms is a logical task that the enterprise needs to solve in order to ensure the sustainability of the development of the production system. Each of the selected stages requires the creation of a unique instrument that takes into account industry-specific features. At the same time, the differentiation of approaches to managing the sustainability of the development of the production system of various technological levels is emphasized primarily by these algorithms.

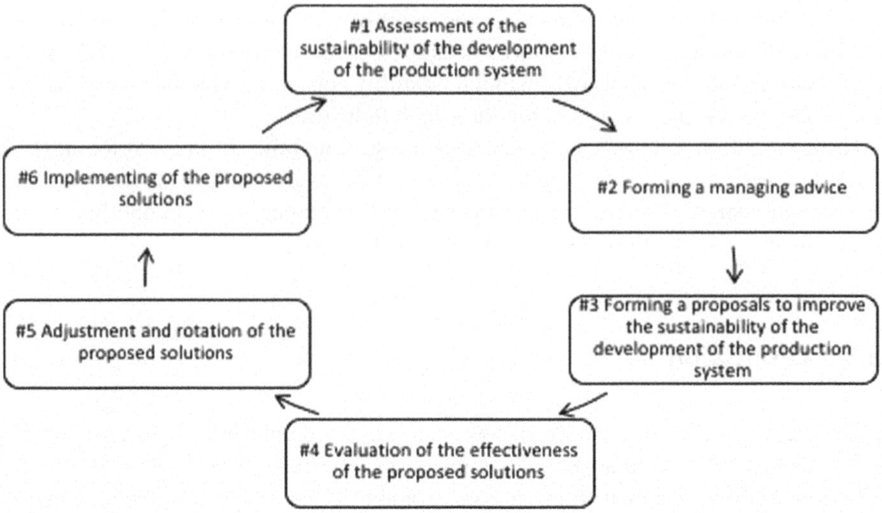

Fig. 28.2 Simplified algorithm of managing the sustainability of the development of low-tech production system

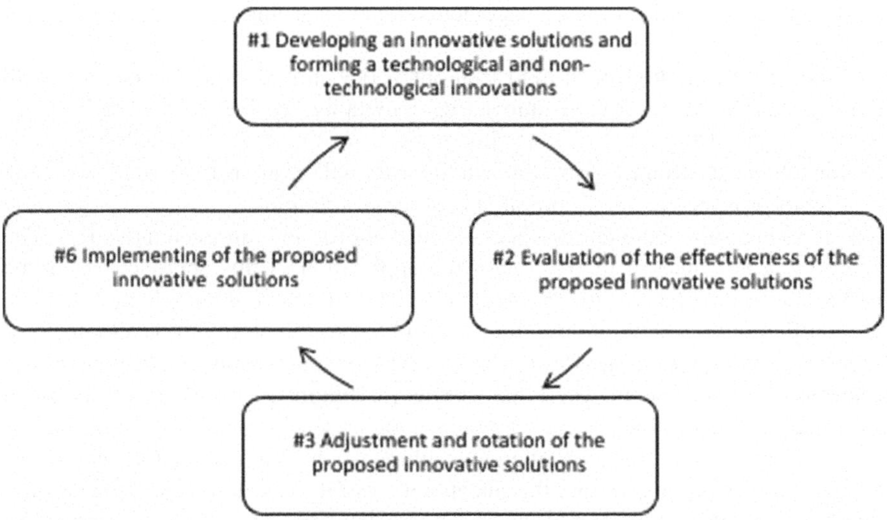

Fig. 28.3 Simplified algorithm of managing the sustainability of the development of high-tech production system

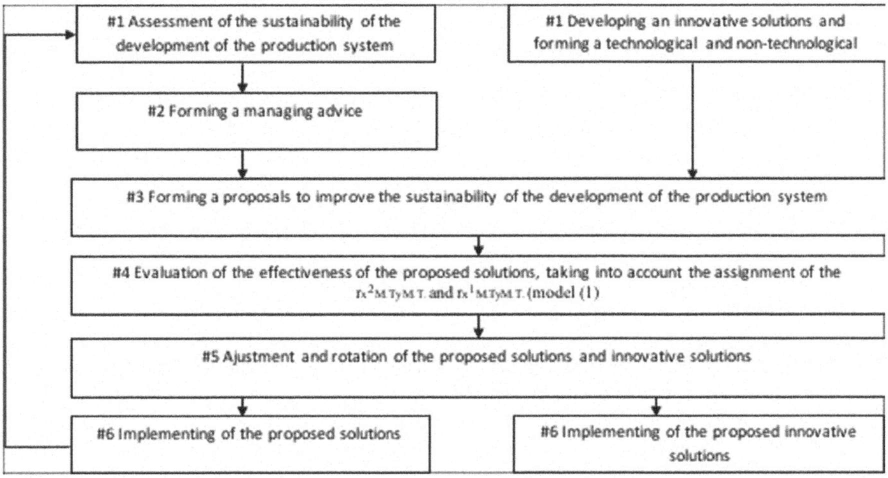

Fig. 28.4 Simplified algorithm of managing the sustainability of the development of medium-tech production system

28.6 Conclusion

The arguments given in this study are necessary form a statement that the source of sustainability of the development of production systems of low-tech industrial enterprises is their financial sustainability, not implying qualitative changes in the production process, while the source of sustainability of the development of production systems of high-tech industrial enterprises are promising innovative solutions leading to quality change. As part of further research, the authors intend to offer a competitive instruments for the implementation of each of the stages of the algorithms. Thus, the key vector for the development of this study is the digitalization of algorithms.

Acknowledgements The research is carried out with the financial support of the grant from the Program Competitiveness Enhancement of the Peter the Great St. Petersburg Polytechnic University, Project 5-100-2020.

References

1. Afanasieva N, Rodionov D, Vasilev Y (2018) System of indicators of coal enterprise competitiveness assessment. Espacios 39(36)
2. Aksenova VV (2015). Modern approaches to the definition of sustainable enterprise development: the ratio of growth and development. Sci Bull Volgogr Branch RANEPS Ser: Econ. [Nauchnyj vestnik VF RANHIGS. Ekonomika], 1:21–26 (in Russian)

3. Idrisova JI, Myasnikov VN, Uljanov AI et al. (2018). Increasing the efficiency of labor protection in the enterprise. In: International conference on information networking, 586–588 Jan 2018
4. Klochkova E, Evdokimov K, Klochkov Y et al. (2018) Methodology for reducing risk of underperformance of personnel functions. Eng Rural Dev 17:1213-1222
5. Klochkov Y, Gazizulina A, Golovin N, et al. (2018)Information model-based forecasting of technological process state. In: 2017 international conference on infocom technologies and unmanned systems: trends and future directions, ICTUS 2017, 709–712 Jan 2018
6. Kuchmin AY, Karheily S (2017) Automatic 3D human body modelling. In: Studies in Systems, Decision and Control, vol 95 pp 83–94
7. Kutovaya AS (2012) Analysis of approaches to the definition of the concept of "sustainable enterprise development". Bull Saratov State Soc Econ Univ [Vestnik Saratovskogo gosudarstvennogo sotsialno-ekonomicheskogo universiteta], 5(44):39–43 (in Russian)
8. Malizia EE, Feser EJ (1999) Understanding local economic development. New, Brunswick, N, J: Center for Urban policy research, pp: 314
9. Novokshenov A, Nemov A, Mamchits D, et al. (2017) Integrated system as a tool for implementation of simulation—and optimization-based design methodology. Mater Phys Mech 34(1):76–81
10. Popovich AA, Sufiiarov VS, Borisov EV et al. (2018) Design and manufacturing of tailored microstructure with selective laser melting. Mater Phys Mech 38(1):1–10
11. Pravdyuk VN (2009) Economic development as a subject of theoretical economics. Econ Sci [Ekonomicheskie nauki] 2(51):74–76 (in Russian)
12. Rudskoi AI, Bogatov AA, Nukhov DS et al. (2018) On the development of the new technology of severe plastic deformation in metal forming. Mater Phys Mech 38(1):76–81
13. Rudskaia I (2017) A regional innovation system: formation features and growth areas (case study: St. Petersburg). In: The proceedings of the 30th international business information management association conference, IBIMA 2017—Vision 2020: sustainable economic development, innovation management, and global growth, pp 541–547
14. Rudskaya I, Rodionov D (2018) Comprehensive evaluation of Russian regional innovation system performance using a two-stage econometric model. Espacios 39(4)
15. Schumpeter JA (1947) Theoretical problems: theoretical problems of economic growth. J Econ History Econ Hist Assoc. 7, supplement: economic growth: a symposium (1947) 1–9
16. Todaro MP, Smith StC (2008) Economic development, 10 edn. Pearson Education, p 891
17. Zobacheva AY, Nemov AS, Borovkov AI, et al. (2017) Design and simulation of additive manufactured structures of three-component composite material. Mater Phys Mech 34(1):51–58

Chapter 29
Reliability Assessment of Multi-release Software System Under Imperfect Fault Removal Phenomenon

Vibha Verma, Sameer Anand, and Anu Gupta Aggarwal

29.1 Introduction

The necessity for high-quality software product has increased manifolds in almost every domain. Software developer aims to improve reliability as the testing progresses and deliver a qualitative product to users. In order to assess the reliability, many time-dependent Software Reliability Growth Models (SRGMs) [1] have been developed in the literature. These models mathematically define the fault detection and removal process. All these models assess the reliability growth for single version software. But their failure curves cannot explain the reliability growth of multiple version software because the probability of failure is higher in case of feature enhancement or addition. Hence the firms release the multiple versions of the software that accommodates in it the additional features and also helps to remove the weaknesses of the previous releases.

Researchers have studied SRGMs based on perfect and imperfect debugging scenarios. In 1985, Goel [2] initiated the idea of imperfect debugging. Detected faults are not always removed completely, i.e., fault removal efficiency need not always be 100%. The ability to remove faults is known as fault removal efficiency (FRE). Considering this factor Zhang, Teng and Pham [3] first introduced a FRE model considering both types of imperfect debugging conditions.

V. Verma (✉) · A. G. Aggarwal
Department of Operational Research, University of Delhi, Delhi, India
e-mail: vibhaverma.du.aor@gmail.com

A. G. Aggarwal
e-mail: anuagg17@gmail.com

S. Anand
Department of Computer Science, Shaheed Sukhdev College of Business Studies, University of Delhi, Delhi, India
e-mail: sananddu@gmail.com

P. K. Kapur et al. (eds.), *Decision Analytics Applications in Industry*, Asset Analytics,
https://doi.org/10.1007/978-981-15-3643-4_29

With time research has contributed to the reliability growth models but many realistic situations have not been considered such as the difference between faults removed and failures experienced. This is modelled as "Fault Reduction Factor (FRF)" introduced by Musa [4]. During run-through process of software, FRF is either constant or less than unity. In further studies, it has been considered that FRF is affected by many factors, for example, dependence of faults, code complexity, etc.

In this paper, we have considered the various factors such as error generation, FRF, FRE to profile the reliability growth model for software released in multiple versions. The developed model is validated on the Tandem computer dataset [5]. The unknown parameters are estimated in SPSS software and model performance is tested using some descriptive and predictive performance measures. Reliability growth curves and boxplots for each release based on the model estimates have been plotted.

The remainder of the paper is: Sect. 29.2 gives brief study of the related concepts while in Sect. 29.3; we have presented the mathematical formulation of the developed model. Further in Sect. 29.4, it is validated on the four release fault Dataset followed by evaluation of the model performance. Further in the next section we assess the impact of incorporating various factors into the model considering some special cases. Finally, in the last section, the conclusion of the work is presented followed by the future scope.

29.2 Related Work

This section discusses the literature work of the basic concepts used in the study.

Software reliability growth modelling characterizes the reliability growth process of the software system as mathematical relation between testing time and detected or removed faults. Yamada, Ohba and Osaki [1] described the SRGM considering time gap in the failure occurrence and removal of the fault causing failure.

Musa [4] defined FRF as ratio of faults removed to the failures experienced. The various forms of FRF have been proposed in the literature. Musa assumed FRF to be proportional to the hazard rate function. FRF is defined as constant as well as the time variable. Considering that scenario Hsu, Huang, and Chang [6] defined it to be constant, decreasing and increasing for software with a single release. Pachauri, Dhar, and Kumar [7] considered Inflexion S-shaped FRF for the multiple versions of the software system under perfect and imperfect debugging environment. Recently Anand, Verma, and Aggarwal [8] proposed a two-dimensional Multi-Release (M-R) model incorporating FRF. Tandon, Aggarwal, Gandhi and Verma [9] discussed an SRGM with FRF under imperfect debugging conditions for software released in multiple versions.

The models developed incorporating the imperfect debugging scenario is more realistic and accurate in fault prediction. Debugging process is said to be imperfect when it is not possible to remove fault in a single attempt or when removal of a fault leads to the introduction of some new faults known as "error generation" [10]. The functional form of imperfect debugging is considered to be linearly or exponentially increasing with time. In later years Kapur and Younes [11] introduced the imperfect debugging SRGM considering faults removal process to be exponentially distributed.

Abundant number of models has been developed to study failure phenomenon of single release software systems and improve their reliability. To keep pace with the fast-evolving technological developments developers need to develop upgraded versions of the software product. The upgraded software consists of old features with some modifications and also some newly added features. The main goal of the software organization behind the multiple versions is to increase its customer size by retaining the old ones and attracting the potential ones. To release new version the developer is concerned about the newly added faults plus the leftover faults from the last versions. Therefore the updated versions are complex to handle but, on the other hand, it is one of the most effective ways for the firms to remain competitive in the market. In literature, several researchers have discussed the fault removal process for software released in multiple versions. One of the basic M-R models was given by Kapur, Aggarwal, and Kaur [12]. They introduced the multi-up-gradation SRGM supposing that the faults in current release depend upon faults in last releases.

The detected faults are not always completely corrected. A proportion of them remains in the system. This signifies that the debugging process lacks some efficiency in removing faults. This measure helps the development team to analyze the effectiveness of fault removal process and estimate the further efforts required for debugging. Higher is the efficiency higher will be the fault removal rate and vice versa. Jones [13] defined FRE as the proportion of faults removed by testing the software, reviewing and inspecting the software. Liu, Yang and Qu [14] proposed an NHPP based SRGM where FRE was defined as the function of debugging time. They assumed the bell-shaped fault detection rate. During the debugging process, the developers are successfully able to remove p % of the detected faults.

29.3 Modelling Reliability Growth

Here, we develop the mathematical expressions of the model. The notations for developing the proposed multi-release model for ith release ($i = 1, 2 \ldots n$) are listed below:

$m_i(t)$	Mean value function for fault removed during (0, t)
$b(t)$	Fault detection rate
α_i	Error Generation rate
a_i	Initial Fault content
p_i	Fault Removal Efficiency
β_i	Proportionality constant
$r_i(t)$	Fault reduction Factor (Time-dependent)
t_i	Time of ith release

The proposed model has been formulated by making following assumptions:

1. Failure process follows NHPP.
2. Fault removal process is not 100% efficient.
3. A fault may not get removed immediately after its detection.
4. Number of failures occurrences at time t is not same as the number of faults detected up to that time, i.e. there is some delay between the failure occurrences and fault detection process.
5. The fault removal process is not perfect.
6. The software product is released with multiple upgradations.

The failure phenomenon of a software system that includes the above-mentioned assumptions is,

$$\frac{\mathrm{d}m(t)}{\mathrm{d}t} = b(t)(a(t) - pm(t)) \tag{29.1}$$

Since we have considered the debugging process as imperfect therefore the fault content of the software is linearly related to the number of faults removed ($m(t)$). Moreover $b(t)$ represents the fault detection rate. Mathematically it is as follows:

$$a(t) = a + \alpha m(t) \text{ and } b(t) = \beta r(t) \tag{29.2}$$

Using (29.2) in Eq. (29.1) makes it:

$$\frac{\mathrm{d}m(t)}{\mathrm{d}t} = \beta r(t)(a + \alpha m(t) - pm(t)) \tag{29.3}$$

where $r(t) = \frac{b^2 t}{1+bt}$ (Delayed S-Shaped model) and $\alpha > 0, \beta > 0, p > 0$

Using initial condition (at $t = 0$, $m(t) = 0$) and solving differential Eq. (29.3), $m(t)$ is computed as:

$$m(t) = \frac{a}{p-\alpha}(1 - ((1+bt)^{\beta(p-\alpha)} e^{-bt\beta(p-\alpha)})) \tag{29.4}$$

Here $\alpha < p$ i.e., error generate rate is less than FRE. If this condition is violated then as the testing continues the fault content of the software will become infinitely large. (29.4) is the general equation for a software system with single version incorporating all mentioned factors. The MVF of different releases is as follows:

Release 1: In the first release the software is introduced to the market for the first time. Before launching it to the market the developers try to minimize the number of faults latent in the system. The mathematical expression for Fault Removal Phenomenon (FRP) for the first release is,

$$m_1(t) = \frac{a_1}{p_1 - \alpha_1} F_1(t) \quad 0 \le t < t_1 \tag{29.5}$$

where $F_1(t) = \left[1 - \left((1 + b_1 t)^{\beta_1(p_1-\alpha_1)} e^{-b_1\beta_1(p_1-\alpha_1)t}\right)\right]$

Release 2: In the 2nd release, the organization provides new features to the users by modifying the first version. During this release, the testing and debugging are done to remove the faults due to new features added in the software along with the faults that could not be detected or removed in the first release. The expression for FRP of this release is as follows:

$$m_2(t) = \left(\frac{a_2}{p_2 - \alpha_2} + \frac{a_1}{p_1 - \alpha_1}(1 - F_1(t_1))\right) F_2(t - t_1) \quad t_1 \le t < t_2 \tag{29.6}$$

In general for *i*th release: For subsequent releases, the expression can be given as follows:

$$m_i(t) = \left[\frac{a_i}{p_i - \alpha_i} + \left(\frac{a_{i-1}}{p_{i-1} - \alpha_{i-1}}(1 - F_{i-1}(t_{i-1}))\right)\right] F_i(t - t_{i-1}) \quad t_{i-1} \le t < t_i \tag{29.7}$$

29.4 Model Validation

Now, here we evaluate the performance of developed model on a real-life fault dataset. Further, we estimate the unknown parameters and thus find the cumulative number of faults estimated through the model. The four Release fault Dataset of Tandem computer [7] from literature is used for model analysis. Table 29.1 gives a brief summary of the four release dataset.

The Mathematical Equation for four Releases based on expression (29.7) is summarized in Table 29.2.

The unknown parameters of the developed model are estimated through Non-Linear Least Square Estimation Method. The parameters for each release are estimated using SPSS and are given in Table 29.3.

The estimated values curve obtained after the non-linear regression analysis for the proposed model are compared with observed values and is shown graphically in Fig. 29.1a–d. Figure 29.1a–d are the goodness-of-fit curves and they show the plots

Table 29.1 Dataset statistics

Release	Testing time (weeks)	No. of faults detected/removed
1	20	100
2	19	120
3	12	61
4	19	42

Table 29.2 Mathematical expression for four releases

Release	Expression corresponding to the release
1	$m_1(t) = a_1^* F_1(t); 0 \le t < t_1;\ a_1^* = \dfrac{a_1}{p_1 - \alpha_1},$ $F_1(t) = \left[1 - \left((1 + b_1 t)^{\beta_1(p_1-\alpha_1)} e^{-b_1\beta_1(p_1-\alpha_1)t}\right)\right]$
2	$m_2(t) = a_2^* F_2(t - t_1); t_1 < t \le t_2;\ a_2^* = \left[\frac{a_2}{p_2-\alpha_2} + \left(\frac{a_1}{p_1-\alpha_1}(1 - F_1(t_1))\right)\right]$
3	$m_3(t) = a_3^* F_3(t - t_2); t_2 < t \le t_3;\ a_3^* = \left[\frac{a_3}{p_3-\alpha_3} + \left(\frac{a_2}{p_2-\alpha_2}(1 - F_2(t_2))\right)\right]$
4	$m_4(t) = a_4^* F_4(t - t_3); t_3 < t \le t_4;\ a_4^* = \left[\frac{a_4}{p_4-\alpha_4} + \left(\frac{a_3}{p_3-\alpha_3}(1 - F_3(t_3))\right)\right]$

Table 29.3 Estimated parameters

Release	a	p	α	β	b	R^2
1	99.245	0.922	0.025	0.237	0.757	0.978
2	120.62	0.894	0.004	0.423	0.453	0.994
3	63.125	0.996	0.001	0.982	0.292	0.992
4	42.116	0.992	0.005	0.292	0.590	0.995

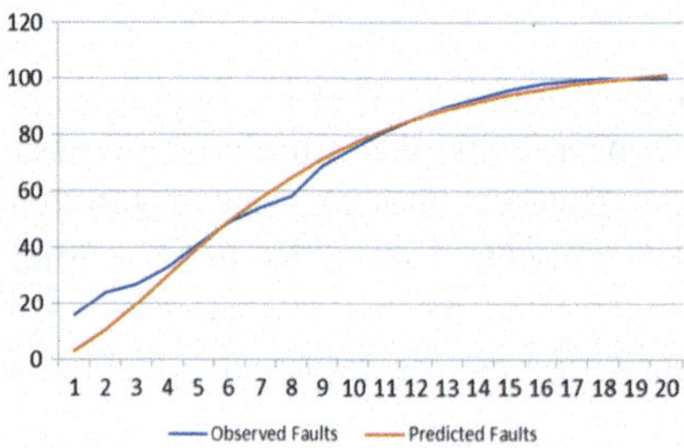

(a) Release 1

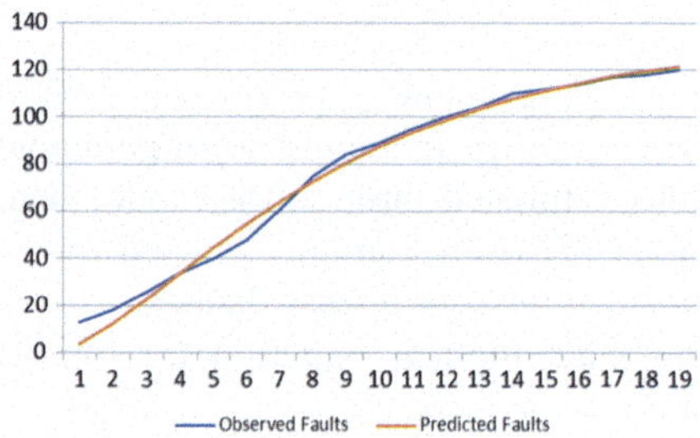

(b) Release 2

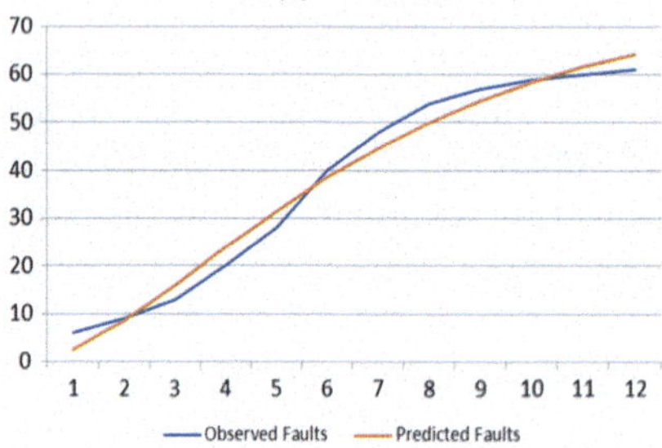

(c) Release 3

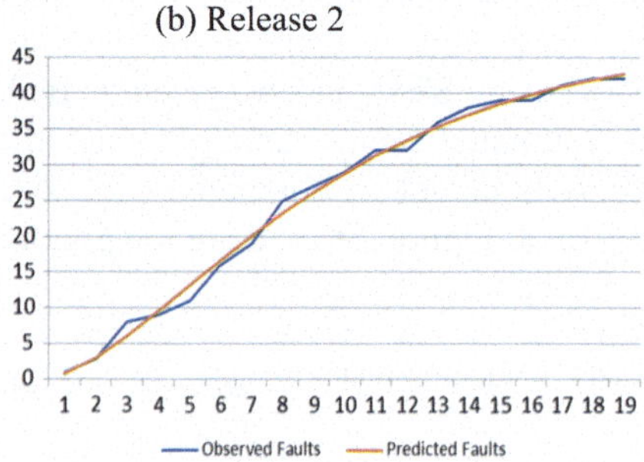

(d) Release 4

Fig. 29.1 Goodness-of-fit curves

corresponding to estimated values and observed values. Closer the two values better is the model fitting.

Further to evaluate the model performance boxplots are plotted. Boxplot can be defined as the exploratory graphic that divides the data into quartiles and thus helps in the analysis of variability/spread of the data at a glance. It is also termed as whiskers plot since whiskers are present at upper and lower ends of the box representing the top and bottom 25% of values respectively. The data point at the lowest end is the minimum value while the topmost represents the maximum value. This data excludes the outliers that may be present in the data which are plotted separately as the single data points. The interquartile range i.e. the area covered by the box consists of 50% data with median lying within the box depicted using a line. In this study, we have plotted the boxplots for absolute errors between the observed faults during testing and the predicted faults using the proposed model for each calendar time.

The boxplot for Release 1 (Fig. 29.2) shows that the data points are concentrated at the lower end and the box formed is also short. This implies that the difference between the observed and predicted faults is less except for the four outlier's present. The proposed model is successful in predicting the number of faults. The data is rightly skewed because of data accumulation at the lower end. The range for this boxplot is 3 whereas the interquartile range is around 2.50. Thus we can conclude that the data points have less spread.

The boxplot for Release 2 (Fig. 29.3) has one outlier with most of the values being close to each other. A smaller range of values due to observed and predicted faults being close has led to the short box. Top 25% of data has more variability than in the case of the bottom 25% of data. Median is lying nearer to the lower end of the box

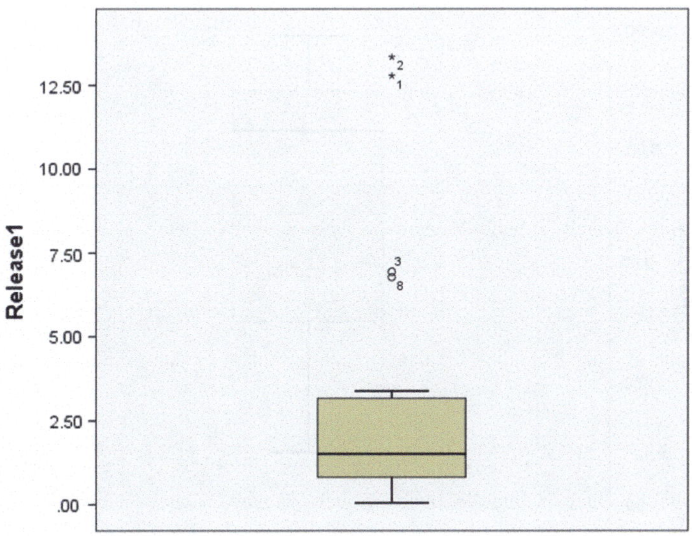

Fig. 29.2 Boxplot for Release 1

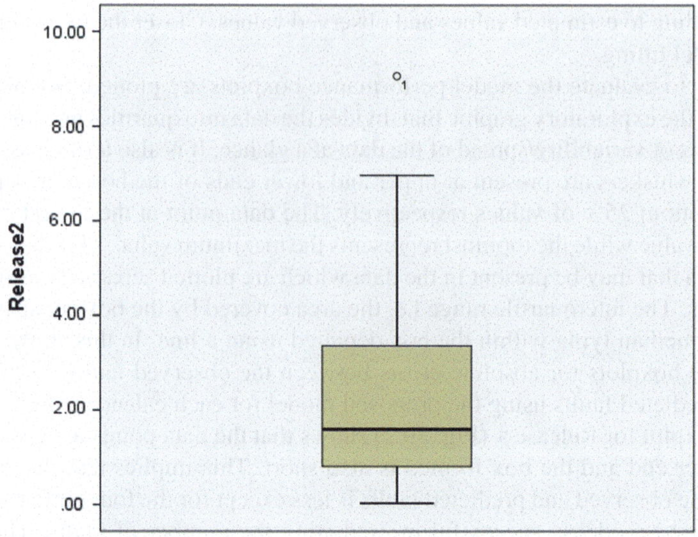

Fig. 29.3 Boxplot for Release 2

thus it is rightly skewed. The interquartile range lies between 1 and 3 while the range is 7. So we can say that there is a little bit of spread in the data of absolute error.

Observing the boxplot for Release 3 (Fig. 29.4) we conclude that the variability is more in comparison to other releases. Median lies at upper end of the left-skewed

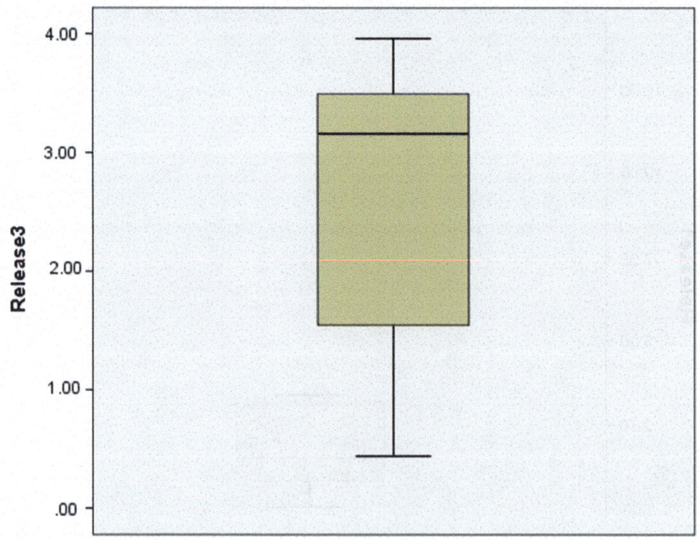

Fig. 29.4 Boxplot for Release 3

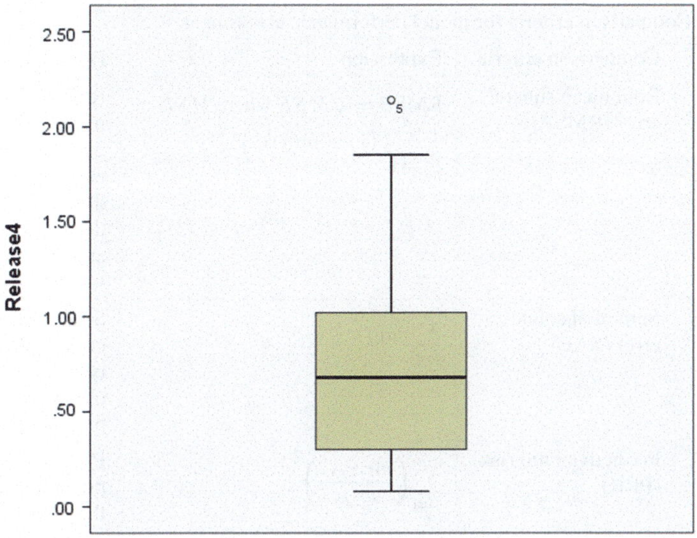

Fig. 29.5 Boxplot for Release 4

box suggesting that most of the data points lying at upper end. Box range is 6 while the interquartile range is 2. Also, there are no outliers present. The variability and spread in the data are large as observed from the length of the box.

Release 4 Boxplot (Fig. 29.5) also has a short box whose interquartile range is 0.75 while the range is 2 suggesting less spread in the data. The median of the data lies in the middle of the box, so it is symmetric in nature. This implies that most of the data points are in the middle only. We can also see that the maximum value is about 2 while the minimum is around 0.20. There is only one outlier present in the data. Boxplots analysis concludes that the developed model predicts the faults with much accuracy. Overall it may be concluded that proposed SRGM fits the dataset quite well.

29.5 Model Comparison

To judge the performance of the proposed model we need to evaluate it on the basis of a number of performance criteria. The performance Criteria used for assessing the Model are listed in Table 29.4 with their mathematical expressions and definitions. In Table 29.4 $\widehat{m}(j)$ represents estimated cumulative number of faults, y_j represents observed faults and n is the number of data points. The last two measures also consider the outliers present in the dataset and help in the determination of the model accuracy. The Lower values for all the performance criteria except coefficient of

Table 29.4 Comparison criteria for model performance assessment

S. no.	Comparison criteria	Expression	Definition
1	Root mean square error (RMSE)	$\text{RMSE} = \sqrt{MSE}$ where $MSE = \frac{1}{n}\sum_{j=1}^{n}\left(\left(y_j - \widehat{m}(j)\right)^2\right)$	RMSE is absolute measure of goodness of fit that evaluates the difference between the predicted values and the observed values
2	Sum of absolute error (SAE)	$\sum_{j=0}^{n}\left\|\widehat{m}(j) - y_j\right\|$	SAE is the absolute error or difference between the observed and predicted faults
3	Predictive ratio risk (PRR)	$\sum_{j=1}^{n}\left(\frac{\widehat{m}(j)-y_j}{\widehat{m}(j)}\right)^2$	PRR shows the predictive power of the model
4	Predictive power (PP)	$\sum_{j=1}^{n}\left(\frac{\widehat{m}(j)-y_j}{y_j}\right)^2$	This is also one of the predictive power comparison technique
The above-mentioned criteria do not take into consideration the outlier's present in the data. For that, we use the following two criteria's			
5	Mean absolute percentage error (MAPE)	$\frac{100}{n}\sum_{j=1}^{n}\left\|\frac{y_j-\widehat{m}(j)}{y_j}\right\|$	It is also known as mean absolute percentage deviation (MAPD) and it measures the accuracy of prediction done using the derived model
6	Mean absolute scaled error (MASE)	$= \frac{1}{n}\sum_{j=1}^{n}\frac{\left\|e_j\right\|}{\frac{1}{n-1}\sum_{j=2}^{n}\left\|y_j - y_{j-1}\right\|}$ $= \frac{\sum_{j=1}^{n}\left\|e_j\right\|}{\frac{n}{n-1}\sum_{j=2}^{n}\left\|y_j - y_{j-1}\right\|}$ where forecast error is $e_j = y_j - \widehat{m}(j)$ and $\widehat{m}(j) = y_{j-1}$	MASE is used for comparative analyses of prediction. The forecasting value is taken as the actual value at the previous step

determination $\left(R^2\right)$ imply better fitting of the model whereas in case of R^2 (coefficient of determination), higher is the value, better is the model fit.

Now, we will see how the presence of parameters impacts the model performance. We consider the following Special cases for comparing the performances of the model with respect to different parameters.

Case 1 ($\alpha = 0,\ p > 0$): Firstly we consider the case that no faults are generated or added i.e. $\alpha = 0$ in the four release problem of the proposed model. Substitute $\alpha = 0$ in Eq. (29.7).
Case 2 ($0 < \alpha < 1,\ p = 1$): Secondly we consider the case when the development team is 100% efficient in removing the faults and thus $p = 1$ in the Eq. (29.7) for all the four releases.
Case 3 ($\alpha = 0,\ p = 1$): In the last case we consider both types of perfection i.e. no introduction of new faults in the system during the process and the debugging efficiency is also 100%. Mathematically we substitute $\alpha = 0$ and $p = 1$ in Eq. (29.7) for all the four releases.

Further, we run least square estimation in SPSS on Tandem Computers dataset for four releases to obtain the unknown parameters of the new models obtained. Table 29.5 shows the model estimates of proposed model and three Cases considered for each of the four releases. Additionally Table 29.6 shows the release wise model performance of the models including proposed model and special cases.

From Table 29.6 it can be observed that R^2 values are maximum for the proposed model for all the 4 releases. This is so because; inclusion of factors such as error

Table 29.5 Estimated parameters

Model	a	α	p	b	β	R^2
Release 1						
Proposed model	99.24	0.025	0.922	0.757	0.237	0.978
Case 1	102.54	–	0.934	0.635	0.277	0.977
Case 2	96.61	0.12	–	0.657	0.294	0.977
Case 3	109.05	–	–	0.586	0.304	0.976
Release 2						
Proposed model	120.62	0.004	0.894	0.453	0.423	0.994
Case 1	129.02	–	0.967	0.728	0.19	0.990
Case 2	118.34	0.132	–	0.645	0.239	0.990
Case 3	129.39	–	–	0.545	0.286	0.989
Release 3						
Proposed model	63.125	0.001	0.996	0.292	0.982	0.992
Case 1	68.138	–	0.942	0.424	0.386	0.989
Case 2	72.678	0.01	–	0.439	0.11	0.990
Case 3	69.546	–	–	0.314	0.865	0.981
Release 4						
Proposed model	42.116	0.005	0.992	0.59	0.292	0.995
Case 1	40.231	–	0.879	0.435	0.341	0.993
Case 2	39.452	0.142	–	0.366	0.437	0.993
Case 3	49.567	–	–	0.523	0.209	0.992

Table 29.6 Model comparisons results

Model	R^2	RMSE	SAE	PRR	PP	MAPE	MASE
Release 1							
Proposed model	0.978	4.877	56.12	7.439	0.771	20.041	0.533
Case 1	0.977	4.983	56.51	8.594	0.813	21.399	0.536
Case 2	0.977	4.983	56.51	8.594	0.813	21.399	0.536
Case 3	0.976	5.074	56.81	9.602	0.847	22.523	0.539
Release 2							
Proposed model	0.994	3.154	37.83	0.983	0.303	8.63	0.298
Case 1	0.990	4.177	55.59	3.431	0.525	14.596	0.438
Case 2	0.990	4.291	57.55	4.223	0.569	15.991	0.544
Case 3	0.989	4.928	57.89	5.549	0.648	18.154	0.570
Release 3							
Proposed model	0.992	2.427	15.06	11.169	0.741	34.35	0.226
Case 1	0.989	2.766	21.67	12.766	1.596	36.278	0.325
Case 2	0.990	2.753	21.40	12.768	1.597	36.265	0.321
Case 3	0.981	3.762	31.15	15.741	1.449	37.674	0.468
Release 4							
Proposed model	0.995	1.127	13.76	0.121	0.166	6.281	0.310
Case 1	0.993	1.333	18.78	0.148	0.182	6.508	0.423
Case 2	0.993	1.313	18.62	0.128	0.130	6.867	0.420
Case 3	0.992	1.452	21.07	0.158	0.194	6.946	0.475

generation and Fault Removal Efficiency makes the model more accurate and practical. The model is able to explain the failure process very well using the available information. Similarly, other performance criteria values given in Table 29.6 signify that proposed model as better performance. When a parameter is dropped from the model the accuracy decreases.

From the Table 29.6, it can be observed that for Release 1 the coefficient of variation R^2 is maximum for the proposed model whereas it is minimum for special case 3 where we have considered a perfect debugging scenario. A similar trend is observed for the other comparison a criterion's used in the analysis. Likewise, for Release 2, 3 and 4 also proposed model also better fits preceded by special cases 1, 2 and 3. Hence it may be noted from the values of performance criteria that the inclusion of crucial factors like FRF, FRE in the SRGM largely affects the accuracy and predictive behavior of the reliability growth models.

29.6 Conclusions and Future Scope

In this article, an NHPP-based growth model is developed for a software system with multiple versions considering error generation, fault removal efficiency and Delayed S-Shaped FRF. The tandem computer's four release failure dataset has been used for validating the model using goodness-of-fit, predictive and descriptive performances. Several comparison criteria's and boxplots have been used to evaluate model performance. In addition to this, the reliability curve for each release has been plotted. The curves are of increasing nature. Initially, the increase is higher but as the testing progresses the increase gets stagnant and eventually stable. At this stage, further improvement in reliability is less visible. The observed results of the proposed model and its comparison with the special cases suggest that the model better fits the failure data and is also capable of predicting faults better. This model will help the developer to make precise fault predictions and thus accordingly make reasonable decisions for software development.

Furthermore, the model can be used for modelling of the optimal release time problems, software development cost optimization problem, Maximizing reliability subject to cost constraint etc. More accurate models can be developed by considering multiple change points and related release time problem may be formulated.

References

1. Yamada S, Ohba M, Osaki S (1984) S-shaped software reliability growth models and their applications. IEEE Trans Reliab 33(4):289–292
2. Goel, AL (1985) Software reliability models: assumptions, limitations, and applicability. IEEE Trans Softw Eng SE-11(12):1411–1423
3. Zhang X, Teng X, Pham H (2003) Considering fault removal efficiency in software reliability assessment. IEEE Trans Syst, Man, Cybern-Part A: Syst HumS 33(1):114–120
4. Musa JD (1975) A theory of software reliability and its application. IEEE Trans Software Eng 3:312–327
5. Wood A (1996) Predicting software reliability. Computer 29(11):69–77
6. Hsu C-J, Huang C-Y, Chang J-R (2011) Enhancing software reliability modeling and prediction through the introduction of time-variable fault reduction factor. Appl Math Model 35(1):506–521
7. Pachauri B, Dhar J, Kumar A (2015) Incorporating inflection S-shaped fault reduction factor to enhance software reliability growth. Appl Math Model 39(5–6):1463–1469
8. Anand S, Verma V, Aggarwal AG (2018) 2-Dimensional multi-release software reliability modelling considering fault reduction factor under imperfect debugging. Ing Solidar 14(25):1–12. ISSN online 2357-6014 ISSN print 1900-3102
9. Aggarwal AG, Gandhi N, Verma V, Tandon A (2019) Multi-release software reliability growth assessment: an approach incorporating fault reduction factor and imperfect debugging. Int J Math Oper Res (Accepted for publication)
10. Yamada S, Tokuno K, Osaki S (1992) Imperfect debugging models with fault introduction rate for software reliability assessment. Int J Syst Sci 23(12):2241–2252
11. Kapur P, Younes S (1996) Modelling an imperfect debugging phenomenon in software reliability. Microelectron Reliab 36(5):645–650

12. Kapur P, Aggarwal AG, Kaur G (2010) Testing-resource dependent flexible reliability growth model for software with multiple releases. In: Proceedings of international conference on development and applications of statistics in emerging areas of science and technology (ICDASEAST)
13. Jones C (1996) Software defect-removal efficiency. Computer 29(4):94–95
14. Liu H-W, Yang X-Z, Qu F, Shu Y-J (2005) A general NHPP software reliability growth model with fault removal efficiency

Chapter 30
Understanding Guest Experience Using Online Reviews

Sanchita Aggarwal, Ajay Jaiswal, and Anu Gupta Aggarwal

30.1 Introduction

In this era of technological advancement where worldwide network has perforated all walks of lives and is being incorporated in almost every field, service provision through electronic medium has taken its own place in the market. Customer as well as the service providers are more inclined towards E-Services due to its ease of access, transparency of transactions, increased market reach and time-saving factors to name a few [1]. This has given a boost to Online Travel Agencies. The hike in OTA market resulted in opening up of its dimensions for evaluation. Lodging industry occupies major part of the hospitality sector and has an immense scope of improvement.

Guest Experience is one dimension, which is of great interest for the researchers as well as the practitioners. One of the factors to study guest experience is through E-WOM. Online Reviews are perceived to be more uncensored, honest and firsthand view of the service and are found to be more insightful [2]. With Web 2.0 in the picture, user can easily share their views, opinions and experiences with the masses through web portals, forums and websites [3]. Review of product/service has always been a part of traditional (feedback and WOM) as well as the electronic (ratings and e-WOM) commerce [1]. An estimated growth of 800% has been observed in recent past in popularity of reviews [4]. Research has shown that these can be utilized to understand customers' vision, perception, demand and preference [5].

S. Aggarwal (✉)
University of Delhi, Delhi, India
e-mail: sanchitaaggarwal.du.aor.17@gmail.com

A. Jaiswal
Shaheed Sukhdev College of Business Studies, Delhi, India
e-mail: ajayjaiswal@sscbsdu.ac.in

A. G. Aggarwal
Department of Operational Research, University of Delhi, Delhi, India
e-mail: anuagg17@gmail.com

P. K. Kapur et al. (eds.), *Decision Analytics Applications in Industry*, Asset Analytics,
https://doi.org/10.1007/978-981-15-3643-4_30

Online reviews are of great importance for travel planners as they are experiences of fellow travelers in their own words rather than the content provided by the service providers. Subsets of the crowd contributes in creation of online content as "Wisdom of Crowds" which helps the travel fraternity in making informed travel decisions [6]. In hospitality and tourism sector, reviews and ratings can be a rich source of information to extract guest experience/satisfaction factors. In past, eWOM has been studied to draw out dimensions of loyalty, repeat purchase, sales, etc. [7, 8].

Guest satisfaction and guest experience have become a hot topic for the hotel industry. Guest Satisfaction has been recognized as how many parameters of customer expectations have been fulfilled. Satisfaction in the form of ratings and reviews is the empirical representation of their feelings and emotions associated with their experience which affects buying behavior of other travelers. Customer Satisfaction is a multidimensional construct which is decided by different aspects somewhat similar to service quality [9]. Customer reviews have been studied since 1970s to understand different dimensions of satisfaction. Prior studies employed traditional methods to draw these dimensions which provided less accurate results. Assessment of Guest satisfaction through text analysis is a fairly new concept. User rating to measure different aspects of service has been widely read and studied in the past but review analysis is a comparatively new field [7].

In this study, we aim to find out the major features that have an impact on guest satisfaction with the help of text analysis technique to deeply analyze customer provided reviews. Text analysis was employed to extract the features of importance and machine learning techniques were employed to study joined effect of reviews and ratings on guest satisfaction. Machine learning has been scarcely utilized in tourism sector. Also, we are using both text analysis through Term Frequency-Inverse Document Frequency (TF-IDF) and machine learning using support vector machine (SVM) for our study which is an untapped combo in the hospitality sector.

The paper proceeds in the following manner. Introduction of the paper is followed by a literature review on guest satisfaction. Further, text analysis and machine learning techniques employed are discussed. Methodology section details the data collection process. It also specifies the text analytical and machine learning techniques used. Further, obtained results are presented and discussed. This paper closes on the conclusion drawn from this study along with limitations and future study prospects of the same.

30.2 Theoretical Background

Guest satisfaction is a state of mind of the customer pertaining to hospitality industry. This is a complex human trait which is being studied since as early as 1970s [10]. Different definitions of guest satisfaction have emerged over the years. Satisfaction has been considered as an evaluation of what customer has experienced is at least as good as it is supposed to be [11]. Importance of hotel location on determinants of guest satisfaction have also been studied [12]. Understanding customer experience through content analysis of online reviews is a transpiring field in hospitality industry. Guest satisfaction has been widely studied to find out the dimensions it impacts. Relative importance of hotel factors for overall satisfaction and likelihood of subsequent return has been studied [13]. Prior research found that customer satisfaction along with a favorable image can positively impact customer loyalty [8]. Further, it has also been studied that Guest return intention which is strongly connected to financial performance is an indicator of unit level sales of a hotel. Since guest satisfaction is an antecedent of return intention, it indirectly affects the sales [10]. Since, it is evident that a satisfied customer is the one who actually impacts all the aspects of business, considering guest satisfaction to be one of the most important aspect cannot be proved wrong.

In the proposed framework, we utilize text analytics techniques: converting textual content into a data representation and mining it for insights, trends or patterns. Prior studies have majorly employed surveys, exploratory case studies or travel blog analysis to extract key factors influencing satisfaction [10, 13]. A few text analytics techniques have also been utilized to mine meaning out of online review. Text analytics has been used to understand satisfied and unsatisfied customers [14]. A short-text sentiment topic model was also studied to identify topics and sentiments from an online review [15]. LSA was applied to conclude that determinants of customer satisfaction or dissatisfaction are different and specific to particular hotels. They contributed to the literature by providing clues to hoteliers to enhance satisfaction and alleviate dissatisfaction by improving service (which are different for different types of hotels) [14].

In this study we are using TF-IDF to extract features out of a corpus of user generated content. Further, we are employing SVM technique to classify on the basis of extracted features and validate our model. SVM is a technique which is comparatively unstudied in marketing environment. Its utility has been well establishes in other sectors like statistics, computer science, engineering, agriculture [16]. this study differs from previous studies in the following ways. First, here we attempt to examine online reviews and customer ratings together. Second, we are creating a combination of feature extraction through text analysis technique of tf-idf and machine learning technique of SVM. SVM draws its strength from its ability of avoiding over fitting and no local minima.

30.3 Proposed Approach

The entire scheme of guest satisfaction classification which is proposed in this paper can be curtailed as follows: The detailed methodology of the proposed model is described in Fig. 30.1 and the subsequent steps are itemized below.

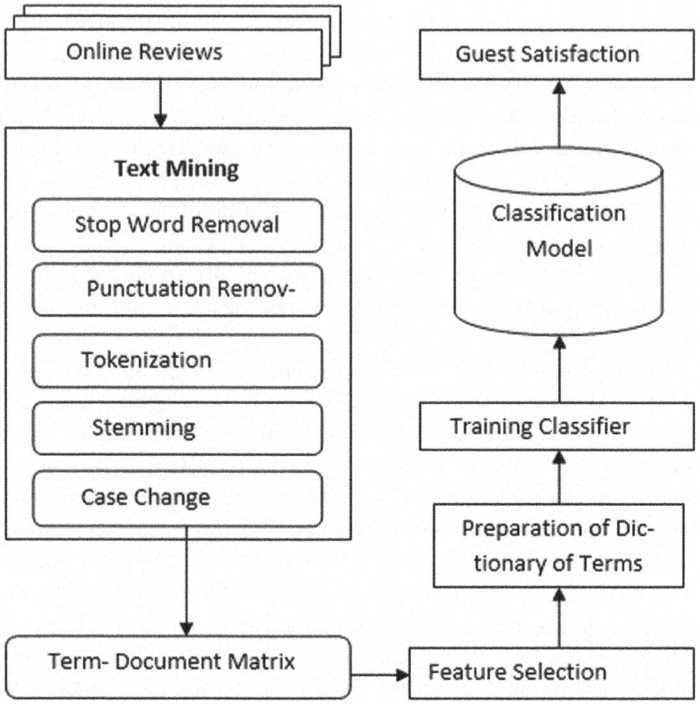

Fig. 30.1 Proposed model

30.3.1 Data Acquisition

For this study, our target citizenry is customers from hospitality industry, particularly the lodging sector. For our probationary setting, we choose consumer review website TripAdvisor. A total of 69,861 reviews spread across 280 hotels were downloaded for the year 2015. The consumer review enclosed both the textual experience as well as the overall rating provided by the customer to the hotel. The description of dataset is presented in Table 30.1.

Of the amassed 69,861 reviews, which comprised of 3 different categories namely hotels, restaurants and attractions, only 36,560 hotel reviews were shortlisted for the present study. These were further categorized as satisfied and unsatisfied customers

Table 30.1 Dataset information

Categories	No. of reviews
Satisfied customers	33,526
Unsatisfied customers	3034
Total no. of hotel reviews (n)	36,560
Total no. of reviews	69,861

on the cutoff condition of 60% satisfaction. This gave us an aggregate of 33,526 satisfied and 3034 unsatisfied customers.

30.3.2 Text Pre-processing

The CSV file of hotel reviews is taken as an input to our python code. On its textual content, standard preprocessing steps are executed [17].

(a) Stop Word Removal: It is the task of removing frequently occurring words with no specific meaning to reduce the data. They generally comprise of prepositions, conjunctions and articles.
(b) Punctuation Removal: It is the task of removing spacing and symbols to further clean the textual data.
(c) Tokenization: It is the process of detaching a string of textual data into pieces such as words, keywords, phrases and other elements called tokens. These become fabricated inputs for another process.
(d) Stemming: It is the task of reducing a word variants to its root form. We used standard porter stemmer to conduct the stemming process given by Porter [17].
(e) Case Change: This process encompassed changing all the tokens to lowercase to eliminate chances of having multiple copies of a word.

30.3.3 Feature Extraction

After the preprocessing step, a term-document matrix is created, in which each column depicts the terms occurring in documents (reviews) and each row depicts an individual review. Cells of the matrix contain respective TF-IDF score, which is zero if a particular word is absent from a given review. TF-IDF score is generally being used to give weight (importance) to each term. It is determined by the product of inverse document frequency and term frequency and is given by

$$TF * IDF = n_c^r \log_2 \frac{R}{R_t} \tag{30.1}$$

where t = count of term t in the review r, R_t = no. of reviews comprising of term t and R = review volume.

Utilizing TF-IDF scores, top 30 features were identified by our python code as shown in Table 30.2. These features were further categorized into five categories based on a logical connection between the terms. This process was first carried out by one researcher and then confirmed by a second researcher. Top words belonging to each category are depicted in Table 30.3.

Table 30.2 Top 30 extracted feature

Recommended	Free	Month	Buffet	Block
Water front	Bed	Office	Situated	Overlooking
Excellent	Internet	Right	Express	View
Directly	Easy	Staying	Airport	Limited
Food	Modern	Menu	Well	Access
Comfortable	Station	Getting	Look	Couple

Table 30.3 Classification of top terms

Features	Terms
Staff	Excellent, modern, well, look, easy
Service	Recommended, food, buffet, office, bed, menu, getting, overlooking, block, limited
Location	Access, directly, situated, staying, station, airport
Amenities	Waterfront, express, free, internet
Hybrid	Comfortable, couple, month, right, view

TripAdvisor facilitates its customers to rate their overall experience in the form of star ratings, with 1 star hotel rating being the least satisfaction and 5 star being the most satisfied category (see Fig. 30.2). A cutoff of 60% was applied to divide the population into satisfied and not satisfied customers, i.e. those who awarded star rating 3 or above were categorized as satisfied customers and others were categorized to be unsatisfied customers. It was established that 91.7% of the reviewers (33,526 out of 36,560) were satisfied with their hotel experience and 8.3% of the reviewers (3034 out of 36,560) were not so happy with their experience.

Fig. 30.2 Hotel ratings and review distribution on the basis of satisfaction level

Table 30.4 Performance measures

Precision	$\frac{TP}{TP+FP}$	Out of data points that were found to be relevant, actually were relevant
Recall	$\frac{TP}{TP+FP}$	Ability to find out all the relevant occurrences in the data
F1-Measure	$2 \times \frac{\text{precision} \times \text{recall}}{\text{precision}+\text{recall}}$	A measure to seek a balance between precision and recall
Accuracy	$\frac{TP+TN}{TP+TN+FP+FN}$	To measure effectiveness of a classifier in prediction

Table 30.5 Prediction analysis of guest satisfaction

	Precision	Recall	F1-measure	Accuracy
Satisfied customer	0.31	0.31	0.31	0.88544
Unsatisfied customer	0.94	0.94	0.94	

30.3.4 Classifiers and Evaluation Matrix

Selected features of importance are fed into SVM as machine learner. The model is tutored using training data and performance is gauged using test data. Training and testing data are divided into 7:3 ratio of the available dataset.

Guest satisfaction lies in the range of 0–1. Satisfaction threshold value is an important parameter of guest satisfaction prediction model. In accordance with our work, its default value is set at 0.60 on the basis of prior studies and the proven efficiency therein [18]. In Table 30.4, experiment is done using standard 10-fold cross validation method and performance evaluation of the model is conducted using F-measure, recall and precision.

30.3.5 Support Vector Machine

SVM was proposed using statistical learning theory [19]. It can illustrate a discriminative classifier, which is defined by a dividing hyperplane and discovers an optimal hyperplane which separates the dataset belonging to two different classes. Data sample present on hyperplane are a representation of support vectors. SVM generates optimal hyperplane using a Lagrangian optimization problem which is developed in dual space and considers kernels. Linear kernel can be calculated using the dot product of input variable x_i and x_j

$$K\left(x_i, x_j\right) = 1 + \text{sum}\left(x_i * x_j\right) \tag{30.2}$$

Table 30.6 Confusion matrix using proposed classifier

Actual class	Predicted class	
	Satisfied	Not Satisfied
Satisfied	236	522
Not Satisfied	525	7857

30.4 Experimental Results

A user can share their hotel experience on TripAdvisor via two major way i.e. through textual review and hotel star rating. We used these two factors to find out the major factors that affect customer satisfaction. Text analysis was applied using TF-IDF method to extract top features that most reviewers are talking about (refer Table 30.2) Based on the overall rating provided by consumer to the hotel, users were categorized into satisfied and unsatisfied customers. Machine learning algorithm, SVM was used to classify guest satisfaction on the basis of extracted features. Firstly, a confusion matrix was created as shown in Table 30.6. Then other performance parameters were computed as shown in Table 30.5. Satisfied customer class delivers precision, recall and f1 value of 31%. Similarly, unsatisfied customer class delivers a value of 94% for precision, recall and f1 measures. Overall accuracy of the model comes out to be 88.5%.

30.5 Conclusion

In this paper, an amalgamated approach of text analytics and machine learning technique was employed to scrutinize cognitive characteristics of consumer in order to identify guest satisfaction factors. Customer provided reviews were filtered to diminish the prospect of misclassification and to enhance the performance of satisfaction prediction. TF-IDF technique was used for the extraction of top influential features and SVM technique was utilized for machine learning classification. Precision, Recall and F1-measure was studied to detect accuracy of the model. The prediction model was found to be considerably accurate. This study had its own set of limitations as it focused only on the hotel reviews and filtered out other travel-related industries. Further, the study focused only on top few features and rarely or infrequently occurring terms were disregarded. Each of the above mentioned limitation along with better computing techniques can be a rich avenue for future research.

References

1. Aakash A, Aggarwal AG, Aggarwal S (2020) Analyzing the impact of e-WOM text on overall hotel performances: a text analytics approach. In: Exploring the power of electronic word-of-mouth in the services industry. IGI Global, pp 240–264
2. Fernandes T, Fernandes F (2018) Sharing dissatisfaction online: analyzing the nature and predictors of hotel guests negative reviews. J Hosp Mark Manag 27(2):127–50
3. Bertola F, Patti V (2016) Ontology-based affective models to organize artworks in the social semantic web. Inf Process Manage 52(1):139–62
4. De Ascaniis S, Borrè A, Marchiori E, Cantoni L (2015) Listen to your customers! How hotels manage online travel reviews. The case of Hotels in Lugano. In: Information and communication technologies in tourism 2015. Springer, Cham, pp 59–72
5. Aggarwal AG, Aakash (2018) Multi-criteria-based prioritisation of B2C e-commerce website. Int J Soc Syst Sci 10(3):201–22
6. Surowiecki J (2005) The wisdom of crowds. Anchor
7. Ye Q, Law R, Gu B (2009) The impact of online user reviews on hotel room sales. Int J Hosp Manag 28(1):180–2
8. Kandampully J, Suhartanto D (2000) Customer loyalty in the hotel industry: the role of customer satisfaction and image. Int J Contemp Hosp Manag 12(6):346–51
9. Tellis GJ, Johnson J (2007) The value of quality. Mark Sci 26(6):758–73
10. Wu CH, Liang RD (2009) Effect of experiential value on customer satisfaction with service encounters in luxury-hotel restaurants. Int J Hosp Manag 28(4):586–93
11. Hunt JD (1975) Image as a factor in tourism development. J Travel Res 13(3):1–7
12. Yang Y, Mao Z, Tang J (2018) Understanding guest satisfaction with urban hotel location. J Travel Res 57(2):243–59
13. Susskind AM, Kacmar KM, Borchgrevink CP (2018) The relationship of service providers' perceptions of service climate to guest satisfaction, return intentions, and firm performance. Cornell Hosp Q 59(4):390–6
14. Berezina K, Bilgihan A, Cobanoglu C, Okumus F (2016) Understanding satisfied and dissatisfied hotel customers: text mining of online hotel reviews. J Hosp Mark Manag 25(1):1–24
15. Xiong S, Wang K, Ji D, Wang B (2018) A short text sentiment-topic model for product reviews. Neurocomputing 5(297):94–102
16. Cui D, Curry D (2005) Prediction in marketing using the support vector machine. Mark Sci 24(4):595–615
17. Porter MF (1980) An algorithm for suffix stripping. Program 14(3):130–7
18. Krishnamoorthy S (2015) Linguistic features for review helpfulness prediction. Expert Syst Appl 42(7):3751–9
19. Vapnik V (2013) The nature of statistical learning theory. Springer Science & Business Media

Chapter 31
e-Epidemic Model for the Analysis of Antidotal and Quarantine with Impact of Malicious Objects in Computer Network

Aditya Kumar Singh

31.1 Introduction

Coercively ceasing the cycling course of Time's chariot is a vain essay as it crushes down moldy odds with evolving changes. The World of Cyber that emerged from this evolution brought a sea-change in the societal scenario. Since the 1980's, the internet technology has been continuously offering multiple functionalities and facilities. By transforming the Globe into a hamlet over time the ever-augmenting efficiency of cyber technology has widened its arch for swift exchanging of information. It, in turn, has conferred coziness to our causes and made the world accessible to the touch of our clutch. But it would be a Utopian Universe if its pros could overpower its cons. It is jeopardous that the cyber world continuing ceaseless combat to shielded its vulnerability from penetrative malicious objects in terms of worms, virus and Trojan horses. These alluded arch enemies have immense influence in crumpling computer network. Pro-tempore, email has become a ceaseless causeway for these intruding malicious objects. Epidemicity of these intruders is analogous to biological epidemic diseases. Recurrent complexity to subjugate those malicious objects in computer network currently is the chief cause of cephalalgia to researchers. In order to curb the malicious object, we propose a SIQRAS model. Some of the steps to curb the transmission of the malicious objects could be isolation and some nodes equipped with strong anti-malicious software called antidotal, which are easy and can equally play a pivotal preface in constipating the transmission of these malicious objects. The word 'Quarantine', symbolizes force isolation. The biological world has experienced radical changes in the last decades with the successful implementation of quarantine to reduce the transmission of human ailments such as leprosy, plague, smallpox, etc. The same panacea has been adopted in the cyber world to dilute this debacle.

A. K. Singh (✉)
Ramgarh Engineering College, Ramgarh, Jharkhand, India
e-mail: adityanil25@gmail.com

P. K. Kapur et al. (eds.), *Decision Analytics Applications in Industry*, Asset Analytics,
https://doi.org/10.1007/978-981-15-3643-4_31

The impact of malicious objects throughout a network can be studied by using epidemiological models for disease propagation [1–9]. Based on the Kermack and McKendrick SIR classical epidemic model [10, 11], dynamical models for malicious objects were proposed, providing estimations for temporal evolutions of infected nodes depending on network parameters considering topological aspects of the network [12–17]. The SEIR model proposed by the authors Yan and Liu [18] assumes that recovery hosts have a permanent immunization period with a certain probability, which is not consistent with the real situation. In order to overcome the limitation, Mishra and Saini [1] present a SEIRS model with latent and temporary immune periods which can reveal common worm propagation. Recently, more research attention has been paid to the combination of virus propagation models and antivirus countermeasures to study the prevalence of the virus, for example, virus immunization [19–22] and quarantine [23]. Extending the SAIR model of Piqueira and Arujo [24], a new compartment exposed has been introduced and its effect has been analyzed in this paper. We considered a network in which some machines have been equipped with anti-virus software and for those machines, a new compartment A has been formed. After the incorporation new compartment, A of antidotal population, the SIR model is converted to SAIR (Susceptible-Antidotal-Infected-Removed) model and that was studied in [7, 8, 25].

In the SIQRAS model for the infections that do not confer permanent immunity, susceptible nodes go to infectious class, thereafter some nodes remain in the infected class while they are infectious and then move to the recovered class after the run of anti-malicious software and some nodes which are equipped with strong anti-malicious software go to antidotal compartment. Other most infected nodes are transferred into the quarantined class while they are infectious and then move to the recovered class after their recovery. The model here has a variable total population size because they have recruitment into the susceptible class by the inclusion of some new nodes and they have crashing of nodes due to the malicious objects and reason other than the attack of malicious codes.

31.2 SIQRAS Model for the Transmission of Malicious Objects

The model considered here is an extension of the SIAR model. We have included an antidotal population compartment (A) which represents nodes of the network equipped with the effective anti-malicious program and quarantine class in which most infected nodes are isolated. The total population of T is divided into five compartments. Compartment S represents nodes susceptible to the attack of malicious objects; compartment A represents uninfected computers equipped with strong anti-malicious software, compartment I for the infectious class, compartment R for the recovered class and compartment Q for quarantined nodes (Fig. 31.1).

Here S denotes the number of susceptible computers at any t.

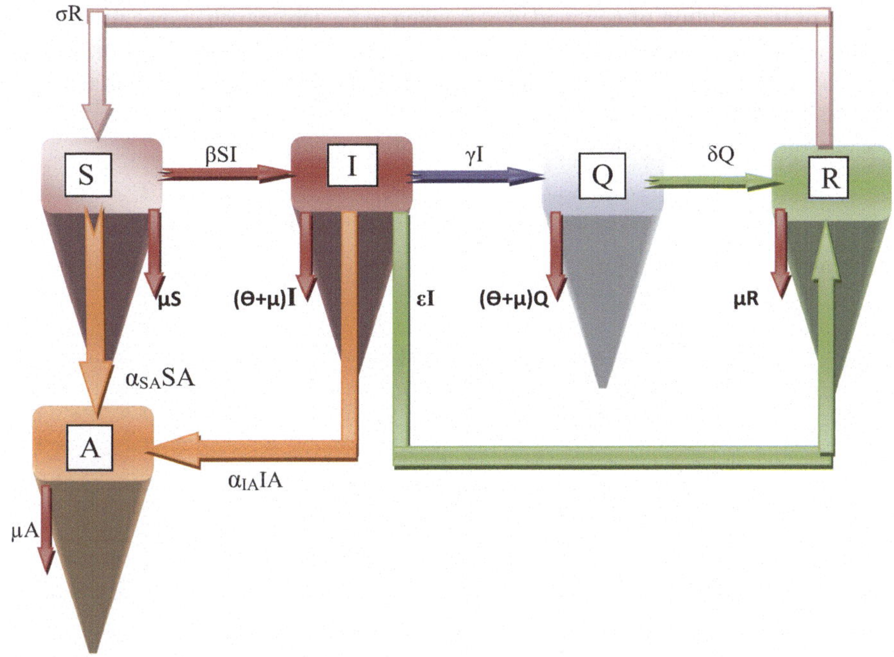

Fig. 31.1 Schematic diagram for the flow of malicious objects in the computer network

$$
\begin{aligned}
\frac{\mathrm{d}S}{\mathrm{d}t} &= N - \alpha_{SA}\,SA - \beta SI - \mu S + \sigma R \\
\frac{\mathrm{d}I}{\mathrm{d}t} &= \beta SI - \alpha_{IA} IA - (\gamma + \mu + \theta + \varepsilon)I \\
\frac{\mathrm{d}Q}{\mathrm{d}t} &= \gamma I - (\mu + \delta + \theta)Q \\
\frac{\mathrm{d}R}{\mathrm{d}t} &= \delta Q - (\mu + \sigma)R + \varepsilon I \\
\frac{\mathrm{d}A}{\mathrm{d}t} &= \alpha_{SA}\,SA + \alpha_{IA}\,IA - \mu A
\end{aligned}
\tag{31.1}
$$

where N be the influx rate that is incorporation of new nodes to the network and each node is assumed to be susceptible, β is the per capita contact rate, which is the average number of effective contact with other nodes per unit time, denotes death rate not owing to attack of malicious objects (that is crashing of nodes due to technical reasons) and θ be the death rate due to malicious objects. γ be the rate of transmission of quarantined nodes from the infectious class, δ and ε be rate of recovery from the quarantined and infectious class, respectively. α_{SA} and α_{IA} rate of conversion from susceptible to antidotal and rate of conversion from infection to antidotal, respectively. σ be the rate of transmission of nodes from recovered class to susceptible class.

Rate of the addition of new nodes N is taken to be zero which shows that during the spread of malicious objects is faster than the network expansion. The same reason justifies the choice of $\mu = 0$, considering that the machine desuetude time is larger than the time of the virus action.

Consequently, we get the system of differential equation as

$$\frac{dS}{dt} = -\alpha_{SA} SA - \beta SI - \mu S + \sigma R$$
$$\frac{dI}{dt} = \beta SI - \alpha_{IA} IA - (\gamma + \theta + \varepsilon + \mu)I$$
$$\frac{dQ}{dt} = \gamma I - (\theta + \delta + \mu)Q$$
$$\frac{dR}{dt} = \delta Q - (\sigma + \mu)R + \varepsilon I$$
$$\frac{dA}{dt} = \alpha_{SA} SA + \alpha_{IA} IA - \mu A \tag{31.2}$$

It has been taken that $T = S + I + Q + A + R$, i.e. total population remains constant. One of the equations can be expressed as a linear combination of the other three.

31.2.1 Malicious Objects Free Equilibrium Point

For the infection-free equilibrium state, i.e. $I = 0$ (absence of malicious objects), we get the following two points from the system of Eq. 31.2.

$$P1 = (S, I, Q, R, A) = (0, 0, 0, 0, T)$$

$$P2 = (S, I, Q, R, A) = (T, 0, 0, 0, 0)$$

Considering the equilibrium point $P1$, the corresponding linear system around it has a Jacobian given by

$$J_{P1} = \begin{bmatrix} -\alpha_{SA} T - \mu & 0 & 0 & \sigma & 0 \\ 0 & -(\gamma + \theta + \varepsilon + \mu) - \alpha_{IA} T & 0 & 0 & 0 \\ 0 & \gamma & -(\delta + \delta + \mu) & 0 & 0 \\ 0 & \delta & 0 & -(\sigma + \mu) & 0 \\ \alpha_{SA} T & \alpha_{IA} T & 0 & 0 & 0 \end{bmatrix}$$

The Eigen values of

$$J_{P1} = (-\alpha_{SA}\,T - \mu,\ -(\gamma + \theta + \varepsilon + \mu) - \alpha_{IA}T,\ -(\delta + \theta + \mu),\ -(\sigma + \mu),\ -\mu)$$

Since all the parameter values are real and negative, implying that P1 is asymptotically stable. For the infection free equilibrium point P_2, we construct the following Jacobian matrix J_{P2}

$$J_{P2} = \begin{bmatrix} -\mu & -\beta T & 0 & \sigma & -\alpha_{SA}\,T \\ 0 & \beta T - (\gamma + \theta + \varepsilon + \mu) & 0 & 0 & 0 \\ 0 & \gamma & -(\delta + \theta + \mu) & 0 & 0 \\ 0 & \varepsilon & \delta & -(\sigma + \mu) & 0 \\ 0 & 0 & 0 & 0 & \alpha_{SA}\,T - \mu \end{bmatrix}$$

The eigenvalues of

$$J_{p2}\ \text{are} : (-\mu,\ \beta T - (\gamma + \theta + \varepsilon + \mu),\ -(\theta + \delta + \mu),\ -(\mu + \sigma),\ \alpha_{SA}\,T - \mu)$$

If these two eigenvalues, $\beta T - (\gamma + \theta + \varepsilon + \mu) > 0$ and $\alpha_{SA}\,T - \mu > 0$ then P_2 is unstable if these eigenvalues are negative, this implies that P_2 is stable.

The stability analysis performed above indicates that this model presents two malicious objects free equilibrium points, one system P_2 is unstable if $\beta T - (\gamma + \theta + \varepsilon + \mu) > 0$, $\alpha_{SA}\,T - \mu > 0$, otherwise both the system is asymptotically stable. As the linear approximation method is local, it is not sure whether any initial condition drives the network to the asymptotically stable malicious objects free equilibrium.

31.2.2 Endemic Equilibrium Points

We assume that the infected nodes exist in the network ($I \neq 0$) this implies that R is not equal to zero. Consider that the network has no antidotal node; there is an endemic equilibrium point, $P_3 = (S, I, Q, R, 0)$ given by,

$$S = \frac{\gamma + \theta + \mu + \varepsilon}{\beta};\ Q = \frac{\gamma I}{\theta + \delta + \mu};\ R = \frac{(\gamma + \theta + \varepsilon)I}{\sigma};\ I = \frac{T - \frac{\gamma + \theta + \mu + \varepsilon}{\beta}}{(1 + \frac{\gamma}{\theta + \delta + \mu} + \frac{\gamma + \theta + \varepsilon}{\sigma})};\ A = 0$$

Consequently, the condition $T < \frac{\gamma + \theta + \mu + \varepsilon}{\beta}$ avoids the existence of this point. Suppose the existence of this equilibrium point, the Jacobian J_{P3} is given by,

$$J_{P3} = \begin{bmatrix} -\beta I - \mu & -(\gamma + \theta + \mu + \varepsilon) & 0 & \sigma & -\alpha_{SA}\frac{(\gamma + \theta + \mu + \varepsilon)}{\beta} \\ -\beta I & 0 & 0 & 0 & \alpha_{IA}\,I \\ 0 & \gamma & -(\mu + \delta + \theta) & 0 & 0 \\ 0 & \varepsilon & \delta & -(\sigma + \mu - \varepsilon) & 0 \\ 0 & 0 & 0 & 0 & \alpha_{SA}\frac{\gamma + \theta + \mu + \varepsilon}{\beta} + \alpha_{IA}\,I \end{bmatrix}$$

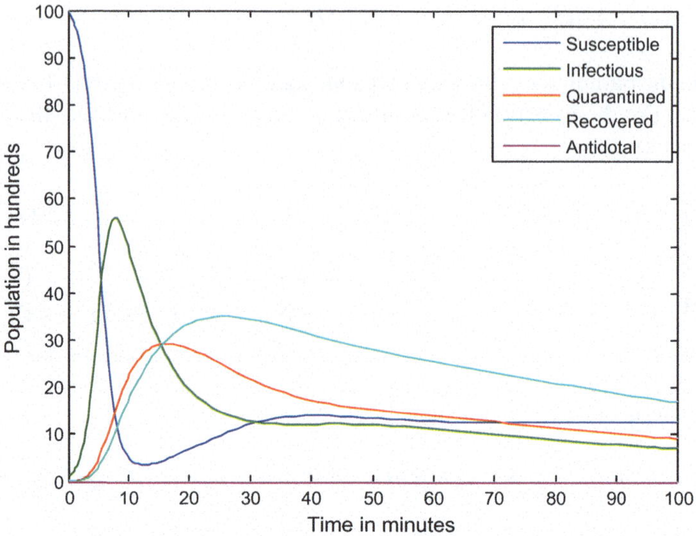

Fig. 31.2 Dynamical behavior of S, I, Q, R and A with time t for the system 1

where $I = \dfrac{T - \frac{\gamma+\theta+\mu+\varepsilon}{\beta}}{(1+\frac{\gamma}{\theta+\delta+\mu}+\frac{\gamma+\theta+\varepsilon}{\sigma})}$

The characteristic equation of the above matrix is

$$|J_{P3} - \lambda I| = \left| \begin{array}{ccccc} -\beta\left(\frac{T-\frac{\gamma}{\beta}}{1+\frac{\gamma}{\delta}+\frac{\gamma}{\sigma}}\right)-\lambda & -\gamma & 0 & \sigma & -\alpha_{SA}\frac{\gamma}{\beta} \\ \beta\left(\frac{T-\frac{\gamma}{\beta}}{1+\frac{\gamma}{\delta}+\frac{\gamma}{\sigma}}\right) & -\varepsilon-\lambda & 0 & 0 & \alpha_{IA}\left(\frac{T-\frac{\gamma}{\beta}}{1+\frac{\gamma}{\delta}+\frac{\gamma}{\sigma}}\right) \\ 0 & \gamma & -\delta-\lambda & 0 & 0 \\ 0 & \varepsilon & \delta & -\sigma-\lambda & 0 \\ 0 & 0 & 0 & 0 & -\lambda+\alpha_{SA}\frac{\gamma}{\beta}+\alpha_{IA}\left(\frac{T-\frac{\gamma}{\beta}}{1+\frac{\gamma}{\delta}+\frac{\gamma}{\sigma}}\right) \end{array} \right|$$

Using the Routh–Hurwitz Criterion, we observe that the roots have negative real parts indicating that the system is stable. Runge–Kutta Fehlberg method of order 4–5 has been employed to solve the system of Eq. (31.2) and using the real parametric values, the system has been simulated and is depicted in Fig. 31.2.

31.3 Conclusion

Compartmental SIQRAS model has been developed for a given total number of nodes, T, of a network and a new parameter, death due to malicious object has been added. Depiction is that if the eigenvalues of all possible parameters are real and negative, the malicious objects free equilibrium state is asymptotically stable, which will give us a guarantee for a good performance of the network. If any roots are non-negative eigenvalue, the equilibrium state is unstable. Endemic equilibrium has also been discussed. To solve and simulate the system of differential equations, four and fifth Runge–Kutta Method of Numerical Analysis are employed and the behavior of the susceptible, infectious, quarantined, antidotal and recovered nodes with respect to time are observed which are depicted in Fig. 31.2. From Figs. 31.2, 31.3, 31.4 and 31.5 we observed that the system is asymptotically stable.

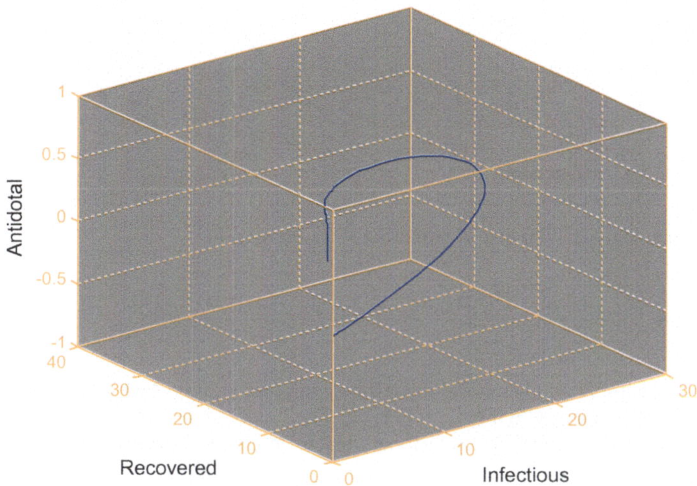

Fig. 31.3 Dynamical behavior of *I*, *R* and *A* for the system 1

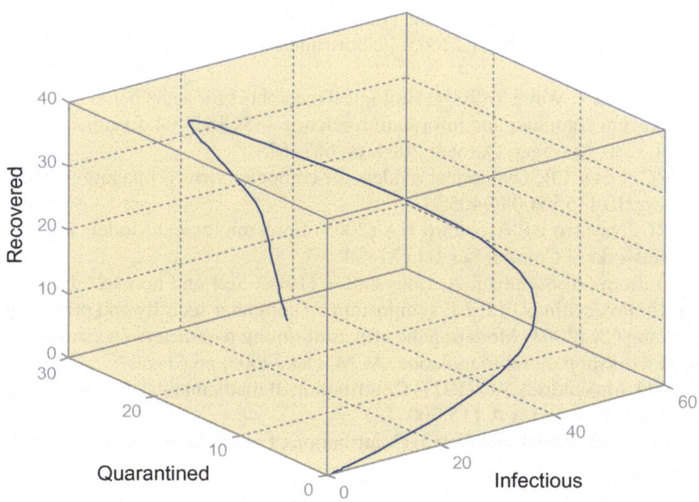

Fig. 31.4 Dynamical behavior of *I*, *Q* and *R* for the system 1

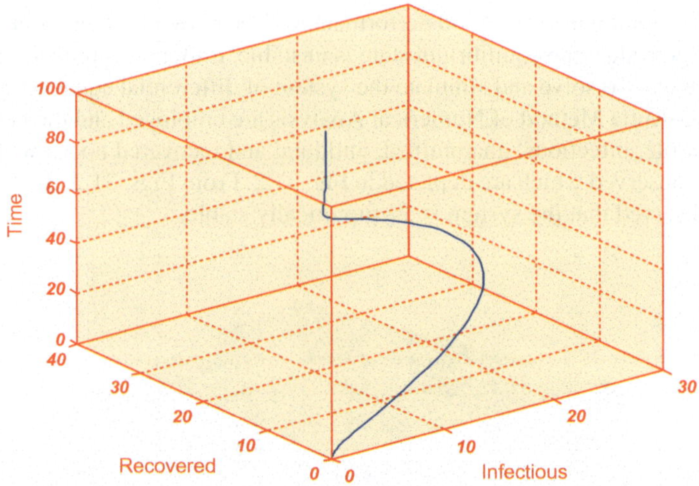

Fig. 31.5 Dynamical behavior of I, R with time T for the system 1

References

1. Mishra BK, Saini D, Mathematical models on computer viruses. Appl Math Comput. https://doi.org/10.1016/j.amc.2006.09
2. Mishra BK, Saini D (2007) Mathematical models on computer viruses. Appl Math Comput 187(2):929–936
3. Mishra BK, Jha N (2007) Fixed period of temporary immunity after run of anti-malicious software on computer nodes. Appl Math Comput 190(2):1207–1212
4. Gelenbe E (2007) Dealing with software viruses: a biological paradigm. Inform Security Technical Rep 12(4):242–250
5. Gelende E (2005) keeping viruses under control. In: 20th international symposium on computer and information sciences—ISCIS 2005. Lecture notes in computer science, vol 3733. Springer, October 2005
6. Gelenbe E, Kaptan V, Wang Y (2004) Biological metaphors for agent behavior. In: 19th international symposium computer and information science—ISCIS 2004. Lecture notes in computer science, vol 3280. Springer, October 2004, pp 667–675
7. Piqueira JRC, Cesar FB, Dynamical models for computer virus propagation. Math Prob Eng. https://doi.org/10.1155/2008/940526
8. Piqueira JRC, Navarro BF, Monteiro HA (2005) Epidemiological Models applied to virus in computer networks. J Comput Sci 1(1):31–34
9. Forest S, Hofmeyr S, Somayaji A, Longstaff T (1994) Self and non-self discrimination in a computer. In: Proceedings of IEEE symposium on computer security and privacy, pp 202–212
10. Wang Y, Wang CX (2003) Modeling the effects of timing parameters on virus propagation. In: 2003 ACM workshop on rapid malcode, ACM, Oct., 2003, pp 61–66
11. Kermack WO, Makendrick AG (1927) Contribution of mathematical theory to epidemics. Proc Royal Soc London—Series A 115:700–721
12. Kermack WO, Makendrick AG (1932) Contribution of mathematical theory to epidemics. Proc Royal Soc London—Series A 138:55–83
13. Kermack WO, Makendrick AG (1933) Contribution of mathematical theory to epidemics. Proc Royal Soc London—Series A 141:94–122

14. Zou CC, Gong WB, Towsley D, Gao LX (2005) The monitoring and early detection of internet viruses. IEEE/ACM Trans Interwork 13(15):961–974
15. Kephart JO, White SR, Chess DM (1993) Computers and epidemiology. IEEE Spectr 20–26
16. Keeling MJ, Eames KTD (2005) Networks and epidemic models. J Roy Soc Interf 2(4):295–307
17. Williamson MM, Leill J (2003) An epidemiological model of virus spread and cleanup. http://www.hpl.hp.com/techreports/
18. Yan P, Liu S (2006) SEIR epidemic model with delay. J Aust Math Soc Series B-Applied Math 48(1):119–134
19. Newman MEJ, Forrest S, Balthrop J (2002) Email networks and the spread of computer virus. Phys Rev E 66:035101–035101-4
20. Hethcote HW, Lewis MA, Van Den Driessche P (1989) An epidemiological model with delay and a nonlinear incidence rate. J Math Bio 27:49–64
21. Hethcote HW (2000) The mathematics of infectious diseases. SIAM Rev 42:599–653
22. Kephart JO, Hogg T, Huberman BA (1989) Dynamics of computational ecosystems. Proc Phys Comput 40(1):404–421
23. Zou CC, Gong W, Towsley D (2003) Worm propagation modeling and analysis under dynamic quarantine defense. In: Proceedings of the ACM CCS workshop on rapid malcode, ACM, pp 51–60
24. Piqueira JRC, Araujo VO (2009) A modified epidemiological model for computer viruses. Appl Math Comput. https://doi.org/10.1016/j.amc.2009,03,023
25. Piqueira JRC, de Vasconcelos AA, Gabriel CECJ, Arauja VO (2008) Dynamic models for computer viruses. Comput Secur 27(7–8):355–359

Chapter 32
Study of Identity-Based Encryption for Cloud Data Security

Pragya Mishra and Vandani Verma

32.1 Introduction

In the present era, cloud computing is the next-generation highly efficient distributed computing platform in which resources of computing are dealt as a service, leveraging internet technologies and virtualization. Cloud computing is a paradigm that gives a giant computation capacity as well as huge memory space at a lower cost. 'Utility computing' and 'On-demand' are attracting key features of cloud computing for which it is being used massively. There exist three cloud-based delivery models presently. One model is 'Software as a Service (SaaS)' model. In this model, the user uses the application services without having control of the hardware, network infrastructure, operating system or Platform as a Service (PaaS) is the second one, where the users provide a domain for their applications as well as control the applications too. However, they don't control the network infrastructure, operating system or hardware being used by them. The final model is Infrastructure as a Service (IaaS), in which the user is authorized to access basic resources of computing such as CPU, storage, memory and middleware.

The concept of storage of applications and relevant data on the system infrastructure as well as services that are out of their control doesn't seem safe to users and dealing organizations. In addition, shifting of workloads on a shared infrastructure pacifies the illicit gain of sensitive data and its disclosure as well. Cloud-based systems require to be steady regarding identity management, authentication, access-related technologies and compliance which are growing very important day by day.

P. Mishra (✉) · V. Verma
AIAS, Amity University, Sector-125, Noida, India
e-mail: pragya.login@gmail.com

V. Verma
e-mail: vandaniverma@yahoo.com

P. K. Kapur et al. (eds.), *Decision Analytics Applications in Industry*, Asset Analytics,
https://doi.org/10.1007/978-981-15-3643-4_32

In a true sense, cloud computing stands for services brought over the internet from any person, any place or any device. Hence, cloud computing allows dramatic flexibility in processing decisions world wide. Shared resources feature of cloud environment provides faster computation and improved operational agility at a lower cost. Although cloud computing offers a number of benefits including economical aspects, it also carries risks of security issues too.

In this paper, we have studied an IBE scheme for cloud data security. This scheme requires less computations and hence is very efficient. We have organized this paper as follows. Section 32.2 presents concepts involved in cloud data security. In Sect. 32.3, we have made a literature survey based on various developments made in the IBE scheme to achieve cloud data security and many more. In Sect. 32.4, we have presented our identity-based encryption scheme for cloud data security and Sect. 32.5 holds the comparative study of the proposed model. In Sect. 32.6, we have made a conclusion and future scope.

32.2 Definitions

32.2.1 Bilinear Pairing

Let G_1 be a cyclic additive group of prime order p, G_2 be a cyclic multiplicative group of same order p. g is a generator of G_1. A bilinear pairing is defined as a mapping e: $G_1 \times G_1 \rightarrow G_2$ with the properties as under:

1. Bilinearity—$e(mu, nv) = e(u, v)^{mn}$, for all $u, v \in G_1$ and $m, n \in Z_p^*$
2. Non-degenerate—$e(g, g)$ is generator of G_2 only if g is generator of G_1.
3. Computability—If $a, b \in G_1$ then $e(m, n)$ will be computable easily.

32.2.2 Identity-Based Encryption

Identity-Based Encryption (IBE) is based on public-key encryption mechanism where public key is a string of arbitrary length like unique name, email address, telephone number, etc. The only authority private key generator is one who creates the private key for the user using its own master secret key.

32.2.3 Cloud Security

Identity-based access to cloud data should achieve the following security goals:

32.2.3.1 Data Confidentiality

The plaintext of the shared data in an original form stored on the cloud server should not be accessible to unauthorized users.

32.2.3.2 Forward Security

Forward security means when the secret key of the user gets compromised or authority of user is run out, the data accessed by a revoked user would not be able to access the plain text of data shared by now. Here, under the identity of her/his, the subsequently shared data of plain text is still encrypted.

32.2.3.3 Backward Secrecy

Backward security means when the secret key of the user gets compromised or authority of user is run out, the data accessed by the revoked user previously would not be accessible to him to get the plain text of data shared. Here, under the identity of her/his, the subsequently shared data of plain text is still encrypted.

32.2.3.4 Cloud Storage

Cloud storage stands for "the storage of data online on a cloud" where "cloud" is some collection of data centres which are maintained as well as owned by a third party.

32.3 Literature Survey

Cloud computing is a very fast computation technique that is easily accessible to users from anywhere. As it comes with data sharing outlooks and associated combative models and vulnerabilities, security threats have an open invitation for this. Thus, to secure client identity and data, IBE would be a vital solution. The foundation of IBE is public-key Infrastructure (PKI). The PKI uses human coherent identities, i.e. name, phone number, IP address, email address, etc. to streamline the key authorities. This feature of IBE is very useful because it does not require certificate management which is a rigorous task. As mostly the cloud data is in nature of public cloud, the security of cloud data is a more concerned aspect of study in the present scenario.

First the IBE scheme was constructed by Shamir [1]. The first identity-based encryption scheme practically implemented by Boneh and Franklin [2] is a proven secure in random oracle model. However, first, fully secure IBE without random oracle model was proposed by Gentry [3]. Subsequently, Canetti et al. [4] proposed a provenly secure public-key encryption (PKE) scheme under selective-ID security in the standard model. Hanaoka et al. [5] presented a new scheme in which users need to update their private keys periodically without interacting with PKG. In this scheme, there was a concept of a mediator who helps the user to decrypt the cipher text every time. In case of revocation of an identity, the mediator stops helping the user. Boldyreva et al. [6] constructed a new revocable IBE scheme in 2008 which was based on fuzzy IBE primitive [7]. This scheme [6] was based on the idea of deploying a binary tree data structure to store identities of users on the leaf nodes. Thus, it reduces the key-update efficiency of PKG up to logarithmic scale in the total number of system users. Li et al. [8] who introduced for the first time the outsourced computation into ID-based encryption and constructed a revocable ID-based encryption scheme to achieve the solution of identity revocation in the server aided setting. The key-update efficiency of this scheme is constant and hence independent of a number of system users. They achieved this goal by employing a hybrid private key for each user by using AND gate for connecting and bounding the identity and time component. Libert and Verngnaud [9] improved the scheme of Boldyreva et al. [6] to achieve adaptive-ID security. In an identity-based encryption scheme, only PKG is responsible for the generation of private keys. To reduce this load of PKG, a hierarchical architecture for cloud computing (HACC) was introduced by H. Li et al. [10]. They also proposed an IBE scheme and an IBS scheme for hierarchical architecture for cloud computing. Further, they constructed an Authentication Protocol for Cloud Computing (APCC) using IBE, HACC and IBS. This APCC meets the demand for cloud computing deftly. In recent years, various studies have been done to deal with the realistic troubles in the cloud served model to get a junction point into outsourcing computation and cloud computing.

32.4 Identity-Based Encryption Scheme for Cloud Data Security

Our proposal of cloud data security is based on the following encryption scheme [11].

This scheme contains the following steps:

Set up:

In this step, PKG runs the algorithm in the following ways:

- Select two cyclic groups G_1 and G_2, where G_1 is an additive cyclic group with generator p. Then bilinear mapping is $e : G_1 X G_1 \rightarrow G_2$.
- It randomly chooses $s, t \in z_p^*$; two secret values, where t is master time key and s is master identity key. Then it computes $P_{\text{pub}} = sP$, $P_t = tP$. It keeps s secret and forwards t to the CSP
- The three hash functions defined are

$$H_0 : \{0,1\}^* \rightarrow G_1,\ H_i : \{0,1\}^* \rightarrow G_1,$$

$$H_j : G_2 \rightarrow \{0, 1\}^*,$$
$$H_k : \{0, 1\}^* X \{0, 1\}^* \rightarrow Z_p$$

- We publish the public parameters $PP = (G_1, G_2, p, P_{\text{pub}}, e, H_i, H_j, H_k)$

Initial Key Generation:
PKG sets the following for a user with the identity ID-$Q_{ID} = H_0(ID)$, $S_{ID} = sQ_{ID}$ and send this initial key S_{ID} through a secure manner to the user.

Time-Key Updation:
When CSP receives a request for key-update from the user with identity ID at the time period T_i, it computes $Q_{ID}, T_i = H_i(ID, T_i)$, $T_{ID,T_i} = tQ_{ID,T_i}$ and sends back T_{ID,T_i} to the authorized user.

Encryption:
If a user wants to encrypt his/her message m for identity ID, using his/her private key S_{ID}, public key Q_{ID} and system parameters, selects a random $\sigma \in \{0, 1\}^n$ and computes

$$r = H_k(\sigma, m), C_1 = rP,$$
$$C_2 = \sigma \oplus H_j((g_1.g_2)^r), C_3 = m \oplus H_j((g_1.g_2)^r)$$

$$where, g_1 = e(Q_{ID}, P_{Pub}), g_2 = e(Q_{ID,T_i}, P_t)$$

Then publish the ciphertext as CT $= < C_1, C_2, C_3 >$.

Decrypt:
To decrypt given ciphertext CT, for a period i, the receiver uses his/her identity S_{ID} and time update key T_{ID,T_i} to compute $\sigma = C_2 \oplus H_j(e(S_{ID} + T_{ID,T_i}, C_1))$, message $m = C_3 \oplus H_j(e(S_{ID} + T_{ID,T_i}, C_1))$, $r = H_k(\sigma, m)$.

The correctness of the decryption algorithm follows since

$$\begin{aligned} & C_3 \oplus H_j(e(S_{ID} + T_{ID,T_i}, C_1)) \\ & \quad = m \oplus H_j((g_1.g_2)^r) \oplus H_j(e(S_{ID} + T_{ID,T_i}, C_1)) \\ & \quad = m \oplus H_j((g_1.g_2)^r) \oplus H_j(e(g_1^r.g_2^r) \\ & \quad = m \end{aligned}$$

Where

$$\begin{aligned} & H_j(e(S_{ID} + T_{ID,T_i}, C_1)) \\ & \quad = H_j(e(S_{ID}, C_1).e(T_{ID,T_i} C_1)) \\ & \quad = H_j(e(sQ_{ID}, rP).e(tQ_{ID,T_i}, rP)) \\ & \quad = H_j(e(Q_{ID}, sP)^r.e(Q_{ID,T_i}, tP)^r) \\ & \quad = H_j(g_1^r.g_2^r) \end{aligned}$$

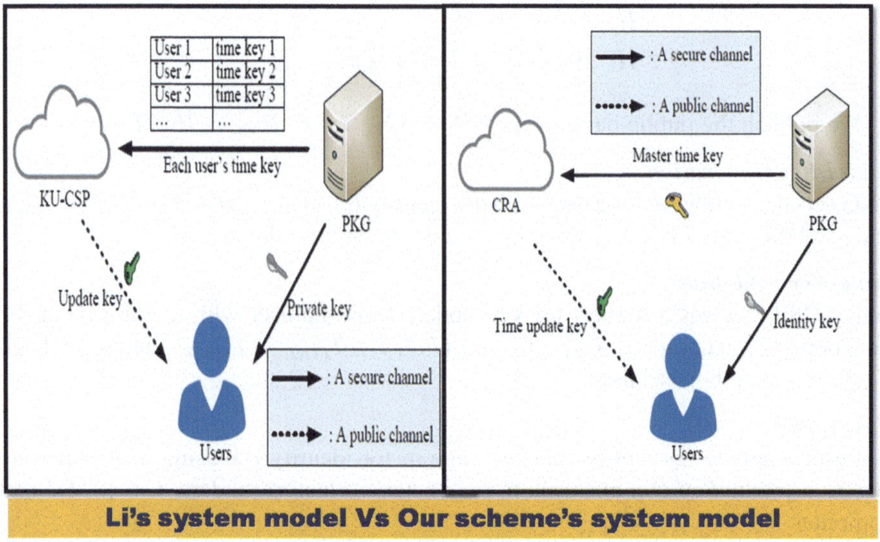

Fig. 32.1 Comparison of a system model of both the schemes

32.5 Comparative Study of Proposed Model

Considering the security of cloud data and identity concerns of users, we have performed a comparative study of our proposed model with previously developed Li's model [8]. As our scheme holds only one master key for all users, Li's scheme holds separate and unique key for every user. This single-key security approach of this proposed scheme offloads the burden of a large number of keys, to be shared for the secrecy of data. If there are a large number of users, then it becomes a herculean task to allot each user a separate key and manage it efficiently and safely. Our schemes definitely overcome this problem. A comparative system model of both the schemes are shown in Fig. 32.1.

Computation comparison is also studied for both the scheme and shown in Table 32.1. By going through the table, we can see that our proposed scheme is more efficient than Li's scheme [8].

32.6 Conclusion and Future Scope

We studied previously proposed revocable identity-based encryption schemes which were contributed to data security on the cloud. In this paper, we have constructed an IBE scheme based on [11] for cloud data security which requires less pairing and computation than Li's scheme [10] and hence it is much more efficient. For future studies, a security model with a double-key approach may be considered to avoid the

Table 32.1 Computation comparison

S no.	Computational comparison		
	Computation type	Li et al.'s scheme [10]	Our scheme
1	Computations in time update key	$H + 3\exp$	$H + m$
2	No. of key stored in the cloud authority	n	1
3	Computation involved in encryption	$P + 2H + m + 4\exp$	$2P + 3H + m + \exp$
4	Computation involved in decryption	$4p + 4m$	P

where H = Hash function; exp = exponentiation; P = Bilinear pairing; m = scalar multiplication in G_1

identity revocation and data security issues in the cloud. In this model, key-holding power may be delegated to CSPs equivalent to PKGs without revealing the identity of the user to CSP which ensures data security in that condition also when one of the keys gets compromised. Moreover, by using the forward security and double layering of keys, this scheme would require additional computation and complexity and will make cloud data more secure.

References

1. Shamir A (1985) Identity-based cryptosystems and signature schemes. In: Advances in cryptology, pp 47–53
2. Boneh D, Franklin M (2003) Identity-based encryption from the Weil pairing. SIAM J Comput 32:586–615
3. Gentry C (2006) Practical identity-based encryption without random oracles. In: Advances in cryptology-EUROCRYPT. Springer, pp 445–464
4. Canetti R, Halevi S, Katz J (2003) A forward-secure public-key encryption scheme. In: Biham E (ed) Advances in cryptology EUROCRYPT 2003. ser. LNCS, vol 2656. Springer, Berlin/Heidelberg, pp 646–646
5. Hanaoka Y, Hanaoka G, Shikata J, Imai H (2005) Identity based hierarchical strongly key-insulated encryption and its application. In: Roy B (ed) Advances in cryptology-ASIACRYPT. ser. LNCS, vol 3788. Springer, Berlin/Heidelberg, pp 495–514
6. Boldyreva A, Goyal V, Kumar V (2008) Identity based encryption with efficient revocation. In: Proceedings of 15th ACM conference on computer and communications security ser. CCS'08, ACM, New York, USA, pp 417–426
7. Sahai A, Waters B (2005) Fuzzy identity-based encryption. In: Cramer R (ed) Advances in cryptology EUROCRYPT 2005. ser. LNCS, vol 3494. Springer, Berlin/Heidelberg, pp 557–557
8. Li J, Li J, Chen X, Jia C, Lou W (2015) Identity based encryption with outsourced revocation in cloud computing. IEEE Trans Comput 64(2)
9. Libert B, Verngnaud D (2009) Adaptive id-secure revocable identity based encryption. In: Fischlin M (ed) Topics in cryptology CT-RSA 2009. ser. LNCS, vol 5473. Springer, Berlin/Heidelberg, pp 1–15

10. Li H, Dai Y, Dong B (2011) Identity-based cryptography for cloud security. IACR Cryptol e Print Arch 169–169
11. Verma V, Gupta D (2016) An efficient signcryption algorithm using bilinear mapping. In: Proceedings of 3rd international conference on computing for sustainable global development (INDIACom). IEEE Xplore Digital Library, pp 988–990

Chapter 33
An Assessment of Research Process Management Software Usage in Indian Higher Education

Sachin Kumar

33.1 Introduction

Scientific analysis and research output come out of the regress process which involves asking a question and setting hypothesis, literature survey, experiment design, experiment conduction, data collection and visualization, result analysis, conclusion, peer review process and publications, etc. [15]. After this process, knowledge generated is called validated knowledge and truth. It comes out after the debate for the reliable, validated systems of knowledge which is acceptable and consistent. This process results in the whole range of experiments and discussions. This led to the development of scientific methods based on research, evaluation and validation through empirical methods [20]. In addition to this more and more development took place for improvement in methods of validity and verification in different domains and measurement of accuracy came into being. With extensive work and applicability of the domain experts, researchers and philosophers started the concept of the systematic study and its verifiability. This process is also called the research cycle. The research cycle is defined as a sequence of the steps and phases in the research process starting from asking questions and getting the answers after the systematic development and scientific methods of study [6]. Over the period of the development of scientific discipline and technological advancement, especially in the area of information and communication technology (ICT), has developed some of the tools for making researchers' life easy which we use in daily life [4]. This has also some of the development in the research domain. As it is understood that there are two important perspectives of the research domain. One of them is finding the answers to the questions and testing the hypothesis. The second one is how to manage the complex processes spanned over many tasks and activities of research phases in different domains in different ways. Several studies have suggested that tools can help

S. Kumar (✉)
Cluster Innovation Centre, University of Delhi, Delhi, India
e-mail: officialid.sachin@gmail.com

P. K. Kapur et al. (eds.), *Decision Analytics Applications in Industry*, Asset Analytics,
https://doi.org/10.1007/978-981-15-3643-4_33

in minimization of the efforts and contribution to the actual research being done. Due to the complexity of the process and having many phases with several task and activities, it requires proper management of the tasks and proper sequencing. If some tools or software can help that is much better. In some of the studies, it has been established that researchers and young researchers lose a significant amount of time in such research processes and activities without proper management and this creates stress and frustration due to deadline and pressure for coming with something novel in the given time frame [5, 13]. So there is a need to develop efficient and productive processes with the integration of the research process management (RPM) tools and software [2]. These research software and tools are helpful and are used in a different set of activities and phases of the research process cycle. There are some stages of the research process such as setting the right questions, doing a literature survey, developing experiments and data collection, data analysis and data visualization and research analysis, writing a publication, storing the research documents and conclusions. In these steps of the research study, the software can help in managing the task efficiently and effectively. For example, Latex [16], Mendeley, Endnote [24], R Studio, Python, Pandas, Google, Google Scholar, Overleaf, Sharelatex, etc. [9] can contribute positively. Indian academic research output is not appreciable in terms of quality and quantity. There might be reasons related to RPM. This needs investigation and assessment. There is a need reason to understand and quantify the patterns of their uses in the Indian universities and among academics and researchers. Are these softwares part of the practice of the researchers and well integrated into the research process in universities and research institutions? This is the question worth asking. In addition to this, there is a need to understand the working habits and patterns of the researchers. Researchers in the west have employed such software and there is proper exposure of such tools to the researchers and professors. In Indian circumstances, the state of Indian higher education is not so well to be appreciated and need more significant changes. There are a clear-cut need and research gap in research process management. In some of the best performing institute which has foreign faculty and exposure, they might be using such tools and software but it is being observed that most of the Indian academicians and researchers are not using them and does not understand the research process management (RPM). So there is a need to validate this hypothesis and quantify the patterns and state of the present-day situation of RPM in Indian academic institutions with some exploratory research approach. This research uncovers the scenarios and patterns of the RPM state in researchers, region-wise, university-wise, subject-wise and according to their skills set knowledge, etc. This study proposed the assessment and quantification of the scenarios of the modern practice of RPM in Indian academicians.

This study is designed as follows. Section 33.2 describes a review of the literature. Section 33.3 states the specific problems and the research objectives. Section 33.4 introduces the methodology and experiment design. Section 33.5 deals with the research results, findings of and discuss them extensively. Section 33.6 concludes the study.

33.2 Literature Survey

Researchers across the world have been doing research to optimize the process-related activities so that productivity and efficiency can be improved. Moreover, researchers have done some specific studies to explore the problems and also suggested the solutions to make the process of research pipeline and process more conducive and less complex and easy for the researchers. Such research is called in the domain of research process management (RPM). There are many studies in the RPM and they have taken many different dimensions to understand the process so that efficiency and productivity can be improved. Some researchers have taken the study of the emotional aspect and the psychological problem faced by researchers and professors. In this process, Trace et al. [22] explored the information management in the humanities. Scholarship, processes and tools have also been explored and suggested some remedies. Lubke et al. worked on suggesting the way to improve the literature process. This helped in the development of the research process in a better way [18]. Aguinis worked on the managerial aspects of the methodology and proposed the different models and solutions in methodology for promoting transparency in management research [1]. Hart extensively worked in the area of literature review and explained the innovative ways of writing a literature survey using imaginations [12]. Joyner et al. proposed several ways of writing the best ways of the thesis in stepwise fashion [14]. In academic writing, plagiarism is bad practice and must be stopped to promote ethical and moral conscience of the community. In this respect, Harris mentioned various ways of using sources effectively [11]. Parija et al. discuss the writing and publishing of a scientific research paper [19]. Similarly, other researchers such as Grech and Cuschieri discussed the ways of writing scientific papers and also associated it with the career aspect-oriented skill [10]. Chambers, J. M mentioned that graphical models are better ways to communicate the results in scientific writing.

Turabian et al. wrote the manual for writing the research publications including papers, theses and dissertations [23]. Nardi et al. has worked extensively to gives systematic and efficient ways of doing survey research in the context of quantitative methods [3] Routledge. Dowling and Wilson studied the available online tools for the better management of exploratory study of Ph.D. candidates' use of online tools [8]. Levecque et al. study the problems of the Ph.D. students and researchers and developed the framework that gave insights about their problems leading to mental health problems [17]. There are several works which suggested the ways of managing the research projects as well though tools and new approaches such as Polonsky et al. Odena, Oscar et al. describe the way of doctoral students and graduates research and innovation experiences and strategies for their thesis writing learning process: a qualitative approach [21]. Devos et al. worked in the mental and psychological aspects of the researchers and Ph.D. students and study the effects of the completion of attrition on their level of sense, progress and distress [7]. After going through the literature survey, it is observed that the RPM-related study has not been conducted in Indian circumstances. So this study tries to address the research gap of the RPM in Indian academic and research institutions about their working pattern and use of research tools and software.

33.3 Problems Formulation

As has been discussed and explained in the introduction that due to several factors related to administration, autonomy, freedom, funding, exposure, Indian academic institutions are facing many problems. Academic institutions are being directed to increase their research output through several guidelines and promote the quality of research and innovation. But there are gaps in the approaches being used and their implementations. Here we have proposed a study to know about the awareness of the researchers and professors and young innovators in their respective domains in terms of their knowledge of managing their research process through tools and software to improve productivity and efficiency. The main aim of this study is to understand and analyze the working of the researchers and professors in managing their tasks and activities in the research cycle. The study tries to understand the patterns of use of specific softwares or tools in research by researchers and young innovators to increase their efficiency and productivity. Moreover, the following are the objectives: To know the status of knowledge of research process management. To know the pattern of use of software in several pages of research process management. To quantify the knowledge of the software used by researchers of different fields and their expertise in that software Location of the Institute of employer effects on the patterns of research process management and its software uses. To know about the quantification of software profession wise and their use pattern in the researcher process management. To know whether the institute or supervisors provide such courses to make the research easy and productive for the students.

33.4 Experimentation and Methodology

According to the established objectives and research gaps, in this project the collection of data is through survey methodology. This study basically deals extensively based on quantitative approaches for making the decision and asking the question relevant to establish facts and to gain the status of the working of researchers in terms of the research process. A detailed and deep questionnaire has been prepared to address the objectives and hypotheses to collect data in a proper manner. Sequentially, initially, the hypothesis is established to be tested or to explore something in quantification form. In this case, the target is to know about the status of working of the research domain and their management by the academic and research community. The discussion about the question and their responses has been prepared which has given us a total of two segments of the questions. One is based on personal and institutional information and the second is related to the RPM. In both of them, there are a total of 22 questions. Information about the researchers and professor is taken from the total 12 questions and the rest of the questions belong to the research process management. Some of the questions may have more than one answer and some of them are single-answer multiple-choice questions and some are representing true and false. The list of the question and their expected answer are given below. After that, the questions list have been created using Google forms so that more data can be collected and shared across the globe. The data was collected from several Facebook groups having different domain researchers and academicians such as Delhi University Teachers, CPDHE Participants WhatsApp groups and Professor in India to get the diverse data that makes the data rich in the context. After collection of data, the research analysis is done for each question and their correlation with other factors have been taken and analyzed in depth and discussions were made to come to the very fruitful and surprising answers and the state of working of academic and researchers. The following is the list of the questions and their responses.

33.5 Experiment Results and Discussion

This section covers the results after the data collection and their analysis. In this section, results are discussed and analyzed in the following manner. The study has taken a quantitative approach to get the state and the general trend in the researchers and academicians in India. The study has taken the location of the employers of the researchers and professors to know about the regional trends in the knowledge of RPM. The study also takes the country of the researchers to compare the trends of the Indian researchers with the rest of the world in RPM. In the research study, people from all across India participated. Education qualification wise, the statistics are given in Fig. 33.1, which states that most of the researchers are Ph.D. holders with 59.1% followed by 36.4% postgraduate and 4.5% having postdoctoral experience.

Steam-wise information is also required which needs to be taken into consideration of research process management. Science streams cover 63.6%, humanities with 13.6% and arts having the participation of 22.7% in the study. Pictorial representation is shown in Fig. 33.2.

Presently, most of the researchers are from India and they are from each region of the country such as North, South, North East, East, West and Central India. In the study, the teaching experience of the participants has also been incorporated. Total teaching experience varies from 1 year to 25 years of teaching experience and the majority of having 5 years of teaching experience. This says that most of the researchers are young and entering into the education and research system. When asked about whether they know about the RPM, their responses are given in Fig. 33.3 which states that 59.1% says Yes and 40.9% say No. This means that researchers understand the importance of the RPM. But the question is the execution of the managing and integration of it into real-life working of the researchers which can be understood from their working which has been explored from the next set of questions and data.

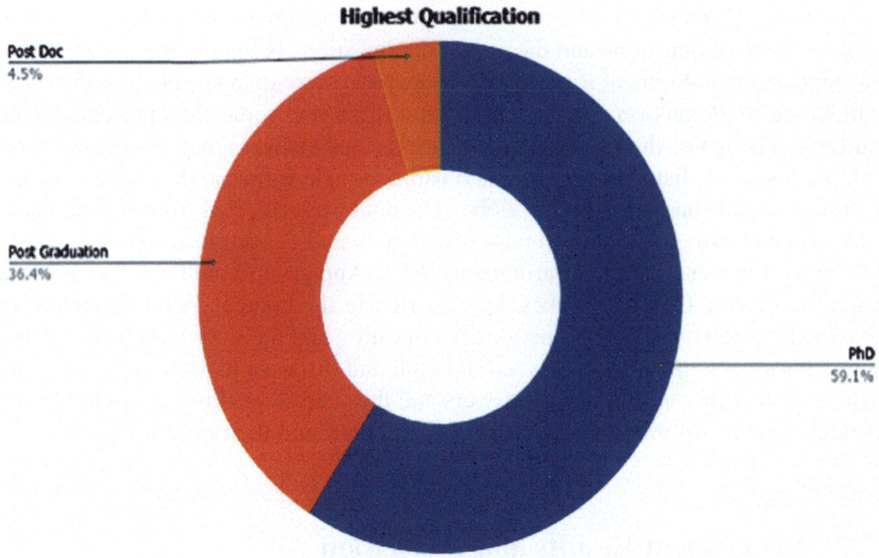

Fig. 33.1 Distribution of qualification of the researchers participated in the study

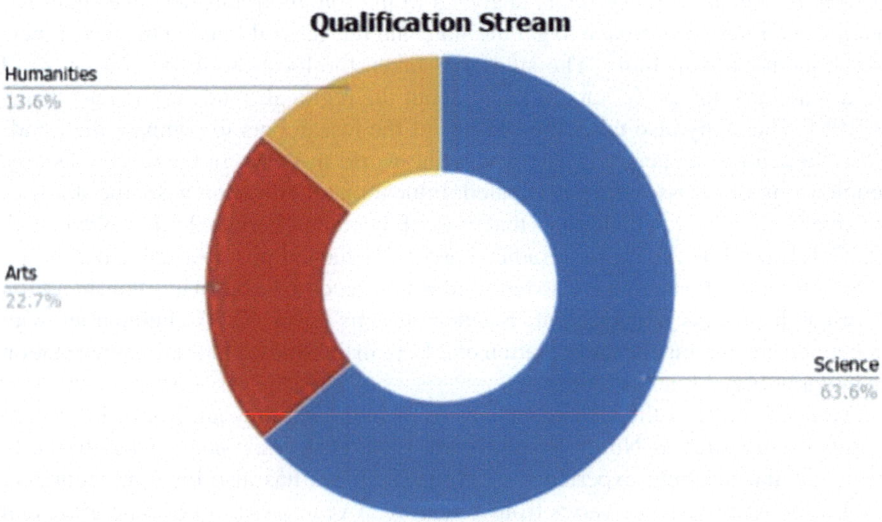

Fig. 33.2 Streams wise distribution

When asked about do they know any of the following software/tools for managing the research process and productivity improvement? The following responses have come as depicted in Fig. 33.4.

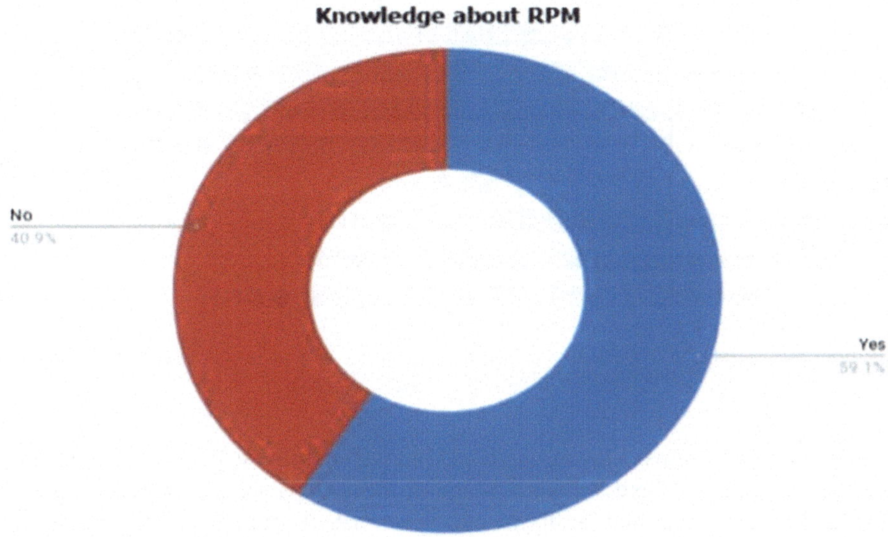

Fig. 33.3 Knowledge about RPM

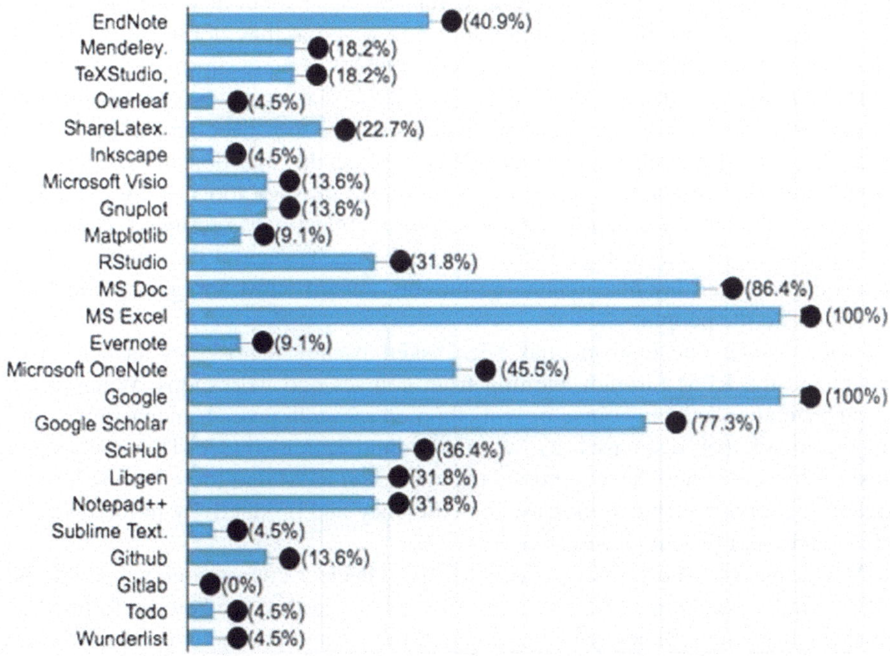

Fig. 33.4 Knowledge about the software

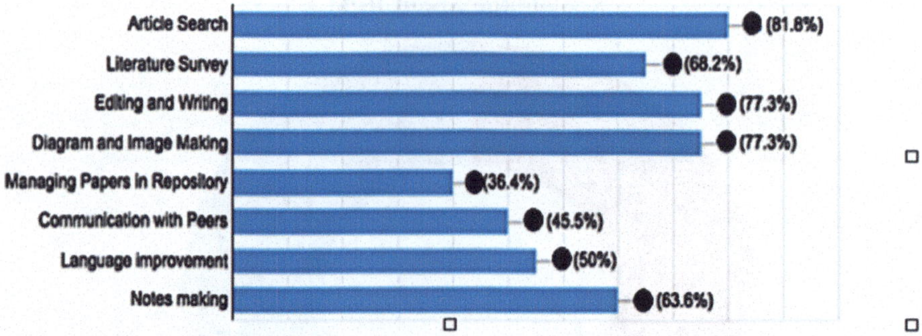

Fig. 33.5 Use of the software in different phases of research

There are many steps in the research and innovation process that can be managed through research tools and software. Some of these steps are the literature survey, editing phase, diagram making tools, etc. When asked about in which phases the participants use the tools or software? Their response was as given in Fig. 33.5. This means that maximum use is in the article search followed by the diagram-making and image-making and then literature search. Some of them also use the tools in notes making too. But the very important thing was that they don't use or lest use the software for the repository of research data which is quite surprising.

There is a need to understand if the researchers are providing some form of courses or workshops to the incoming students so that they become competent and productive in a research environment. When asked the question of whether you provide the information and hands-on experiments on such tools to your students doing research under your guidance, the response collection is presented in pictorial fashion in Yes and No in Fig. 33.5. Here 86% says Yes and 13% say No, which means that present-day researchers are moving toward RPM and giving hands-on experience to the people.

In this study, one question has been asked if they themselves have tried to understand the RPM through learning from courses and workshops. When asked if you have attended any course or workshop on research process management and tools/software, their response is 27.3% said Yes and 72.7% said No. This implies that most of the researchers have not done the course on RPM and there is little knowledge of the tools which can increase the efficiency and productivity in the research process and in different phases (Fig. 33.6).

Above results and discussion states that there is clear-cut research gap about the RPM in Indian academic and research institutions and that needs to be fulfilled as it will improve the efficiency of the research output and productivity of individual research or institution and will help in making the country stand tall in research output about in comparison of other countries. This is the best practice and must be adopted in Indian academic institutions through research methodology courses in postgraduate and Ph.D. courses and should be included in and orientation courses (Fig. 33.7).

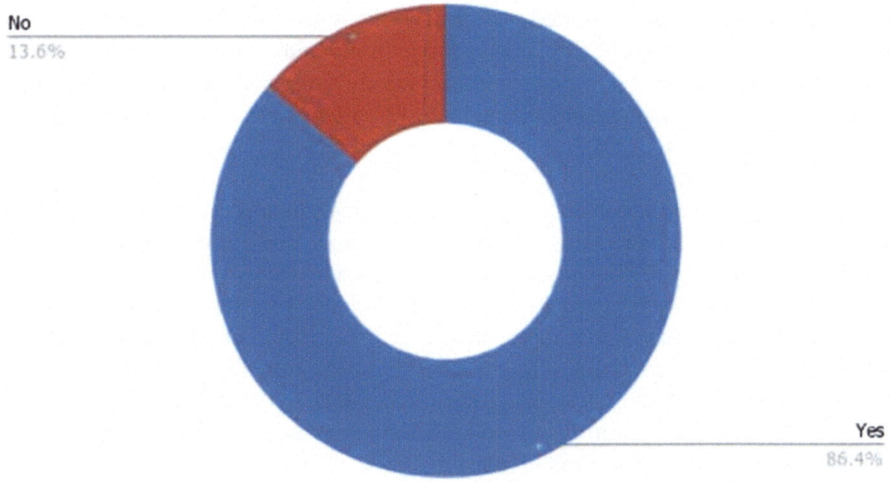

Fig. 33.6 Pattern of providing information to the student on RPM and Software

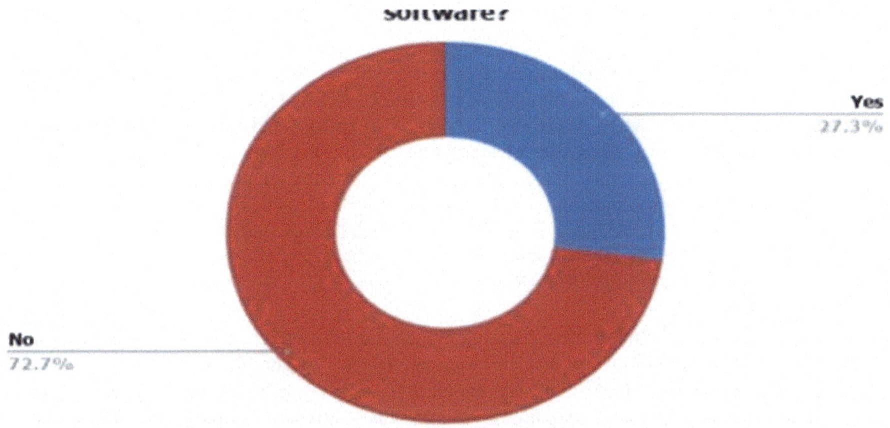

Fig. 33.7 Results showing attending workshop on RPM

33.6 Conclusion

This study explores the concept of RPM in the context of Indian circumstances and uses the quantitative approach based on surveys and questionnaires. Results obtained suggest that there is a need to address the research gap among Indian researchers, academician and young innovators. The study presents the status of knowledge of RPM and the types of software tools used by the researchers and their patterns. This demonstrates that most of the research is using simple tools such as Excel and Word for editing and image-making and for literature Google, not the Google Scholar. Some qualifications have been presented that most of the researchers does not understand the RPM and software tools. This makes productivity and efficiency very less in the research publication output. Science researchers are using more tools while in some science, they are not being used. People and researchers employed at distant from metropolitan cities have clearly shown the gap and lack of understanding about the processes and tools that must be fulfilled. The location of the institute clearly plays an important role in information access and technological advancement as stated by this study. Some supervisor provides such information and courses but that is maximum 50% which should be 100% and some policy decision on the part of AICTE and UGC must be taken to make it mandatory for the researches in refresher and orientation programs and research methodology courses. There is a limitation of the study. Right now this study is in the starting phase and this requires more data from all the states and other educational institutes. For comparative study in the world, there is a need to have data from other countries as well. This study in the next phase is designed to look after these aspects and limitations.

References

1. Aguinis H, Ramani RS, Alabduljader N (2018) What you see is what you get? Enhancing methodological transparency in management research. Acad Manag Ann 12(1):83–110
2. Benner MJ, Tushman ML (2003) Exploitation, exploration, and process management: the productivity dilemma revisited. Acad Manag Rev 28(2):238–256
3. Brace I (2018) Questionnaire design: how to plan, structure and write survey material for effective market research. Kogan Page Publishers
4. Camarinha-Matos LM, Afsarmanesh H (2005) Collaborative networks: a new scientific discipline. J Intell Manuf 16(4–5):439–452
5. Chen X, Anderson RC (2008) Reflections on becoming a successful researcher. Educ Psychol Rev 20(1):65–70
6. De Villiers MR (2005) Interpretive research models for informatics: action research, grounded theory, and the family of design-and development research. Alternation 12(2):10–52
7. Devos C, Boudrenghien G, Van der Linden N, Azzi A, Frenay M, Galand B, Klein O (2017) Doctoral students experiences leading to completion or attrition: a matter of sense, progress and distress. Eur J Psychol Educ 32(1):61–77
8. Dowling R, Wilson M (2017) Digital doctorates? an exploratory study of PhD candidates use of online tools. Innov Educ Teach Int 54(1):76–86
9. Falagas ME, Pitsouni EI, Malietzis GA, Pappas G (2008) Comparison of PubMed, Scopus, web of science, and Google scholar: strengths and weaknesses. FASEB J 22(2):338–342
10. Grech V, Cuschieri S (2018) Write a scientific paper (wasp)-a career-critical skill
11. Harris RA (2017) Using sources effectively: strengthening your writing and avoiding plagiarism. Routledge
12. Hart C (2018) Doing a literature review: releasing the research imagination. Sage
13. Herbert DL, Coveney J, Clarke P, Graves N, Barnett AG (2014) The impact of funding deadlines on personal workloads, stress and family relationships: a qualitative study of australian researchers. BMJ Open 4(3):e004462
14. Joyner RL, Rouse WA, Glatthorn AA (2018) Writing the winning thesis or dissertation: a step-by-step guide. Corwin Press
15. Kothari CR (2004) Research methodology: methods and techniques. New Age International

16. Lamport L (1994) LATEX: a document preparation system: user's guide and reference manual. Addison-Wesley
17. Levecque K, Anseel F, De Beuckelaer A, Van der Heyden J, Gisle L (2017) Work organization and mental health problems in PhD students. Res Policy 46(4):868–879
18. Lubke J, Paulus TM, Britt VG, Atkins DP (2017) Hacking the literature review: opportunities and innovations to improve the research process. Ref User Serv Q 56(4):285–295
19. Parija SC, Kate V (2017) Writing and publishing a scientific research paper. Springer
20. Patton MQ (2005) Qualitative research. Encyclopedia of statistics in behavioral science
21. Polonsky MJ, Waller DS (2018) Designing and managing a research project: a business student's guide. Sage Publications
22. Trace CB, Karadkar UP (2017) Information management in the humanities: scholarly processes, tools, and the construction of personal collections. J Assoc Inf Sci Technol 68(2):491–507
23. Turabian KL (2018) A manual for writers of research papers, thesis, and dissertations: Chicago style for students and researchers. University of Chicago Press
24. Yamakawa EK, Kubota FI, Beuren FH, Scalvenzi L, Miguel PAC (2014) Comparativo dos softwares de gerenciamento de referências bibliográficas: Mendeley, EndNote e Zotero. Transinformação 26(2):167–176

Chapter 34
Conical Conformal Antenna Array Using *SIW* Backed Feeding Technique for X Band Applications

Rehan Ahmed Siddiqui, Bishal Mishra, and Malay Ranjan Tripathy

34.1 Introduction

The Earth Observation satellites will serve advance services and more complex sensing techniques, which required the high data rate in downlink transmission [1]. But data transmission using the current data link in x band is (8.02–8.40 GHz) which is not sufficient. Also, the adjoining frequency bands are reserved for some sensitive space station receivers which must be protected against any interferences [2]. So that we can use the upper part of the x band, i.e. from 10 to 11 GHz for earth exploration satellite [3].

Antenna design for space application is different from the other applications. Different considerations are taken into account such as the atmosphere in which antenna is exposed, safety measurement, regulatory requirements and surface on which antenna will be installed. Antenna designed for this application must withstand static and dynamic stress. For high-speed aircraft, antennas must be designed aerodynamically to tackle physical stress on antenna [4].

In the past, many microstrip fed antenna arrays were studied. But the major problem with this configuration is that it suffers very high losses as frequency range increases [5]. Substrate integrated waveguide (SIW) technique was proposed which provides low-cost substitution for the higher frequency range [6]. SIW provides properties of rectangular waveguide. It also offers other advantages like compact size, low cost, lightweight and easy manufacturing using PCB. Microstrip and co-planar waveguide can also integrate with SIW for wideband transmission [7].

R. A. Siddiqui (✉) · B. Mishra · M. R. Tripathy
Department of Electronics and Communication, Amity University, Noida, India
e-mail: rsiddiqui07@gmail.com

B. Mishra
e-mail: bishal071294@gmail.com

M. R. Tripathy
e-mail: mrtripathy@amity.edu

P. K. Kapur et al. (eds.), *Decision Analytics Applications in Industry*, Asset Analytics,
https://doi.org/10.1007/978-981-15-3643-4_34

Table 34.1 The parameters of the proposed conformal tapered antenna array are given in this table

Parameter	Numeric value
Hight (h)	110 mm
Bottom radius	150 mm
Upper radius	80 mm
Diameter of SIW Via (d)	1.2 mm
Period between Via (p)	1.6 mm
Permittivity (ϵ_r)	2.2
Solution frequency	10.5 GHz

In this proposed design, the characteristics of design like the gain, return loss and radiation pattern with and without the SIW configuration for the conical conformal shaped antenna array is investigated. The design procedure involved carefully designing the SIW metallic vias so that we get the better performance of the conformal antenna array. The simulated results are carefully analyzed and explained so that this proposed antenna array can be used in satellite and space applications (Table 34.1).

34.2 Antenna Design

First, a rectangular planer patch antenna having dimension 45 × 35 mm is taken as the reference antenna. Then curved this patch on a conical platform having height 110 mm bottom and upper radius of the conical platform of 80 mm and 150 mm, respectively. The ground is taken as aluminum. Commercially available Roger RT/duroid 5880 ($\epsilon_r = 2.2$) having a thickness 5 mm is used as a substrate.

The proposed work includes the very fundamental design of a Simple Rectangular Waveguide in a conformal shape and its equivalent Substrate Integrated Waveguide (SIW). The most prolific constructional parameters are as follows:

- **p** as the period between vias
- **d** as the diameter of vias
- **a** width of rectangular waveguide
- $\mathbf{a_s}$ the width of the 'dielectric field-based metallic waveguide'.

where the reference patch antenna is as shown in Fig. 34.1. Structure of Rectangular Substrate Integrated Waveguide along with these considerations, the values of 'p' along 'd' are calculated as follows.

For a rectangular waveguide, cut-off frequency of arbitrary mode is found by the following formula:

$$f_c = \frac{c}{2\pi}\sqrt{\left(\frac{m\pi}{a}\right)^2 + \left(\frac{n\pi}{b}\right)^2} \tag{34.1}$$

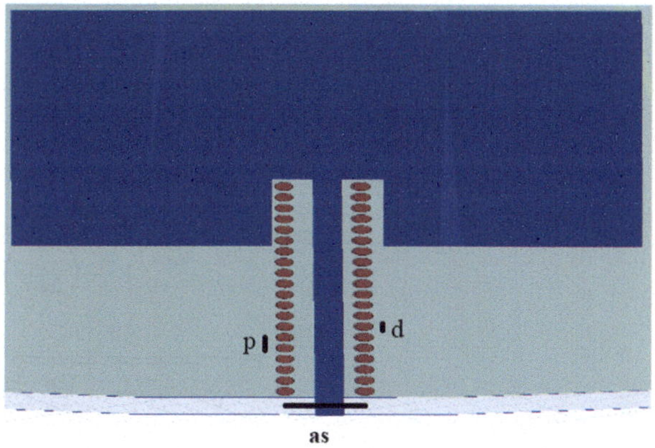

Fig. 34.1 Reference patch antenna

where

c speed of light
m, n mode numbers
a, b dimensions of the waveguide

For TE_{10} mode, the much-simplified version of this formula is

$$f_c = \frac{c}{2a} \tag{34.2}$$

For DFW with same cut off frequency, dimension "a_d" is found by

$$a_d = \frac{a}{\sqrt{\varepsilon_R}} \tag{34.3}$$

Having determined the dimension "a" for the DFW, we can now pass to the design equations for SIW.

$$a_s = a_d + \frac{d^2}{0.95p} \tag{34.4}$$

where

d diameter of the via.
p pitch (distance between the vias).

In this paper, for the design of SIW, the following two conditions are required:

$$d < \frac{\lambda_g}{5} \tag{34.5}$$

$$p > 2d \tag{34.6}$$

where the guided wavelength is

$$\lambda_{\text{guide}} = \frac{\lambda_{\text{freespace}}}{\sqrt{1 - \left(\frac{\lambda_{\text{freespace}}}{\lambda_{\text{cutoff}}}\right)^2}}$$

$$\lambda_{\text{guide}} = \frac{c}{f} \times \frac{1}{\sqrt{1 - \left(\frac{c}{2a \cdot f}\right)^2}} \tag{34.7}$$

$$\lambda_g = \frac{2\pi}{\sqrt{\frac{\varepsilon_R (2\pi f)^2}{c^2} - \left(\frac{\pi}{a}\right)^2}} \tag{34.8}$$

Copper metallic SIW via having diameter 1.2 mm and a center to center distance 1.6 mm are carefully chosen to get good matching and lower the return loss S_{11} (Fig. 34.2).

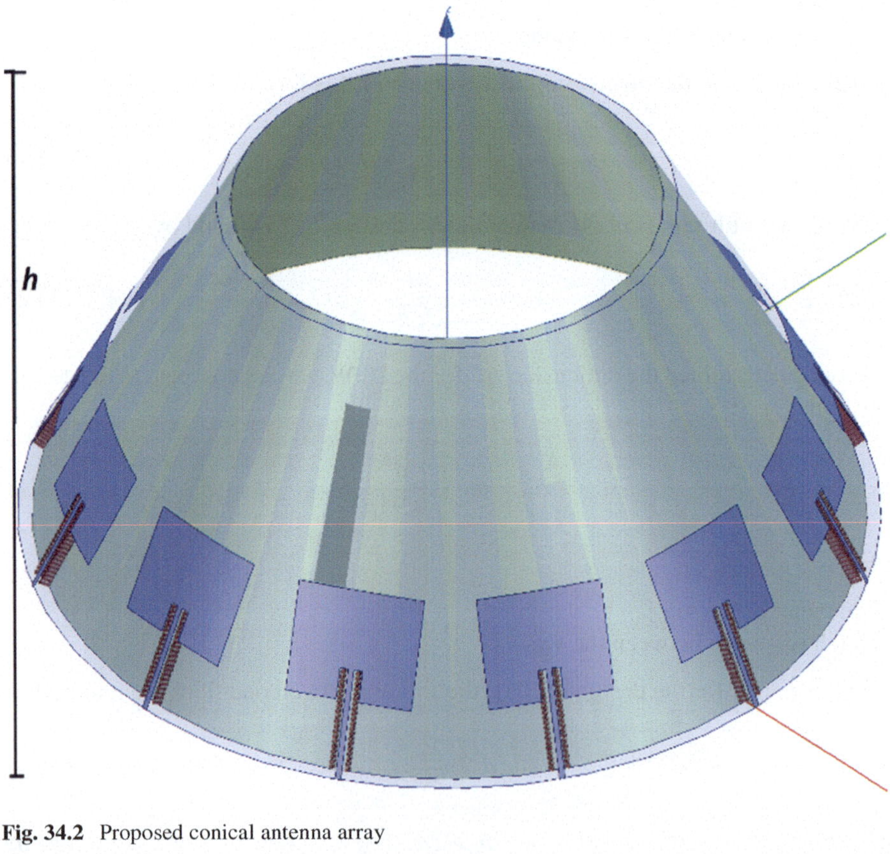

Fig. 34.2 Proposed conical antenna array

34.3 Result and Discussion

34.3.1 Impedance Bandwidth

Figure 34.3 shows the return loss of proposed design without SIW, the <-10 dB bandwidth is not obtained and the losses in this configuration are very high.

Figure 34.4 shows the return loss of the proposed antenna with the SIW backed feed line. <-10 dB bandwidth is 238.6 MHz from 10.378 to 10.616 GHz. The minimum return loss is −25.14 at 10.505 GHz. As shown in Fig. 34.4 good matching is obtained using the SIW at 10.5 GHz.

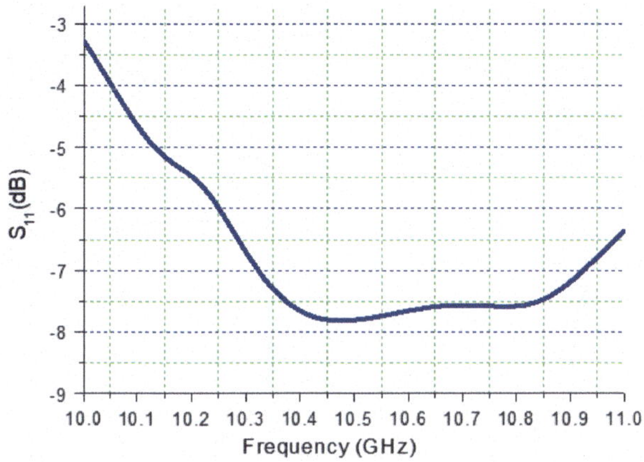

Fig. 34.3 Return loss S_{11} without SIW

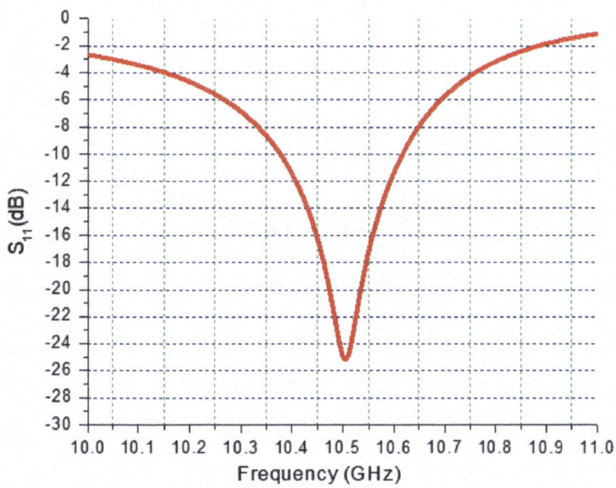

Fig. 34.4 Return loss S_{11} with SIW

34.3.2 Radiation Pattern and Efficiency

Figure 34.5 shows the radiation pattern of the proposed design without the SIW configuration. It is seen that the radiation pattern is in multiple directions with a maximum gain of 0.39 dB.

Figure 34.6 shows the radiation pattern of the proposed conical conformal antenna design at 10.5 GHz. The radiation pattern is more directional with the SIW configuration. The maximum peak gain of antenna is 7.20 dB.

Figure 34.5 shows the 3D radiation pattern of the proposed antenna array with SIW. It clearly is seen from Fig. 34.7 that the radiation pattern of the antenna is directional in the elevation plan.

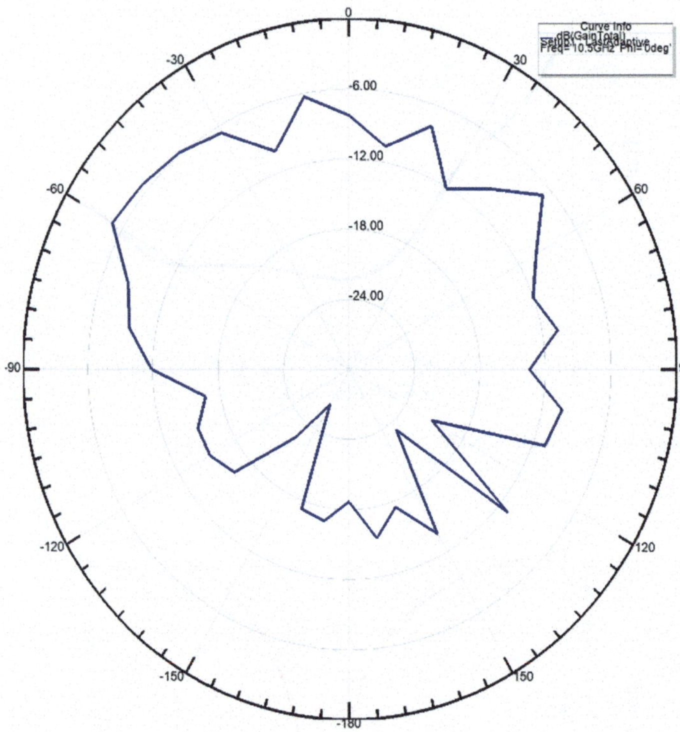

Fig. 34.5 Radiation pattern at 10.5 GHz without SIW

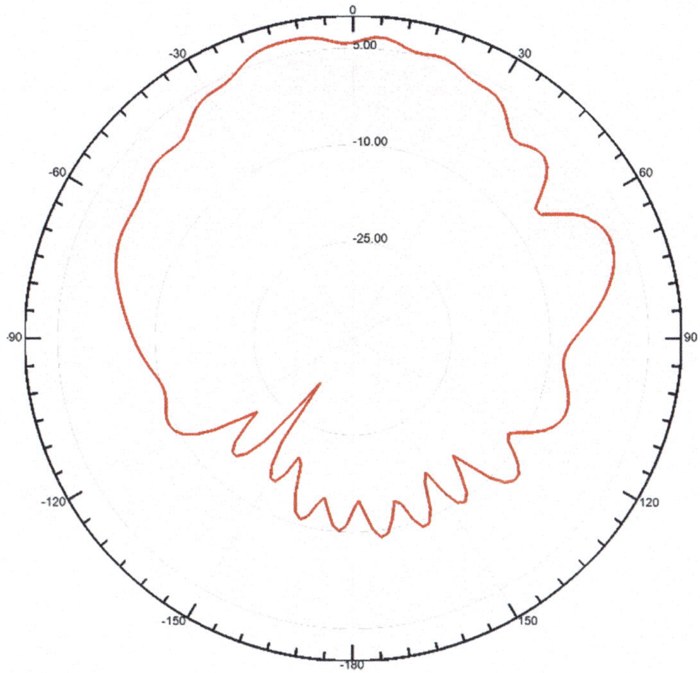

Fig. 34.6 Radiation pattern at 10.5 GHz

Figure 34.8 shows the radiation efficiency of the proposed design. At 10.5 GHz 91% efficiency is obtained. From 10.25 to 10.90 GHz the radiation efficiency of the proposed design is more than 85%.

34.4 Conclusion

The simulated results clearly show that the proposed conical conformal antenna array demonstrates good performance characteristics like increased gain and improved return loss at 10.50 GHz. This conformal array can be used in space applications like Earth exploration satellite and radar application.

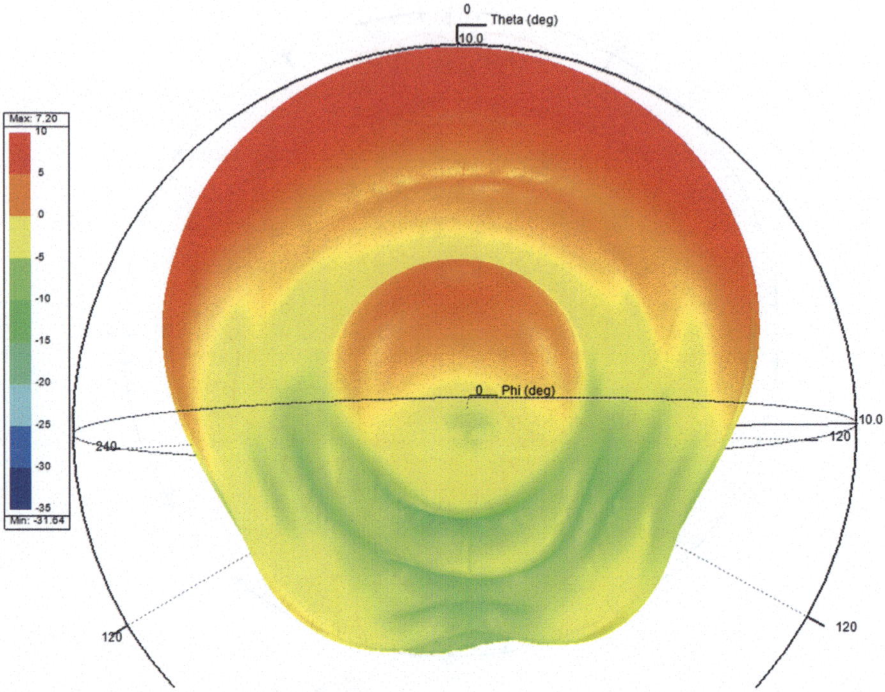

Fig. 34.7 3D radiation pattern at 10.50 GHz

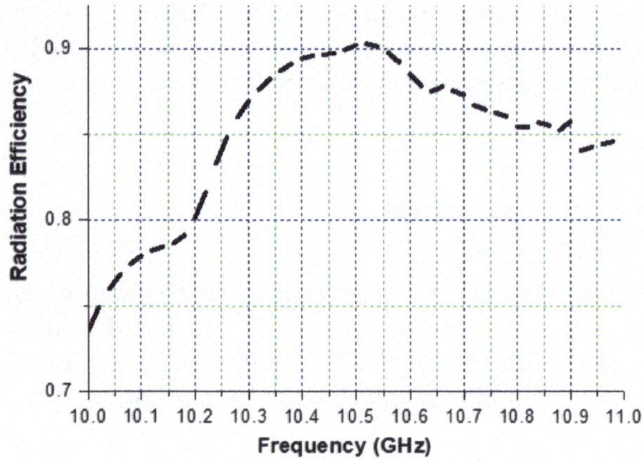

Fig. 34.8 Radiation efficiency

References

1. Shimada M, Tadono T, Rosenqvist A (2010) Advanced land observing satellite (ALOS) and monitoring global environmental change. Proc IEEE 98(5):780–799
2. Use of the frequency band 8025–8400 MHz by EESS Electronic Communications Committee (ECC) within the European conference of postal and telecommunications administrations (CEPT), ECC Rep. 115, February 2008
3. FCC online table of frequency allocation. http://transition.fcc.gov/oet/spectrum/table/fcctable
4. James JR, Hall PS (eds) (1989) Handbook of microstrip antennas. Peter Peregrinus, London
5. Wu K (2008) Substrate Integrated Circuits (SICs) for low-cost high-density integration of millimeter-wave wireless systems. In: Proceedings of RWS2008, January 2008, pp 683–686
6. Deslandes D, Wu K (2001) Integrated microstrip and rectangular waveguide in planar form. IEEE Microw Wirel Comp Lett 11:68–70
7. Yang S (2008) Antennas and arrays for mobile platforms direct broadcast satellite and wireless communication PhD dissertation, University of Tennessee, Knoxville

Chapter 35
Effect of Preservation Technology on Optimization of Two Warehouse Inventory Model for Deteriorating Items

Anjali Harit, Anurag Sharma, and Shivraj Singh

35.1 Introduction

For the last few years, a large number of researchers show interest in the effect of permissible delay in payment on optimum models. Since enterprises purchase a higher amount of goods than the volume available to carry goods in their OW. So, the enterprises stored excess quantity of goods in additional storage space, the RW. The rented warehouse charged higher units holding cost then OW. Many items deteriorate during the shortage period. It is generally considered in the inventory model that the management-owned warehouse is sufficient for the purchase of inventory to stock. However, it may be attractive to purchase more quantity of goods that can be kept in reserved owned warehouse and the rest of the goods are stored in rented storage. RW is situated far from OW. The management works on this application either when acquisition cost is more than the holding charges in RW or does it when there is an attractive amount discount for maximum purchases. Consumers are only served from OW inventory generally charged in RW is more than OW to diminish the holding shares of cost RW is first approved. This type of model was initially developed by Hartley [1]. He adopted a computational process to quantify the optimum order quantity under this hypothesis that the price of moving one item from RW to OW is not very high. Sarma [2] discussed a model in which he considers the cost of taking one unit from RW to OW and also presenting a solution process for determining an inventory model with two levels of storage. Agrawal et al. [5]

A. Harit (✉) · A. Sharma
DN College, Meerut, India
e-mail: anjaliharit94@gmail.com

A. Sharma
e-mail: sharmaadr@gmail.com

S. Singh
CCS University, Meerut, India
e-mail: shivrajpundir@gmail.com

P. K. Kapur et al. (eds.), *Decision Analytics Applications in Industry*, Asset Analytics,
https://doi.org/10.1007/978-981-15-3643-4_35

proposed an EOQ model with two warehouses having lifetime deteriorating items. They considered that the demand for model depends exponentially on time with a linearly dependent holding cost. The model is considered under inflation with shortage which is fulfilled by the backlogging rate which is taken as exponential function of time. Harit et al. [7] proposed a model with stock-dependent demand with the effect of preservation techniques for deteriorating goods under inflation. Omar and Zulkipli [8] proposed a model based on just-in-time inventory in which they consider single raw supplier and a single vendor who produces the products in batches at a finite rate and delivered the products to the buyer's warehouses in different shipments before they are shifted to the display area. Chakrabarty et al. [6] extended Liao and Huang [4] model. In this article, we extended Chakrabarty et al. [6] model by considering linear holding cost and preservation technology for deteriorating items in a rented warehouse which is a more realistic phenomenon. The article is organized structured as follows. Section 35.2 explains the assumptions and notations which have been used in this paper. Section 35.3 demonstrates the mathematical interpretation of the proposed model. Sections 35.4 and 35.5 model is illustrated by a numerical example and sensitivity analysis, respectively. Section 35.6 is devoted to concluding remarks and references.

35.2 Notation and Assumption

Notations and assumptions of this model are based on Chaudhari et al. [6] model.

35.2.1 Notation

C_o	Just-In-Time setup cost
$D(t)$	Demand rate at time
β	Constant deterioration rate
W	Volume of own-warehouse
T	Length of replenishment rotation
S	Selling cost at t time
C	Purchasing cost per unit time
$I_R(t), I_O(t)$	Inventory at t (units) time in RW and OW respectively
h_o	Storage cost charged by OW
h_r	Storage cost per unit time charged by RW
t_w	Time in which Inventory level in RW reaches zero
M	Credit period set by the supplier
I_c	Capital opportunity cost (as a percentage)
I_e	Earned interest rate (as a percentage)
H	Length of planning horizon
r	Discount rate

$R = (r - i)$	Constant discount rate of inflation
t_j	Time in which jth replenishment drops to zero
T_j	Total time including the jth replenishment rotation, $T_0 = 0$
$m(\xi)$	The preservation technology $\xi > 0$
τ_p	Resultant deterioration rate $\tau_p = (\theta - m(\xi))$
C_s	Shortage cost at time
N	Total Replenishments during the planning horizon, $N = H/T$
Q	Item's order level

Decision Variable

F Fraction of replenishment rotation where net stock is greater than zero.

35.2.2 Assumptions

1. The inventory planning horizon is finite. With the end of planning horizon, inventory level reaches zero.
2. The demand rate is dependent on time $D(t) = ae^{bt}$: $a, b > 0$.
3. The OW has fixed volume of W units and the RW has no specified volume to store items.
4. First, items kept in RW are consuming after then items in OW will be consumed.
5. Shortage is allowed and dissatisfied demand is partially backlogged. During the stock-out period, the backlogging rate depends on the length of time waiting until the next lot arrives. In this situation, the backlogging rate is defined as $e^{-\delta(T-t)}$, $\delta > 0$.
6. The number of replenishment is restricted to natural one.
7. The inventory costs in OW are lesser than those in RW.
8. To satisfy the shortage, the last order of last period will be placed.
9. Deterioration rate in Ow is considered constant. No replacement is allowed for deteriorating items.
10. Holding Cost in RW is linearly dependent on time while in OW it is constant.
11. Preservation technology 'τ_p' is used in RW only to reduce deterioration rate.
12. RW is near to OW s.t. travelling charge is negligible.

35.3 Mathematical Formulation

Here, we have discussed a situation of retailers on how they know whether to hire a RW or not to keep goods. We have order quantity $Q = I(0) = \frac{\alpha}{\alpha+b}\left(e^{(\alpha+b)t_1} - 1\right)$ and if the order quantity Q $\leq$ W then it is not necessary to hire any warehouse. Thus, $\frac{\alpha}{\alpha+b}\left(e^{(\alpha+b)t_1} - 1\right) \leq W$ implies $t_1 \leq \frac{\alpha}{\alpha+b}\left(e^{(\alpha+b)t_1} - 1\right)$. Let $t_a = \frac{\alpha}{\alpha+b}\left(e^{(\alpha+b)t_1} - 1\right)$.

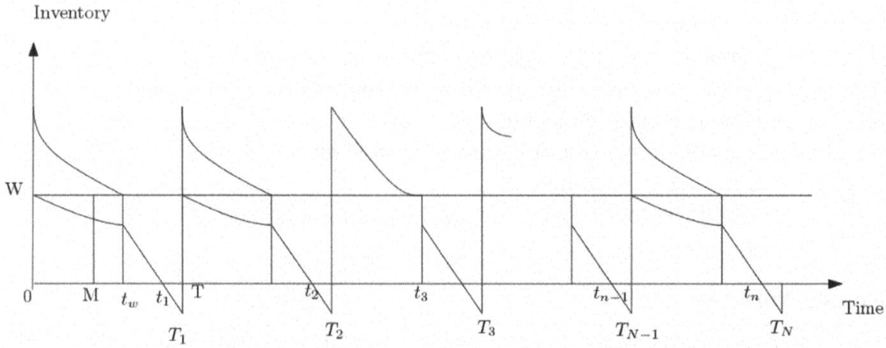

Fig. 35.1 Graph of inventory cycle

Therefore, the inequality Q > W holds iff $t_a < t_1$ which implies that there are W units of items stored in OW and the rest are shipped in RW and then the condition of two warehouse systems will arise.

Modelling ($t_a < t_1$)

The system includes two warehouse models with Q units. The first W units are stored in OW and placed in the RW. Therefore, during (0, t_w), the items in RW decrease because of the exponential demand rate and reaching zero levels. Deterioration rate is being stable by given conservation technique ($\theta - m(\xi)$). During the [0, t_w], the level of the inventory on the OW decreases by deterioration. Then during the [t_w, t_1], the level of item demand and decline in OW reached level zero. In the interval [t_1, T] commodity starts decreasing, the demand is partly with the backlog rate $e^{-\delta(T-t)}$ and after completing the demand the new replenishment cycle begins. Inventory status is shown in Fig. 35.1 given below.

Therefore, the inventory level for both the warehouse that can be described by the help of differential equation over [0, T] is given by:

$$\frac{dI_R(t)}{dt} + \tau_p I_R(t) = -ae^{bt} \quad 0 \le t \le t_w \tag{35.1}$$

$$\frac{dI_O(t)}{dt} = -aI_O(t); \quad 0 \le t \le t_w \tag{35.2}$$

$$\frac{dI_O(t)}{dt} + \alpha I_O(t) = -ae^{bt}; \quad t_w \le t \le t_1 \tag{35.3}$$

$$\frac{dI_O(t)}{dt} = -ae^{bt}e^{-\delta(T-t)}; \quad t_1 \le t \le T \tag{35.4}$$

With condition

$$I_R(t_w) = 0 = I(t_1), I_O(0) = W \tag{35.5}$$

The solution of above Eqs. (35.1)–(35.4) subject to the conditions (35.5) are given by

$$I_R(t) = \frac{a}{b+\tau_p}(e^{bt_W+\tau_p(t_W-t)} - e^{bt}); \quad 0 \le t \le t_w \tag{35.6}$$

$$I_O(t) = We^{-\alpha t}; \quad 0 \le t \le t_w \tag{35.7}$$

$$I_O(t) = \frac{a}{b+\alpha}(e^{bt_1+\alpha(t_1-t)} - e^{bt}); \quad t_W \le t \le t_1 \tag{35.8}$$

$$I(t) = \frac{ae^{-\delta T}}{b+\delta}(e^{(b+\delta)t_1} - e^{(b+\delta)t}); \quad t_1 \le t \le T \tag{35.9}$$

The extreme level of inventory in the starting of each rotation:

$$I_m = I_R(0) + I_O(0) = \frac{a}{b+\tau_p}(e^{(b+\tau_p)t_W} - 1) + \mathrm{W} \tag{35.10}$$

The extreme level of shortage in the starting of each rotation:

$$I_s = -I(T) = \frac{a}{b+\delta}(e^{bT} - e^{bt_1+(t_1-T)\delta}) \tag{35.11}$$

35.3.1 JIT Setup Cost

Since there is N replenishment, therefore, the Just-In-Time setup cost is on the planning horizon with inflation

$$C_{SET} = \sum_{J=1}^{N} C_O \frac{a(e^{bT}-1)}{bI_m} e^{-JRT} = C_O \frac{a(e^{bT}-1)}{bI_m}\left(\frac{e^{-RH}-1}{e^{\frac{-RH}{N}}-1}\right) \tag{35.12}$$

35.3.2 Holding Cost

Holding cost in RW

$$C_{RW} = \int_0^{t_1} (h_r + \lambda t) I_R(t) \mathrm{d}t = \frac{ah_r}{a+\tau_p}\left(-\frac{e^{bt_w}}{\tau_p} + \frac{e^{(b+\tau_p)t_w}}{\tau_p} - \frac{e^{bt_w}}{b} + \frac{1}{b}\right)$$

$$+\frac{a\lambda}{b+\tau_p}\left(-\frac{t_w e^{bt_w}}{\tau_p}-\frac{e^{bt_w}}{\tau_p^2}-\frac{t_w e^{bt_w}}{b}-\frac{e^{bt_w}}{b^2}+\frac{e^{(b+t_w)}\tau_p^2}{\tau_p^2}-\frac{1}{b^2}\right)$$

Holding cost in OW

$$C_{OW}=h_o\left(\int_0^{t_w} I_O(t)dt+\int_{t_w}^{t_1} I_O(t)dt\right)$$
$$=\frac{h_0 W}{\alpha}(1-e^{\alpha t_w})+\frac{ah_o}{b+\alpha}(\frac{1}{\alpha}(e^{(b+\alpha)t_1-\alpha t_w}-e^{bt_1})+\frac{1}{b}(e^{bt_w}-e^{bt_1}))$$

Total holding cost

$$C_{HC}=C_{RW}+C_{OW}$$
$$=\frac{ah_r}{a+\tau_p}\left(-\frac{e^{bt_w}}{\tau_p}+\frac{e^{(b+\tau_p)t_w}}{\tau_p}-\frac{e^{bt_w}}{b}+\frac{1}{b}\right)$$
$$+\frac{a\lambda}{b+\tau_p}\left(-\frac{t_w e^{bt_w}}{\tau_p}-\frac{e^{bt_w}}{\tau_p^2}-\frac{t_w e^{bt_w}}{b}-\frac{e^{bt_w}}{b^2}+\frac{e^{(b+t_w)}\tau_p^2}{\tau_p^2}-\frac{1}{b^2}\right)$$
$$+\frac{h_0 W}{\alpha}(1-e^{\alpha t_w})+\frac{ah_o}{b+\alpha}\left(\frac{1}{\alpha}(e^{(b+\alpha)t_1-\alpha t_w}-e^{bt_1})+\frac{1}{b}(e^{bt_w}-e^{bt_1})\right) \tag{35.13}$$

Total holding cost over planning horizon with inflation

$$C_{HCI}=\sum_{J=1}^{N-1}(C_{RW}+C_{OW})e^{-JRT}=(\frac{ah_r}{a+\tau_p}\left(-\frac{e^{bt_w}}{\tau_p}+\frac{e^{(b+\tau_p)t_w}}{\tau_p}-\frac{e^{bt_w}}{b}+\frac{1}{b}\right)$$
$$+\frac{a\lambda}{b+\tau_p}\left(-\frac{t_w e^{bt_w}}{\tau_p}-\frac{e^{bt_w}}{\tau_p^2}-\frac{t_w e^{bt_w}}{b}-\frac{e^{bt_w}}{b^2}+\frac{e^{(b+t_w)}\tau_p^2}{\tau_p^2}-\frac{1}{b^2}\right)$$
$$+\frac{h_0 W}{\alpha}(1-e^{\alpha t_w})+\frac{ah_o}{b+\alpha}(\frac{1}{\alpha}(e^{(b+\alpha)t_1-\alpha t_w}-e^{bt_1})$$
$$+\frac{1}{b}(e^{bt_w}-e^{bt_1})))(\frac{e^{-RH}-1}{e^{\frac{-RH}{N}}-1}) \tag{35.14}$$

Putting $T=\frac{H}{N}$ and $t_1=\frac{FH}{N}$

$$C_{HCI}=(\frac{ah_r}{a+\tau_p}\left(-\frac{e^{bt_w}}{\tau_p}+\frac{e^{(b+\tau_p)t_w}}{\tau_p}-\frac{e^{bt_w}}{b}+\frac{1}{b}\right)$$
$$+\frac{a\lambda}{b+\tau_p}\left(-\frac{t_w e^{bt_w}}{\tau_p}-\frac{e^{bt_w}}{\tau_p^2}-\frac{t_w e^{bt_w}}{b}-\frac{e^{bt_w}}{b^2}+\frac{e^{(b+t_w)}\tau_p^2}{\tau_p^2}-\frac{1}{b^2}\right)$$

$$+\frac{h_0 W}{\alpha}(1-e^{\alpha t_w})+\frac{ah_o}{b+\alpha}(\frac{1}{\alpha}\left(e^{(b+\alpha)\frac{FH}{N}-\alpha t_w}-e^{b\frac{FH}{N}}\right)$$

$$+\frac{1}{b}\left(e^{bt_w}-e^{b\frac{HF}{N}}\right)))(\frac{e^{-RH}-1}{e^{\frac{-RH}{N}}-1}) \tag{35.15}$$

35.3.3 Shortage Cost

Average reduction S should be determined first

$$S=\frac{1}{2}\int_{t_1}^{T}\frac{a}{b+\delta}\left(e^{bT}-e^{\delta(t_1-T)+bt_1}\right)\mathrm{d}t$$

$$=\frac{1}{2}\frac{a}{b+\delta}\left(e^{bT}-e^{\delta(t_1-T)+bt_1}\right)(T-t_1) \tag{35.16}$$

Total shortage cost over planning horizon with inflation

$$C_{SCI}=\sum_{J=1}^{N-1}SC_Se^{-JRT}=\frac{1}{2}\frac{a}{b+\delta}\left(e^{bT}-e^{\delta(t_1-T)+bt_1}\right)(T-t_1)\left(\frac{e^{-RH}-1}{e^{\frac{-RH}{N}}-1}\right)$$

Putting $T=\frac{H}{N}$ and $t_1=\frac{FH}{N}$

$$=\frac{1}{2}\frac{a}{b+\delta}\left(e^{\frac{bH}{N}}-e^{\delta(\frac{FH}{N}-\frac{H}{N})+\frac{bFH}{N}}\right)\left(\frac{H}{N}-\frac{FH}{N}\right)\left(\frac{e^{-RH}-1}{e^{\frac{-RH}{N}}-1}\right) \tag{35.17}$$

35.3.4 Purchasing Cost

Purchasing cost of the jth cycle C_{pj} is calculated as

$$C_{\mathrm{pj}}=CI_m+CI_se^{-T}$$

$$=\frac{Ca}{b+\tau_p}(e^{(b+\tau_p)t_W}-1)+CW+\frac{Ca}{b+\delta}(e^{bT}-e^{bt_1+(t_1-T)\delta})e^{-T} \tag{35.18}$$

Total purchasing cost over planning horizon with inflation is

$$C_{PUR}=\sum_{J=1}^{N-1}C_{pj}e^{-JRT}$$

$$= (\frac{Ca}{b+\tau_p}(e^{(b+\tau_p)t_W} - 1) + CW + \frac{Ca}{b+\delta}(e^{bT} - e^{bt_1+(t_1-T)\delta})e^{-T})\left(\frac{e^{-RH}-1}{e^{\frac{-RH}{N}}-1}\right)$$

$$\text{Putting } T = \frac{H}{N} t_1 = \frac{FH}{N}$$

$$= (\frac{Ca}{b+\tau_p}(e^{(b+\tau_p)t_W} - 1) + \mathrm{CW}$$

$$+ \frac{Ca}{b+\delta}(e^{b\frac{H}{N}} - e^{b\frac{FH}{N}+(\frac{FH}{N}-\frac{H}{N})\delta})e^{-\frac{H}{N}})\left(\frac{e^{-RH}-1}{e^{\frac{-RH}{N}}-1}\right) \quad (35.19)$$

35.3.5 *Interest Earned and Charged* ($M < T_W < T_1$)

Since the retailer sells the goods and collects the sales revenue continuously and 0 to M. During the period of the period, the interest earns with the rate. Again, for the retailer's stock, he starts compensating interest for the goods in M time with I_e rate.

$$I.E = I_e S\left(I_s M + \int_0^M Dtdt\right)$$

$$= I_e S(\frac{a}{b+\delta} M(e^{bT} - e^{bt_1+(t_1-T)\delta} + \frac{a}{b^2}(Mbe^{Mb} - e^{Mb} + 1)) \quad (35.20)$$

Total Interest received over the scheduling horizon with inflation is

$$TIE = \sum_{J=1}^{N-1} I.Ee^{-JRT}$$

$$= I_e S(\frac{a}{b+\delta} M(e^{bT} - e^{bt_1+(t_1-T)\delta} + \frac{a}{b^2}(Mbe^{Mb} - e^{Mb} + 1))\left(\frac{e^{-RH}-1}{e^{\frac{-RH}{N}}-1}\right)$$

$$\text{Putting } T = \frac{H}{N} \text{ and } t_1 = \frac{FH}{N}$$

$$= I_e S(\frac{a}{b+\delta} M(e^{b\frac{H}{N}} - e^{b\frac{FH}{N}+(\frac{FH}{N}-\frac{H}{N})\delta}$$

$$+ \frac{a}{b^2}(Mbe^{Mb} - e^{Mb} + 1))\left(\frac{e^{-RH}-1}{e^{\frac{-RH}{N}}-1}\right) \quad (35.21)$$

Interest payable is

$$\mathrm{I.}C = I_c C \int_M^{t_1} I(t)\mathrm{d}t = I_c C\left(\int_M^{t_W} I_R(t)dt + \int_M^{t_W} I_o(t)dt + \int_{t_W}^{t_1} I_o(t)dt\right)$$

Total interest payable over planning horizon with inflation

$$TIC = \sum_{J=1}^{N-1} I.Ce^{-JRT} = I_c C(\frac{a}{b+\tau_p}\left(-\frac{1}{\tau_p}\left(e^{bt_w} - e^{bt_w+\tau_p(t_w-M)}\right) - \frac{1}{b}\left(e^{bt_w} - e^{bM}\right)\right)$$
$$-\frac{1}{b}\left(e^{bt_w} - e^{bM}\right) - \frac{W}{\alpha}\left(e^{-\alpha t_w} - e^{-\alpha M}\right) + \frac{a}{b+\alpha}(\frac{1}{\alpha}(e^{b\frac{FH}{N}+\left(\frac{FH}{N}-t_w\right)\alpha} - e^{b\frac{FH}{N}})$$
$$-\frac{1}{b}\left(e^{b\frac{FH}{N}} - e^{t_w}\right)))(\frac{e^{-RH}-1}{e^{\frac{-RH}{N}}-1}) \tag{35.22}$$

Now, total cost function is

$$TC(F) = C_{SET} + C_{HCT} + C_{SHT} + C_{PUR} + TIC - TIE$$
$$= C_O \frac{a\left(e^{bT}-1\right)}{bI_m}(\frac{e^{-RH}-1}{e^{\frac{-RH}{N}}-1}) + (\frac{ah_r}{a+\tau_p}\left(-\frac{e^{bt_w}}{\tau_p} + \frac{e^{(b+\tau_p)t_w}}{\tau_p} - \frac{e^{bt_w}}{b} + \frac{1}{b}\right)$$
$$+\frac{a\lambda}{b+\tau_p}\left(-\frac{t_w e^{bt_w}}{\tau_p} - \frac{e^{bt_w}}{\tau_p^2} - \frac{t_w e^{bt_w}}{b} - \frac{e^{bt_w}}{b^2} + \frac{e^{(b+t_w)}\tau_p^2}{\tau_p^2} - \frac{1}{b^2}\right) + \frac{h_0 W}{\alpha}(1-e^{\alpha t_w})$$
$$+\frac{ah_o}{b+\alpha}(\frac{1}{\alpha}\left(e^{(b+\alpha)\frac{FH}{N}-\alpha t_w} - e^{b\frac{FH}{N}}\right) + \frac{1}{b}\left(e^{bt_w} - e^{b\frac{HF}{N}}\right)))(\frac{e^{-RH}-1}{e^{\frac{-RH}{N}}-1})$$
$$+\frac{1}{2}\frac{a}{b+\delta}\left(e^{b\frac{H}{N}} - e^{\delta\left(\frac{FH}{N}-\frac{H}{N}\right)+b\frac{FH}{N}}\right)\left(\frac{H}{N} - \frac{FH}{N}\right)\left(\frac{e^{-RH}-1}{e^{\frac{-RH}{N}}-1}\right)$$
$$+(\frac{Ca}{b+\tau_p}(e^{(b+\tau_p)t_W} - 1) + CW + \frac{Ca}{b+\delta}(e^{b\frac{H}{N}} - e^{b\frac{FH}{N}+\left(\frac{FH}{N}-\frac{H}{N}\right)\delta})e^{-\frac{H}{N}}\left(\frac{e^{-RH}-1}{e^{\frac{-RH}{N}}-1}\right)$$
$$+I_c C\left(\frac{a}{b+\tau_p}\left(-\frac{1}{\tau_p}\left(e^{bt_w} - e^{bM}\right)\right) - \frac{1}{b}\left(e^{bt_w} - e^{bM}\right) - \frac{W}{\alpha}\left(e^{-\alpha t_w} - e^{-\alpha M}\right)\right.$$
$$\left.+\frac{a}{b+\alpha}\left(\frac{1}{\alpha}\left(e^{b\frac{FH}{N}+\left(\frac{FH}{N}-t_w\right)\alpha} - e^{b\frac{FH}{N}}\right) - \frac{1}{b}\left(e^{b\frac{FH}{N}} - e^{bt_w}\right)\right)\right)\left(\frac{e^{-RH}-1}{e^{\frac{-RH}{N}}-1}\right)$$
$$-I_e S\left(\frac{a}{b+\delta}M(e^{b\frac{H}{N}} - e^{b\frac{FH}{N}+\left(\frac{FH}{N}-\frac{H}{N}\right)\delta} + \frac{a}{b^2}\left(Mbe^{Mb} - e^{Mb} - 1\right)\right)\left(\frac{e^{-RH}-1}{e^{\frac{-RH}{N}}-1}\right)$$

Since TC(F) is very large and complicated to differentiate by hand, Mathematica is used to solve it numerically. So, the sufficient condition $\frac{\partial^2 TC(F)}{\partial F^2} > 0$ must be satisfied.

35.4 Numerical Example

Consider an example: Let us take parameter values in the inventory system as $h_o =$ \$3, $h_r =$ \$1, $M = 1/12$, $\tau_p =$ \$0.4, $\alpha = 0.05$, $C_o =$ \$35, $C_s =$ \$15, R = .1, $\delta = 0.6$, H = 2, $I_e =$ \$0.35, $I_c =$ \$6.5, N = 2, a = 100, b = 4, s = 7, $t_w = .05$ and $\lambda = .02$ in suitable units. The optimum value TC(F) = 864346.2, F* = 0.332219 (Fig. 35.2).

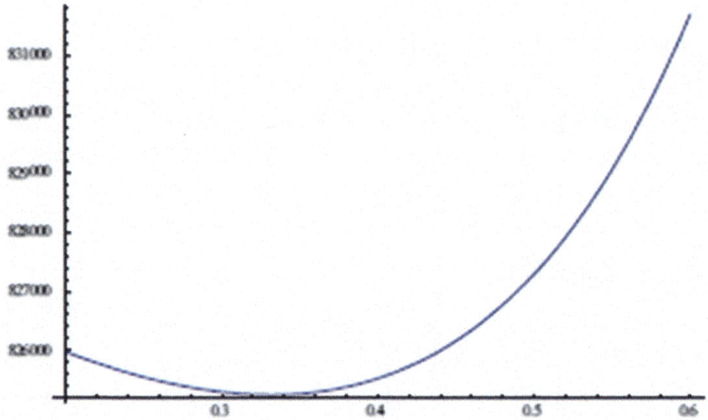

Fig. 35.2 Total cost function TC versus F

35.5 Sensitivity Analysis of the Above Example

Parameter	Change rates	F*	% effect in TC(F*)
h_o	+50	0.321522	0.0319
	+20	0.321522	0.0319
	−20	0.321522	0.0319
	−50	0.321522	0.0319
h_r	+50	0.321521	0.0316
	+20	0.321521	0.0317
	−20	0.321521	0.0319
	−50	0.321521	0.0320
λ	+50	0.321628	0.0312
	+20	0.321565	0.0315
	−20	0.321476	0.0321
	−50	0.321406	0.0326
τ_p	+50	0.321521	41.79
	+20	0.321521	14.95
	−20	0.321521	−12.91
	−50	0.321521	−29.11
C_s	+50	0.393128	0.629
	+20	0.380737	0.629
	−20	0.283503	−0.230
	−50	0.207192	−0.659
C_o	+50	0.321521	0.039
	+20	0.321521	0.035
	−20	0.321521	0.029
	−50	0.321521	0.025

(continued)

(continued)

Parameter	Change rates	F*	% effect in TC(F*)
R	+50	0.321521	−2.29
	+20	0.321521	−0.909
	−20	0.321521	9.92
	−50	0.321521	2.47
δ	+50	0.306866	−0.074
	+20	0.315522	−0.0112
	−20	0.327710	0.0762
	−50	0.337339	0.137
I_c	+50	0.250262	0.137
	+20	0.288989	0.082
	−20	0.362359	0.046
	−50	0.470482	−0.23
I_e	+50	0.321085	0.0062
	+20	0.321347	0.022
	−20	0.321696	0.042
	−50	0.321958	11.64
b	+50	0.466665	−81.54
	+20	0.391209	−79.31
	−20	0.237334	−87.31
	−50	0.021598	−81.01
a	+50	0.320410	−84.99
	+20	0.329065	−87.99
	−20	0.322357	−19.97
	−50	0.324874	−49.99

With this sensitivity table, it can be easily justified that the parameters show more effect on decision variable and total cost function. The sensitivity is taken into consideration by change rate of each parameter by −50%, −20%, +20% and + 50%. While finding the value of decision variable and total cost function for each changing rate, consider one parameter at one time change its value by applying the changing rate and keep other parameter fixed at that time. Repeat this process to get values for every parameter. With this, it can be observed that some parameters are more sensitive and some are less. Total cost TC(F) is directly proportional to the parameters $h\text{-}_o$, h_r, C_o, R, λ, I_c and $I_{e.}$ Other parameters are extremely sensitive to total cost as compared to these parameters. Hence the maximum and minimum value of total cost depends on these parameters.

35.6 Conclusion

In practice, we see that credit policy is an attractive feature in inventory management to purchase more items than RW capacity. The present study incorporates the concept of warehousing and conservation techniques. Analytical results suggest that there is a unique and better optimal lot size, which maximizes the expected rate per unit. Some realistic features such as inflationary effects, preservation technology on deteriorating items and lack of money and time are also taken into account. This model can be used for different types of products like food and cosmetics, etc. With the help of this study, we also demonstrate the retailer's optimum replenishment policies which reduce the total cost. Finally, the model has been illustrated by a numerical example and sensitivity analysis.

References

1. Hartely RV (1976) Operations research-a managerial emphasis. Good Year Publishing Company, California, pp 315–317
2. Sharma KVS (1983) A deterministic inventory model with two levels of storage and an optimum release rule. Opsearch 20:175–180
3. Liao JJ, Huang KN (2010) Deterministic inventory model for deteriorating items with trade credit financing and capacity constraints. Comput Ind Eng 59:611–618
4. Agarwal A, Sangal I, Singh SR, Rani S (2018) Two-warehouse inventory model for life time deterioration and inflation with exponential demand and partial lost sales. Int J Pure Appl Math 118(22):1253–1265
5. Chakrabarty R, Roy T, Chaudhari KS (2018) Two warehouse inventory model for deteriorating items with capacity constraints and back-ordering under financial considerations. Int J Appl Comput Math 58:1–16
6. Harit A, Sharma A, Singh SR (2019) Effect of preservation technology on optimization of two warehouse inventory model for deteriorating items with stock dependent demand under inflation. Int J Interdiscip Res Innov 7(2):587–600
7. Omar M, Zulkipli H (2019) An integrated just-in-time inventory system with stock-dependent demand. Bull Malaysian Math Soc Series 2, 37(4):1085–1098

Chapter 36
Techniques of Test Case Prioritization

Arpit Rajput, Anirudh, and Sheetal Joshi

36.1 Introduction

Software testing and prioritization are two coherent terms associated with software development. Software testing detects the bugs and flaws in the software module. It then validates and authenticates the software to meet the stakeholders' needs. It is a prominent step in the SDLC (Software Development Life Cycle) [1], which detects all the required functions that work properly or not and specifies the area of improvements. Prioritization is conducted for the test cases to determine faults and reduce test cases to maximize the sore objective function [10].

36.1.1 The Factors Affecting Prioritization Are as Follows

A. Customer requirements dependant techniques. Such prerequisite based strategies are techniques to straightforwardly organize test cases from necessity specified. The higher the values denotes the need for the required prioritization of test cases.
B. Coverage-based techniques, these are inclusion-based methods are auxiliary white-box testing procedures. "Coverage" in itself denotes the testing process like coverage of total, additional requirements.

A. Rajput (✉) · Anirudh · S. Joshi
ASET, Amity University, Noida, Uttar Pradesh, India
e-mail: arpitrajput666@gmail.com

Anirudh
e-mail: aniirudd@gmail.com

S. Joshi
e-mail: ssharma31@amity.edu

P. K. Kapur et al. (eds.), *Decision Analytics Applications in Industry*, Asset Analytics,
https://doi.org/10.1007/978-981-15-3643-4_36

C. Cost effective based techniques are strategies for organizing experiments dependent on just cost components, like the investigating cost, regression cost, analysis cost and cost of prioritization.
D. History-based techniques, on test execution history factors. Such factors can include past achievements which can act as a boon or a bane for future performances which will be relevant in progressing sessions [1–3].

36.2 Theoretical Prospects and Big Data Analytics

36.2.1 Study of Test Case Prioritization Technique Using AFPD

Factors that are considered for prioritization. In this technique, we consider two factors that are given below:

1. **Fault Detection Rate**:

$$\text{FRi} = (\text{Number of faults/Time}) * 10 \tag{36.1}$$

2. **Impact of fault**

In this technique, we give severity values to each fault on the basis of fault impact and type [9] (Fig. 36.1).

1. SM for complex: 9–10	2. SM for moderate: 6
3. SM for low: 4	4. SM for very low: 2
SM: Severity Measure	

Here, we assign a severity value to faults on the basis of its type. In a test, if a fault found in function, assigns 9 severity values to that test or if a fault found in data, assigns 3 severity value to that test. Severity values of various faults are [5] (Fig. 36.2, Table 36.1):

GUI 1	Environment 6
Checking 2	Assignment 7
Data 3	Unknown 8
Algorithm 4	Function 9
Interface 5	Timing/serialization 10

TSFD is defined as

$$\text{TSFD} = \sum i = 1 \text{ to n (SM)} \tag{36.2}$$

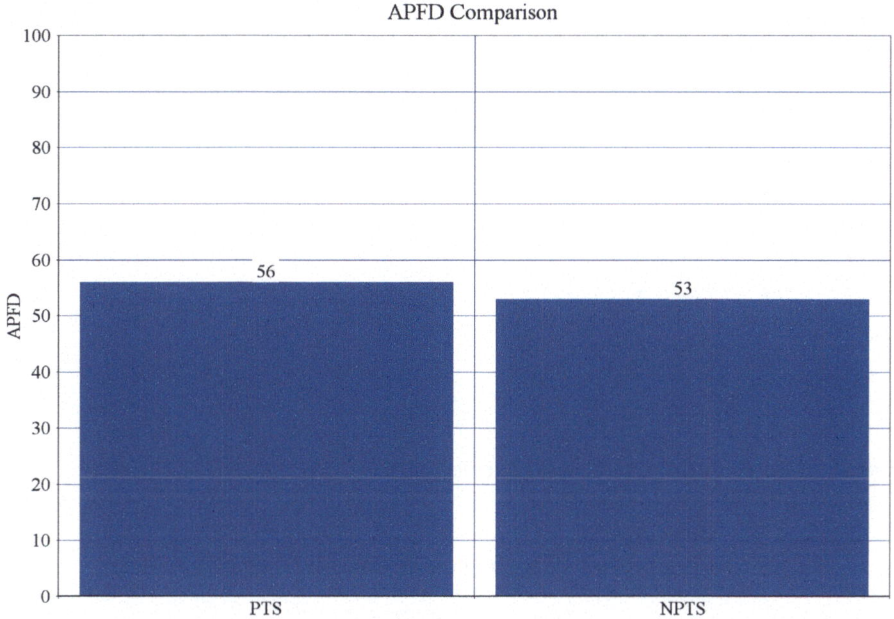

Fig. 36.1 Comparison based on APFD PTS: Prioritized Test Suit NPTS: Non-Prioritized Test Suit

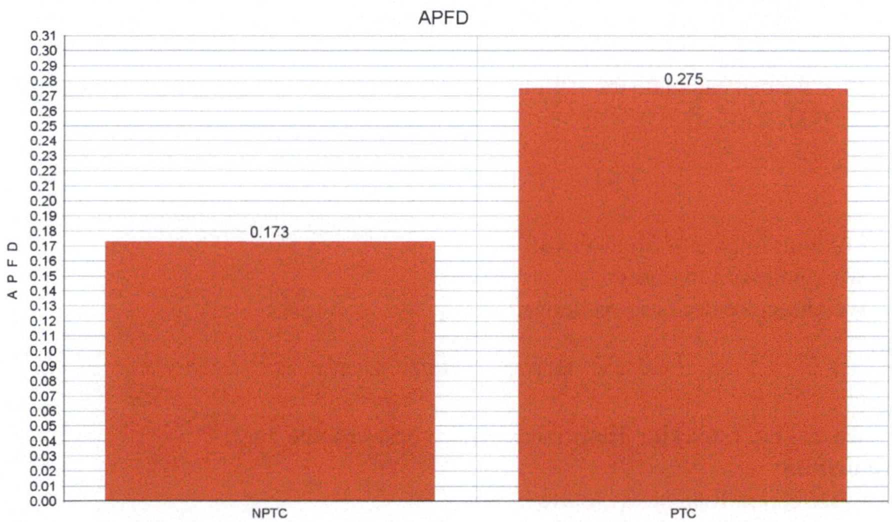

Fig. 36.2 Comparison based on APFD PTC: Prioritized Test Case NPTC: Non-Prioritized Test Case

Table 36.1 Calculating the severity of the fault

Test case	Assign severity value									
	1	2	3	4	5	6	7	8	9	10
CI	–	–				–				
CII					–			–		–
CIII			–						–	
CIV		–					–			
CV	–			–						
CVI		–							–	
CVII			–				–			–
CVIII					–					

TSFD: Total Severity of Faults Detected
n: number of faults
SM: Severity Measure
For the test cases severity values can be calculated by using Eq. (36.2)

$$\begin{array}{ll} TC1 = 9 & TC2 = 23 \\ TC3 = 12 & TC4 = 9 \\ TC5 = 5 & TC6 = 11 \\ TC7 = 20 & TC8 = 5 \end{array}$$

Now we can calculate the fault impact of ith test case using the above values [9] (Fig. 36.3):

$$FIi = (TSFD/MAX(TSFD) * 10 \quad (36.3)$$

FIi: fault impact of *i*th test case
Weightage of test cases:
Weightage of test cases is defined as [9]:

$$WTCi = FRTi * FIi \quad (36.4)$$

We use Eq. (36.4) for prioritizing the test cases (Table 36.2).
Algorithm:
Fault Detection Rate are as follows:

1. FRT1 = (3/2)*10 = 15 2. FRT2 = (1/3)*10 = 3.3
3. FRT3 = (1/2)*10 = 5 4. FRT4 = (2/3)*10 = 6.6
5. FRT5 = (1/8)*10 = 1.25 6. FRT6 = (1/9)*10 = 1.11
7. FRT7 = (3/7)*10 = 4.28 8. FRT8 = (2/9)*10 = 2.22

Fault impacts are as follows:

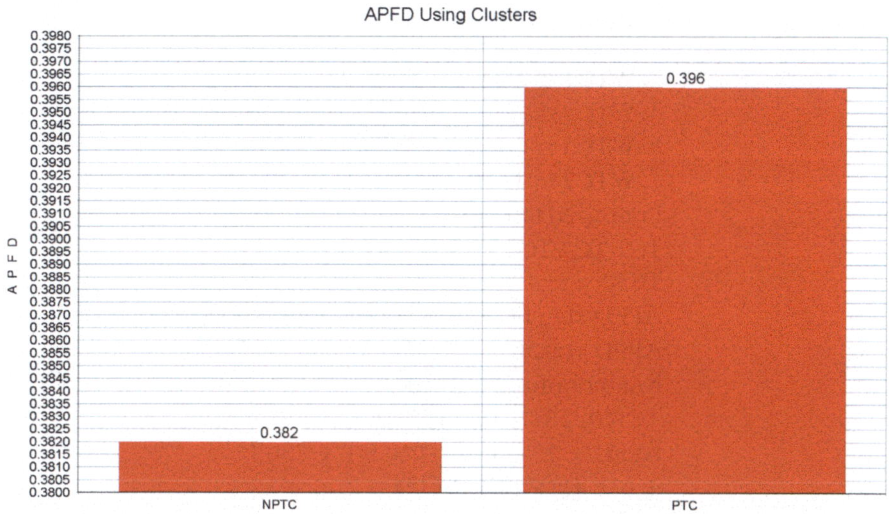

Fig. 36.3 APFD using clusters PTC: Prioritized Test Case NPTC: Non-Prioritized Test Case

Table 36.2 Fault matrix

Faults	Test cases							
	TC1	TC 2	TC3	TC 4	TC5	TC6	TC7	TC8
1	–		–		–			–
2		–						
3				–				
4	–				–			
5							–	
6	–		–					
7						–		–
8				–				
No. of faults	3	1	1	2	1	1	3	2
Time	2	3	2	3	8	9	7	9
Severity	9	23	12	9	5	11	20	5

1. FI1 = (9/23)*10 = 3.9
2. FI2 = (23/23)*10 = 10
3. FI3 = (12/23)*10 = 5.2
4. FI4 = (9/23)*10 = 3.9
5. FI5 = (5/23)*10 = 2.17
6. FI6 = (11/23)*10 = 4.78
7. FI7 = (20/23)*10 = 8.69
8. FI8 = (5/23)*10 = 2.17

Weightage of test cases are as follows:

1. WTC1 = 58.5 2. WTC2 = 33
3. WTC3 = 26 4. WTC4 = 25.74
5. WTC5 = 2.7 6. WTC6 = 5.25
7. WTC7 = 37.19 8. WTC8 = 4.81

Prioritized test order:
TC1 TC7 TC2 TC3 TC4 TC6 TC8
TC5
APFD=1−(1+3+5+1+2+1+6+5)/(8*8)+1/(2*8)
APFD = 0.56
Non-prioritized test order:
TC1 TC2 TC1 TC4 TC7 TC1 TC6 TC4
APFD = 1 − (1 + 2 + 4 + 1 + 7 + 1
+ 6 + 4)/(8*8) + 1/(2*8)
APFD = 0.53

36.2.2 Effectiveness of Prioritization Techniques Using APFD Techniques Metric

"The triangle program takes three integers x, y and z as input. These are taken to be the sides of the triangle. The output of the program can be a Scalene Isosceles, Equilateral Triangle or "Not a Triangle".

Testing techniques used here include scheduling by prioritizing the test cases to improve the regression techniques which will reduce errors and improves the quality of the software. The goal is to formulate the APFD. A metric through which it is determined by considering the weight on average of the percentage of faults obtained through the execution of the test suite [6]. The value of APFD will vary till 100. A higher value denotes that the fault detection rate is faster which are in turn better in technique. APFD can be calculated as follows [13]:

$$\text{APFD} = 1 - (\text{Tf1} + \text{Tf2} + \ldots + \text{Tfm})/\text{nm} + 1/2\text{n} \tag{36.5}$$

m = total number of faults

n = number of test cases and

TFi = position of the first test in T that exposes fault i

To demonstrate this technique, we take a program as given below (Table 36.3).

We set priority by calculating the rate of fault detection, here the rate is

$$\text{RTi} = \text{fault/time} \tag{36.6}$$

Table 36.3 Fault matrix

Faults	Test cases						
	TC 1	TC 2	TC 3	TC 4	TC 5	TC 6	TC 7
F1			f(0.13881)			f(0.1239375)	
F2		f(0.1338525)	f(0.1437675)	f(0.1140225)	f(0.1338525)	f(0.14128875)	f(0.13881)
F3	f(0.1536825)	f(0.13633125)	f(0.188385)	f(0.13881)		f(0.188385)	
F4	f(0.1635975)						f(0.16111875)
F5	f(0.18590625)	f(0.16111875)	f(0.188385)	f(0.16111875)	f(0.14128875)		f(0.1635975)
F6	f(0.188385)	f(0.17103375)			f(0.16111875)	f(0.188385)	
F7					f(0.188385)		f(0.188385)
No of faults	4	4	4	3	4	4	4
Time	0.69157125	0.60233575	0.6593475	0.413951	0.624645	0.64199625	0.65191125

$$RT1 = 4/.69157125 = 5.78393014 \quad RT5 = 4/.624645 = 6.4036372$$
$$RT2 = 4/.60233575 = 6.6408145 \quad RT6 = 4/.64199625 = 6.2305659$$
$$RT3 = 4/.6593475 = 6.0666037 \quad RT7 = 4/.65191125 = 6.135804529$$
$$RT4 = 3/.413951 = 7.2472345$$

Greater RTi means greater the priority of the test case. Therefore, we arranged the test cases priority order in descending order of RTi.
Prioritized order:
TC4 TC2 TC5 TC6 TC7 TC3 TC1

In the above program, the value of m is 7 and the value of n is 7. We calculate the APFD by putting these values into the following equation.

$$m = 7, n = 7$$
$$APFD = 1 - (Tf1 + Tf2 + \ldots + Tfm)/(nm) + 1/2n$$
$$APFD = 1 - (6 + 2 + 7 + 7 + 7 + 7 + 3)/(7 * 7) + 1/2 * 7$$
$$APFD = 0.275$$

Non-prioritized order:
TC7 TC6 TC5 TC4 TC3 TC2 TC1

$$APFD = 1 - (Tf1 + Tf2 + \ldots + Tfm)/(n * m) + 1/2n$$
$$APFD = 1 - ((5 + 6 + 7 + 7 + 7 + 7 + 3)/(7 * 7) + 1/2 * 7$$
$$APFD = 0.173$$

2.1 Study Of Test Case Prioritization Technique using clustering approach-

1. **Prioritized Test Suite S**
2. **Begin**
3. set S' as empty
4. **for each** case sS'
5. **do**
6. **Cluster =∑ i =1 to k Si**
7. calculate Product M as Rs*Fc
8. end **for loop**
9. Sort S in decreasing value of M for each value of test case
10. let us take S' as S
11. **end**

This algorithm is used to calculate the product metric where Rs is the statement coverage and Fc is the number of factions for every test case. Therefore, an algorithm is used to sort M the product in decreasing order. Sorting techniques may include quicksort, mergesort and heapsort [2] (Table 36.4).

Table 36.4 Clustering matrix

Clusters	C1	C2
Test cases	TC1	TC4
	TC2	TC5
	TC3	TC6
		TC7

Non-prioritized

$C1 = (TC1 + TC2 + TC3)/3$

$C1 = (0.69157125 + 0.60233575 + 0.6593475)/3$

$C1 = .6510848333$

$C2 = (TC4 + TC5 + TC6 + TC7)/4$

$C2 = 0.5831174375$

$(C1 + C2)/2 = 0.61710111354$

$Ci = 1 - 0.61710111354$

$Ci = 0.3828988646$

Prioritized

$C1 = (T4 + T2 + T5)/3 = 0.54697725$

$C2 = (T6 + T7 + T3 + T1)/4 = 0.661198125$

$(C1 + C2)/2 = 0.6040876875$

$Ci = 1 - 0.6040876875$

$Ci = 0.3959123125$

36.2.3 Optimization Using Ant Topology Techniques

This type of technique is based on the basic instinct of ants to search for food in their surroundings [1]. Ants are blind but still, they communicate and travel through teamwork without knowing the actual path. It achieves this by secreting a chemical substance called "Pheromone". The other ants in the army follow these traces by smelling the odor left behind by other ants [11]. ACO thus has two significant procedures which are deposition of pheromones and its evaporation. The former is based on random choice and the later on the equation. Equation to determine the path which can be related to the pheromone trail to beginning from a node.

Table 36.5 Comparison

Technique	Advantages	Disadvantages
Clustering algorithm	Simple to execute With an expansive number of factors, K-Means may be computationally speedier than different hierarchical procedures	Difficult to envision K-value It doesn't function admirably with (in the original data) different size and density
	Optimal answer to a problem can be found using cluster which can produce tighter results. The clusters are globular	With the global cluster, it didn't function admirably
Ant-topology	It can be used to find the smallest possible path for fault detection	Prioritization of the test cases through the statement coverability method has not been established
	Optimal answer to a problem can be found using artificial intelligence which has proposed searching codes and algorithms to get the desired solution	Past experience or knowledge about the subject is required for further go in analyzation
APFD (Advanced Percentage Fault Detection)	Requires certain criteria and procedures for better and accuracy of the result	The technique based solely on assumptions
	Equal and the same severity is required	Can provide unsatisfactory results is conditions not met
	All test cases used should have the same amount of its cost	Requires calculation which can cause time and cost delay

36.3 Result

By using the above-given technique, we get the following result, which shows that the test case suite for higher value of APFD covers the maximum faults and takes less time to check the faults in order to improve the software testing process. It has also come to notice that other empirical techniques used required a set of experience and knowledge [10] (Table 36.5).

36.4 Conclusion

Investigation improved the situation in organized and non-organized cases with the assistance of APFD [2]. It is demonstrated that after the organized instances are run at that point end result is extra talented. In the destiny, experiment prioritization can be completed by utilizing more factors and investigated by means of PTR, Weighted Defect Density (WDD), Defect Removal Effectiveness (DRE), Defect Removal Efficiency (DRE), Weighted Percentage Based on Fault Severity (WPFS) and chance measurements [13]. Usefulness of testing may be progressed in the least time and assets. This can improve a software program product. Organizing experiments will restrict the time, attempt and fee of large reimbursement together with the tester's labour which in flip reduces the complexities.

References

1. Vegas, S, Basili V (2005) A characterization schema for software testing techniques. Empir Softw Eng 10(4) 437–466
2. Sethi N, Rani S, Singh P (2014) Ants optimization for minimal test case selection and prioritization as to reduce the cost of regression testing. Int J Comput Appl
3. (2001) Prioritizing test cases for regression testing. IEEE Trans Softw Eng 27(10)
4. Broekman B, Notenboom, E (2003) Testing embedded software. Addison-Wesley
5. Srivastava PR (2008) Test case prioritization. J Theor Appl Inf Technol 178–181
6. Ammann P, Black PE, Ding W (2002) Model checkers in software testing technical report, National Institute of Standards and Technology
7. Nita M, Notkin D (2009) White-box approaches for improved testing and analysis of configurable software systems. IEEE
8. Janardhanudu G (2009) White box testing
9. Kapfhammer GM (2008) Software testing. In: ACM. [7] Oltsik J (2008) Black box testing and codenomicon DEFENSICS
10. Do H, Mirarab S, Tahvildari L, Rothermel G (2008) An empirical study of the effect of time constraints on the cost-benefits of regression testing. In: ACM
11. Saha GK (2008) Understanding software testing concepts. In: ACM
12. Kaur P Test case prioritization using ant colony, Thapar University
13. Badwal J (2013) Test case prioritization using clustering. In: International press conference, INPRESSO, vol 3, no 2
14. Pradeepa R (2013) Effectiveness of test case prioritization using APFD. In: ICRTCT-2013

Chapter 37
Condition Monitoring and Reliability Prediction of I&C Cables for Use in Probabilistic Safety Assessment of NPPs

T. V. Santhosh, V. Gopika, A. K. Ghosh, B. G. Fernandes, and J. Chattopadhyay

37.1 Introduction

The instrumentation and control (I&C) cables in nuclear power plants (NPPs) are considered to be long-lived components as they are not generally replaced during the entire design life. However, their performance assessment and remaining life estimation is an important aspect for safe and reliable operation of the plant. The assessment of long-term ageing performance has been practiced for years through accelerated life testing; however, correlating the laboratory accelerated ageing data with the actual field ageing data has still not been understood properly. The use of accelerated thermal ageing from the Arrhenius principles has many limitations and a few researchers have observed that a particular dose rate can have a different impact on the performance [1, 2]. A few studies have also reported that simulating the operating conditions and predicting the degradation is more challenging. The problem becomes even more complicated when manufacturers use their proprietary formulations with respect to antioxidants, flame retardants, filler materials, and important

T. V. Santhosh (✉) · V. Gopika · J. Chattopadhyay
Reactor Safety Division, Bhabha Atomic Research Centre, Mumbai, India
e-mail: santutv@barc.gov.in

V. Gopika
e-mail: vgopika@barc.gov.in

J. Chattopadhyay
e-mail: jchatt@barc.gov.in

A. K. Ghosh
Health Safety & Environment Group, Bhabha Atomic Research Centre, Mumbai, India
e-mail: ashok.ghosh@gmail.com

B. G. Fernandes
Department of Electrical Engineering, Indian Institute of Technology, Mumbai, India
e-mail: bgf@ee.iitb.ac.in

P. K. Kapur et al. (eds.), *Decision Analytics Applications in Industry*, Asset Analytics,
https://doi.org/10.1007/978-981-15-3643-4_37

chemicals to enhance the performance withstanding properties, and also to enhance the physical and mechanical properties. The type of geometry, design, and the fabrication process also affect the overall characteristics of ageing in insulation materials. Also, the combined effects due to thermal and radiation ageing have made difficult to understand this topic even further [3, 4]. A firm approach to modelling the performance of cable materials is, generally, from the condition monitoring (CM) techniques namely, the insulation resistance (IR) measurement, elongation-at-break (EAB) measurement, oxidation induction time (OIT), Fourier transform infrared spectroscopy, density measurement, etc. [5–7]. Although there are many remaining useful life estimation and CM techniques available in the literature, a well-established approach for incorporating the cable ageing into PSA is still not available. Therefore, the aim of this research work is to develop an integrated approach to assess the insulation degradation by subjecting to various ageing mechanisms and to eventually determine the time-dependent reliability for incorporating the cable ageing in PSA of NPPs.

IR measurement of four similar type I&C cables from various manufacturers was carried out prior to accelerated thermal and radiation ageing and their stability was measured. Due to slow kinetics, the exact oxidation peak could not be established from differential scanning calorimeter (DSC) heat flow signals under OIT measurement even after 8 h of continuous heating at constant isothermal temperatures in the range between 180 and 220 °C. However, a similar set of samples from the same manufacturers have shown a clear oxidation peak in DSC heat flow signals under oxidation onset temperature (OOT) measurement [8]. Hence, the isothermal OITs were derived from the OOT from the thermodynamic principles [9]. Fourier transform infrared spectroscopy (FTIR) was also performed on unaged and irradiated cable samples to study the structural damage in the insulation. The results obtained from OIT/OOT and FTIR studies are in good agreement with the EAB results. A large free volume fraction observed from positron annihilation lifetime spectroscopy (PALS) corresponding to 5 Mrad dose also suggests good correlation with other performance parameters. The micrographs from scanning electron microscopy (SEM) also indicate significant damage due to ageing, and strongly support the results from other CM approaches. This paper illustrates incorporating the cable ageing in PSA from an integrated framework proposed on the basis of employing CM approaches and Weibull distribution.

37.2 Experimental

Performance assessment of I&C cables involves measurement of two or more performance parameters which are correlated to the functional performance of the cable on which the measurements have been made [10, 11]. It is also possible to link the measured performance indicators with an independent parameter such as time or cycles in order to identify trends in the degradation state of the cable. Currently, various

CM approaches such as indenter modulus, OIT, OOT, infrared spectroscopy, density measurement have been found to be effective techniques for studying insulation degradation [12, 13].

37.2.1 Insulation Resistance Measurement

The low voltage cables have been selected for the performance assessment due to their extensive use in I&C applications. Four specimen cables for IR measurement are shown in Table 37.1. The insulation material is of type flame retardant low smoke (FRLS) PVC and sheath of type heat resistant (HR) PVC for all the cables.

The IR measurement was performed using a 10 kV high current IR tester (Megger S1-1052/2). Since all the selected cables are above 1100 V rated, the test voltage chosen was 1000 VDC for all the cables. The insulation materials were PVC with flame retardant additives in both insulation and sheath. The IR test was performed on a minimum five cores with respect to ground (remaining cores were grounded) on each multicore cable, and IR and polarization index (PI) were measured. PI is the ratio of IR measured corresponding to 10 min to the IR measured corresponding to 1 min. In order to compare the measured values against recommended values, the minimum recommended IR values as per IEC 60364-6 [14] are shown in Table 37.2 and the typical values of PI as per IEEE 43-2000 [15] are shown in Table 37.3.

Table 37.1 Cables for IR measurement

Cable	Specifications	Insulation	Sheath
Manufacturer 1-control cable	10 core, 1.5 mm^2, 1100 V	FRLS PVC	HR PVC
Manufacturer 2-signal cable 1	37 core, 1.5 mm^2, 1100 V	FRLS PVC	HR PVC
Manufacturer 2-signal cable 2	19 core, 1.5 mm^2, 1100 V	FRLS PVC	HR PVC
Manufacturer 3-control cable	37 core, 1.5 mm^2, 1100 V	FRLS PVC	HR PVC

Table 37.2 Minimum IR values

Nominal voltage (Vac)	Test voltage (Vdc)	Insulation resistance (MΩ)
Extra low voltage	250	≥0.5
Up to 500 V	500	≥1.0
Above 500 V	1000	≥1.0

Table 37.3 Typical PI values

PI	Insulation condition
<2	Problematic
2–4	Good
>4	Excellent

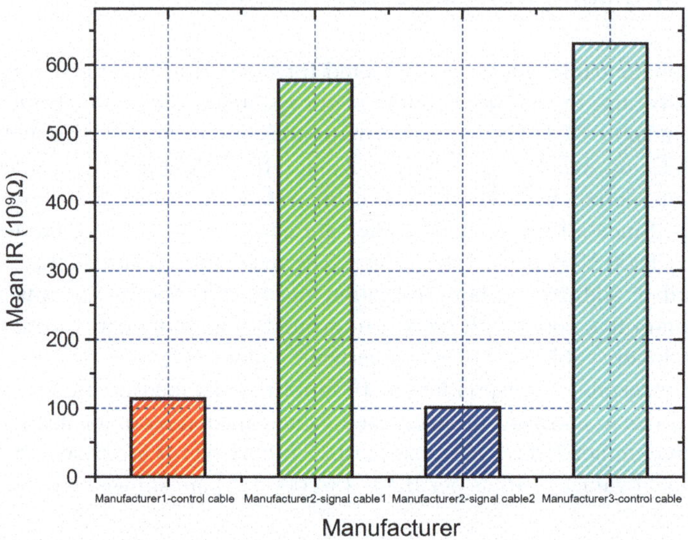

Fig. 37.1 IR comparison

The IR and PI values have been compared amongst the manufacturers for studying the manufacturing process variation. The bar charts of IR and PI comparison are shown in Figs. 37.1 and 37.2, respectively.

It is observed from Figs. 37.1 and 37.2 that the control cable from Manufacturer 3 demonstrates relatively high IR and PI values. Though the IR of signal cable 1 from Manufacturer 2 is significantly high, its PI is much too low. It means that the signal cable 2 from Manufacturer 2 has significant steady-state losses. It is also evident from Fig. 37.1 that the IR values of all cables are well above the minimum recommended values. However, the PI values of all cables are below typical values. It can be noted here that the PI values given in Table 37.3 are considered typical but not minimum recommended values, unlike IR.

As insulation materials in all the cables are of type PVC, comparison between signal and control cable was not envisaged. Also, there was no clear demarcation of signal and control cables from the results except a minor increase in PI values for control cables. As PI measurement is insensitive to temperature and considered to be an accepted index for assessing the quality of the cable, the Manufacturer 1 control cable and Manufacturer 2 signal cable 1 have conservatively been selected for radiation and thermal ageing analysis and subsequent reliability prediction.

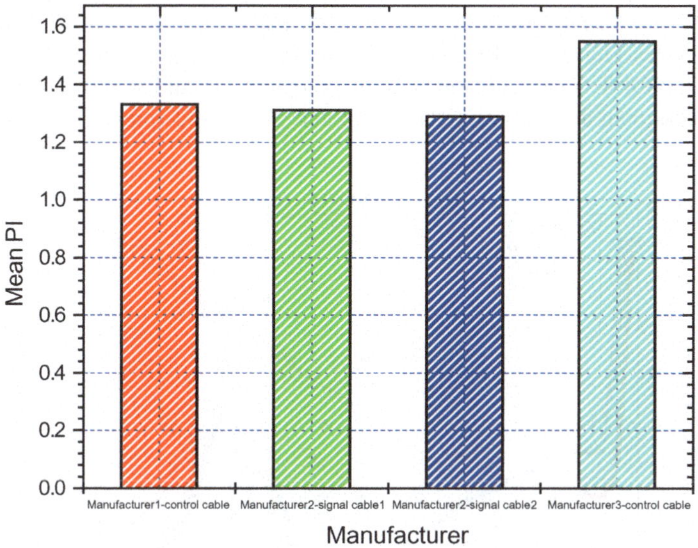

Fig. 37.2 PI comparison

37.2.2 Accelerated Radiation Ageing

The specification of the control cable selected for accelerated radiation ageing experiment is shown in Table 37.4.

The radiation ageing was performed according to IEC 544 [16–18] under aerated condition at room temperature with dose rate of equal to 0.17 Mrad/h in a gamma chamber having Cobalt-60 source. Experiment was conducted on samples made from insulation and sheath materials by exposing to doses starting from 2.5 to 50 Mrad in order to simulate the various doses normally expected during normal and accidental conditions of NPP.

37.2.2.1 Measurement of OIT and OOT

The OIT measurements were carried out between 180 and 220 °C, and up to 8 h on unaged and irradiated materials according to ASTM D3895-07 standard [19] using SETARAM DSC131 DSC. However, the exact peak corresponding to an exothermic

Table 37.4 Specification of control cable for accelerated radiation ageing

Cable	Specifications	Insulation	Sheath
Manufacturer 1-control cable	10 core, 1.5 mm^2, 1100 V	FRLS PVC	HR PVC

reaction could not be determined from the DSC heat flow signals due to slow kinetics in these samples. Hence, OOT measurement was employed to determine the OIT from the measured OOT. The heat flow signals of insulation and sheath obtained under OOT measurement are shown in Figs. 37.3 and 37.4 respectively. It is evident from

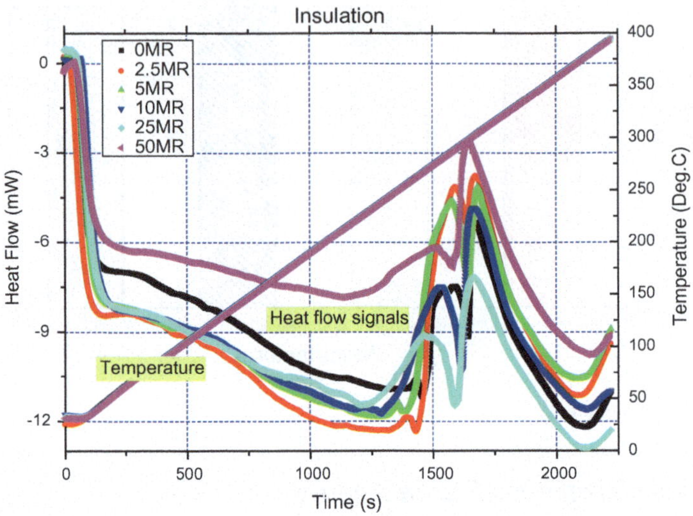

Fig. 37.3 Heat flow signals of insulation

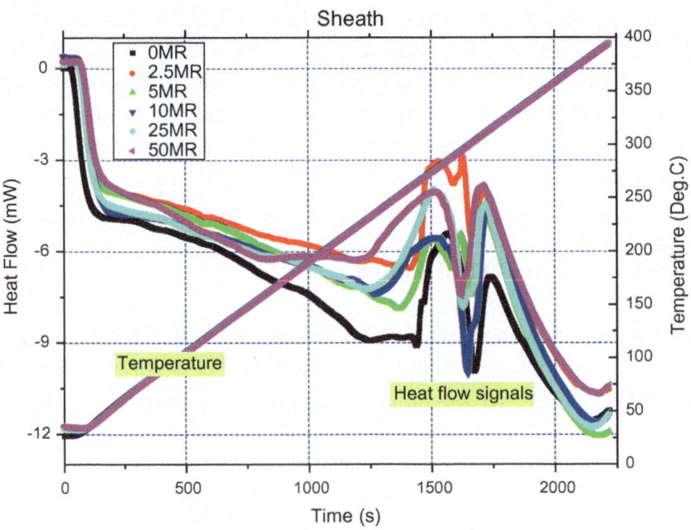

Fig. 37.4 Heat flow signals of sheath

the heat flow signals that a distinct exothermic peak exists between 200 and 280 °C in both insulation and sheath at various dose conditions.

Since the insulation materials are composed of several other additives in addition to base polymers (in this case the base polymer is PVC), it is possible that there could be several other peaks that are insignificant with respect to the strong oxidation of the sample.

37.2.2.2 Elongation-at-Break Measurement

The elongation-at-break measurements were carried out using Tinius Olsen Universal Testing Machine accompanying a load cell with 10 kN according to ASTM D638 [20]. The experiments were carried out at a strain rate of 25 mm/min with a gauge length of 25 mm. The experimental data was then analyzed to study the ageing effect in both insulation and sheath in order to assess the degradation and subsequent reliability estimation. The EAB as a function of radiation dose for insulation and sheath are shown in Figs. 37.5 and 37.6 respectively. The coefficient of determination for the exponential fit for insulation and sheath respectively are of 0.58 and 0.97.

From Figs. 37.5 and 37.6, it is evident that though the ageing phenomenon is exponential in both insulation and sheath, the degradation rates are entirely different. The variation is mainly due to the antioxidant and other additives present in each material. It is also evident in the case of sheath that the kinetics of the exothermic reaction is slow compared to insulation. This justifies the use of many other physical and mechanical strength improvement additives in sheath. The life-stress relationship

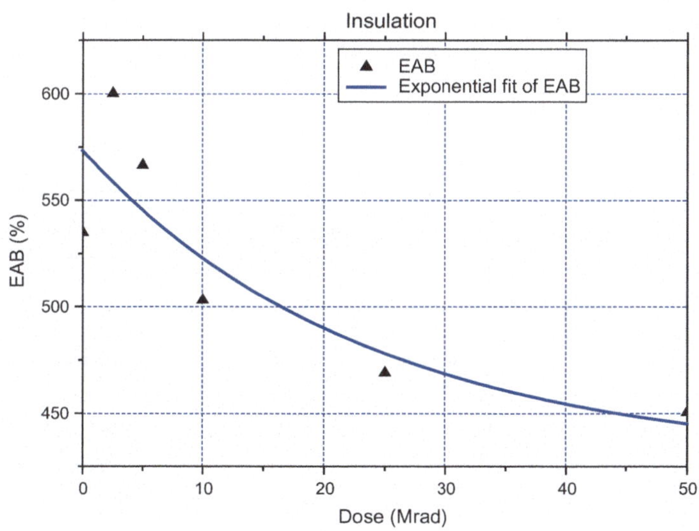

Fig. 37.5 EAB versus dose for insulation

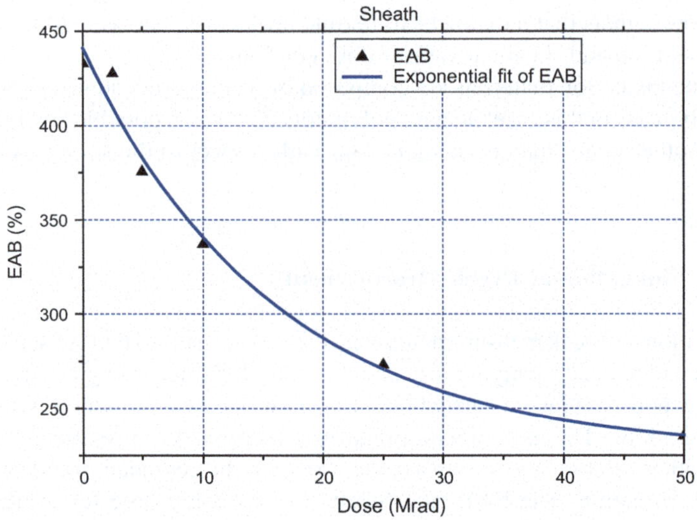

Fig. 37.6 EAB versus dose for sheath

obtained by analyzing the degradation data for both insulation material and sheath is given by:

$$e = \mathrm{A} + Be^{-\mathrm{C}d} \tag{1}$$

where e is the EAB, A, B and C are model parameters of the life-stress relationship and d is the radiation dose. Model parameters along with standard error for insulation material and sheath are given in Tables 37.5 and 37.6, respectively.

Table 37.5 Insulation parameters

Parameter	Value	Standard error
A	427.94	87.22
B	145.45	79.74
C	0.043	0.06

Table 37.6 Sheath parameters

Parameter	Value	Standard error
A	226.41	17.13
B	214.50	16.92
C	0.06	0.014

37.2.2.3 Determination of OIT from OOT

The approach developed by Gimzewski [9] has been employed to determine the OIT from the measured OOT. The thermally induced reaction that occurs in DSC is given by:

$$\mathrm{RH} + \mathrm{O_2} \rightarrow \mathrm{R} \cdot + \mathrm{HO_2} \cdot \tag{2}$$

Here, RH is undamaged polymer chain, R· is a polymer free-radical, and $HO_2\cdot$ is a hydroperoxide free-radical. Considering that the reaction in Eq. (2) follows an Arrhenius process [9], the corresponding rate equation is given by:

$$\frac{\mathrm{d}}{\mathrm{d}t}[\mathrm{RH}] = -[\mathrm{RH}][\mathrm{O_2}]\mathrm{A}\mathrm{e}^{-\frac{\mathrm{E_a}}{\mathrm{kT}}} \tag{3}$$

Here, A is a constant with respect to the initial polymer concentration, E_a is the activation energy, k is Boltzmann's constant, and T is the absolute temperature. The direct measurement of OIT is generally performed in the isothermal mode of DSC. In this mode letting $T = T_{\mathrm{iso}}$, Eq. (3) is then integrated until antioxidant is completely consumed as given by:

$$\frac{\mathrm{d}[\mathrm{RH}]}{[\mathrm{RH}][\mathrm{O_2}]\mathrm{A}} = -\mathrm{e}^{-\frac{\mathrm{E_a}}{\mathrm{kT}}}\mathrm{d}t = -(\mathrm{OIT})\mathrm{e}^{-\frac{\mathrm{E_a}}{\mathrm{kT_{iso}}}} \tag{4}$$

While OIT cannot be obtained from Eq. (4), this is used in OOT measurement for obtaining the isothermal OIT. This gives a measure of OOT; that is the temperature at which the material undergoes an exothermic reaction due to continuous heating under oxygen atmosphere. The ramp rate, $\frac{\mathrm{d}T}{\mathrm{d}t}$ is introduced into Eq. (3) as shown below:

$$\frac{\mathrm{d}}{\mathrm{d}T}[\mathrm{RH}]\left(\frac{\mathrm{d}T}{\mathrm{d}t}\right) = -[\mathrm{RH}][\mathrm{O_2}]\mathrm{A}\mathrm{e}^{-\frac{\mathrm{E_a}}{\mathrm{kT}}} \tag{5}$$

The integrating Eq. (5) from initial T_0 to OOT gives:

$$\int \frac{d[RH]}{[RH][O_2]A} = -\frac{1}{\alpha}\int_{T_0}^{OOT} e^{-\frac{E_a}{kT}}dt \tag{6}$$

Here, ramp rate is denoted by α. The integrals on the LHS of Eqs. (4) and (6) are equal as both equations are integrated until the entire antioxidant is consumed. Hence, the RHS of Eqs. (4) and (6) can simply be set equal, and this enables for determining the isothermal OIT as given by:

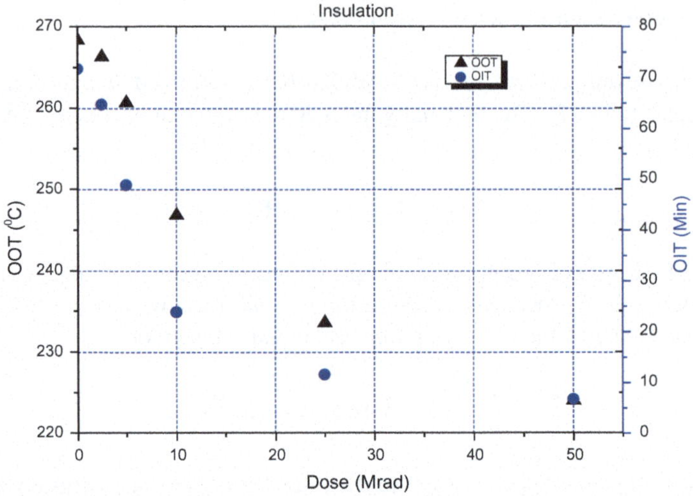

Fig. 37.7 OIT for insulation

$$\mathrm{OIT} = \frac{1}{\alpha} \mathrm{e}^{\frac{\mathrm{E_a}}{\mathrm{kT_{iso}}}} \int_{\mathrm{T_0}}^{OOT} \mathrm{e}^{-\frac{\mathrm{E_a}}{\mathrm{kT}}} \mathrm{d}t \tag{7}$$

Thus, Eq. (7) can be numerically evaluated from the measured OOT and E_a. In general, the activation energy for PVC insulation materials is typically between 0.92 and 1.25 eV [21]. Comparing Figs. 37.5 and 37.6 for the initial margin in the EAB for insulation and sheath, a factor of 1.27 times more EAB is seen for the insulation in comparison with sheath. With this experimental evidence, 0.92 eV for sheath material and 1.18 eV of activation energy for insulation is assumed for determining the OIT from OOT. Figures 37.7 and 37.8 show the OITs obtained from the OOT with respect to radiation dose for insulation and sheath, respectively.

37.2.2.4 Correlation Between OIT and EAB

As EAB is an accepted CM technique for assessing the polymeric insulation materials, a correlation between the determined OITs and EAB needs to be established for degradation assessment and remaining useful life prediction. Figures 37.9 and 37.10 shows correlation plots with 0.99 and 0.93 as a coefficient of determination insulation and sheath, respectively.

It is observed from Figs. 37.9 and 37.10 that the correlation amongst EAB and OIT for both the materials is seen to be exponential for the radiation ageing. It is important to note that degradation rates are different. This is mainly due to the chemical changes in the polymer chain by chain scission, cross-linking, etc. when subjected to gamma radiation, and also due to various antioxidant agents present in insulation materials.

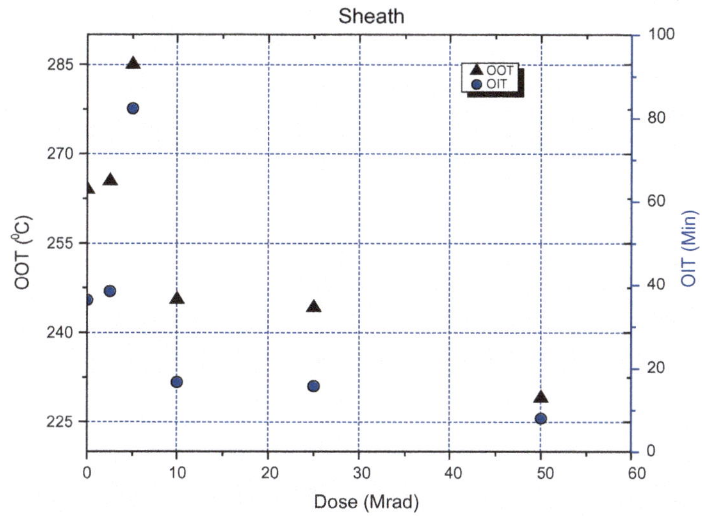

Fig. 37.8 OIT for sheath

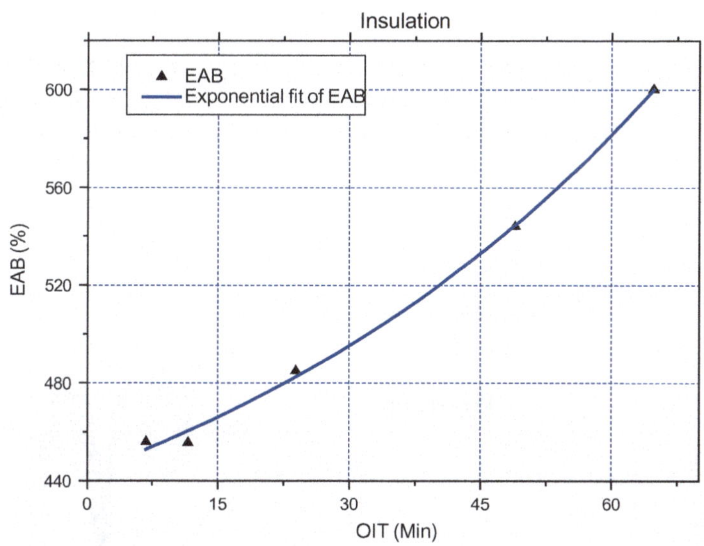

Fig. 37.9 EAB versus OIT for insulation

37.2.2.5 Infrared Spectroscopy

Fourier transform infrared spectroscopy is a technique to monitor the changes in the polymer structure when subjected to ageing environments [22]. As infrared energy is absorbed in specific wavelengths by chemical bonds, the structure of elements can

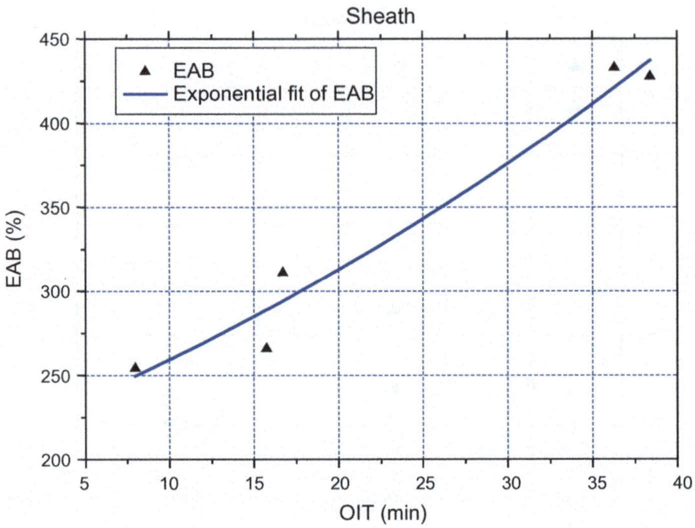

Fig. 37.10 EAB versus OIT for sheath

be identified by spectral locations of their infrared transmittance or absorption [23]. The infrared spectroscopy was performed using Jasco FTIR 660Plus Series spectrometer. Figures 37.11 and 37.12 shows the mean infrared radiation transmittance corresponding to specific chemical bonds at various doses for insulation and sheath respectively.

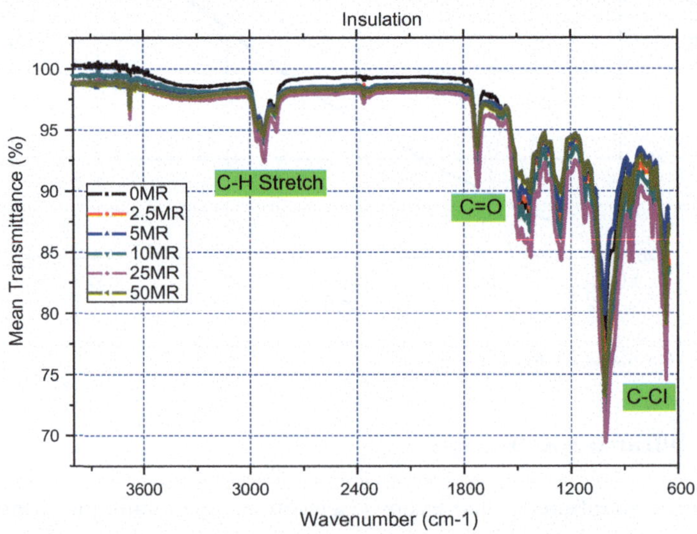

Fig. 37.11 Mean spectra of insulation material

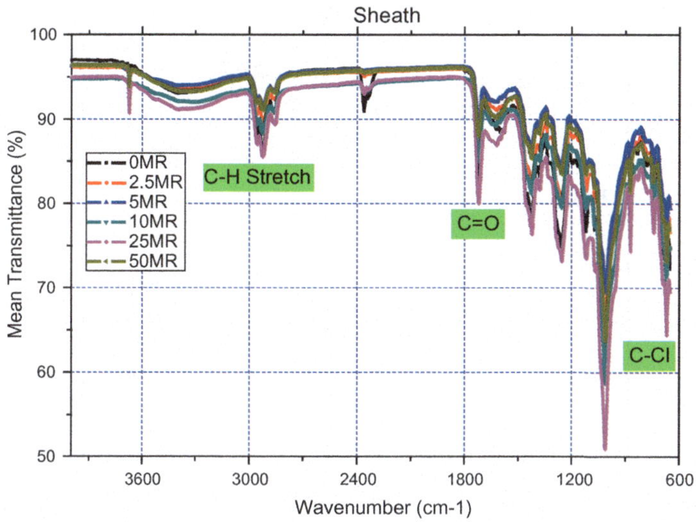

Fig. 37.12 Mean spectra of sheath material

It is observed from FTIR spectra that a noticeable change in the infrared transmittance in both insulation and sheath has occurred at specific bonds due to gamma radiation. This change mainly due to various additives and other filler materials present in the insulation and sheath.

37.2.2.6 Correlation Between Infrared Transmittance and EAB

In order to build confidence in the findings obtained from FTIR analysis, the correlation was established amongst infrared transmittance and EAB. The plots showing the correlation are given in Figs. 37.13 and 37.14 for insulation and sheath respectively. It is evident from the correlation plots that a close correlation exists amongst infrared transmittance and EAB for the insulation. However, the correlation differs from the sheath corresponding to smaller doses. This difference in correlation is expected to occur in case of sheath as manufacturers add several additives and other filler materials in addition to antioxidant to the base polymer to improve the physical and mechanical properties.

37.2.2.7 Free Volume Analysis

Positron annihilation lifetime spectroscopy is a tool for studying the nanostructure and analyzing the movement of molecular in polymers [24, 25]. The analysis was carried out on unaged and irradiated sheath and the free volume fraction was estimated. The initial estimate of free volume fraction in sheath was estimated by assuming all

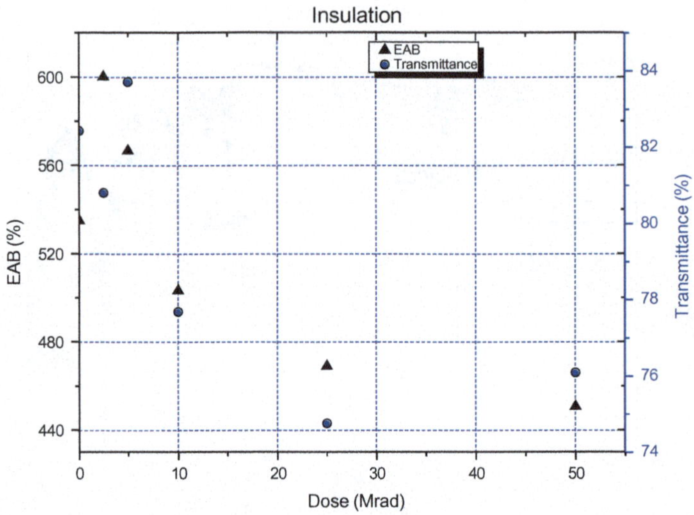

Fig. 37.13 EAB and transmittance for insulation

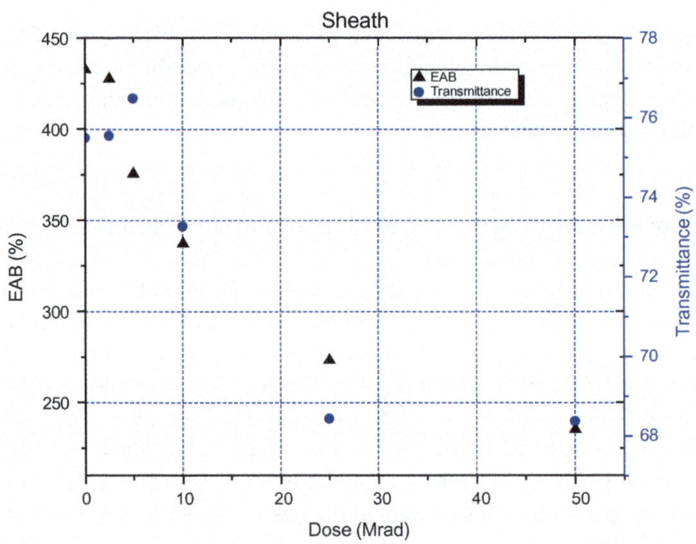

Fig. 37.14 EAB and transmittance for sheath

free volume holes to be of same size and is spherical in shape [26]. The free volume fraction obtained from PALS for sheath of the cable is compared with EAB, and the corresponding plot is shown in Fig. 37.15. In general, gamma radiation causes the chain scission in the polymer structure thereby generating free volume in the bulk material [27]. This process is found to be predominant at lower dose levels

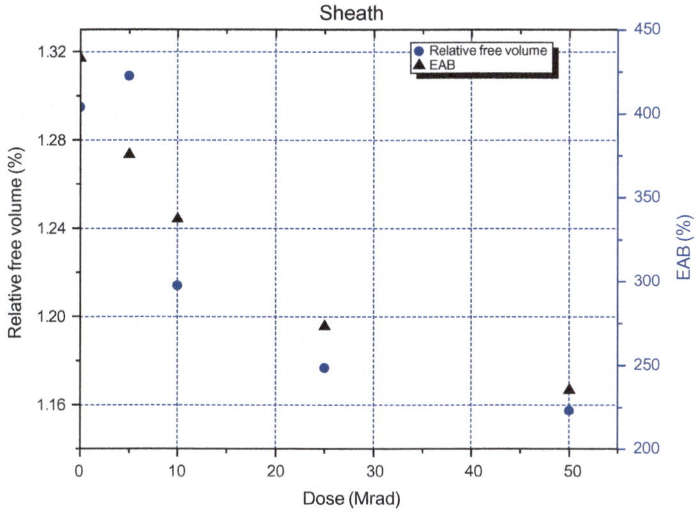

Fig. 37.15 Relative free volume fraction versus dose

up to 5MR in this case. As more free volume is available in the polymer structure, the adjacent bonds undergo cross-linking thereby forming a new links that make the polymer chain rigid resulting reduction in elongation. The effect of such chain scission process was observed in DSC and FTIR analysis as well at lower dose levels.

37.2.2.8 Scanning Electron Microscopy

The scanning electron microscopy is an efficient technique for studying the small topographic details as well as cross-section of an object. This examination yields information related to the topography, composition, morphology, and crystallographic details of the materials [28, 29]. To study the state of unaged and irradiated materials, the samples as shown in Table 37.4 were examined under field emission SEM. A fine coating of gold (1.5–3.0 nm) was done initially on these polymeric samples for making them current conducting. The micrographs of unaged and irradiated materials as shown in Figs. 37.16, 37. 17, 37. 18 and 37.19 indicate changes in the geometry and structure indicating material becoming brittle at high dose levels.

It is evident from the Figs. 37.16, 37. 17, 37. 18 and 37.19 the findings from SEM analysis support the degradation process observed from other CM techniques.

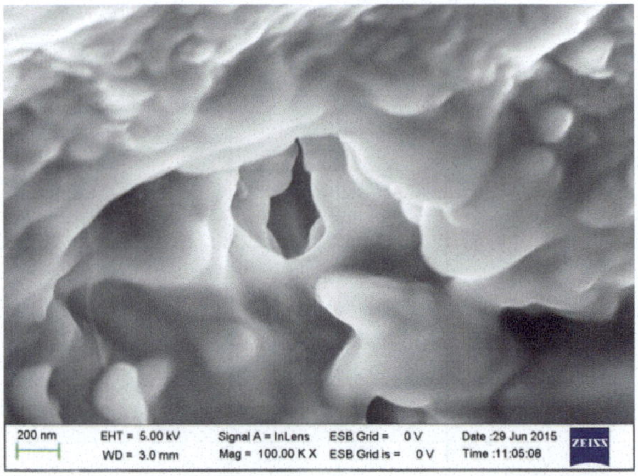

Fig. 37.16 Micrograph of fresh insulation

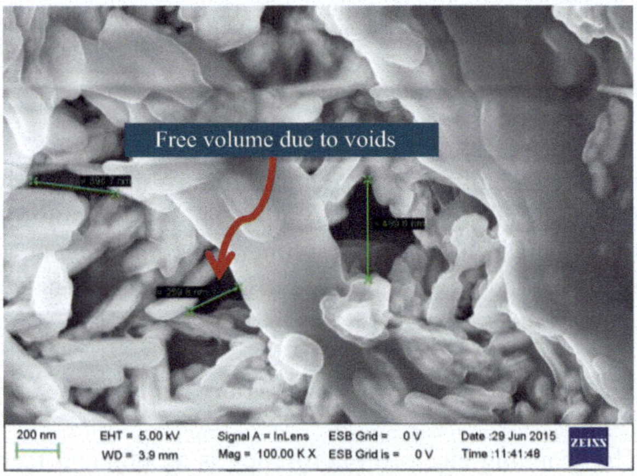

Fig. 37.17 Micrograph of insulation at 25MR ageing

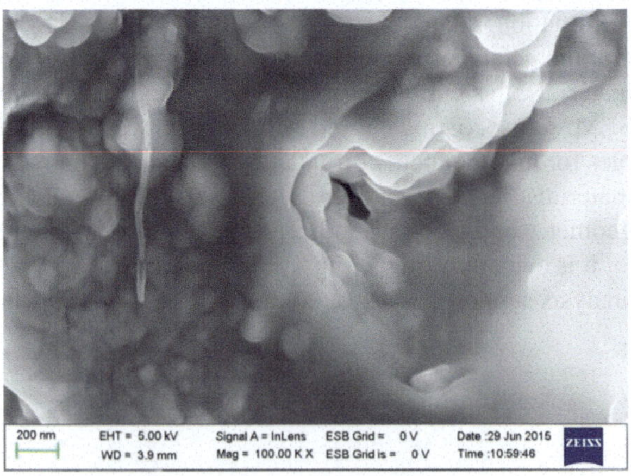

Fig. 37.18 Micrograph of fresh sheath

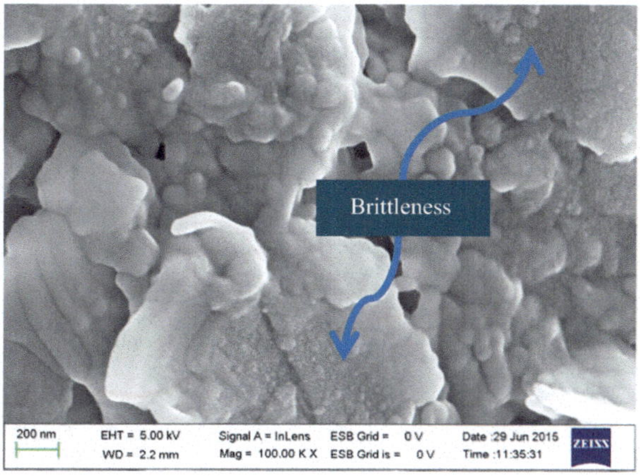

Fig. 37.19 Micrograph of sheath at 25MR ageing

Table 37.7 Details of the cable for thermal ageing

Cable	Specifications	Insulation	Sheath
Manufacturer 2-signal cable 1	37 core, 1.5 mm^2, 1100 V	FRLS PVC	HR PVC

37.2.3 Accelerated Thermal Ageing

The details of the signal cable selected for thermal ageing are given in Table 37.7.

Accelerated thermal ageing of insulation materials was carried out at accelerated ageing temperatures 110, 135 and 150 °C for simulating the field condition IEC 60216 [30]. The periodic measurement of performance indicators and visual examination revealed that the samples simulated at 150 °C condition became hard within 12 days and they could not be useful EAB and other measurements.

37.2.3.1 OIT and OOT Measurement

OIT experiments were performed on unaged and aged samples in the range from 180 to 220 °C, and up to 8 h. However, a distinct exothermic peak could not be determined from the DSC signals due to slow kinetics during the heating; hence OOT measurements were performed. The DSC signals corresponding to 110 and 135 °C are given in Figs. 37.20, 37.21, 37.22 and 37.23 for the samples aged up to 119 days.

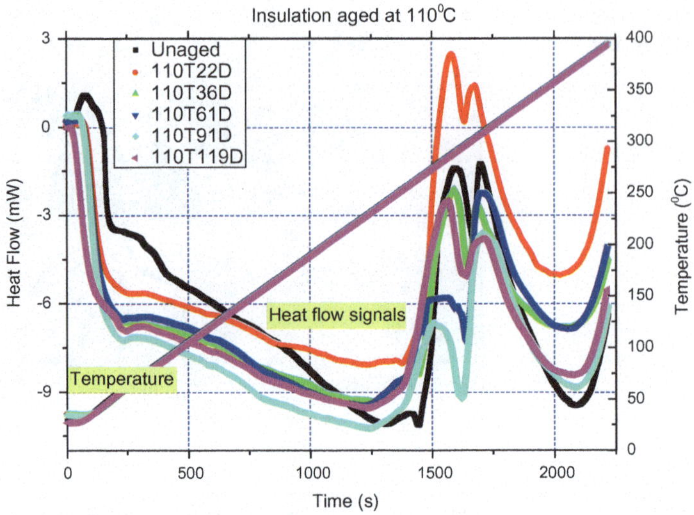

Fig. 37.20 Heat flow signals of insulation aged at 110 °C

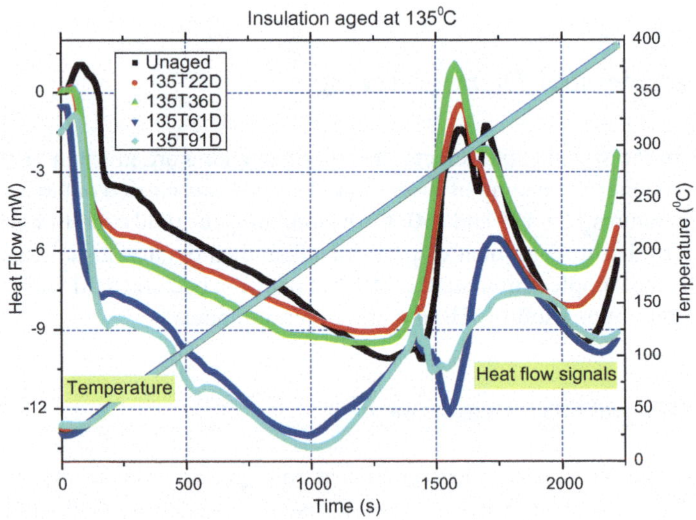

Fig. 37.21 Heat flow signals of insulation aged at 135 °C

37.2.3.2 EAB Measurement

The tensile elongation of thermally aged insulation materials was measured using a Tensile Testing Machine at a strain rate of 25 mm/min with a gauge length of 25 mm. The EAB results corresponding to 110 and 135 °C are given in Figs. 37.24 and 37.25 for insulation and sheath respectively.

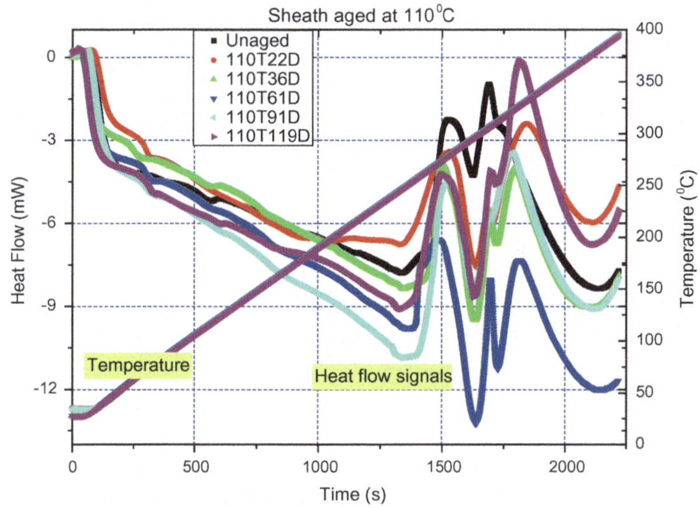

Fig. 37.22 Heat flow signals of sheath aged at 110 °C

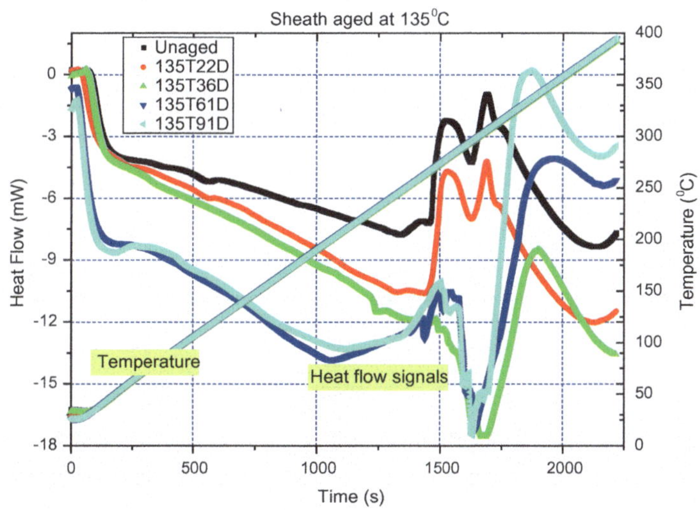

Fig. 37.23 Heat flow signals of sheath aged at 135 °C

37.2.3.3 Determination of OIT from the Measured OOT

The procedure employed for determining OIT from the measured OOT in radiation ageing is adopted for thermal ageing data as well for determining isothermal OIT. The estimated OITs corresponding to the measured OOT corresponding to 110 and 135 °C are given in Figs. 37.26 and 37.27 respectively for insulation and sheath.

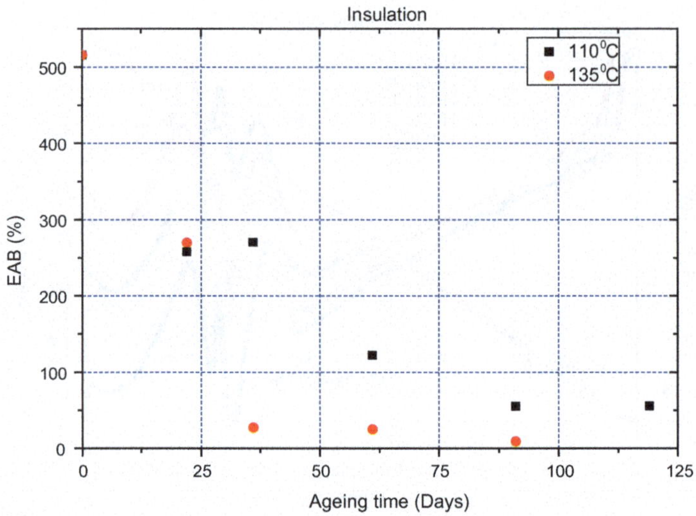

Fig. 37.24 EAB versus ageing time for insulation

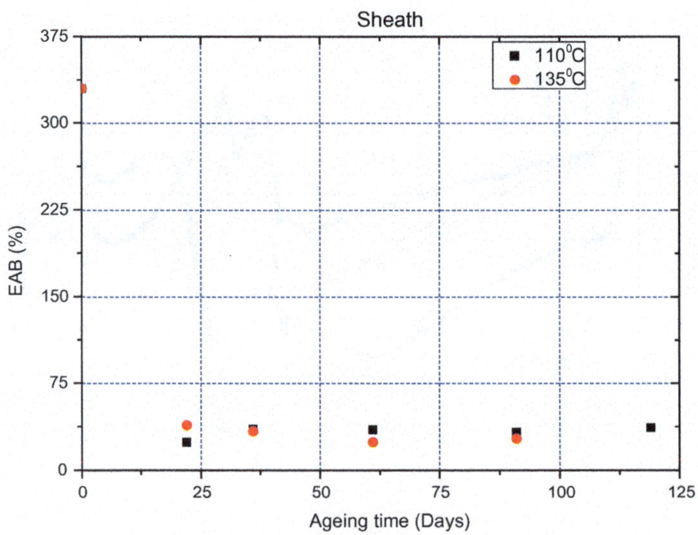

Fig. 37.25 EAB versus ageing time for sheath

37.2.3.4 Correlation Between OIT and EAB

The data was analyzed to study the correlation between OIT and EAB. The resultant fit with adjusted R^2 of 0.76 for insulation and 0.98 for sheath both aged at 110 °C are shown in Figs. 37.28 and 37.29, respectively.

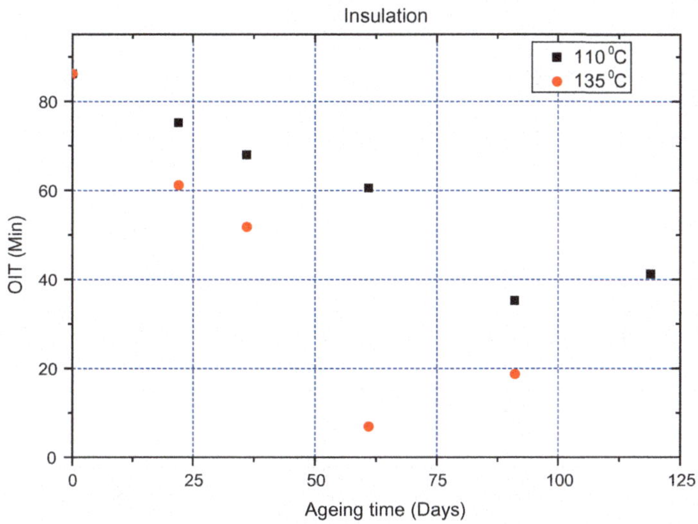

Fig. 37.26 OIT versus ageing time for insulation

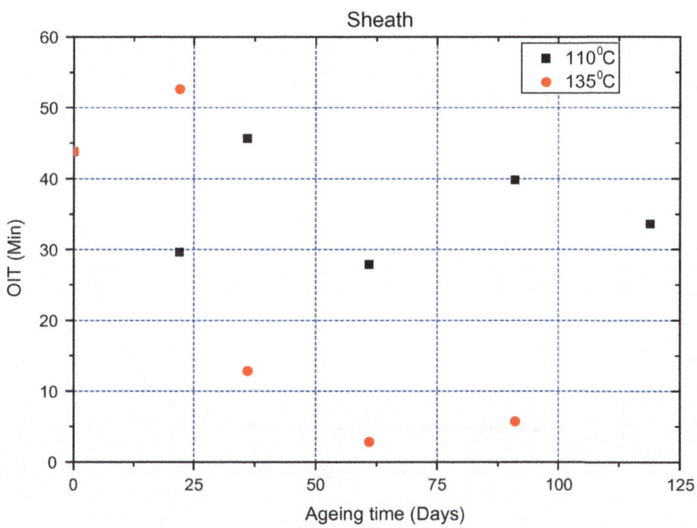

Fig. 37.27 OIT versus ageing time for sheath

The empirical model of the data for EAB is given in Eq. (8).

$$e = A + Be^{C*OIT_{acc}} \tag{8}$$

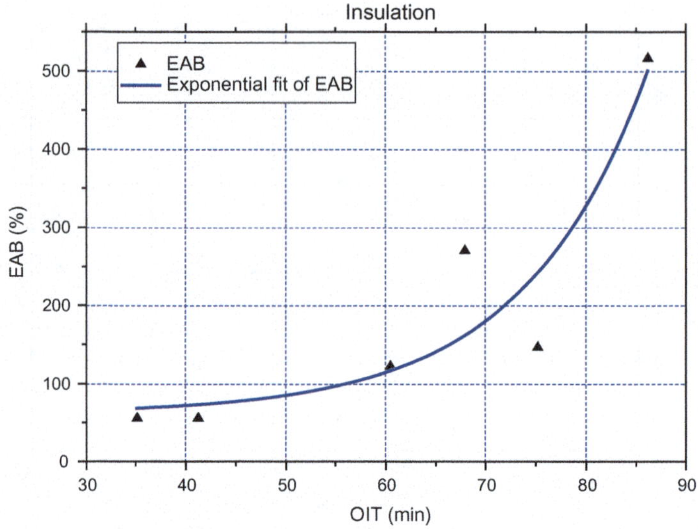

Fig. 37.28 EAB versus OIT for insulation

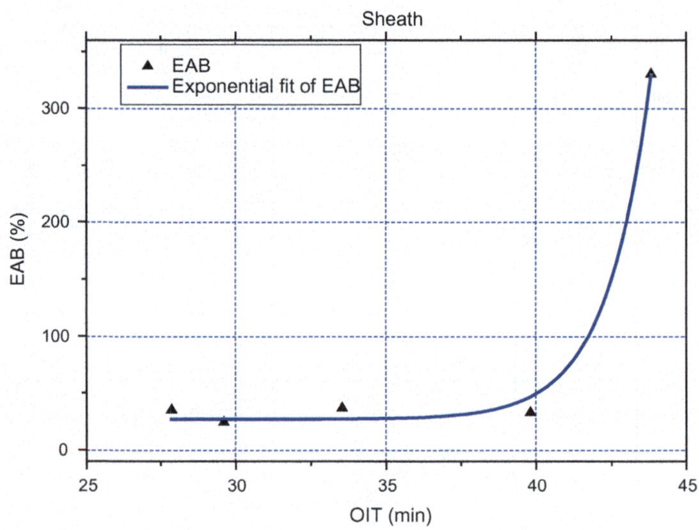

Fig. 37.29 EAB versus OIT for sheath

where, e is the EAB, A, B and C are the parameters of the model and OIT_{acc} refers to OIT calculated at an isothermal temperature of 200 °C. The parameters of the model and standard error are shown in Tables 37.8 and 37.9 for insulation and sheath respectively. It is observed from the correlation plots that under thermal ageing the samples have exhibited relatively an expected correlation [31].

Table 37.8 Parameters for the insulation

Parameter	Value	Standard error
A	62.5	73.1
B	0.40	1.69
C	0.08	0.04

Table 37.9 Parameters for the sheath

Parameter	Value	Standard error
A	29.89	6.54
B	4.18e-11	4.23e-10
C	0.59	0.2

It is apparent from Figs. 37.9, 37.10, 37.28 and 37.29 that a near exponential degradation process occurs under both radiation and thermal ageing of insulation materials. It is evident that due to changes in the polymer structure caused by radiation and thermal ageing, a significant reduction in thermo-oxidative stability and mechanical properties of insulation material is seen.

37.2.3.5 Scanning Electron Microscopy

The SEM analysis was performed on the sample shown in Table 37.7. The images of SEM analysis for unaged and aged materials are shown in Figs. 37.30, 37.31, 37.32 and 37.33.

The SEM micrographs of thermally aged samples reveal a significant degradation in the insulation material. It is also evident that the material has become brittle in about 22 days of ageing at relatively high-temperature condition and noticeable cracks can be found in the samples that support the findings observed from other techniques and

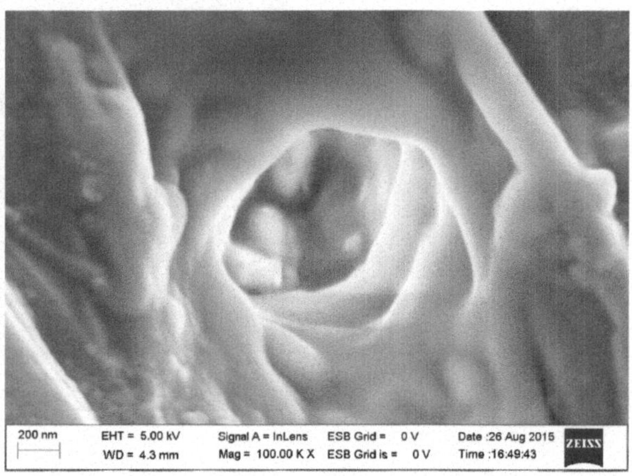

Fig. 37.30 SEM micrograph of fresh insulation

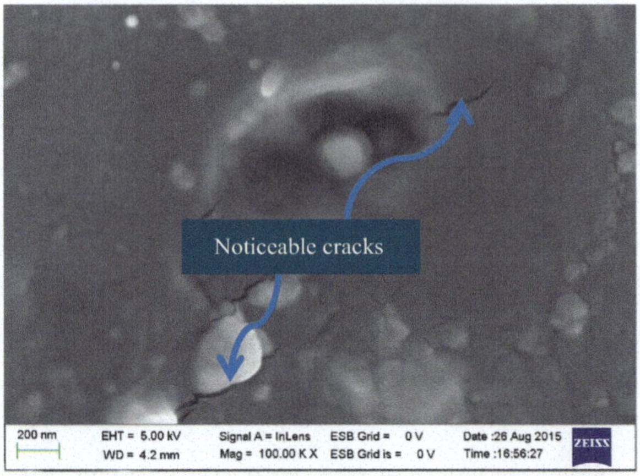

Fig. 37.31 SEM micrograph of insulation after 22 days at 135 °C

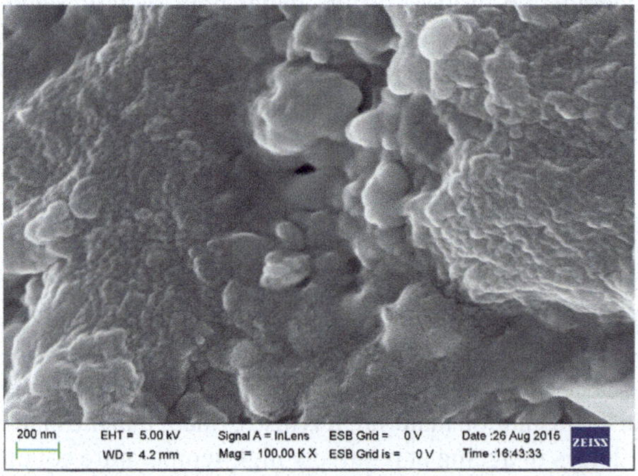

Fig. 37.32 SEM micrograph of fresh sheath

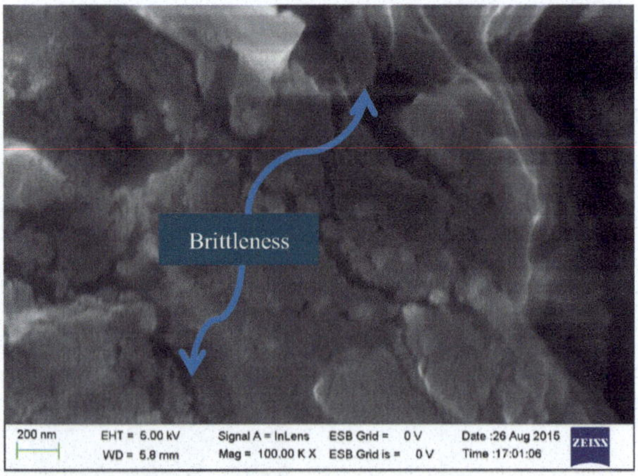

Fig. 37.33 SEM micrograph of sheath after 22 days at 135 °C

their correlation. As the OITs determined from the OOT measurement technique are in good correlation with the benchmark CM techniques, the OITs predicted from OOT are used for predicting the reliability and its incorporation in PSA. The approach for reliability prediction from the predicted performance indicators is discussed below.

37.3 Reliability Prediction

This section presents an overview of how a performance indicator is used in estimating the reliability of I&C cables for use in PSA of NPPs. Generally, the OIT decreases exponentially when subjected a sample to increasing isothermal temperature and this follows an Arrhenius process [9]. Hence, the OIT determined from isothermal temperature may be directly related to an Arrhenius equation as given in Eq. (9).

$$\boldsymbol{OIT} = \boldsymbol{A}\boldsymbol{e}^{\frac{\Phi}{kT_{OIT\cdots}}} \tag{9}$$

where, T_{OIT} is the temperature corresponding to isothermal condition, A is a model constant, Φ is the energy corresponding to reaction and k is Boltzmann's constant. In Eq. (9), OIT is equivalent to as characteristic life. From the acceleration factor relationship given by Eq. (10):

$$\boldsymbol{AF} = \frac{\boldsymbol{t}_{\text{use}}}{\boldsymbol{t}_{\text{acc}}} = \frac{\boldsymbol{OIT}_{\text{use}}}{\boldsymbol{OIT}_{\text{acc}}} \tag{10}$$

Using Eqs. (9) and (10), the use of condition OIT is estimated from Eq. (11):

$$\boldsymbol{OIT}_{\text{use}} = \boldsymbol{OIT}_{\text{acc}}\boldsymbol{e}^{\frac{\Phi}{k}\left[\frac{1}{T_{\text{use}}} - \frac{1}{T_{\text{acc}}}\right]} \tag{11}$$

From Eq. (8), the OIT with respect to 50% absolute EAB is obtained as 8.4 min under 200 °C isothermal temperature when insulation is subjected to ageing of 110 °C. The normal environmental temperature in NPPs is around 40 °C. Therefore, the OIT corresponding to a normal operating temperature of 40 °C with activation energy, Φ = 1.15 eV is estimated as 29 years from Eq. (11). For estimating the reliability a suitable life distribution is required to best explains the failure characteristics of components. The most widely used distribution is Weibull distribution [32–34]. The Weibull reliability function is given by:

$$\boldsymbol{R}(\boldsymbol{t}) = \boldsymbol{e}^{-\left(\frac{t}{\eta}\right)} \tag{12}$$

where, η is the scale parameter referred to as characteristic life, β is the parameter explaining the shape of the distribution, and t refers to the component operating time. As OIT is equivalent to characteristic life and can be related to scale parameter of the

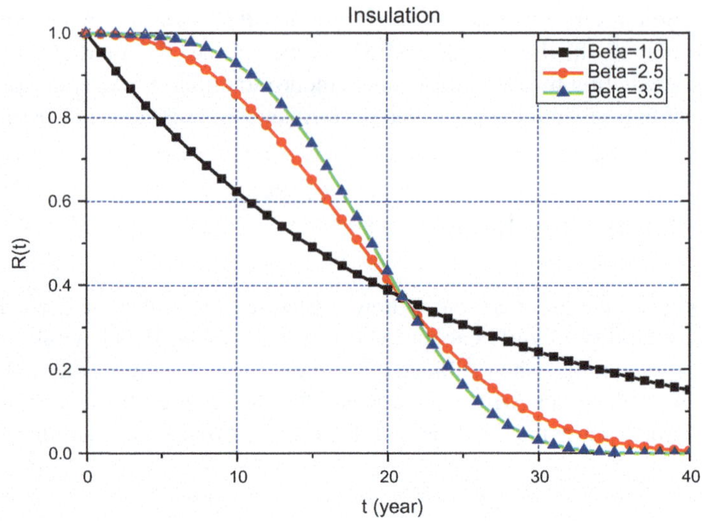

Fig. 37.34 Reliability versus time

Weibull distribution [9], the time-dependent reliability is estimated from Eq. (12). Considering time t from 0 to 40 years (typical design life of NPP) with $\eta = \text{OIT}$ equal to 29, the reliabilities predicted over 40 years of service life. The time-dependent reliabilities for various values of β for the thermally aged cable insulation are given in Fig. 37.34.

It is observed from Fig. 37.34 that when times-to-failure of the insulation material follows an exponential distribution, the reliability is substantially high as failures are treated as random failures. Study also demonstrates that under thermal and radiation ageing both insulation and sheath materials exhibited an exponential degradation process; hence it is appropriate to consider the exponential failure rates for these insulation materials. As samples were subjected to accelerated ageing, other failure distributions were also considered in this study to demonstrate the reliabilities from non-exponential failures as well. The reliability predicted in this study is used in PSA for accounting the cable failures due to ageing.

The general approach for determining the reliability of I&C cable based on the performance indicators and by employing Weibull distribution is summarized as follows:

- use DSC to measure OIT at an isothermal temperature (OIT_{acc})
- derive relationship between OIT and EAB at a particular stress level (e.g. 110 °C)
- determine the OIT corresponding to 50% absolute EAB as minimum acceptable value
- use the acceleration factor (Eq. 10) to calculate OIT for temperature(s) at use condition (OIT_{use})

- calculate the reliability using Eq. (12) where the shape parameter β equal to 1 (i.e. exponential case) and OIT at use is used as η, and
- make an assessment of time-dependent reliability over a design life to ensure the reliability targets.

37.4 Conclusions

The radiation and thermal stability of I&C cable insulation materials were studied using several CM methods by subjecting to accelerated thermal and radiation ageing. The findings from DSC, FTIR, PALS suggest a good correlation with EAB. The SEM image analysis also supports the correlation among various CM techniques. The difference in degradation in insulation and sheath is mainly due to the fact that the manufacturers add several additives and filler materials in addition to antioxidant to improve the properties. Using the results from various CM techniques and from Weibull distribution, a general approach for determining reliability of I&C cables has been proposed and time-dependent reliabilities have been predicted. As samples were subjected to accelerated ageing, other failure distributions were also considered in this study to demonstrate the reliabilities from non-exponential failures as well. The predicted reliabilities are useful in reliability analysis of safety and control systems of NPPs to account for cable failures in PSA.

Acknowledgements The authors gratefully thank Dr. Kallol Roy for his constant help and suggestions during the course of this research work. Authors would also like to thank Dr. K. A. Dubey and Dr. K. Sudarshan for their help in conducting a few experiments.

References

1. Lofaro RJ, Grove E, Soo P (2000) An evaluation of condition monitoring techniques for low-voltage electric cables, BNL-NUREG-67389, New York
2. IAEA-TECDOC-932 (1997) Pilot study on the management of ageing of instrumentation and control cables, International atomic energy agency
3. Anandakumaran K, Seidl W, Castaldo PV (1999) Condition assessment of cable insulation systems in operating nuclear power plants. IEEE Trans Dielectr Electr Insul 6(3):376–384
4. Jacobus MJ (1992) Ageing, condition monitoring, and loss-of coolant accident tests of class 1E electrical cables: Ethylene propylene rubber cables, NUREG/CR-5772, SAND91-1766/2, vol 2
5. Ekelund M, Edin H, Gedde UW (2007) Long-term performance of poly (vinyl chloride) cables. Part 1 Mechanical and electrical performances. Polym Degrad Stab 92:617–629
6. Burnay SG (2001) An overview of polymer ageing studies in the nuclear power industry. Nucl Instrum Methods Phys Res B 185:4–7
7. Clavreul R (1999) Ageing of polymers in electrical equipment used in nuclear power plants. Nucl Instrum Methods Phys Res B 151:449–452

8. Mason LR, Reynolds AB (1998) Comparison of oxidation induction time measurements with values derived from oxidation induction temperature measurements for EPDM and XLPE polymers. Polym Eng Sci 38(7):1149–1153
9. Focke W, van der Westhuizen I (2010) Oxidation induction time and oxidation onset temperature of polyethylene in air. J Thermal Anal Calorim 99:285–293
10. NUREG/CR-6704 (2001) Assessment of environmental qualification practices and condition monitoring techniques for low voltage electrical cables, Brookhaven National Laboratory
11. Schmid M, Ritterand A, Affolter S (2006) Determination of oxidation induction time and temperature by DSC: results of round robin tests. J Therm Anal Calorim 83(2):367–371
12. Hsu YT, Chang-Liao KS, Wang TK, Kuo CT (2007) Correlation between mechanical and electrical properties for assessing the degradation of ethylene propylene rubber cables used in nuclear power plants. Polym Degrad Stab 92:1297–1303
13. Gordon GV, Shen C, Shaw MT (1992) A method for correlating natural ageing data and its application to insulation ageing in nuclear power plants. In: Proceedings of 1992 annual report: conference on electrical insulation and dielectric phenomena, pp 694–699
14. IEC 60364-6:2006 (2006) Low-voltage electrical installations—Part 6: Verification
15. IEEE Std. 43-2000 (R2006) (2006) IEEE recommended practice for testing insulation resistance of rotating machinery
16. IEC 544-1 (1995) Guide for determining the effects of radiation on insulating materials, Part 1: radiation interaction and dosimetry, Geneva
17. IEC 544-2 (1991) Guide for determining the effects of radiation on insulating materials, Part 2: procedures for irradiation and test, Geneva
18. IEC 544-4 (1985) Guide for determining the effects of radiation on insulating materials, Part 4: classification system for service hi radiation environments, Geneva
19. ASTM D 3895-07 (2007) Test method for oxidative induction time of polyolefins by thermal analysis, ASTM International
20. ASTM D638-10 (2010) Standard test method for tensile properties of plastics, ASTM International
21. Dalal SB, Gorur RS, Dyer ML (2005) Ageing of distribution cables in service and its simulation in the laboratory. IEEE Trans Dielectr Electr Insul 12(1):139–146
22. Anandakumaran K, Barreca S, Seidl N, Castaldo PV (2001) Nuclear qualification of PVC insulated cables. IEEE Trans Dielectric Electric Insul 8(5)
23. Rajini V (2011) Experimental evaluation and comparison of performance of polymeric insulating materials in radiation environment, Electrical engineering. Elixir Elec Eng 40:5554–5562
24. López-Castañares R, Olea-Cardoso O, Vázquez-Moreno F, Lizama-Soberanis B, Camps-Carvajal E (2002) Positron annihilation for characterizing polymeric materials. Bulg J Phys 29:55–178
25. Kobayashi Y, Zheng W, Meyer EF, McGervey JD, Jamieson AM, Simhall R (1989) Free volume and physical ageing of poly(viny1 acetate) studied by positron annihilation. Macromolecules 22:2302–2306
26. Dai G, Deng Q, Liu J, Shi H, Huang C et al (1993) Free-volume hole distributions of polymers by positron annihilation lifetime spectroscopy. J Phys IV, 03 (C4), C4-233–C4-239
27. Liao KS, Chen H, Awad S, Yuan JP, Hung WS, Lee KR, Lai JY, Hu CC, Jean YC (2011) Determination of free-volume properties in polymers without orthopositronium components in positron annihilation lifetime spectroscopy. Macromolecules 44:6818–6826
28. Reimer L, Hawkes PW (2010) Scanning electron microscopy: physics of image formation and microanalysis. Springer-Verlag, Berlin (DE)
29. Deepalaxmi R, Rajini V (2014) Performance evaluation of gamma irradiated SiR-EPDM blends. Nucl Eng Des 273:602–614
30. IEC 60216 (1994) Guide for the determination of thermal endurance properties of electrical insulating materials, 4th issue
31. Bell RM, Doyle TE, Reynolds AB (1990) Correlation of oxidative induction time with thermal and radiation ageing of EPR and XPLE cable insulation. In: IEEE international symposium on electrical insulation, Toronto, Canada, June 3–6

32. Dissado LA (2002) Predicting electrical breakdown in polymeric insulators from deterministic mechanisms to failure statistics. IEEE Trans Dielectr Electr Insul 5:860–875
33. Fothergill JC, Dissado LA (1992) Electrical degradation and breakdown in polymers. P. Peregrinus for IEE, London
34. Tavner PJ, Ran L, Penman J, Sedding H (2008) Condition monitoring of rotating electrical machines. Power and Energy Series (56), IET

Chapter 38
Industry 4.0 Optimizing Logistics Processes and Supply Chain Management in Industries

Ashok G. Matani

38.1 Introduction

Implementing Information and Communication Technologies (ICT) in industries has become unavoidable, mainly by being vital for increasing the organizational efficiency and its level of competitively—resulting in the promotion of adopting ICT in most of the industry activities especially in logistics and production operations. This technological evolvement is implemented by most organizations in Enterprise Resource Planning (ERP), Warehouse Management Systems (WMS), Transportation Management Systems (TMS), Intelligent Transportation Systems (ITS). The information available in real-time and in-line, gathered through VANET Systems, drone points and business intelligence systems, sensor networks will improve and enhance the decision-making efficiency of management and become more and more flexible and efficient into the near future [1, 2].

In an Industry Internet of Things (IIoT) context, logistics challenges will be requiring high need for transparency of supply chain visibility; integrity control of right products, at the right time, place, quantity, condition and at the right cost of the supply chain; dynamic reconfigurability of supply networks, especially by re-examining service-level agreements with upstream and contracted suppliers; supply network design, towards achieving lean, agile, resilient and green supply chains [3]. The Logistics 4.0 purpose is not to replace humans in their works, rather it is to avoid inaccuracies and to have faster processes shared effortless and in real-time. It will be always needed the involvement of people controlling the processes and taking control of any system failure [4].

Nowadays the rapid development of internet has resulted in a huge amount of information produced and collected on a daily basis that their processing and analysis are beyond the capabilities of traditional tools. Big Data facilitates to quickly

A. G. Matani (✉)
Mechanical Engineering Department, Government College of Engineering, Amravati, India
e-mail: ashokgm333@rediffmail.com; dragmatani@gmail.com

P. K. Kapur et al. (eds.), *Decision Analytics Applications in Industry*, Asset Analytics,
https://doi.org/10.1007/978-981-15-3643-4_38

and efficiently manage and use this constantly growing thanks to reaping information from many different sources database [5]. This technology allows analysis and separation of the important from the less important—helping to draw conclusions and support effective transfer of knowledge to carry out business objectives.

According to Forrester's definition, Big Data consists of four dimensions namely:

- Value—value data—the main objective is to isolate the whole mass of information to what is most important for us—this is why it is so important that the results reflect the actual conditions and led to the most favourable business activities [7].
- Variety—variety of data—Big Data comes from a variety of sources namely: transactional systems, social networking sites or the internet. These data change dynamically and are much unstructured, which means that they are not suited to traditional forms of analysis including images, video and content from social networking sites.
- Volume—amount of data—McKinsey Global Institute identifies the concept of Big Data referring to datasets whose size exceeds the capacity of ordinary tools for collection, storage, management and analysis—it is connected with the technological capabilities to manage these data [6].
- Velocity—the speed of generation of new data and analysis—data analysis is carried out on Big Data in near real-time, as the correct conclusions from the constantly flowing and changing data need to be implemented on an ongoing basis.

Big Data makes it possible to analyze the data at a more advanced level than traditional tools allowed. With this technology, even data that has been collected in various mutually incompatible systems, databases and websites are processed and combined to give a clear picture of the situation in which there is a specific company or person [8] (Tables 38.1, 38.2).

38.2 DHL Effectively Implementing Big Data Technologies in the Logistics

The use of Big Data technologies in the area of logistics is effectively implemented by DHL named "Resilience360"—an instrument designed to manage risk in the supply chain. The company provides customers with information on potential interference of their respective supply chains. It is through the collection and evaluation of data that it is possible not only to protect but also to improve the efficiency of the supply chain. Resulting in no interruption in operations and permanently achieving customer satisfaction [9].

DHL utilizes the use of Big Data analytics towards increasing operational efficiency while providing the opportunity to explore new business models. DHL "Resilience360" contains two elements that are associated with the risk assessment analysis, as well as tools to monitor the supply chain that works in almost real-time. The

Table 38.1 Typical applications of IoT in various countries

Industries/companies and countries implementing	**Modifications/improvements expected**
Smart community, Canada and China	Value-added services, e.g. utility management and social networking
A cloud implementation using Aneka, Australia	A seamless independent IoT working architecture
Healthcare and social applications, USA	Increased processing capabilities
Machine-to-machine measurement, Ireland and France	Enhanced interpretation from users
Smart cities, Padova, Italy	Reduced traffic congestion, improved energy efficiency
IoT Gateway system, China	Improved functions such as data display, topology, etc.
IoT application framework, India and France	Improved collaborative work, improved productivity of stakeholders
IoT-enabled energy management, Italy and Spain	Enhanced energy data analysis, integrated energy data management
IoT-enabled real-time information capturing and integration framework, China	Real-time information capturing

Table 38.2 Strategic transformations via digital supply network

Supply chain transformations	Implementing tactics and strategies
Design process optimization	Rapid prototyping, virtual design simulation
Planning and inventory efficiency	Dynamic inventory fulfilment, sensor-driven forecasting
Product optimization	Using 3 D printing
Risk prevention and mitigation	Proactive risk sensing, proactive quality sensing
Sales optimization	Inventory-driven dynamic pricing, sensor-driven replenishment pushes
Supplier collaboration	Blockchain enabled transparency, cloud/control tower optimization
Aftermarket sales and services	Augmented reality enabled customer support, end to use transparency to customers
Operations efficiency	Automated production, predictive maintenance
Logistics optimization	Automated logistics, driverless trucks, dynamic routing

strength of the chain and associated revenue losses depend on whether a break occurs in the production, and this should be less prone to failures. DHL is in the pilot phase on the model of the forecast number of packages DHL, which has also been taken in connection with the analysis of Big Data [10].

Big Data enables service providers to optimize logistics processes, improve customer service, and presents a promising starting point for developing new business models. Big Data suggests some instruments operating in the field of geomarketing for small and medium-sized companies and enterprises. Another model "DHL Geovista" allows detailed analysis and evaluation of very complex geographic data to be obtained, which greatly facilitates the logistics service providers to anticipate the multiplicity of sales, which generates small and medium-sized enterprises. Among the sources from the supply system is information from retailers, transport, invoices and more. Data from customer profiles, social networking profiles, orders, market forecasts and geographical schemes also plays a role. Using customer data to analyze information from the delivery system, retailers can meet the expectations of customers by anticipating their purchasing behaviour [11].

38.3 The Internet of Things Enabling Effective Supply Chain Management Operations

- Smart environment is a category of solutions, Internet of Things, which from the daily consumer perspective are the least visible. These are the basis for the safe operation of the entire anthropogenic environment manmade, e.g. Urban, industrial areas, agricultural areas that make the ecosystem friendly to economic development and the functioning of societies [12].
- Smart water management, a wide range of issues related to the administration and management of key resources for the functioning of the environment. This category includes regulation of rivers and protection against floods, the impact of water resources on the environment, their use and protection deficits, waterways, hydropower or security [13].
- Smart industry is entering the area of the Internet of Things in solutions related to particular sectors of the national economy.
- Smart production as well as intelligent industry includes solutions in specific sectors of the economy. Examples include issues related to agriculture—temperature control and irrigation to prevent drought or the formation of fungi, breeding—monitoring living conditions and grazing livestock, and control of production lines—readers, sensors, video surveillance—useful in the management and inspections as well as control of the rotation of products on store shelves and in warehouses [14].
- Smart transport includes issues namely—the location of transported goods—checking routes of hazardous, delicate or precious materials and control of the conditions of transport or storage conditions.
- Smart energy includes a number of solutions that enable the management of utilities. These include the monitoring of individual consumption, as well as the processes for its production and use of solar systems, windmills and water management.

- Smart cities are another area in which the Internet of Things can play an increasingly important role. Its capabilities promise a lot of applications—from the organization of pedestrians and traffic—monitoring traffic congestion, parking spaces, intelligent roads, providing information about the state of roads, traffic problems, monitoring of weather or accidents on the road, the diagnosis of safety threats—vibrations and strength of materials in buildings, bridges, historic buildings, noise, lighting—adaptive to the level of cloud cover and waste management—filling level of containers [15].
- Smart buildings are a whole range of facilities, which can be used both at the individual as well as industrial level including monitoring the property—fences, windows and doors, motion sensors, smart irrigation, earning thermostats.
- Smart apartments are a category of equipment, which are typically for individual applications covering refrigerators—informing content, shelf life, the need to replenish, remote machines—allowing the use of energy at lower tariffs, cookers—for remote setting of the oven [16].
- Smart health covers a wide range of applications used in the monitoring of health and physical activity of the elder people, vitality—people active in sport, patient safety both in hospital and at home.
- Smart life is a whole range of consumer solutions aimed at comfort and safety.

The IoT offers new possibilities in the area of performance. Road transport trucks and luggage carrying vehicles can be automatically controlled to the specification of hosts, which will allow them to operate in predefined intervals and with standard speed, so as to maximize fuel economy. The Daimler Group has invested in the development of mobile services such as car2go, myTaxi or moovel; General Electric, has invested in systems to operate equipment and factories use a system called industrial design (internet industry); LG is preparing for smart homes, producing televisions and household appliances which can connect to the internet, enriching the offer of related services. With the IoT, it is possible to supervise the process of travel of packages and letters. In case of delay, the customer can be informed in advance of complications. Analysis associated with the development of the connected fleet can help predict failure and automatically plan moves aimed at improving the supply chain. In the case of storage in warehouses, intelligent shelving and pallets will become the driving force of modern inventory management. In respect of the carriage of goods, tracking and tracing becomes faster, more precise, predictable and safe [17].

According to research conducted in 2014 by Forrester Consulting on behalf of Zebra Technologies:

- Nearly 90% of companies from the logistics and transport sector have implemented IoT solutions [18]
- More than 50% of respondents expect that the Internet Things will improve the supply chains
- 40% of respondents expect that the IoT will help companies increase their level of safety and cost-effectiveness

- Key technologies in the implementation of the Internet of Things are assumed to be Wi-Fi connectivity, security sensors, NFC communications (Near Field Communications)
- Nearly 40% of respondents expressed their concerns about the privacy and security of information as the biggest obstacle to the implementation of IoT solutions [19]
- 38% indicate a high degree of complexity of these solutions, and as such a high risk of implementation.

These results indicate emerging challenges for the Internet of Things for the logistics and transport sector. Solutions in this field can provide operational data on the location and monitoring of the condition of things. With this information, it is possible to improve customer service by shortening the cycle of logistics processes and optimize their cost [20].

38.4 Logistics 4.0 Using Five Technological Applications

Smart Logistics is a logistics system enhancing the flexibility, the adjustment to the market changes and will make the company be closer to the customer needs thereby improving the level of customer service, the optimization of the production and helps in lowering the prices of storage and production. As the Smart Logistics will change accordingly to the actual technology-driven, it has a time dependency. This new paradigm is the result of the increased use of internet that enables the communication between each other machines and humans in real-time and the implementation of advanced digitalization [21].

An efficient and strong Logistics 4.0 implements the following technological applications:

(1) Resource Planning,
(2) Warehouse Management Systems,
(3) Transportation Management Systems,
(4) Intelligent Transportation Systems and
(5) Information Security.

38.4.1 Resource Planning

The resource planning management procedures and the implementation of Cyber-physical systems (CPS) will enhance flexibility and agility to the changes that might occur in the supply chains thereby enhancing the overall productivity. The proper alignment and integration between the main actors of the supply chain, and the increasing level of visibility and transparency will result in an adequate utilization of resources—people, materials, equipment, therefore optimizing resources/processes, the time to market alignment and raise the asset employment [22].

38.4.2 Warehouse Management Systems

In today's economic climate the warehouses also need to serve as a key source of competitive advantage for logistics providers. The applications of the Industry 4.0 parameters will result in remarkable changes in the way warehouse works these days. Especially, the introduction of smart management throughout the proper adoption and implementation of Warehouse Management Systems (WMS) which will transform the warehouse activities into the future requirements of the inbound logistics according to the Industry 4.0 paradigm [23].

38.4.3 Transportation Management Systems

A transportation management system is part of supply chain management cantered on transportation logistics. A transportation management system enables interactions between an order management system and distribution centre and a warehouse. With the advancements in implementing the transportation management system these systems help companies to control and manage ever-higher freight costs; integrate with other supply chain technologies, warehouse management systems and global trade management systems; and handle electronic communications with customers, trade partners, and carriers. With the increased offers in cloud services and cloud computing, transportation management system cloud-based is becoming popular as the standard parameter [24].

38.4.4 Intelligent Transportation Systems

Intelligent transportation system is a novel field which is interoperating in various fields of transportation systems—transportation management, control, infrastructure, operations, policies and control methods. Intelligent transportation system covering new technologies such as data processing, computing hardware, positioning system, sensor technologies, telecommunications, virtual operation and planning techniques. The idea of virtual technologies integration in the transportation field plays a vital part to overcome the issues in global world. Intelligent transportation system are important for increasing safety and reliability, travel speeds, traffic flow and reducing risks, accidents rate, carbon emissions and air pollution. An intelligent transportation system provides solutions for cooperation and reliable platform for transport.

Electronic Toll Collection (ETC), Highway Data Collection (HDC), Traffic Management Systems (TMS), Vehicle Data Collection (VDC), Transit Signal Priority (TSP), Emergency Vehicle Preemption (EVP) is some applications of the intelligent transportation system. Intelligent transportation system is not limited for vehicular traffic, it also provides other services implemented in navigation systems, air transport systems, water transport systems and rail systems [25].

38.4.5 Information Security

The proliferation of the internet-based applications promoted by the emergence of cloud-based systems, the Internet of Things (IoT), Big Data, Industry 4.0, BYOD (Bring Your Own Device) and CYOD (Choose Your Own Device) trends have revolutionalized the way organizations conduct their businesses. Organizations are especially interested in finding new technological initiatives at low operating cost, in order to offer better and innovative services and thus gain competitive advantage.

38.5 KPMG International and Alibaba Cloud Focusing on Data Security, Regulatory Compliance and Seamless Integration

KPMG International and Alibaba Cloud—the cloud computing arm of Alibaba Group had plans to form a global alliance to provide digital transformation to businesses of all sizes and across multiple industries initially in Asia-Pacific and Europe. The alliance focuses on digital transformation, new retail, and IT strategy consulting and cloud-based digital solutions. Alibaba's strong cloud computing and artificial intelligence capabilities, together with KPMG's global network, global enterprise solutions, business acumen and knowledge of global industries with many Fortune 1000 companies will complement each other in helping clients to expand into new markets. The purpose of this global network is to build a strong global ecosystem with cloud services and infrastructure, in serving organizations with multi-cloud strategies including small- to medium-sized enterprises. As one of the world's leading global infrastructure-as-a-service (IaaS) providers, and China's largest provider of public cloud services, Alibaba Cloud will be the latest global cloud provider to join KPMG's alliance community, allowing KPMG to offer businesses with multi-cloud strategies a single, independent advisor for cloud- and platform-based digital services and solutions. Under this alliance, KPMG and Alibaba Cloud will form a strategy team comprised of KPMG and Alibaba Cloud software and systems architects and engineers to develop cloud infrastructure solutions for clients expanding into new markets, with a focus on data security, regulatory compliance and seamless integration. KPMG specialists will receive training and certification in Alibaba

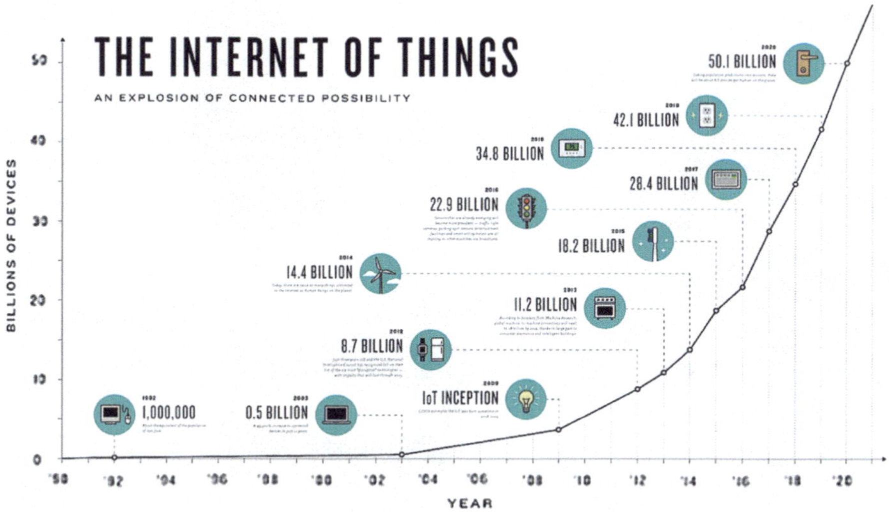

Fig. 38.1 Impact of internet of things and connected possibility

Cloud enterprise internet architecture, cloud computing and security, big data, and robotic process automation (RPA) [12] (Fig. 38.1).

38.6 Satellites Used to Identify Potential Air Quality Hot Spots

Satellite data can fill the coverage gaps in the existing network to support routine monitoring. They could also be used to identify potential air quality hot spots. It has been found that using satellite data is more economical than setting up and operating a number of fixed stations. For instance, most of the NO_2 in the atmosphere is contained within 1 km from the surface of the earth. That makes satellite measurements a useful representation of surface NO_2 concentrations. For PM assessment, too, satellite-based assessment of aerosol optical depth is used. Both the magnitude and spatial extent of sources can be determined and exposure can be assessed with the help of satellites. Satellite observations can change emissions and concentrations observed through ground-based measurements. Air quality monitoring networks are instrumental to the evaluation and management of air pollution by governments, policy makers and regulatory bodies. Ambient data from monitoring networks, subject to rigorous analysis, can reveal the pollutant concentrations, correlations and trends at measurement locations. Such information is invaluable for estimating the actual effects of social and infrastructure changes, and policy interventions on air quality [1] (Fig. 38.2).

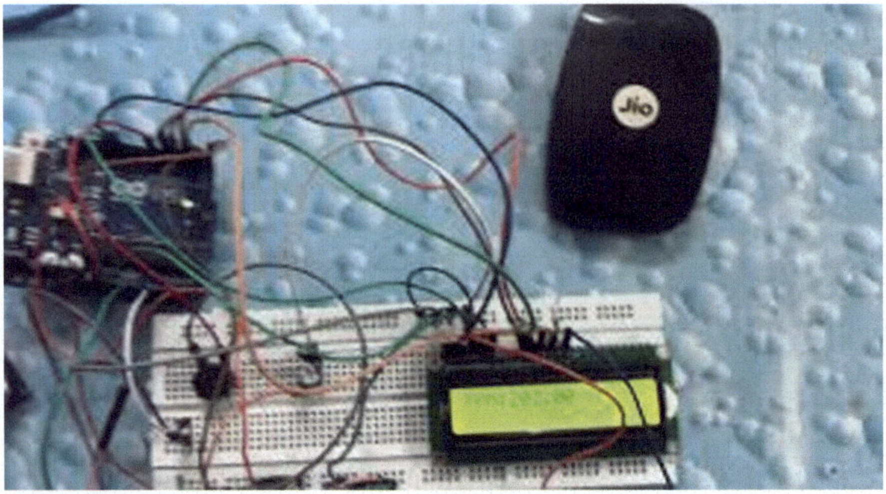

Fig. 38.2 IoT-based air pollution monitoring system

38.7 Conclusions

Technologies and IT infrastructure elements of business intelligence will impact the supply chain activities through cost-reduction opportunities and an increase of the process-transparency, therefore processes will be more digital and technological, where the company's personnel are able to acquire and share information using the business intelligence technology from anywhere around the globe. Procurement processes will be optimized, as suppliers can be fully flexible and autonomously chosen by specific software. Smartphone Apps will have an impact on the organization of the supply chain activities from a technological perspective as well. In future, each employee will be equipped with this kind of mobile device, interact with colleagues, perform time-management and execute specific activities in the manufacturing process by using the Smartphone and Smartphone Apps. Specific Apps will be created to enhance the efficiency of the production processes—track and trace system of specific product components, or by assisting software for the human activities in the company. The medical industry is a leading industry which is already including the smartphone Apps within its supply chains. The term Industry 5.0 has been introduced to the research areas which are considered as the next industrial revolution which is a more systematic transformation that includes impact on civil society, governance and structures, and human identity in addition to solely economic/manufacturing ramifications.

References

1. Cardenas A, Amin S, Sastry S (2008) The 28th International conference on distributed computing systems workshops, Beijing, pp 495–500
2. Reiner A (2014) Industrie 4.0—advanced engineering of smart products and smart production. In: 19th International seminar on high technology, Piracicaba, Brasil, 9th October 2014
3. Ali S, Qaisar SB, Saeed H, Khan MF, Naeem M, Anpalagan A (2015) Network challenges for cyber physical systems with tiny wireless devices: a case study on reliable pipeline condition monitoring. Sensors (Basel) 15(4):7172–7205
4. Valdez CA, Brauner P, Schaar AK, Holzinger A, Ziefle M (2015) Reducing complexity with simplicity—usability methods for industry 4.0. In: 19th Triennial congress of the international ergonomics association (IEA 2015)
5. Matani AG (2001) Managing new product innovations. Ind Eng J 4(1):21–23
6. Matani AG (1999) Strategies for better waste management in industrial estates. J Ind Pollut Control 22(1):67–72
7. Qin E, Long Y, Zhang C, Huang L (2013) International conference on human interface and the management of information-HIMI 2013. Human interface and the management of information. Information and Interaction for Health, Safety, Mobility and Complex Environments 3(1):173–180
8. Shrouf F, Ordieres-Mere J, García-Sánchez A, Ortega-Mier M (2014) Optimizing the production scheduling of a single machine to minimize total energy consumption costs. J Clean Prod 67:197–207
9. Guo ZX, Ngai EWT, Yang C, Liang X (2015) An RFID-based intelligent decision support system architecture for production monitoring and scheduling in a distributed manufacturing environment. Int J Prod Econ 159:16–28
10. Hartenstein H, Laberteaux K (2010) VANET vehicular applications and inter-networking technologies, 1st edn. Wiley Publication, London
11. Carvalho H, Cruz-Machado V (2011) Integrating lean, agile, resilience and green paradigms in supply chain management (LARG_SCM). Supply Chain Management, InTech
12. Jeschke S, Brecher C, Song H, Rawat DB (eds) (2017) Industrial internet of things and cyber manufacturing systems. Springer International Publishing, pp 3–19
13. Macaulay J, Buckalew L, Chung G (2015) Internet of things in logistics. DHL Customer Solutions & Innovation, Germany
14. Nasser J (2014) Cyber physical systems in the context of Industry 4.0. In: 2014 IEEE international conference on automation, quality and testing, robotics. IEEE
15. Witkowski K (2017) Internet of things, big data, industry 4.0—innovative solutions in logistics and supply chains management science direct. Proc Eng 182:763–769
16. KPMG (2016) The factory of the future: industry 4.0—the challenges of tomorrow
17. Barreto L, Amaral A, Pereira T (2017) Industry 4.0 implications in logistics: an overview. Proc Manuf 13(1):1245–1252
18. Heuser L, Nochta Z, Trunk NC (2008) ICT shaping the world: a scientific view. Wiley Publication, London, ETSI
19. Lu YQ, Xu X (2017) A semantic web-based framework for service composition in a cloud manufacturing environment. J Manuf Syst 42:69–81
20. Bartevyan L (2015) Industry 4.0—summary report, L., DLG-expert report
21. McKinsey Digital (2015) Industry 4.0 how to navigate digitization of the manufacturing sector, pp 1–62
22. RY Y, Zhong A, Xu X, Klotz E, Newman ST (2017) Intelligent manufacturing in the context of industry 4.0: a review. Engineering 3:616–630
23. Hasan S, Siddique N, Chakraborty S (2013) Intelligent transport systems: 802.11-based roadside-to-vehicle communications. Springer Publication, New York

24. Shrouf F, Miragliotta G (2015) Energy management based on internet of things: practices and framework for adoption in production management. J Clean Product 235–246
25. Wagner T, Herrmann C, Thiede S (2017) Industry 4.0 impacts on lean production systems. Proc CIRP 63(2):125–131

Chapter 39
Integrating Science of Six Sigma to the Art of Marketing

K. Muralidharan and Neha Raval

39.1 Introduction

Six Sigma as a process improvement approach has lots of significance in a structured manufacturing environment. The growing pace of service industry and the need for making service offerings with more reliability have contributed to the popularity of Six Sigma to new heights. Many companies including GE, Motorola, and Bank of America are experiencing bottom-line benefits by applying Six Sigma to service processes like accounting and finance [40]. However, marketing being an organizational function is slow in the adoption of this methodology [39]. It has been observed that the marketing activity is completely a customer-centric organizational function, and therefore, its integration with Six Sigma could be the formula for all-round success of an organization [13].

Six Sigma Marketing (SSM) is a fact-based, data-driven disciplined approach to growing market share by providing targeted product/markets with superior value [13, 22, 31, 39]. By lauding components of Six Sigma and Marketing, SSM models them into a stalwart discipline designed to achieve lofty financial goals and gross sales. SSM is a disciplined method that focuses on more operative and well-organized marketing activity to attain customer value aims at optimum cost. These features of new Six Sigma approach put it in the cadre of next-generation Six Sigma, which focuses upon achieving monetary benefits by focusing on customer value instead of merely targeting defects reduction. Expanding its horizon from manufacturing, Six Sigma has been adopted by various business processes to improve their functionality. Marketing is a major part of the customer focus business process, it is using different tools to deploy Six Sigma effectively like Voice of Customer (VOC), Quality Function

K. Muralidharan (✉) · N. Raval
Department of Statistics, Faculty of Science, The MSU, Vadodara, India
e-mail: muralikustat@gmail.com

N. Raval
e-mail: njr.sky@gmail.com

P. K. Kapur et al. (eds.), *Decision Analytics Applications in Industry*, Asset Analytics,
https://doi.org/10.1007/978-981-15-3643-4_39

Deployment (QFD), SIPOC (Supplier-Input-Process-Output-Customer), Prioritizing Market opportunity matrix (P/M Matrix), etc. Many other process-specific tools can be used to integrate Six Sigma with marketing. Prime focus of SSM is to understand different elements of the marketing process and thrive for continuous improvement. If Six Sigma is science, then marketing is art. By applying Six Sigma to marketing, organizations can identify leading indicators of growth and become proactive about performance improvement. Major advantage of amalgamation of Six Sigma and marketing is improving decision-making process through fact-based approach instead of gut feelings and guess. Therefore, this new formulation has the ability to change the vision and mission of SSM to a successful venture for any organization.

To get insight into possible integration between Six Sigma and marketing, it is important to understand their key success factors. Section 39.2 of this paper discusses critical success factors of Six Sigma and Sect. 39.3 explain about success factors of marketing. Section 39.4 discusses the integration between Six Sigma and marketing with reference to contemporary interactive marketing. The article ends with a conclusion.

39.2 Components of Successful Six Sigma Implementation

Though Six Sigma has its success story with different companies, still many companies failed miserably to achieve desired results out of Six Sigma implementation [21]. As Moosa and Sajid (2010) mentioned "*The success and failure of Six Sigma programmes largely depend upon their implementation*", it is important to inspect components contributing to the successful Six Sigma deployment in the organization. These components are entrenched as "key elements" for the effective execution of quality initiative in literature. Important elements for effective implementation of Six Sigma are discussed by different authors [4, 12, 18, 19, 24]. The role of top management, process-based approach, and project selection are some of the ingredients of a successful Six Sigma implementation. Below we discuss these concepts in detail from a marketing point of view.

39.2.1 Role of Top Management

This component gains its importance invariably in the majority of Six Sigma literature [10, 11]. According to Best and Neuhauser [7], an organizational system gets its result not because of the individual working in the system, but also because of the top management's involvement in it. The top management plays an important role in designing quality based culture, as they are the custodian of quality movement in the organization. Following dimensions are vital for Six Sigma implementation with reference to top management:

- Adopting Six Sigma as a long term management philosophy, not merely as a tool kit to make quick fixes [23].
- Effective communication of clear organizational goals from shop floor to customer level helps to align efforts of each unit towards aimed end result [6].
- Identification of Six Sigma project based on business needs and thereby making connection between quality improvement efforts and financial benefits to the organization [39].
- Identification of key processes to achieve desired goals and proper allocation of resources for their optimum utilization [37].

Explicit description of possession at each level in the institution is necessary for successful Six Sigma project implementation. This is important for unambiguous communication and concretizes the positioning of authority to take proactive decisions and avoid delays in projects. Successful implementation of Six Sigma project greatly depends upon organizational culture. Positive internal climate created by top management can leverage successful Six Sigma implementation in the organization.

39.2.2 Process-Based Approach

Following a process-based approach to any problem-solving is a salient feature of Six Sigma methodology. As defined by the American Society for Quality (ASQ) process is defined as, "a set of interrelated work activities characterized by a set of specific input and value-added tasks that makeup procedure for a set of specific output". Process plays key role in creating a structure to keep improvement efforts align with goals. Effective identification of non-value-added activities and optimizing different phases of process by measurements, reduced inter-functional conflict and connected interdepartmental flow, better alignment of process towards customer need, employee satisfaction towards work, optimum use of resources, reduced cycle time, association between customer value and organizational process, delivering greater value thereby achieving financial gain are well-reported benefits of process-based approach [16, 20, 21, 28, 29, 42, 43].

Effective process management can be achieved through the following steps:

- *Identification and prioritization of key processes*: Identification of key processes can be done based on "exhaustive" or "high-impact" criteria [14]. Processes tightly matching with organizational vision [15] should be given maximum priority.
- *Mapping processes*: Process mapping is the technique used to examine interrelation and sequencing of processes. It is a visual display of the flow of process showing the relationship between activities, data, and objects involved in the production of output [8]. Damelio (2011) explained process mapping at three levels of performance as Organizational level mapping, Process level mapping and Job level mapping.
- *Measuring processes*: Process measurements are useful to manage the effectiveness and efficiency of the process. Measurements of effectiveness give insight

into the ability of the process to meet customer needs. Establishing an appropriate connection between market requirement and business process performance is the key element of process management [24]. Hence, we can say measurements are required for making process valid to the customer needs. Customer satisfaction score, customer loyalty measurement, repeat purchase intentions are examples of process effectiveness measurements. As Snee (1990) mentioned employment of statistical thinking and measurements in Six Sigma is based on following three principles: (i) all processes occurs in a system of interconnected steps, (ii) variation exists in all process, and (iii) understanding and analyzing variation are key to success.

- *Managing processes*: Six Sigma approach relies on a unique structure to manage processes. Processes are managed by assigning process owner who is responsible for the performance of process assigned to him [15]. Process owner can manage the process effectively by creating action plans, resource allocation and assignment of work. Other team members like Master Black Belts, Black Belts and Green Belts contribute significantly to effectively manage process. Some of the frequently used statistical methods like Control charts, Rolled throughput yield (RTY), process capability indices, DPMO (Defects per Million Opportunities) can be used as efficiency measurements while managing processes [41].

Although the efficiency of a process depends on many of the above-mentioned process components, it is essential to establish good communication between voice of business (VOB) and voice of customer (VOC). Such communication enables marketing people to orient the process capability (sigma level per se) to the sales components. Figure 39.1 shows such a process that results in identifying the variation inherent in the process and the necessary inputs needed to meet the customer requirements. The overall assessment of the process performance becomes a by-product of this situation. Therefore, to enhance the efficiency of the marketing process, the Six Sigma quality professionals and marketing professionals should mutually identify the core areas of expertise, which works best for the betterment of organizational growth.

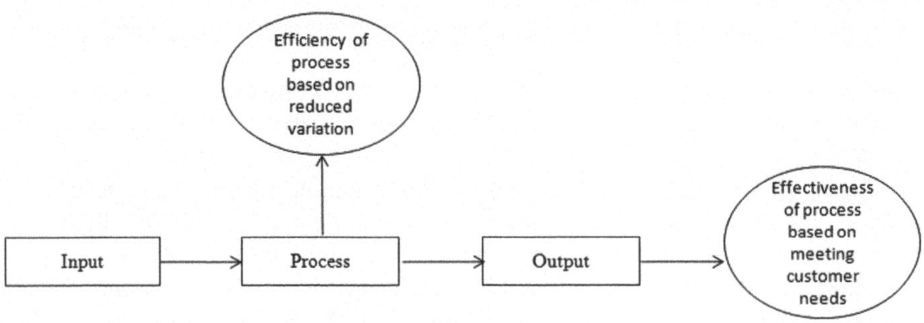

Fig. 39.1 Effectiveness and efficiency of process

39.2.3 Project Selection

To satisfy customer needs projects must be selected and aligned with business objectives [27]. The right selection and prioritization of projects is one of the critical success factors of a Six Sigma Programme [2]. Improper projects, though executed properly, will have a trifling effect on business and organizational goals [36]. "Top-down" and "bottom-top" approaches are proposed for project selection in Six Sigma literature [21]. Top-down approach is based on perpetuating required organizational change for the successful implementation of Six Sigma from top management to down the line [36]. Considering major business issues or improvement opportunities, building an effective team to bring changes at each level of the organization, identifying key processes and establishing their key performance indicators, examining process baseline and identifying opportunities for improvement are the major aspects best addressed through top-down approach.

"Bottom-top" approach can best address issues of operational level of the organization. Initiation of such a project is mainly based on internal voice of customer. Bottom-top projects are generally narrowly focused, not tightly connected with strategic requirements of the organization. For example, production manager addressing local issues related to waste, supplier or engineering needs, bottom-up approach can be used [21]. Bottom-top approach is generally used by the companies in the infancy of Six Sigma implementation and shows the need to develop a data-driven process and they are bound to fail if not getting support from top management.

39.3 Components Critical to Marketing

High commoditization of physical products leads contemporary firms to compete in the market based on peripheral services. These services are bringing a competitive advantage to the firms instead of merely tangible products. Organizations traditionally focusing upon product quality, product advancement are now shifting their focus to their service functions like customer support, finance, human resource and marketing. Marketing as an organizational function has a stand of repute to establish firm-customer communication channel. As Drucker (1954) mentioned, "Business has only two functions- marketing and innovation". Marketing is the key organizational function facilitating customer–product–profit link through effective communication. Inherent creativity required for marketing put it in the category of art. Considering marketing merely as art leads to issues related to the credibility of this function in the organization. The major components of a successful marketing programme are discussed below.

39.3.1 Understanding Customer

Customer focus is the bare important characteristic of marketing. Marketing is responsible for revenue growth, require being attentive towards changing need of customers. As mentioned by Muralidharan [30] customer focus should be considered as major *leading indicator* for marketing to achieve success. The author defined leading indicators as "the indicator that precedes the occurrence of something". Since customer focus is the prerequisite to achieve customer loyalty which paves to increased revenue, lower acquisition and retention cost and increased profitability it is considered as major leading indicator for marketing. Marketing plays a vital role in understating customers based on two major aspects:

- *Understanding customer needs*: Purposeful positioning based on understanding customer's need is one of the major winning dimensions for marketers now a day [5]. As explained by Almquist et al. [1] customer need is the complex combination of price, quality, feature and other dimensions. Hence understanding what customers "value" becomes critical for the companies to survive in the market. Conceptualization of value is based on benefits customers are getting against price they are paying for [25]. As mention by Almquist et al. [1] organizations can compete in the market based on value proposition. Companies proposing value on multiple elements and selecting value elements strategically can outperform competitors.
- *Total experience*: Tough initial marketing model focus upon customer satisfaction as an offering evaluation criteria [32], blurring the difference between tangible offering and services lead to focus upon total experience of the customer. High performing companies are focusing on "total customer experience" which emphasis on creating an emotional connection between customer and brand [31]. Firms now a day are focusing upon understanding what their customer value in their offering. This approach requires intense firm-customer communication channel at each level of product/service consumption. Each touchpoint of firm–customer communication becomes critical in this approach. All these touchpoints offer opportunity for the firms to connect their customers with their offering and create emotional bonding. Enriching "total customer experience" leads to lasting customer loyalty, which is the major source to improve profitability [37]. As shown in Fig. 39.2 customers purchase offer based on expected value and repurchase offering based on their total experience during consumption of offering.

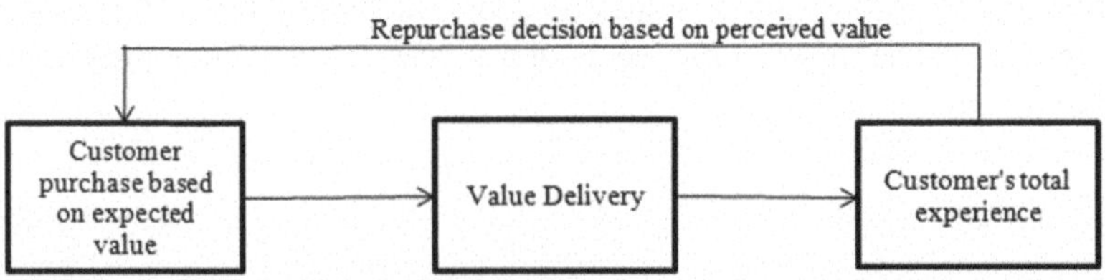

Fig. 39.2 Customer touch points based on "value" aspect

39.3.2 Role of Communication

Need for becoming more flexible to customer's demands required different types of organizational culture. As mentioned by Arons et al. (2014) marketing is covering almost every function of business and hence no longer a discrete department. Schultz (1996) observed the inevitability of integrated marketing communication. As the author mentioned irrespective of marketer's desire consumers integrate messages they received from different channels of organization and this integration may have a negative impact on organization reputation. Unavoidable fervour from customers to integrate marketing communication makes cross-functional communication obligatory [35]. Customer's purview to understand integrated organizational message enkindle need to coordinate communication over different organizational functions. Communication integration is moving from internally planned marketing communication to empowering customers with needed information to make purchase decisions made cross-functional communication inevitable. Inter functional communication being a strategic integration, focus upon common mission and strategy but shared objectives and outcomes become the outcome of the communication [35].

39.3.3 Adoptability of Marketing to Different Context

Marketing as a business function is changing rapidly. This change is ranging from traditional product marketing to service marketing, relationship marketing, total customer experience, network-based marketing and most recently interactive marketing. Marketing has grown as sophisticated filed from its infancy mode called product marketing, where product marketing was focusing upon identification of "marketing mix" for an effective marketing programme. This gave rise to well-known marketing 4P's (product-place-price-promotion) concept. Globalization, technological innovations, increased affluence offers an array of products to the customers. Organizations are shifting their focus to offer compatible services with product. This new service focused environment requires long-term relationships with customers and hence gave rise to relationship marketing [17]. This relationship environment can work efficiently only if all stakeholders like suppliers, employees, customers are working in synchronization. This resulted in network base marketing [34]. Increased pressure to reduce marketing cost, need of maintaining close relationship with customers, addressing the need of customers from any part of the world, active involvement of customers in product and process design, extensive spread of internet lead to the altogether different marketing environment. Marketers are using digital platforms to reach out to the customers, to understand their needs, to maintain a close relationship and to keep them engaged. This contemporary form of marketing is well known as digital marketing.

39.3.4 Role of Data in Current Marketing Management

Data plays an important role to survive and excel in the market. Effective marketing strategy proposes deep insight into customer requirements, customer engagement, competitor's proposition, and cost-benefit analysis and customer satisfaction. The use of these data helps marketing as a function to leverage its effectiveness.

Understanding customer requirements is what differentiates marketing from sales [26]. Instead of *push system* promoted by sales, marketing focus upon *pull system.* Goal of sales is on volume of selling of offerings to make maximum profit. Whereas the effort of marketing is to propose offers that have value to the customer. Instead of changing the customer's mind to fit product, marketing focus on changing the reality as per customer requirements. In order to do that data can be helpful in the following ways:

- To understand what customers really value in their offerings
- For appropriate value proposition which can help marketers to differentiate their offerings from competitors
- Appropriate value proposition helps the customer to make a better choice from the offerings available in the market

In the era of rapid change and digitization lack of engaging customers with your offerings is the ingredient to failure in the market. Technology has changed the way in which we form and maintain the relationship with each other. None of the fields have ever changed as marketing due to technological advancement. Marketers are using digital platforms extensively to keep their customers informed and engaged. These digital platforms are loaded with rich information about customer demographics. They also provide information about customer behaviour on digital platforms. Important data generated through digital platforms like website traffic, likes and share, clicks and navigation on different pages help to design an effective marketing strategy. As mentioned by Cascio and Montealegre [9], big data generated through these digital platforms are transforming the business world by providing an opportunity to serve the customer better and faster.

39.4 Integration Between Six Sigma and Marketing

Recent advancement in Six Sigma and Marketing offers an ingenious environment for integration of these approaches. From the discussions above, it is apparent that communication, digitalization and the data analytics (or Six Sigma per se) plays a vital role in charging the marketing activities to a profit-generating activity. The phase-wise approach of Six Sigma philosophy integrated with marketing activity can pave way for establishing sustainable business in any organization, provided top-level commitment is in place.

39.4.1 Six Sigma Expanding Horizon to Services

Though originated in manufacturing, Six Sigma expanding its horizon to services. Six Sigma has been largely applied into different services like financial, medical, education, accounting, etc. as well as service functions of organization like operation, sales, human resource, etc. Organizations nowadays are competing based on services they provide with their offering. Hence, effective management of services is the need of the time. Poor performance of services leads to higher defects rate resulting into lower sigma level and higher operating cost [3]. Six Sigma as a process improvement method is contributing a lot to improve service quality. The very nature of service is different from product-based approach on the following aspects:

- Services don't exchange parts but they exchange information
- Though process is defined, in-service environment enactment of process is highly variable
- Output measures are vague and perceptual
- Lack of agreed service standards results into hazy approach to defect reduction
- No clear start and endpoint of service delivery.

Six Sigma as a data-based approach can help to overcome the above issues based on:

- Following project-based approach to address quality issues of services
- Identifying the proper matrix to measure vague phenomena
- Following measurement-based process management approach to assess process quality at different stages of progression
- Defining measurements for service defects
- Examine quality of process through a number of defectives and sigma level of the process.

39.4.2 Marketing Big Data

Marketing and marketing activities are changing rapidly. Amalgamation of different elements like increased customer demand, widespread culture of digitization, striving for effectiveness, network-based communication are creating unique opportunities for marketing to become more accountable for investment to it. These items are adding many dimensions to the information, and hence the data becomes huge. Decision-making under these circumstances is the major challenge for scientists and analysts. Hence understanding the environment which provides the scope of integration between structured Six Sigma approach and creative marketing field necessitates within and outside the organization [38]. In Table 39.1, we concise, the components providing scope for integration between Six Sigma and marketing and their joint effect from a marketing perception.

Table 39.1 Integrating components of Six Sigma and marketing

Component	Six Sigma	Marketing	Integrating effect
Essential characteristic	Structured data-driven approach	Seen as creative and artistic field to engage the customer in exchange	Six Sigma providing framework to channelize marketing creativity
Value	Realized by reducing number of defects	Realized by establishing customer–product–financial link	Defining appropriate measure for successful customer-product-financial link
Process	Focusing on reducing number of defect at each process stage	Focusing on making communication more effective at each touchpoint of customer	Making each customer touchpoint more effective through examining sigma level of the process
Data	Used extensively based on matrix structure	Data used to understand customer and market need	Matrix-based approach to understand customer and market need can be used to keep marketing efforts align

Increasing trend of interactive media leads to a new era of marketing called digital marketing. The use of interactive media is changing market place as never before. Interactive marketing possesses unique feature compared to other traditional marketing methods in the following way:

- It is one of the methods of marketing to reach the target market using platforms of technologies such as websites, blogs, social media, paid ads, etc.
- It enables new form of interaction between customers and marketers, e.g. two-way interactivity, seamless transactions, addressability, on-demand availability, customization [33].

Interactive digital marketing differs from traditional marketing in many ways. From business perspectives digital marketing differs based on following aspects compare to traditional marketing:

- Digital platform provides business opportunity irrespective of business size
- Digital platform is less costly compare to other methods of traditional marketing
- Since digital marketing facilitates automatic data generation, it is easier to examine effectiveness of marketing efforts and customer requirements
- At digital platform, refining marketing strategy is very flexible and easy
- Digital marketing provides a platform to encourage customers to visit website, make a purchase decision and give feedback. Hence making the whole process of purchase less time-consuming.

Since an interactive marketing environment is non-instructive in nature, people make choices based on communication. People themselves are searching for information, hence the only thing remaining in the hand of a marketer is providing them

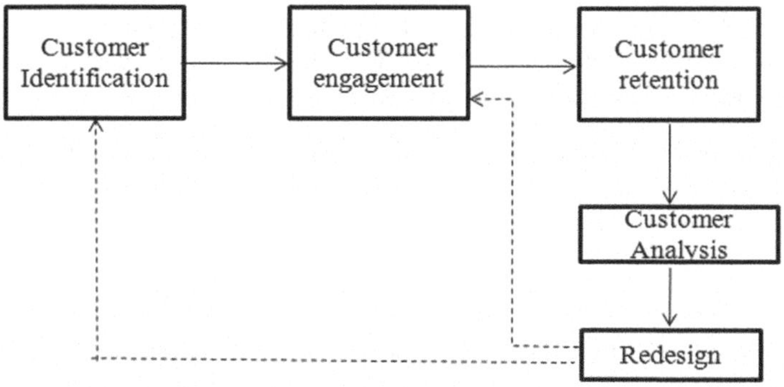

Fig. 39.3 Conceptual framework of interactive marketing

relevant information. Based on this unique conceptual framework of digital marketing one can influence the customer very easily and get their endorsement for the product or service (Fig. 39.3).

Figure 39.3 depicts the customer involvement of interactive marketing at various stages and their interrelationships. Further, the customers can be appraised about the sigma level of each process and their interpretations of effective marketing.

- *Customer identification*: This web-based interactive marketing platform provides rich information about customer segments. Through this platform, it is possible to target customers based on different demographic characteristics, interests, organization, position, etc. Not only that this platform provides different types of discussion forms, groups and community which can be used effectively to identify potential customers, but also, helps to identify those customers who are really interested in your product and services.
- *Customer engagement*: Major focus of interactive marketing is to engage customers with offering. Engaging customers through different modes like web sites, blogs, discussion forums, and social media is important to establish emotional connection between customers and brands. Engaging customers through substantial content on different platforms, making platform interactive, inviting customers to give feedback are the different ways to increase customer engagement.
- *Customer retention*: Once customer visits website or any other web-based application, next critical focus is on bringing customer again to your online platform. Hence interactive marketing is not a onetime process but a continuous process to engage them with business. This can be done by attracting customers through frequent updates on social media and other platforms.
- *Customer analysis*: Digital platform provides an unprecedented opportunity to marketer to learn about their customer demographics, attitude, behaviour, test, preference, etc. This helps the marketer to decide what type of information is worth and how to design an effective marketing strategy based on these information.

- *Redesign*: Redesign is the process of making necessary changes on a digital platform based on customer analysis. Redesigning is important to address the changing need of customers or to address the needs of the new market.

Application of Six Sigma to any function required process-stage approach. With reference to the above conceptual framework, five-step Six Sigma approach for interactive marketing is proposed as Plan–Depict–Reach–Analyze–Optimize (PDRAO), where

- *Plan*: To articulate the business strategy and define marketing goals clearly
- *Depict*: Design modules of digital presence like creating a website, blogs, etc.
- *Reach*: Plan to reach out maximum customer base and design strategy to engage them
- *Analyze*: customer engagement on digital platform and identify their need
- *Optimize*: Make the required changes to keep customers revisit the digital platform.

39.5 Conclusions

Application of Six Sigma to other domains required momentous knowledge of Six Sigma methodology and knowledge of the respective domain. In reference to this, various critical characteristics of Six Sigma and marketing functions are discussed in this paper. Process-based approach; value delivery and use of data are found to be major integrating factors between Six Sigma and marketing. Realizing the importance of the digital era, conceptual framework for interactive marketing is proposed. Finally, the five-step process PDRAO (Plan–Depict–Reach–Analyze–Optimize) is proposed to integrate interactive marketing structure with Six Sigma.

References

1. Almquist E, Senior J, Bloch N (2016) The element of value: measuring and delivering what consumners really want. Havard Bus Rev
2. Antony J (2004) Some pros and cons of six sigma: an academic perspective. TQM Mag 16(4):303–306. https://doi.org/10.1108/09544780410541945
3. Antony J (2006) Six sigma for service processes. Bus Process Manag J 12(2):234–248. https://doi.org/10.1108/14637150610657558
4. Antony J, Banuelas R (2002) Key ingredients for the effective implementation of Six Sigma program. Meas Bus Excel 6(4):20–27
5. de Swaan Arons M, van den Driest F, Weed K (2014) The ultimate marketing machine. Havard Bus Rev
6. Beer M (2006) Why total quality managment programs do not persists: the role of managment quality and implications for leading a TQM trasformation. Decis Sci 34(4):623–642
7. Best M, Neuhauser D (2005) Heroes & Martyrs of Quality and Safety, W Edwards Deming: father of quality management, patient and composer. Qual Saf Health Care 14:310–312

8. Biazzo S (2002) Process mapping techniques and organisational analysis: lessons from sociotechnical system theory. Bus Process Manag J 8(1):42–52
9. Cascio WF, Montealegre R (2016) How technology is changing work and organizations. Annu Rev Organ Psychol Organ Behav 3:375–379
10. Chakrabarty A, Chuan Tan K (2007) The current state of six sigma application in services. Manag Serv Qual: Int J 17(2):194–208. https://doi.org/10.1108/09604520710735191
11. Coronado RB, Antony J (2002) Critical success factors for the successful implementation of six sigma projects in organisations. TQM Mag 14(22):92–99
12. Creveling CM, Hambleton L, McCarthy B (2007) Six sigma for marketing processes: an overview for marketing executives, leaders and managers. Prentice Hall
13. Damelio R (2011) The basics of process mapping, 2nd edn. Productivity Press
14. Davenport TH, Short JE (1990) The new industrial engineering: information technology and business process redesign. Slogan Manag Rev
15. DeToro I, McCabe T (1997) How to stay flexible and elude fads. Qual Prog 30:55–60
16. Drucker PF (1954) The practice of management. Harper and Row, New York
17. Grönroos C (2004) The relationship marketing process: communication, interaction, dialogue, value. J Bus Ind Mark 19(2)
18. Hahn GJ, Hill WJ, Hoerl RW, Zinkgraf SA (1999) The impact of six sigma improvement—a glimpse into the future of statistics. Am Stat 53(3):208–215
19. Hammer M (2002) Process management and the future of six sigma. MIT Sloan Manag Rev 43(2)
20. Hammer M (2007) The Process Audit. Havard Bus Rev
21. Harry MJ, Schroeder R (2005) Six sigma: the breakthrough management strategy revolutionizing the world's top corporations. Currency
22. Hensley RL, Dobie K (2005) Assessing readiness for six sigma in a service setting. Manag Serv Qual: Int J 15(1):82–101
23. Hooper JH (2001) The process approach to QMS in ISO 9001 and ISO 9004. Qual Prog
24. Henderson KM, Evans JR (2000) Successful implementation of six sigma: benchmarking general electric company. Benchmarking: Int J 7(4):260–280
25. Lanning MJ, Michaels EG (1988) A business is a value delivery system. McKinsey staff paper, 41
26. Levitt T (1960) Marketing Myopia. Harvard Bus Rev
27. Lynch DP, Bertolino S, Cloutier E (2003). How to scope DMAIC projects: the importance of the right objective cannot be overestimated. Qual Eng Appl Res
28. McCormack KP, Johnson WC (2001) Business process orientation: gaining the e-business competitive advantage. CRC Press
29. Moosa K, Sajid A (2010) critical analysis of six sigma implementation. Total Qual Manag 21(7):745–759
30. Muralidharan K (2016) Six sigma for organizational excellence: a statistical approach. Springer
31. Mascarenhas OA, Kesavan R, Bernacchi M (2006) Lasting customer loyalty: a total customer experience approach. J Consum Mark 23(7):397–405
32. Parasuraman A, Zeithaml VA, Berry LL (1985) A conceptual model of service quality and its implications for future research. J Mark 49(4):41–50
33. Parsons A, Zeisse M, Waitman R (1998) Organizing today for the digital marketing of tomorrow. J Interact Mark 12(1)
34. Payne A, Ballantyne D, Christopher M (2005) A stakeholder approach to relationship marketing strategy: the development and use of the "six markets" model. Eur J Mark 39(7):855–871
35. Porcu L, del Barrio-García S, Kitchen PJ (2012) How integrated marketing communications (IMC) works? a theoretical review and an analysis of its main drivers and effects. Comunicación y Sociedad 15(1):313–348
36. Ray S, Das P (2010) Six sigma project selection methodology. Int J Lean Six Sigma 1(4):293–309
37. Reichheld FF, Sasser WE Jr (1990) Zero defections: quality comes to services. Havard Bus Rev

38. Reidenbach E (2009) Six sigma marketing: from cutting costs to growing market share. Amer Society for Quality
39. Salzarulo PA, Krehbiel TC (2012) Six sigma sales and marketing: application to NCAA basketball. Am J Bus 27(2):113–132
40. Schultz DE (1996) The inevitability of integrated communications. J Bus Res 37:139–146
41. Škrinjar R, Štemberger MI, Hernaus T (2007) The impact of business process orientation on organizational performance. Paper presented at the informing science and IT education joint conference
42. Škrinjar R, Vukšić VB, Štemberger MI (2010) Adoption of business process orientation practices: Slovenian and Croatian survey. Bus Syst Res 1(1–2):1–50
43. Snee RD (1990) Statistical thinking and its contribution to Total Quality. Am Stat 44(2):116–121

Chapter 40
Quality of Maintenance Activity

Anil Rana

40.1 Introduction

More and more industries these days rely on offloading their major maintenance tasks to external firms for economic reasons. Minor maintenance tasks, such as inspections, lubrication, change of filters etc. are carried out by the in-house skeletal maintenance workforce whereas the major maintenance tasks such as overhauls of diesel engines or gearboxes are offloaded to external agencies. After each such major maintenance tasks the question regarding how good or bad the maintenance task was, remains to be answered. This question can be answered subjectively by the maintenance supervisors/experts based on the system parameters recorded after completion of the maintenance task compared with those recorded when the system was newly commissioned.

One of the problems with this approach is that the parameters that are recorded are often correlated with each other through non-linear functions. On top of that, the assessment of the maintenance experts in themselves is subjective and may differ with one another. If we could create an aggregation of expert opinions, we could then train an adaptive neuro-fuzzy inference system to replicate the decision-making process of experts. We would then have no need for deputing the maintenance supervisors for making an assessment on the quality of maintenance activity carried out by an external firm. This would be particularly beneficial, for instance, for a power generation plant which has several similar diesel engines based systems at various locations in a country. Neuro-fuzzy techniques have been studied and successfully applied to various kind of problems, but the one being studied here involves the aggregation of expert opinions before we can use them [1–11].

A. Rana (✉)
Bhartiya Skill Development University, Jaipur, India
e-mail: anil.rana@ruj-bsdu.in

P. K. Kapur et al. (eds.), *Decision Analytics Applications in Industry*, Asset Analytics,
https://doi.org/10.1007/978-981-15-3643-4_40

40.2 Aggregation of Expert Opinions

40.2.1 Fuzzy Input Parameters of a Diesel Engine

We take a case of a 108 MW power generation plant fitted with four W20V32 engines. Similar engines are also fitted in other locations in the country. A careful selection of system parameters is done that may be crucial for assessing the quality of given major maintenance activity. The selected parameters are SFC (specific fuel consumption), intake temperature, quality of emissions, power produced, overall vibration, system leakages. The measure of some of these parameters is fuzzy and correlated with each other. For example, the specific fuel consumption for the diesel engines is as given below (Table 40.1) by the OEM (original engine manufacturer), however under field conditions, it is not always possible to achieve the design values. Supposing post overhaul, the fuel consumption is 186 g/KW-h at 70% load, can we say that the maintenance performance has been poor? The SFC will also increase as the load decreases; can we then say that the maintenance activity has been poor? The maintenance supervisors would usually use linguistic terms to measure the entire maintenance activity performance, under a given value of SFC and other five parameters.

More important than the non-linear co-related, fuzzy characteristics of the input parameters is the difference in subjective assessments by the supervisors themselves. Hence we need to use a method for arriving at an aggregated measure of their assessment. The process used for such aggregation is given in Fig. 40.1, in the shaded box. Once the aggregated measures are created, we will have with us a set of fuzzy input parameters and an aggregated set of measurements of the quality of the major maintenance activity or outputs. A large collection of such inputs and outputs can then be used to train an adaptive neuro-fuzzy tool (ANFIS) [12] to replicate the decision-making process of maintenance supervisors. Figure 40.1 shows the process used for aggregation and comparison. The errors in comparison of these values are used for adjusting the parameters of ANFIS.

We present an example of a set of input values in fuzzy terms as given in Table 40.2 and will go through the entire process shown as shaded in Fig. 40.1 of the paper including the FMADM process [13, 14]. In the end, we will arrive at a crisp defuzzified score which will give us the measure of maintenance activity performance (in both the fuzzy and crisp terms). Crisp scores so generated (**50 were generated for**

Table 40.1 Specific fuel consumption at different loads

Load in %	SFC (g/KW h)
100	183
85	180
75	180
50	190
30	208

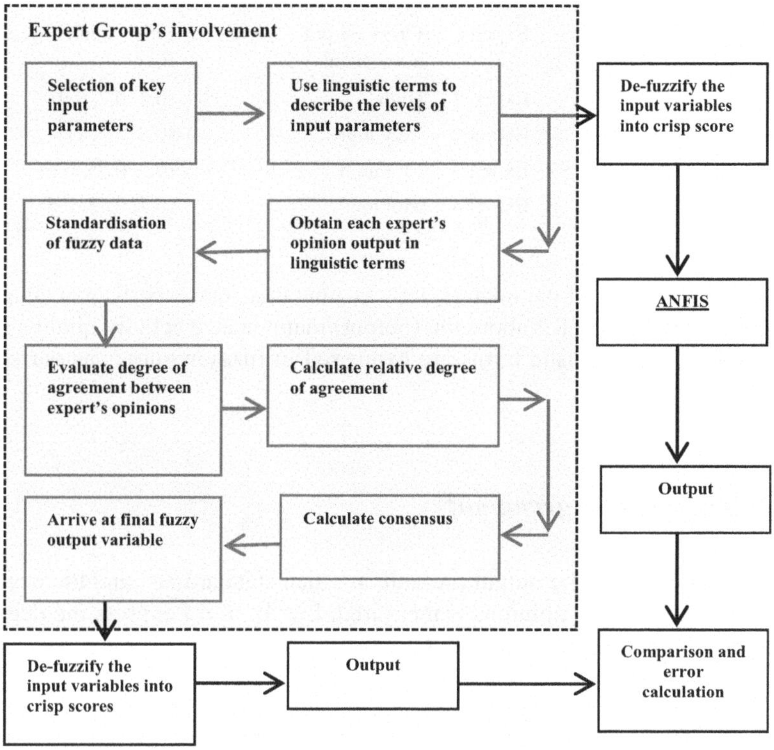

Fig. 40.1 Process: generation of data using expert group's opinion

Table 40.2 Parameters expressed in linguistic terms

Parameters	Fuzzy measure (scale 8) [13]
SFC	Mol low
Ambient Temp	Medium (Fair)
Emission	Low-very low
Leakages	Low-very low
Load	Fairly high
Vibration	Mol high
Performance (output measure)	??

the paper) can then be used to train the Neuro-Fuzzy inference system. Once trained with sufficient data, ANFIS can help measure the maintenance activity quality on any of the power generation plants fitted with the same engines. Since most of the major maintenance activities are these days outsourced to maintenance expert firms, a decision given out by ANFIS will be acceptable as the most unbiased estimate of the quality of their maintenance activity.

Table 40.3 Experts' assessment of maintenance activity effectiveness

Experts	Fuzzy output assessment	Fuzzy number (scale 8) R_i
Expert 1	High	(0.5, 0.7, 0.9)
Expert 2	Very high	(0.8, 0.9, 1)
Expert 3	Mol high	(0.5, 0.55, 0.6)
Expert 4	Mol low	(0.4, 0.45, 0.5)

Given the above set of parameters, let's assume that four experts give an assessment in linguistic terms 'R_i' about the (output) maintenance activity quality shown in Table 40.3. The linguistic terms are expressed in fuzzy numbers as per scale 8 [13].

40.2.2 Degree of Agreement

The expert opinions on the output measure are then standardized and the degree of agreement between their opinions is measured. Let A_{ij}, $i \neq j$ express the degree of agreement between expert 'i' and 'j' then $A_{ij} = A_{ji}$ is given as

$$A_{ij} = 1 - \left[\frac{|a_i - a_j| + |b_i - b_j| + |c_i - c_j|}{3} \right] \tag{40.1}$$

where (a_i, b_i, c_i) are fuzzy numbers expressing the assessment of expert 'i'

The agreement scores deduced using Eq. 40.1, between various experts is given in Table 40.4: (note: $A_{ij} = A_{ji}$).

Table 40.4 Measure of agreement between the experts

Agreement between experts	Scores
A_{12}	0.8
A_{13}	0.85
A_{14}	0.75
A_{23}	0.65
A_{24}	0.55
A_{34}	0.9

Table 40.5 Average agreement and consensus scores

Experts	Average degree of agreement AA_i	Consensus scores CS_i
1	0.8	0.266
2	0.667	0.222
3	0.8	0.266
4	0.733	0.244

40.2.3 *Average Degree of Agreement*

Average degree of agreement for experts denoted by AA_i is calculated for every expert 'i' given by:

$$AA_i = \frac{1}{n-1}\sum_{\substack{i=1 \\ i \neq j}}^{i=n} A_{ij}; \text{ where } n = \text{number of experts} \tag{40.2}$$

40.2.4 *Consensus (Relative) Degree of Agreement*

Once the above measure has been achieved, consensus score for each expert is calculated as:

$$CS_i = \frac{AA_i}{\sum_{i}^{n} AA_i}; \text{ where } n = \text{number of experts} \tag{40.3}$$

Average degree of agreement and consensus measures for experts is given in Table 40.5.

In the end, the final aggregated score 'FS' (output or the maintenance activity effectiveness) in fuzzy score is calculated as follows:

40.2.5 *Final Aggregated Score and Crisp Score Generation*

$$FS = CS_1 \otimes R_1 \oplus CS_2 \otimes R_2 \oplus CS_3 \otimes R_3 \oplus CS_4 \otimes R_4 \tag{40.4}$$

where, R_i is the Fuzzy number expressing the initial assessment of expert 'i' about maintenance effectiveness as shown in Table 40.3 and $\otimes$, $\oplus$ are fuzzy multiplicative and addition operators.

Hence, the FS (fuzzy score), for example, is (0.542, 0.642, 0.743) which is "Fairly Good" in linguistic terms. The fuzzy number can be converted into crisp scores in accordance with Eqs. (40.5)–(40.8). The final crisp score is "0.629". Since ANFIS generates only crisp scores, this score can be compared with the ANFIS output.

$$\mu_{\max}(x) = \begin{cases} x, & \text{for } 0 \leq x \leq 1 \\ 0, & \text{otherwise} \end{cases} \tag{40.5}$$

$$\mu_{\min}(x) = \begin{cases} 1 - x, & \text{for } 0 \leq x \leq 1 \\ 0, & \text{otherwise} \end{cases} \tag{40.6}$$

The right and the left score of an output fuzzy number 'O' can be found as

$$\begin{aligned} \mu_R(O) &= \underset{x}{Sup}[\mu_O(x) \Lambda \mu_{\max}(x)] \\ \mu_L(O) &= \underset{x}{Sup}[\mu_O(x) \Lambda \mu_{\min}(x)] \end{aligned} \tag{40.7}$$

The crisp score can now be given as $\mu_T(O) = [\mu_R(o) + 1 - \mu_L(o)]/2$ (40.8)

40.3 Training of Neuro-Fuzzy System

Once enough number of aggregated outputs (similar to the example as given above) are generated, we can feed the inputs and outputs to an ANFIS system. 50 such sets of input and outputs were generated. 40 sets were used for training and 10 sets for testing the data. Random combination of sets were created to assess the replication capacity of ANFIS. The figures below give the structure, surface views and displays of training and checking data (using Matlab R2016) (Figs. 40.2, 40.3 and 40.4).

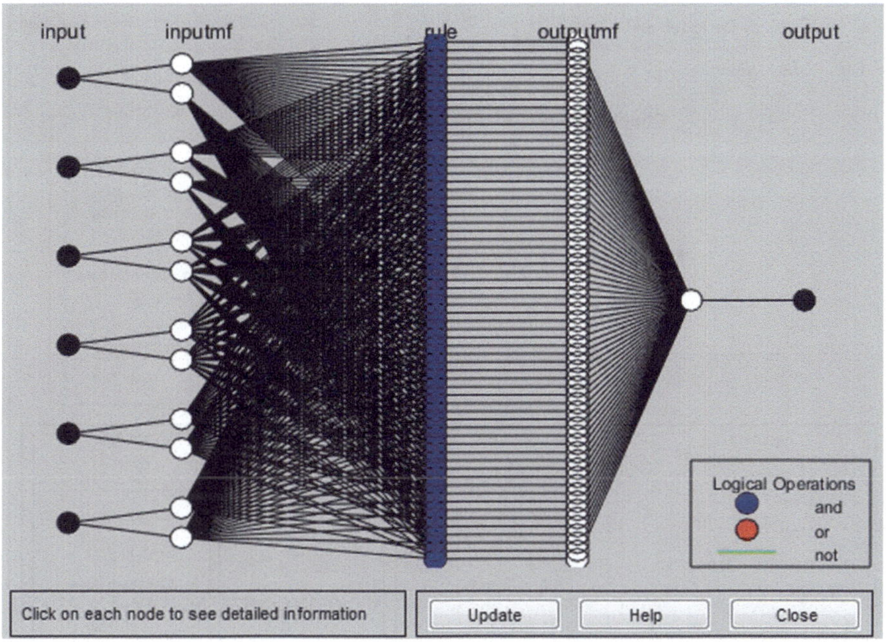

Fig. 40.2 ANFIS structure

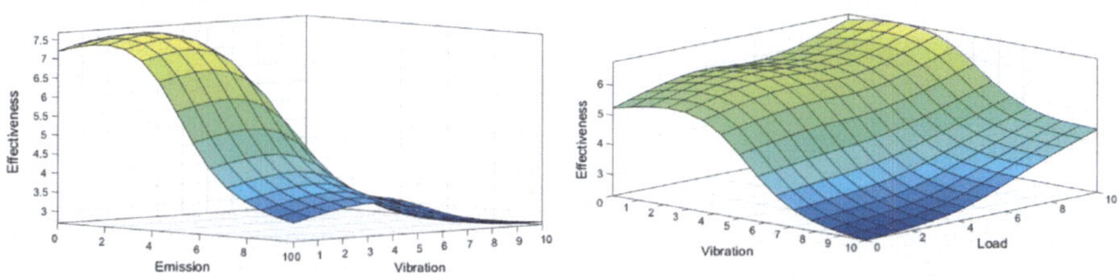

Fig. 40.3 Surface view, two parameters at a time [14]

40.4 Result

The result of an ANSIS trained system matches very closely with the subjective evaluation of the quality of maintenance activity by the supervisors as shown in Fig. 40.4.

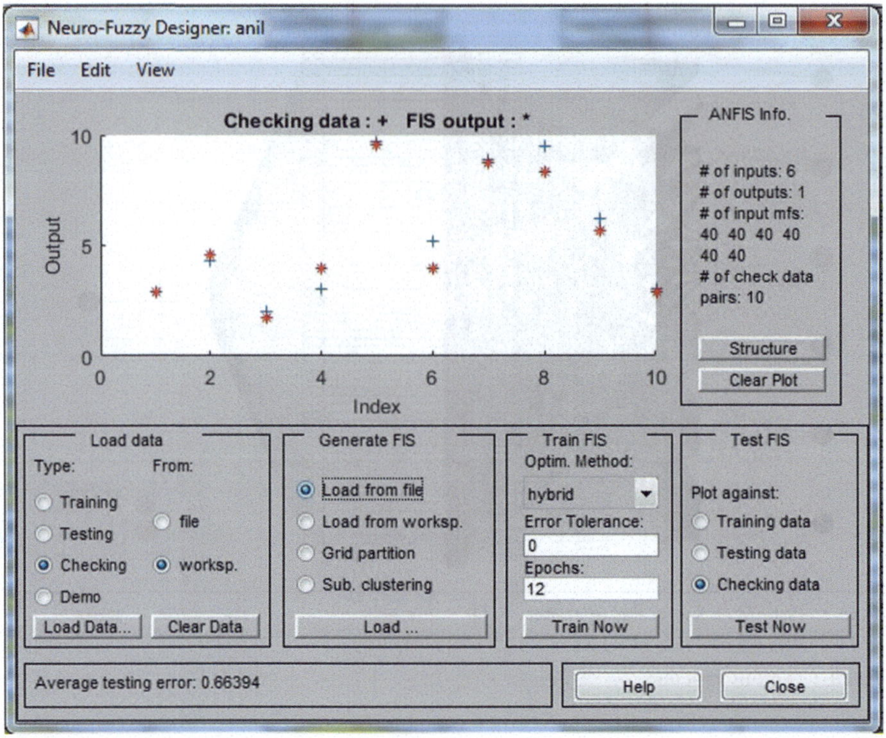

Fig. 40.4 Check data plot in Matlab

References

1. Raza J, Liyanage JP, Al Atat H, Lee J (2010) A comparative study of maintenance data classification based on neural networks, logistic regression and support vector machines. J Qual Maint Eng 16(3):303–318
2. Horikawa S, Furuhashi T, Uchikawa Y (1992) On fuzzy modelling using fuzzy neural networks with the back-propagation algorithm. IEEE Trans Neural Netw 3(5):801–806
3. Tanaka K, Sano M, Watanabe H (1995) Modelling and control of carbon monoxide concentration using a neuro-fuzzy technique. IEEE Trans Fuzzy Syst 3(3):271–279
4. Lin Y, Cunningham G (1995) A new approach to fuzzy-neural system modelling. IEEE Trans Fuzzy Syst 3(2):190–197
5. Zio E, Gola G (2009) A neuro-fuzzy technique for fault diagnosis and its application to rotating machinery. Reliab Eng Syst Saf 94:78–88
6. Du J, Er MJ (2004) Fault diagnosis in air-handling unit system using dynamic fuzzy-neural network. In: Proceeding of the 6th international FLINS conference 2004, pp 483–488
7. Marseguerra M, Zio E, Avogadri P (2004) Model identification by neuro-fuzzy techniques: predicting the water level in a steam generator of a PWR (Pressurized water reactor). Prog Nucl Energy 44(3):237–252
8. Salmalian K, Soleimani M (2011) ANFIS and neural network for modelling and prediction of ship squat in shallow waters. Int J Math Models Methods Appl Sci 5(4):848–856
9. Valcic M, Antonic R, Tomas V (2011) ANFIS based model for ship speed prediction. J BRODOGADNJA (shipbuilding) 62(4):373–382

10. Lin CT (1994) Neural fuzzy control systems with structure and parameter learning. World Scientific, New York
11. Ayoubi M, Isermann R (1997) Neuro-fuzzy systems for diagnosis. Fuzzy Sets Syst 89:289–307
12. Jang JS (1993) ANFIS: Adaptive-network based fuzzy inference system. IEEE Trans Syst Man Cybern 23(3):665–685
13. Chen SJ, Hwang CL (1992) Fuzzy multiple attribute decision making: methods and applications. Springer, New-York
14. Rana A, Koroitamana E (2018) Measuring maintenance activity effectiveness. J Qual Maintenance Eng 24(4):437–448. https://doi.org//10.1108/JQME-11-2016-0061

Chapter 41
Maintenance of Railway Infrastructure Using Cyber-Physical Systems

Adithya Thaduri, Ajit Kumar Verma, and Uday Kumar

41.1 Introduction

Traditionally, industries operate in disintegrated silos with most of the divisions/departments within the organization are not integrated as it should be. Due to this issue, time to take decision in the event of any hazard, vulnerability, loss of profit, etc. is long enough so that it can significantly affect the performance and couldn't meet the ever-growing demands of customers. Hence, industries are considering a way of integrating several systems so that they can increase their productivity and be it in competition.

Cyber-physical systems (CPS) are the upcoming jargon that resembles the integration of physical systems with cyber capabilities such as cloud, networking, computation, etc. It is being in the testing phase in a lot of applications mainly, medical, automation, manufacturing and aviation to take advantage of this integration. It is very important to consider that most of the existing technologies in the capabilities are not fully developed and the expectations from the business are to properly integrate these technologies to meet their demands. Hence, it requires an amalgamation of comprehensive variety of scientific areas, substantial quantity of efforts and research are essential to design, develop and implement CPS methodologies. The assisting technologies that can support in CPS at various stages are IoT (Internet of Things), service-oriented architecture, data analytics, diagnostics and prognostics, context-awareness, etc.

A. Thaduri (✉) · U. Kumar
Luleå University of Technology, Luleå, Sweden
e-mail: adithya.thaduri@ltu.se

U. Kumar
e-mail: uday.kumar@ltu.se

A. K. Verma
Western Norway University of Applied Sciences, Haugesund, Norway
e-mail: AjitKumar.Verma@hvl.no

P. K. Kapur et al. (eds.), *Decision Analytics Applications in Industry*, Asset Analytics,
https://doi.org/10.1007/978-981-15-3643-4_41

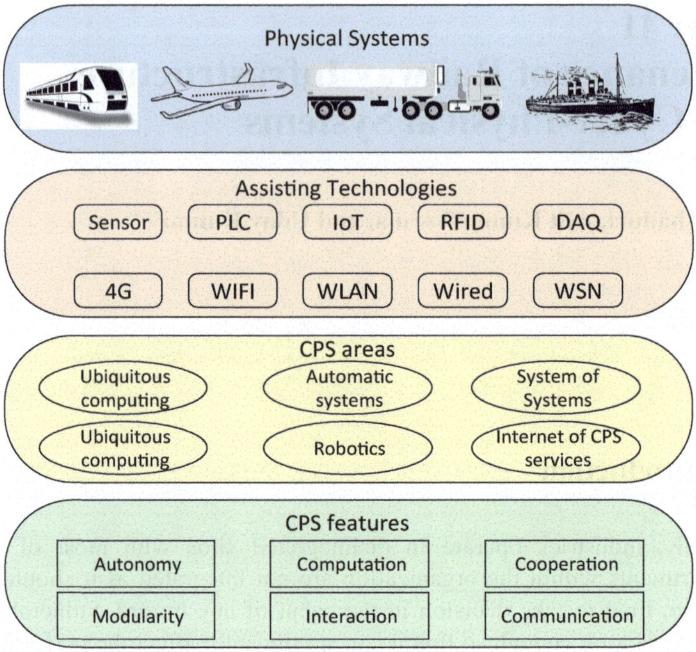

Fig. 41.1 Features and assisting technologies for CPS

CPS considers the computational mechanisms that make use of the shared data, information and knowledge from physical systems with communication and computations to support decisions based on intelligence, approachability and adaptation. The distinguishing feature among all areas is not the dissimilar features but which of them they implement dependent on the requirements as shown in Fig. 41.1 [17]. The assisting technologies in the figure are sensors, programmable logic control (PLC), IoTs, radio frequency identifier (RFID), data acquisition systems (DAQ), 4th generation communication, WiFi, wireless local area networking (WLAN), wired communications and wireless sensor networks (WSN).

In general, the life cycle of a product/system will navigate across different stages, mainly, design, manufacturing, installation, operation, maintenance and disposal. The engineering asset management has attempted to integrate several aspects within product performance within lifecycle perspective. Asset management organizes the alignment of management processes with objectives from the top level to the bottom level to meet the expectations of the customers. Engineering asset management with life cycle management of the product also incorporates several assisting technologies.

To improve the process, planning and control paradigms must be integrated with novel equipment competences. For example, the lifecycle times of equipment in the infrastructures can be improved by incorporating innovative technologies and procedures to function combined with existing equipment to make it more efficient. This integration (convergence) of existing (Operation Technology, OT) with the new technologies (Information Technology, IT) will facilitate the infrastructures in smartness with advanced capabilities to provide an additional layer for life cycle management.

In the context of life cycle management, the role of maintenance should be of the most important means for the operation of the existing infrastructure. The principle purpose of maintenance is to maintain the condition of assets to achieve their specified functions during their entire life cycle. The main purpose is to improve the ecological efficiency of the asset life cycle. Therefore, "life cycle maintenance" is termed to highlight its role from the viewpoint of life cycle management. It is important to predict and control the condition of the assets; estimate the condition under different deterioration/degradation conditions because of change in operational loads and external loads, and the change in demands from the society. This dynamic provides a gap between the required and realized functions. Hence, maintenance is to be performed to compensate for the gaps by means of several capabilities.

In asset and fleet management applications, CPS can provide context-awareness and self-maintenance intelligence. The prognostic capability with the technologies facilitates assets to continuously assess their individual condition and predict probable unplanned failures and decide to implement alternative decisions to improve process or operation. Furthermore, CPS acts as a central point for data and fleet management that recommends health evaluation and assessment with data fusion methods. These are expected to improve asset availability, increase productivity and service quality.

In this paper, integration of cyber-physical systems with the idea of life cycle management for maintenance aspects is introduced and applications of this approach are listed. These applications provide the implementation of CPS with assisted technologies to improve the maintenance management of the assets.

41.2 Cyber-Physical Systems

CPS systems are implemented by the assimilation of physical methods with software and communication with detailed generalizations and design, modelling and analysis techniques for a sub-system, system or systems of systems. The prediction and control of the dynamic behavior of the systems consisting of multidisciplinary areas of computers, networking and physical systems and their interaction in ways that necessitate essentially novel innovative design technologies. A starting point for developing such an architecture is, for example, the five-level architecture of cyber-physical systems (CPS) outlined in see Fig. 41.2. This architecture includes the following basic architectural levels:

Smart connectivity level: Data acquisition using condition monitoring techniques using existing sensors or Internet of Things (IoTs) or unmanned vehicles. The information must be acquired from the item level to the system level depending on the objective of the function expected from CPS architecture. Due to an increase in demands from real-time communication, the data acquired from the systems must be

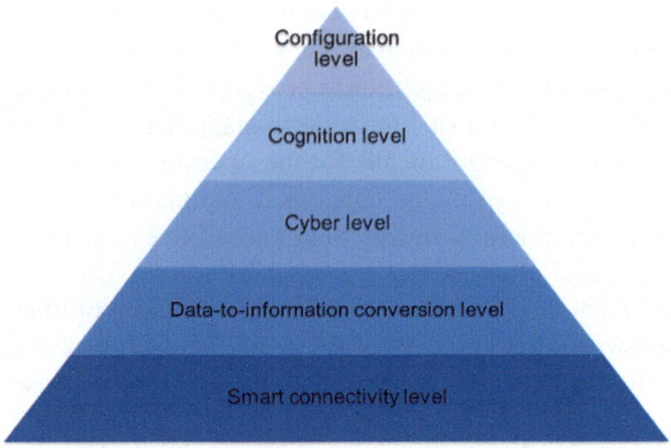

Fig. 41.2 Cyber-physical systems (CPS) architecture of smart systems

stored in the cloud through different communication protocols using service-oriented architecture.

Data-to-information conversion level: The next level in the CPS is to extract the required information from the data obtained in the Smart connectivity level. Sometimes, it will be difficult to extract information due to complexity in nature, accuracy of the data and huge storage. Hence, there is a need to perform advanced data manipulation or pre-processing methods such as data quality, data cleaning, feature extraction to implement efficient data for condition assessment and the prediction. There are also other tools such as data reduction, data normalization and soft sensors to acquire secondary information from the existing data (called precursors or covariates). When wireless sensor systems with limited power harvesting resources are involved this process is particularly challenging and requires the use of energy-efficient computing concepts.

Cyber level: Once information is obtained from every connected component and system is available, modelling and simulation methods are performed to evaluate different operation and maintenance scenarios. Online data analytics tools are implemented to extract useful information to support the above scenarios. Information about different machines of the same type can be compared and used for prediction.

Cognition level: Decision-making is performed by analyzing different maintenance scenarios with optimizing the parameters such as required availability, total cost, total risk and maintainability aspects using efficient genetic algorithms. The main intention of this level is to capitalize on the artificial cognitive systems that can relate with specific objectives to the strategies and actions. At this level, the capabilities to perform detailed simulations at the cyber level is combined with synthesis capabilities enabled by cognitive computation and human collaborative diagnostics and decision making.

Configuration level: The virtual architecture is organized by required entities and operational requirements/indicators defined by business organizations such as Key Performance Indicators (KPIs). The main purpose of the configuration level is to adapt and adopt the strategies according to the dynamic nature of the environment, contextual requirements and business demands by implementing the self-configured

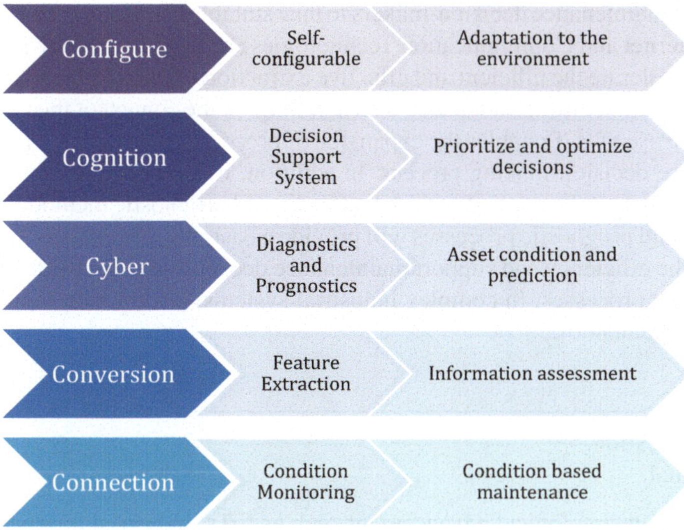

Fig. 41.3 An example of levels depicting CPS

and self-optimizing systems. To implement these systems, the lower systems need to be supported/changed according to the needs of the organized. An example of methods implemented in each level can be shown in Fig. 41.3.

41.3 Assisting Techniques

There are several assisting that are in improvement and upcoming can be incorporated into CPS at several levels in the architecture. These are listed below:

41.3.1 Data Analytics

Decision-making in the industry must be improved to instantaneously detect and resourcefully act, i.e. "the new know" [13]. The new know in maintenance is to emphasis on two main aspects:

- what information can be identified and
- what information must be disseminated,

to allow the maintenance decision-makers to take suitable actions. Enormous applications of Internet and Communication Technologies (ICT) and other developing technologies accelerate the efficient and effective extraction of knowledge from information dissemination through advanced data handling. In the context of the maintenance process, the use of ICT enables the expansion of procedures that aims to provisioning maintenance decision-making process. In addition, ICT can deliver supplementary capabilities that can be applied within diagnostic and prognostic methodologies. The diagnostic and prognostic processes will provide a basis for information logistics that increases the efficiency and support maintenance decision-making by the integration of both these processes. In complex industrial systems, information logistics refers to the management of:

- time
- content
- communication
- contextual.

Due to the technological advancement such as IoTs, industrial internet, big data, Industry4.0, etc., the factories are becoming smart by leveraging these technologies to improve the process of data-information-knowledge chain for efficient and effective operation and maintenance management of the products during the entire life cycle. Karim et al. [13] proposes a concept that can be adapted for Asset Analytics (AA) is based on four phases. The concept aims to enable asset management by improved understanding of data and information. The AA phases are (see Fig. 41.4):

1. Asset Descriptive Analytics, which purposes to answer, "What has happened?"
2. Asset Diagnostic Analytics, which purposes to answer, "Why something has happened?"
3. Asset Predictive Analytics, which purposes to answer, "What will happen in the future?"
4. Asset Prescriptive analytics, which purposes to answer, "What needs to be done?"

Diagnostic detects the abnormal condition (fault detection), determines which component is defected (fault isolation) and estimates the nature and extent of the fault (fault identification). To automatically diagnose the faults of a complex system using machine learning techniques has recently gained more attention. Sub-categories are supervised learning algorithms, e.g. Artificial Neural Networks (ANN), Support Vector Machine (SVM) and unsupervised learning algorithms, e.g. Deep Belief Network and k-Nearest Neighbors.

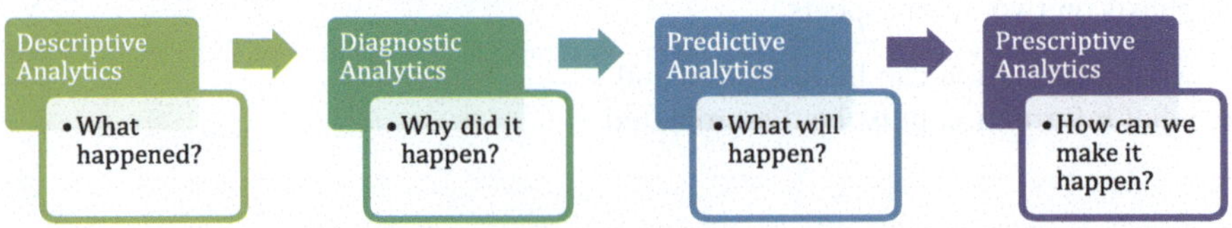

Fig. 41.4 The asset analytics concept, adapted from [13]

41.3.2 Diagnostics and Prognostics

By the adoption of modern software and hardware systems, data acquisition has become comparatively easy and cheap. Data cleaning is an important pre-processing stage for all analyses since condition monitoring data always contain errors. Excess of data is also a problem for data handling to extract the required information and hence need for additional data mining and data filtering algorithms are required to obtain suitable information. As the dimensionality of data increases, new techniques for fault detection needs to be developed. Moreover, relevant features need to be extracted, either in time domain, frequency domain or using a combination of the two. Two typical fault detection indicators in time-domain are root-mean-square (RMS) and kurtosis. Analysis in frequency-domain is a widely used approach for rotating machinery due to the ability to identify and isolate certain gear and bearing frequencies.

Prognostic predicts failure events that are yet to occur and rely on the diagnostic outputs. Several comprehensive reviews on machinery prognostics have been conducted. The estimation of remaining life can be achieved mainly using two approaches: data-driven and model-based. Data-driven approaches rely on past observed data. Different sub-categories are regressions models, Proportional Hazards Modelling, Bayesian models and Artificial Neural Networks. Two common approaches for applying Bayesian network models are Kalman and Particle filtering [22]. Model-based approaches estimate remaining life using deterministic equations or equations derived from extensive empirical data representing the physical behavior of a failure mode. An implementation of physics models discussed in comparison to condition monitoring of bearing in a paper machine has been demonstrated by Mishra et al. [23].

Only a very few studies have been reported on the combination of the data-driven and physics-based models. This calls for the development of a new hybrid approach, which combines the different techniques for remaining life estimation, for condition-based maintenance of critical components in machineries and assets. Through an IoT environment, hybrid modelling can be carried out by utilizing data from the cloud to perform diagnostic and prognostic techniques for assessment and visualization of the health of the asset.

41.3.3 Service-Oriented Architecture

Presently, the industries and governments demand sustainability, flexibility, efficiency and competitiveness because of the dynamic nature of societal and market trends. To attain the above goals, it is important to handle the requirements of multi-stakeholder environment and need for efficient co-operation and collaboration which requires a system-wide architecture. This process also needs to be followed through the value chain and the life cycle of the product and its process [13]. This also required a digitization procedure to be able to connect through different stakeholders in real-time to acquire data, process the information and disseminate the knowledge among the partners.

Considering the perspective of the organization, cooperation and managing the operation assets, the three domains required are stated in Fig. 41.5:

- Product life cycle management (design to support)
- Supply chain management (suppliers to customers)
- Stakeholder integration management (shop floor to business).

These three management domains must work together so that the requirements of multi-stakeholders must be achieved. The dynamic collaboration among each of these domains has the potential possibility of integrated learning among them and transfer of data and information is the key.

To provision these developments in these domains, there are still several gaps in administrative, managerial and technological gaps that need to be addressed where the present state of the art is not enough. Hence, the adaptation of the new technologies will facilitate to an extent to meets the demands of the industries. In addition, the present motive of the organization to move towards automation is also not enough at the larger scale though there were existing systems in automation at a small scale. The adoption in the larger scale is hindered due to the higher cost of operation, higher

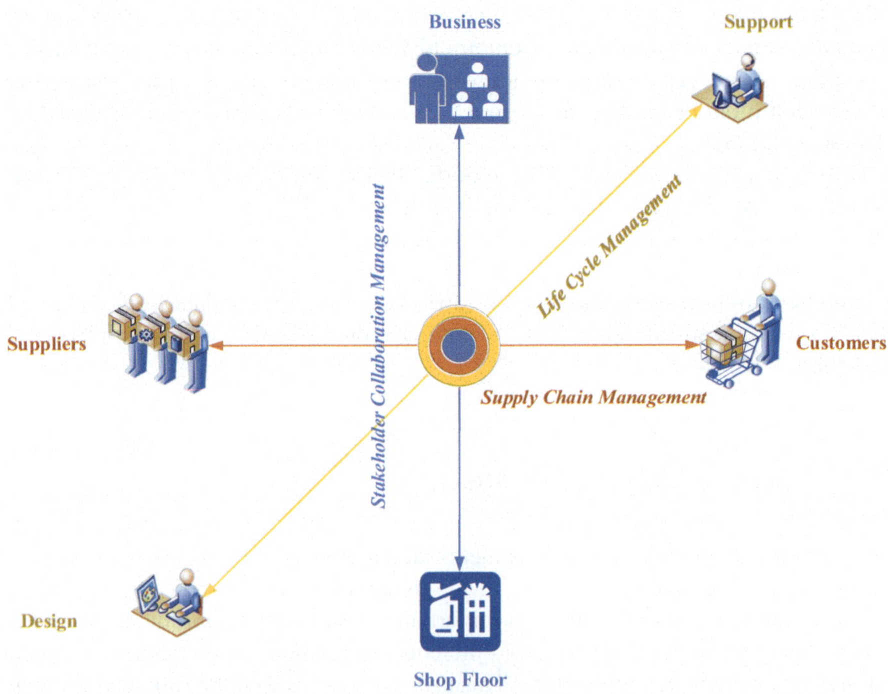

Fig. 41.5 Three important axes for service-oriented architecture

risks and sometimes pressure from the governments to delimit automation to reduce job shortages. The combination of digitization, digitalization and automation could improve competitiveness, flexibility and sustainability.

41.3.4 Internet of Things—IoT

In general, the IoT concept enables Internet access to any type of 'device/thing'. An interesting review on IoT is written by Holler et al. [9]. Each 'thing' will have an Internet Protocol (IP) address and will be addressable using standard Internet technology like Domain Name System (DNS). In an automation context, an IoT will be a device or functionality that can find in standard ISA-95 automation implementations. Automation applications of the Internet of Things typically include Sensors, Actuators, Programmable logic controllers (PLCs) and Control loops. Thus, the Thing can be viewed as a physical device or as a functionality that is computed in a software system executed on any type of device having enough computational resources. Currently, there is no consistent technology that can be attributed to an IoT.

41.3.5 Context-Aware Systems

A context-aware system is a system that acclimates actively and autonomously adapts according to the required function that enables for more relevant information to the users based on information gathered from machines/people's contextual information [31]. The concept of context-aware computing was as "*the ability of a mobile user's applications to discover and react to changes in the environment they are situation*" [27, 28].

Context-aware systems are complex, and they can do different tasks such as data representation, data modelling, data management, reasoning from data and analysis of contextual information [31]. It needed to require collaboration from various domains and from different components in a system. There also exist different context-aware systems that are difficult to provide a generic process though it consists of four main steps as shown in Fig. 41.6 [6].

Context Acquisition: The first step is to select and acquire the necessary data from the sensors. The sensors can be physical sensors such as any wired or wireless sensors depending on the availability. In addition, the information can also be gathered from so-called virtual or soft sensors to get secondary information.

Storing Information: The data will be stored in the repository. Before modelling the data, it is necessary to organize the data in taxonomical order so that the modelling because easier. There are several closed and open standards to store the data from bottom to top level characterizing with different entities. These entities can be defined with failure modes, failure effects, etc.

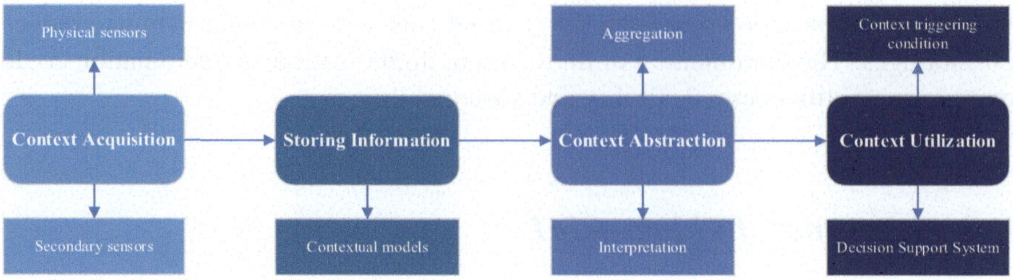

Fig. 41.6 Four steps in context-aware systems

Context abstraction: The context-aware system requires the abstraction level by interpreting or aggregating them which can be useful for data analytics.

Context utilization: At the last step, this data is analyzed with data analytics to provide a decision support system.

41.4 Engineering Asset Management

The engineering asset management (EAM) term was coined by Amadi et al. [2]. He also reviewed state-of-the-art asset management and developed a strategic framework considering comprehensive relationships and mechanisms among asset management processes and activity. Ma [19], stated that the existing diagnostic and prognostic approaches available at that time are not enough for EAM and need further improved techniques and models.

An attempt has been made on an integrated and intelligent EAM system for achieving maintenance, asset management and decision support system [3]. McArther et al. [20] used an agent-based system for monitoring real-time data to spot anomalous states in power systems. Ma et al. [18] proposed an agent-based asset management platform for power plant assets to communicate, coordinate and plan for maintenance services. Wagle et al. [35] created a multipurpose monitoring system that combines information from multiple sensors and equipment with localized information. Frolov et al. [5] exhibited a fundamental asset management ontology within asset management process architecture. The system developed by Trappey et al. [33] involves a mechanism to reinforce the collaborative characteristics of maintenance and distribution of services. The ISO 55000 standard [10] offers an inclusive methodology for implementing an asset management system that aids establishments to attain their business objectives by effective and efficient management of its assets. To rephrase, the interface that connects the cyberspace to the physical space can be accomplished with proper strategy.

Smart analytics including both predictive and prescriptive analytics for attaining such type of intelligence can be used at the specific machine. The influence of failure, operational and maintenance data at the various stages of asset management has been encouraged by the advent of advanced online sensors, the progressive implementation of e-maintenance [7]. The functions of Asset Management are gathering critical data, throughout the life-cycle of the asset by condition monitoring the smart assets systems and acquiring data to conduct intelligent diagnosis and prognosis to provisioning maintenance decision process. To achieve the above goal, condition monitoring research needs to undergo a paradigm shift from its traditional focus that involves researchers from different areas to work collaboratively to develop efficient predictive methodologies for health assessment and effective engineering asset management.

41.5 Life Cycle Management

From the INCOSE Handbook [8], life-cycle model management is defined as '*an organizational process that creates life-cycle models as a basis for common reference to a project's life-cycle*'. To recuperate the effectiveness of the organization, feedback mechanism in the process is implemented for adjusting if there are deviations from the existing process. The International Infrastructure Management Manual (IIMM) studies life-cycle models with attention to the operations and maintenance phases [25]. Life Cycle Costs (LCC) apparently included in the life cycle models for the optimization of decision models for optimal maintenance for savings of the business. Kim et al. [14] demonstrate its application by developing a model to estimate LCC for light rail transit. Asset management embraces the life-cycle management in the better way. Asset management organizations have the structure of its management processes with strategic aims, legal and regulatory necessities and customer hopes [34].

From life cycle management, it is pointed out that operation and maintenance is an important phase considering LCC. The main aim of maintenance is to maintain the condition of products/items to achieve their necessary functions during their entire life cycle. Hence, "life cycle maintenance" to highlight its important role from the viewpoint of lifecycle management [4]. A framework for life cycle maintenance is required to be intervened in the CPS as shown in Fig. 41.7. For satisfying the necessities of life cycle maintenance, the active implementation of a P-D-C-A (plan-do-check-action) cycle is vital. Maintenance strategic planning is an important role for selecting the best maintenance action according to the operation and requirements with further evaluation of the condition assessment of assets with maintenance technologies.

During this entire process of maintenance strategy planning, CPS with the assisting technologies can play a vital part in deciding the best maintenance strategy by efficient communication among peers. These three loops shown in the figure, provide effective instruments for acclimatizing maintenance strategies to various static and dynamic conditions to withstand the successful operation.

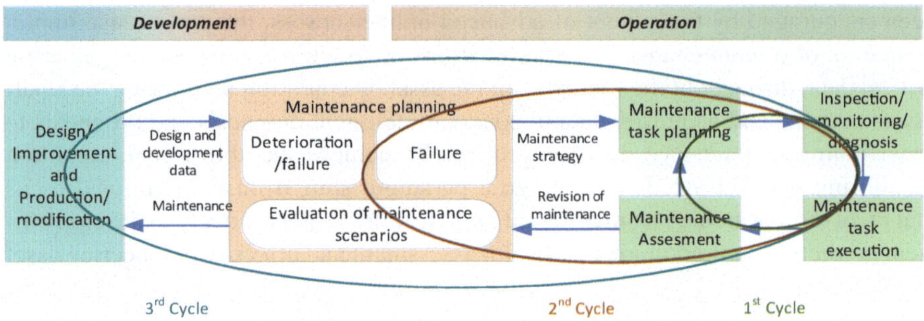

Fig. 41.7 Framework for life cycle maintenance [29]

41.6 Applications

This section details on existing applications of usage of CPS. The main industrial expectations for CPS are

- Industry needs to be more resilient
- Improved capacity by proper planning and operation with flexibility
- Reduced carbon emissions using less stoppages and implementing system of systems approach with less combustions and noise pollutions.

To attain these industrial needs, several consortiums, industries, organizations and universities are working together to meet these goals using CPS.

41.6.1 Cyber-Physical Based Maintenance Strategies

Lee and Behrad [16] proposed a procedure for applying CPS in maintenance applications that is termed as "Time Machine Methodology for Cyber Physical Systems". The information from the physical side like sensory data, work order/maintenance history, operation parameters, weather, etc. are extracted on the cyber side using cyber-physical interface. This interface consists of computation algorithms consisting of prognostics health management (PHM) using physical and data-driven models to evaluate the condition of the assets. The watchdog agent thus converts this information to the virtual space. The virtual prototype for the existing machine can be developed by using virtual synthesis and send this information to cyber space. Cyber space provides whatever the condition in the physical replicates in cyber, the technology known as digital twin.

41.6.1.1 Wind Turbine Health Monitoring System

By implementing the above digital twin approach, the authors applied this approach in the wind turbine to assess the most critical metrics with mechanical component degradation and failure. Supervisory Control and Data Acquisition (SCADA) data and Condition Monitoring data are employed for data acquisition and condition monitoring. Several feature extraction methods have been integrated into predicting component faults, to examine vibration data at various operating scenarios, and diagnose detailed failure modes.

41.6.1.2 Robot Health Monitoring

This was implemented on a fleet of industrial robots that was intended to develop a predictive health monitoring system. Robots were behaving indifferent line speeds and hence a complex multi-scenario CPS based maintenance method has been implemented. The PHM algorithms comprise non-linear relation between speed and torque for controlling the health state of the robot. The feature extraction module in cyber-physical model produces several parameters such as pressure calibration, load ration, gear ratio, tools for robots. These features helped in allocating products to robots from the production line. The operation and maintenance data from all assets in the fleet were processed and stored on the cloud. Several health monitoring procedures were performed on the data to predict the health condition of each robot in the fleet.

41.6.2 MANTIS—EU Project

The MANTIS project's (www.mantis-project.eu/) idea is to deliver a proactive maintenance service platform architecture using CPS that can provide estimation of the future performance. This can be achieved by monitoring the condition, prediction and prevention of the imminent failures and optimize the proactive maintenance. Algorithms need to be custom-built for distributing results at different levels of data processing (see 1). At the lowest level, sensor data collection, data cleaning, data fusion was performed and then several prediction methods such as fuzzy logic, stochastic methods, Kalman filter, rule-based methods were applied. At the highest level, several data-mining methods were performed such as classification, clustering and regression analysis for carrying out data evaluation by anomaly detection of outliers, k-means algorithms, machine learning [24].

The proactive maintenance platform consisted of dispersed processing chains. These chains will proficiently convert raw data into information and knowledge. To achieve the proactive maintenance service with CPS, an integrated domain knowledge system consisting of experts are needed for processing the data. This knowledge is further combined with data-driven approaches such as condition monitoring, connectivity and analytics with self-learning capabilities. There is a need to adopt key technologies such as:

- Smart sensors
- Robust communication systems
- Distributed machine learning
- Cloud-based processing
- Human–machine interface (HMI).

41.6.3 Cyber Physical Interface (CPI) for Automation Systems

Kao et al. [12] proposed CPI for automation systems. The capabilities that were incorporated are; intelligent resilient future system that responds do dynamic environment, learning history and condition of other machines, planning maintenance actions from the other machines, self-assessment of component degradation for quality and learning form human intelligence for fault prevention.

A self-aware and self-maintenance machine structure was developed with the above capabilities that can assess the health on its own, detect and predict the degradation behavior of the assets to support smart maintenance decisions. This concept of Smart analytics can be performed from the individual level to the top level in a system to achieve highest performance.

To form an interface between the cyber space and the physical space, an analytics platform needs to be developed. For automation systems, it is still trivial. While implementing, it is necessary to decide on which aspects such as assets, attributes and requirements which is important for end user that must be replicated from physical to cyber space.

41.6.3.1 Experiments-Ball Screw Case Study

For especially high precision machines that are used in critical systems, the accuracy of prediction ball screw health is important [12]. To improve the accuracy of the prediction, a test rig was built to conduct experiments on life testing of the ball screw by adopting CPS systems. At the smart connectivity level, physical sensors were installed to monitor the critical parameters such as vibration, speed, temperature, etc. By utilizing this data, a health-monitoring model was developed to predict the position error.

For decision making of maintenance actions, there is a need to collect additional data on different failure modes and maintenance data. Because of data acquisition system that collects data for every second, the large data must be processed first before transmitting across to the cloud server. Hence, at the data conversion level, feature extraction was performed that will be feasible to conduct diagnosis and prognosis such as data filtering, data pre-processing and selection of specific features. After the pre-processing level, the specific features or health value was connected to cloud, from physical to cyber.

In the cyber level, a virtual ball screw was simulated by extracting the data analysis model developed from real data using smart connection. The virtual space must simulate and predicts the degradation behavior by considering different scenario of loads during the entire component's life cycle. The virtual model of the ball screw needs to adopt the self-aware and self-predict abilities. These are useful to provide different maintenance actions according to the context. Additionally, at the cognition level, the managers can exploit this information for optimizing maintenance strategies and for providing efficient feedback information at the design level to improve the reliability of the ball screw.

41.6.4 CPS for MRO in Transportation

CPS for Maintenance, Repair and Overhaul (MRO) planning and control was initially discussed among several transportation systems and other MRO companies for airline industry [21, 26]. Knothe and Moos [15] discussed the need for modification from vertical chain to horizontal chain with direct communication by incorporating with an architecture using agent-based technologies. The proposed architecture supported the implementation of the compromise of chain process among four vertical entities t using CPS. These entities consist of specific capabilities, which are as follows:

1. Transportation
   a. Schedule
   b. Resources
   c. platform
2. Navigation
   a. Electronic ID
   b. Schedule plan
   c. platform
3. Capacity
   a. Schedule
   b. Platform
4. Operation
   a. Fight schedule
   b. Predetermined maintenance plan
   c. Platform.

SURFER project led by Bombardier Transportation was designed for developing a CPS system on-line monitoring of door of trains. The proposed architecture is used distinctive features of the holonic principles of recursively and cooperation. Initially, the information is combined and augmented among the various levels of the train. This is helpful for the maintenance center to evade huge unnecessary data and improve reaction timings of the Centre with more accurate information and health assessment. This approach has some limitations such as:

- Doors which are critical for safety.
- No realization by integrating with the maintenance.
- Fleet level was not realized.

41.7 Swedish Railway Infrastructure

The Swedish Transport Administration [32] is divided into seven key functions and five business areas as shown in Fig. 41.8.

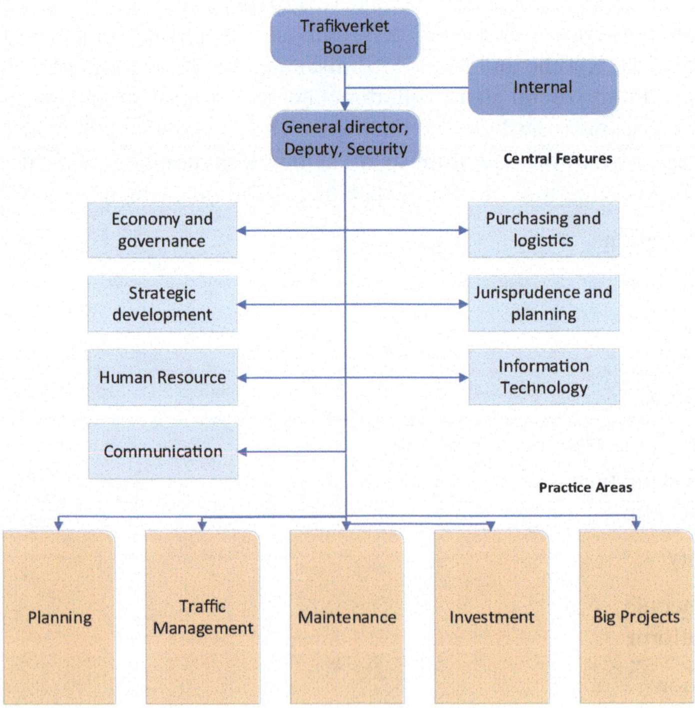

Fig. 41.8 Trafikverket organization

The maintenance is one of the important practice areas and being integrated into the central features of the Swedish Railway organization. Traffic management, planning and maintenance work simultaneously for continuous railway operation. The proposed architecture for CPS for the railway infrastructure is shown in Fig. 41.9.

The knowledge gained from the previous applications and assisting technologies in the previous sections are incorporated into this architecture. Each level is described below:

Smart connection Level: Trafikverket already maintains different sets of data such as asset data, condition data, maintenance work orders, inspection and other additional data as described in Thaduri et al. [30]. In addition, the sensor data from IoTs and RFIDs are also incorporated into the data layer.

Data-to-Information Level: The core approaches are listed in this level. Since the data is always not correct enough, data mining and data cleaning methods are performed for the right decision. A further preprocessing of this data helps to obtain the necessary data for further analysis. By using the feature extractions techniques

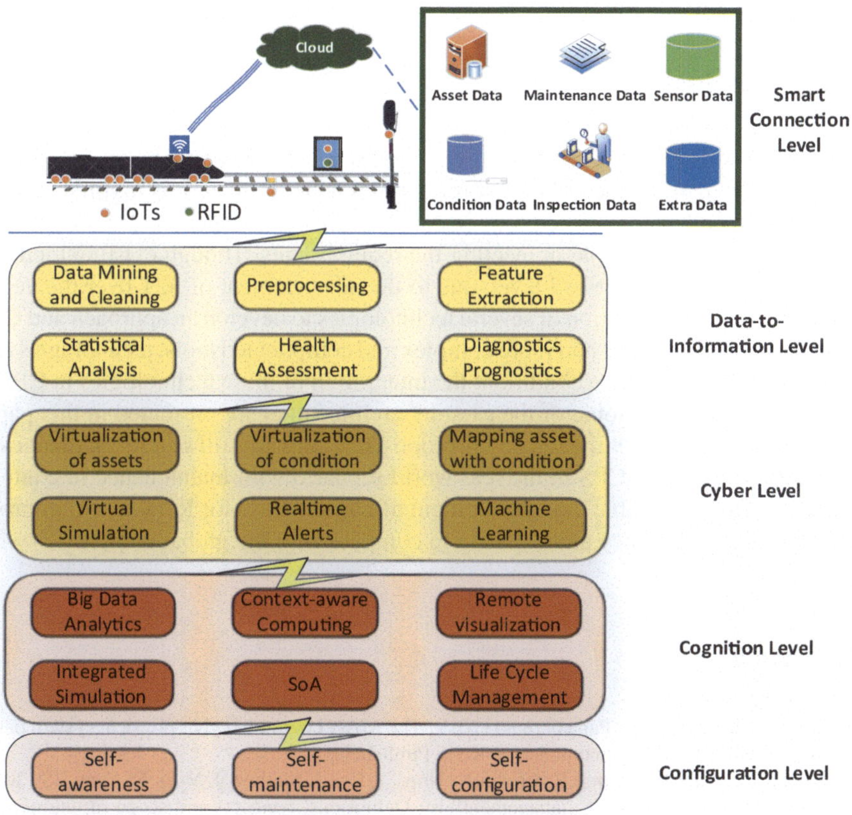

Fig. 41.9 CPS architecture for Swedish Railway

described above, the health of the railway infrastructure is performed using advanced diagnostics and prognostics methods.

Cyber Level: From the above experience, cyber level tends to create a virtual environment. By using the existing data, physical systems are emulated to 3D simulation and to be combined with asset conditions evaluated in the previous level. The proper integration can be carried out using cyber-physical interface using machine learning techniques with locational data.

Cognition Level: This level where most of the data-analytics and computational methods are implemented to develop an interactive atmosphere of railway infrastructure. Life Cycle Management and asset management are conducted in this level for comprehensive information management. Maintenance analytics provided by Kans et al. [11] is also implemented for maintenance decision support. Visualization of the past, present and future status of the railway infrastructure are also represented.

Configuration Level: The self-aware and self-maintenance capabilities proposed by Kao et al. [12] are incorporated to develop real-time and self-adjusted capabilities to maintain the infrastructure operational.

41.8 Conclusion

Due to the industrial requirements, there is tremendous interest increasing over the years in implementing the CPS for several systems. In doing it, it is highlighted that the operation and maintenance phase in the lifecycle management and asset management form a significant role in meeting the requirements. Though existing methods are attaining at the prescribed level, due to the advancement of ICT over the years, there is a need for integration of several technologies to develop an approach and CPS serves this purpose. More and more complex and complicated tools, technologies and algorithms to maintain the condition, the integration of these technologies in CPS is becoming converged. Some of the existing architectures are presented in this paper and CPS architecture for the railway is proposed. There are still several challenges to practically implement CPS in the real-world scenario with maintenance in context. This CPS will also enable the development of Digital Twin for Railways to improve the robustness, efficiency and capacity to compete with the global market.

References

1. Albano M, Jantunen E, Papa G (eds) (2019) The MANTIS book: cyber physical system based proactive collaborative maintenance. River Publishers
2. Amadi-Echendu JE, Willett R, Brown K, Hope T, Lee J, Mathew J, Vyas N, Yang BS (2010) What is engineering asset management? In: Definitions, concepts and scope of engineering asset management. Springer, London, pp 3–16

3. Bangemann T, Rebeuf X, Reboul D, Schulze A, Szymanski J, Thomesse JP, Thron M, Zerhouni N (2006) PROTEUS—creating distributed maintenance systems through an integration platform. Comput Ind 57(6):539–551
4. Donca G, Mihăilă I, Ganea M, HIRłE D, Nica M (2007) Maintenance role in life cycle management. Ann Oradea Univ Fascicle Manag Technol Eng 6(16):2158–2163
5. Frolov V, Mengel D, Bandara W, Sun Y, Ma L (2010) Building an ontology and process architecture for engineering asset management. In: Engineering asset lifecycle management. Springer, London, pp 86–97
6. Galar D (2014) Context-driven maintenance: an eMaintenance approach. Manag Syst Prod Eng
7. Galar D, Kans M, Schmidt B (2016) Big data in asset management: knowledge discovery in asset data by the means of data mining. In: Proceedings of the 10th world congress on engineering asset management (WCEAM 2015). Springer, Cham, pp 161–171
8. Handbook INCOSE (2014) INCOSE systems engineering handbook: a guide for system life cycle processes and activities, version 4. International Council on Systems Engineering
9. Holler J, Tsiatsis V, Mulligan C, Avesand S, Karnouskos S, Boyle D (2014) From machine-to-machine to the internet of things: introduction to a new age of intelligence. Academic Press
10. ISO (2014) ISO 55000: asset management-overview, principles and terminology. Geneva, Switzerland
11. Kans M, Galar D, Thaduri A (2016) Maintenance 4.0 in railway transportation industry. In: Proceedings of the 10th world congress on engineering asset management (WCEAM 2015). Springer, Cham, pp 317–331
12. Kao HA, Jin W, Siegel D, Lee J (2015) A cyber physical interface for automation systems—methodology and examples. Machines 3(2):93–106
13. Karim R, Westerberg J, Galar D, Kumar U (2016) Maintenance analytics-the new know in maintenance. IFAC-PapersOnLine 49(28):214–219
14. Kim GT, Kim KT, Lee DH, Han CH, Kim HB, Jun JT (2010) Development of a life cycle cost estimate system for structures of light rail transit infrastructure. Autom Constr 19(3):308–325
15. Knothe T, Moos F (2014) Systemseitige dezentrale Planung und Steuerung von Wartungs- und Instandsetzungsarbeiten für Verkehrssysteme. Patentantrag. FB 1104-01DE, FB 1104-02US
16. Lee J, Bagheri B (2015) Cyber-physical systems in future maintenance. In: 9th WCEAM research papers. Springer, Cham, pp 299–305
17. Leitão P, Colombo AW, Karnouskos S (2016) Industrial automation based on cyber-physical systems technologies: prototype implementations and challenges. Comput Ind 81:11–25
18. Ma C, Tang WH, Yang Z, Wu QH, Fitch J (2007) Asset managing the power dilemma. Comput Control Eng 18(5):40–45
19. Ma L (2007) Condition monitoring in engineering asset management. In: Kaneko S (ed) Proceedings of APVC2007 12th Asia Pacific Vibration Conference, JSME, CD Rom, pp 1–16
20. McArthur SD, Booth CD, McDonald JR, McFadyen IT (2005) An agent-based anomaly detection architecture for condition monitoring. IEEE Trans Power Syst 20(4):1675–1682
21. Mertins K, Knothe T, Gocev P (2012) Towards CPS based aircraft MRO. In: Collaborative networks in the internet of services, pp 166–173
22. Mishra M, Odelius J, Thaduri A, Nissen A, Rantatalo M (2017) Particle filter-based prognostic approach for railway track geometry. Mech Syst Signal Process 96:226–238
23. Mishra M, Saari J, Galar D, Leturiondo U (2014) Hybrid models for rotating machinery diagnosis and prognosis: estimation of remaining useful life
24. Papa G, Zurutuza U, Uribeetxeberria R (2016). Cyber physical system based proactive collaborative maintenance. In: International conference on smart systems and technologies (SST), Oct 2016. IEEE, pp 173–178
25. Roberts L, Hughson G, Smith R, McIlveen DK, Murray S (2006) International infrastructure management manual. National Asset Management Steering (NAMS) Group, Wellington, New Zealand

26. Sampigethaya K, Poovendran R, Shetty S, Davis T, Royalty C (2011) Future e-enabled aircraft communications and security: the next 20 years and beyond. Proc IEEE 99(11):2040–2055
27. Schilit BN, Theimer MM (1994) Disseminating active map information to mobile hosts. IEEE Netw 8(5):22–32
28. Schilit B, Adams N, Want R (1994) Context-aware computing applications. In: First workshop on mobile computing systems and applications. WMCSA 1994, Dec 1994. IEEE, pp 85–90
29. Takata S (1999) Life cycle maintenance management. In: Computer-aided maintenance. Springer US, pp 209–230
30. Thaduri A, Galar D, Kumar U (2015) Railway assets: a potential domain for big data analytics. Procedia Comput Sci 53:457–467
31. Thaduri A, Kumar U, Verma AK (2014) Computational intelligence framework for context-aware decision making. Int J Syst Assur Eng Manag 1–12
32. Trafikverket (2016) Organisation. http://www.trafikverket.se/om-oss/varverksamhet/trafikverkets-uppdrag/
33. Trappey AJ, Trappey CV, Ma L, Chang JC (2015) Intelligent engineering asset management system for power transformer maintenance decision supports under various operating conditions. Comput Ind Eng 84:3–11
34. Valencia VV, Colombi JM, Thal Jr AE, Sitzabee WE (2011) Asset management: a systems perspective. In: Proceedings of IIE annual conference, Jan 2011. Institute of Industrial and Systems Engineers (IISE), p 1
35. Wagle AM, Lobo AM, Kumar AS, Patil S, Venkatasami A (2008) Real time web based condition monitoring system for power transformers-case study. In: International conference on condition monitoring and diagnosis. CMD 2008, Apr 2008. IEEE, pp 1307–1309

Chapter 42
A Graph-Theoretic Analysis on Functional EEG Network in Igraph R

Gajendra Pratap Singh, Ekta Raphael Anthony, Mamtesh Singh, and Naina

Abbreviations

WHO	World Health Organization
AMA	American Medical Association
MEG	Magnetoencephalography
MRI	Magnetic Resonance Imaging
EEG	Electroencephalography
GBU	General Public Licence Version 2
ARBD	Alcohol-Related Brain Damage
PD	Parkinson's Disease
PD-CN	Parkinson's Disease who are cognitively normal
PET	Positron Emission tomography
fMRI	Functional Magnetic Resonance Imaging
SL	Synchronization Likelihood

G. P. Singh (✉) · Naina
School of Computational and Integrative Sciences, Jawaharlal Nehru University, New Delhi, India
e-mail: gajendra@jnu.ac.in

Naina
e-mail: nainatondak@gmail.com

E. R. Anthony
Integral University, Lucknow, Uttar Pradesh, India
e-mail: anthonyekta16@gmail.com

M. Singh
Department of Zoology, Gargi College, University of Delhi, Delhi, India
e-mail: s.mamtesh@yahoo.com

P. K. Kapur et al. (eds.), *Decision Analytics Applications in Industry*, Asset Analytics,
https://doi.org/10.1007/978-981-15-3643-4_42

42.1 Introduction

Alcoholism is huge intake of alcohol which is an inseparable part of everyday life. World Health Organization (WHO) reported the alcoholism as the third highest risk factor that often results in 2.5 million deaths per year. According to the American Medical Association (AMA) definition, alcoholism is a primary chronic disease with genetic psychosocial and environmental factors influencing its development and manifestation [8]. The main function of the brain is to process information and the primary processing of information is done by neuron. A neuron is a cell of the brain that collects several input signals then processes it and generates an output signal. The signals within neurons are electrochemical in nature. Graph Theory helps in the analysis of the network topology of the neural network both anatomically and functionally [21]. Neural Network possesses small-world property both at the level of anatomically and functionally. In neuroscience, the most challenging task is to understand the network organization of the brain. In the past decade, the development in graph theory has provided many new methods for the analysis of complex neural networks that have already been translated to the characterization of functional and anatomical brain networks. In Graph Theory, brain regions are to be considered as the nodes and edges as the functional connectivity between their nodes. To construct the large-scale functional network, signals are recorded by many neuroimaging techniques like Magnetoencephalography (MEG), Magnetic Resonance Imaging (MRI) and Electroencephalography (EEG). In this paper, the authors used the EEG data set. EEG technique is used to record the electrical activity of the human brain. This technique is inexpensive and convenient to perform [1, 2, 5–8, 10, 13, 14, 25]. Here, we have constructed the 20 individual networks of 20 trial of the Alcoholic subject and 20 individual networks of 20 trial of the Non-Alcoholic subject in Igraph R and then we have done the analysis of 20 Alcoholic and 20 Non-Alcoholic networks in Igraph R by using graph theory measures. Alcoholism is characterized by the following:

(a) frequent, heavy intake of alcohol for a prolong period.
(b) unable to control drinking.
(c) when an individual stops using alcohol there is physical dependence manifested by withdrawal symptoms.
(d) tolerance or the need to use more and more alcohol to achieve the same effect.
(e) due to the consumption of alcohol many legal or social problems arise [8].

EEG is used widely to study neurological disease, mental disorder, neurodegenerative disease. This discussion deals with a study of Alcohol-Related Brain Damage (ARBD).

42.2 Methodology

It consists of two parts, namely, materials and methods.

42.2.1 Material

(i) **Igraph**: It is a software package collection for graph theory and network analysis. Gabur Csardi and Tamas Nepusz developed an Igraph software package. Igraph package source code was originally written in C. It is freely available under GBU (General Public Licence Version 2). Igraph is also written in R and Python Packages. It is widely used in the field of Network Biology, Network science and related fields. It is a good software for network analysis. We have constructed 20 alcoholic and 20 Non-Alcoholic brain networks in Igraph software and also analyzed our networks in Igraph software by using graph theory measures [15].

(ii) **R Programming Language**: R programming language was first created by Robert Gentleman and Ross Ihaka at the University of Auckland, New Zealand. R Development Core Team currently developed R. R is the statistical programming language that is helpful for statistical computing, analysis and graphics. The main core of R is an interpreted computer language that allows functions and looping. We have used the R statistical programming language in Igraph to construct and analyze the 20 alcoholic and 20 Non-Alcoholic networks [20].

(iii) **Data**: The information comes from the EEG data set which is downloaded from 'UCI, Machine Learning Repository, Centre for Machine Learning and Intelligent System'. The EEG data set comes from a large study of Alcoholic and Non-Alcoholic subjects. This data set has only two types of subjects first one is alcoholic and the second one is Non-Alcoholic. Every subject was exposed to 64 electrodes on their scalp to measure or record their electrical activity of the brain. In this data, every subject was exposed to either two stimuli (S1 and S2) or a single stimulus (S1). When subjects are exposed to two stimuli (S1 and S2) then they are represented as either S1 is identical to S2 which is called Matched Condition or S1 is different from S2 which is called as Non-Matched Condition. There are 122 subjects and every subject has completed 120 trials that were exposed to different stimuli. Every electrode was located at standard sites. In 1995, the details of the data collection process were explained by Zhang et al. In this study, we have taken 20 alcoholic EEG data and 20 Non-Alcoholic EEG data of single stimulus (S1) and further constructed networks in Igraph by using R statistical programming language and analyzed network in Igraph by using R statistical programming language by the help of Graph Theory measures [23].

42.2.2 Methods: It Is Divided into Four Parts

i. **Installation of R**: The first and very crucial step of the methodology is the Installation of R on the system. The following commands were given on the terminal of system for the installation the R.

sudo apt-get update

↓

sudo apt –get Install R

ii. **Installation of Igraph**: The second step of the methodology is the installation of Igraph on the system. As we have to use the Igraph R for building the biological network and do network analysis by using graph theory.

Go to igraph in CRAN

↓

Go to package source

↓

Download the files

↓

On the terminal give command "install.packages("filename")"

iii. **Download data from the database**: The data is downloaded from UCI, Machine Learning Repository, Centre for Machine Learning and Intelligent System.

iv. **Network Construction**: All network construction and analysis are done in Igraph software using R statistical programming language. The network is represented as a graph in which there are nodes (64 nodes) and edges (256 edges). Then, scalp electrodes are chosen as the nodes and functional connectivity between channels are chosen as the edges. In this study, we have taken data from UCI Alcoholic and Non-Alcoholic. In Alcoholic, we have taken 20 trials so we have to generate 20 networks of alcoholic trials (*see. Appendix* Figs. 21–40) and the same in case of Non-Alcoholic subject (*see. Appendix* Figs. 1–20) we have to generate 20 networks of 20 trials. In the present study, there is a 64 * 64 correlation matrix C in every trial of Alcoholic and Non-Alcoholic subjects. The elements in Cij in the matrix represent the two corresponding nodes dependent intensity. Then we have applied a threshold to fix the network's edge density. Fix the network edge density to 10% in every 20 trials of Alcoholic and Non-Alcoholic subject of S1 match condition. In this study, we have chosen an appropriate threshold T and binarizing the correlation matrix C results in the adjacency matrix A. If Cij > = T, then Aij = 1, which means the corresponding two nodes have an edge. If Cij < T, then Aij = 0, which means the corresponding two nodes do not have an edge [1]. The human brain network is highly efficient and has low consumption small-world networks [2]. A small world is the basic property of the brain network which is used to fix and confirm the scale of density [1, 21]. *The rule of the small-world network is that the mean degree of every network should be larger than 2log (N), N is the number of vertices. In every network of EEG Data have 64 vertices.*

42.3 Graph Metrics

Different graph metrics were used to analyze EEG Alcoholic and Non-Alcoholic networks. Igraph can calculate the different graph metrics or graph theory measures that help to analyze the EEG Alcoholic and Non-Alcoholic networks. List of formulas and code we used to calculate graph theory measures are given in an appendix. The first graph theory measure used to analyze network is *Mean Degree*, the number of edges incident on particular vertex v called as degree of that vertex v. The mean of all degree in the network called Mean Degree of the network. The network density is closely related to the mean degree [16]. Another important measure is *Degree Centrality*, the number of neighbors a node has. The importance of node increases if it has many neighbors [9]. Another important measure is *Modularity*, it calculates the strength of networks which are divided into modules or subnetworks. A network with high modularity has dense connections within modules [18]. Another important graph measures are *Transitivity*, it measures the extent to which a relationship relates to two nodes in a network is connected by an edge. The transitivity is also known as the Clustering Coefficient [22]. Another important graph measures are *Efficiency*, it helps to measure how efficiently information flows in the network. The efficiency is applied to both local and global networks [12]. Another important graph theory measure is **Assortativity**, it helps to measure the preference of a network's node is connected to other networks which are similar in some way [3]. Another important graph theory measure is the *Average Path Length*, which measures the shortest path between all the pairs of nodes, adding the shortest path and then dividing it by the total number of pairs. It also measures the efficiency of information [4]. Another important graph theory measure is *Network Diameter*, which measures the shortest distance between two most distant nodes [11].

42.4 Results

The following algorithm was used to construct a network from 20 alcoholic and 20 non-alcoholic EEG data on the terminal of LINUX system using R statistical programming language in Igraph.

Read the table of data on the terminal

↓

Find out the dimensions of the data

↓

Read the only fourth column of the table

↓

Find out the correlation of the fourth column

↓

To build adjacency matrix of fourth column, we have to fix the threshold in every graph in a way that the density should be of 10%, if Rij >=T (threshold) then Aij is 1 otherwise 0

↓

Find out the dimension of Aij

↓

Assign the diagonal of Adjacency Matrix as zero

↓

Then after construction of network in igraph then analyse network by using graph theory measures by the help of R statistical programming language.

After the construction of the network several graph theory metrices were used to analyze the constructed 20 alcoholics and 20 non-alcoholic networks. The following Table 42.1 represents the standard deviation values of graph theory metrices in the Alcoholic and Non-Alcoholic network (Figs. 42.1 and 42.2).

Mean Degree—If the mean degree of a network is more then that network nodes is important. More Mean Degree leads to a connection to more nodes that perform more functionality. In our study, Non-Alcoholic mean degree is more than Alcoholic network which means there is more functionality in Non-Alcoholic network.

Degree Centrality—If degree centrality of a node in the network is more, then the importance of that node increases. In this study, degree centrality of Alcoholic network is more than that of Non-Alcoholic network.

Modularity—If modularity increases in a network then small clusters are formed in a network which leads to an increase in robustness and resilience of a network. An increase in robustness, decreases the random and targeted attack in the network.

Table 42.1 Analysis of network

	Alcoholic	Non-Alcoholic
Mean degree	6.56875 ± 0.1595281129	6.5578125 ± 0.1947241138
Degree centrality	0.1504960315 ± 0.0372841039	0.139558523 ± 0.0339150394
Transitivity	0.62813445 ± 0.059064403	0.614576305 ± 0.0358127547
Efficiency	0.29842714 ± 0.0369141504	0.28914014 ± 0.0208561214
Assortativity	0.5237407 ± 0.1680043229	0.48555482 ± 0.1279721646
Average path length	4.1684944 ± 0.642967215	4.1949237 ± 0.5407378086
Network diameter	10.5 ± 2.2022715546	10.85 ± 2.4550967394

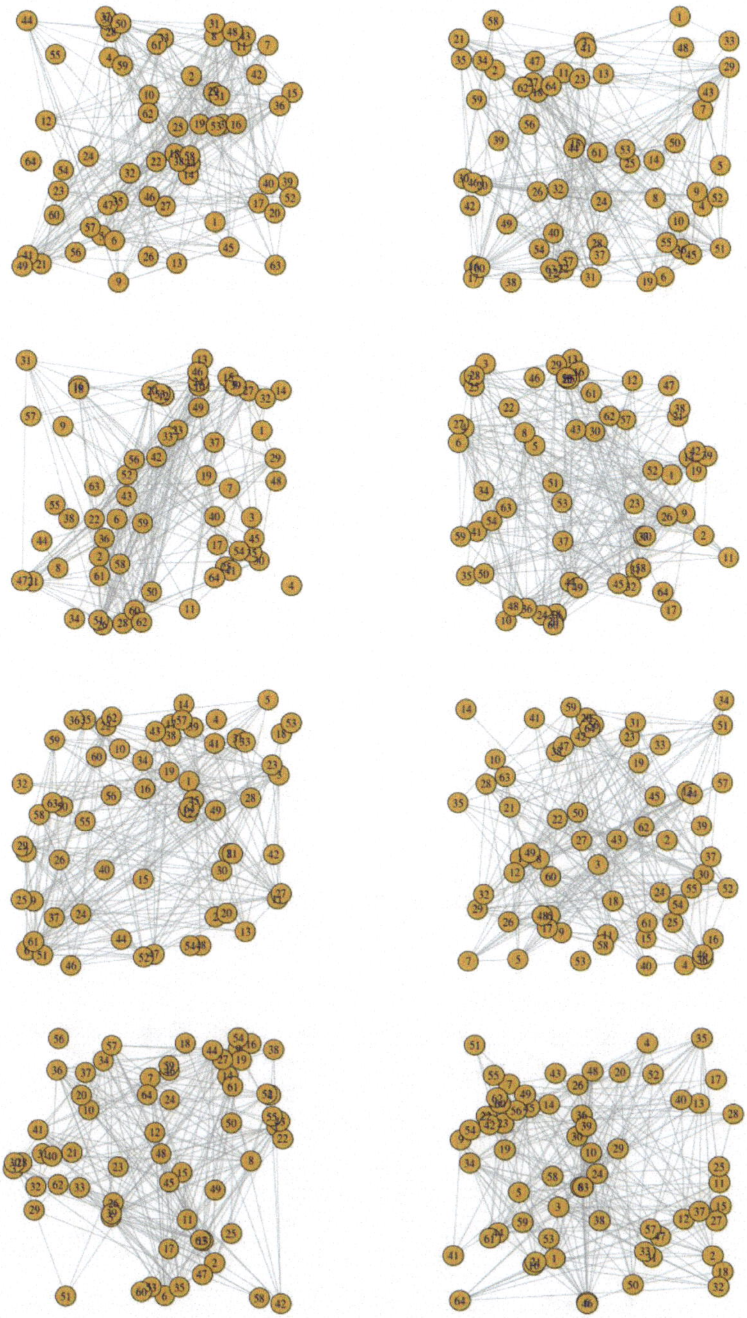

Fig. 42.1 *Non-Alcoholic Graphs*: The following are the graphs generated by EEG Non-Alcoholic data in Igraph R. Further these are analyzed by using graph measures in Igraph R

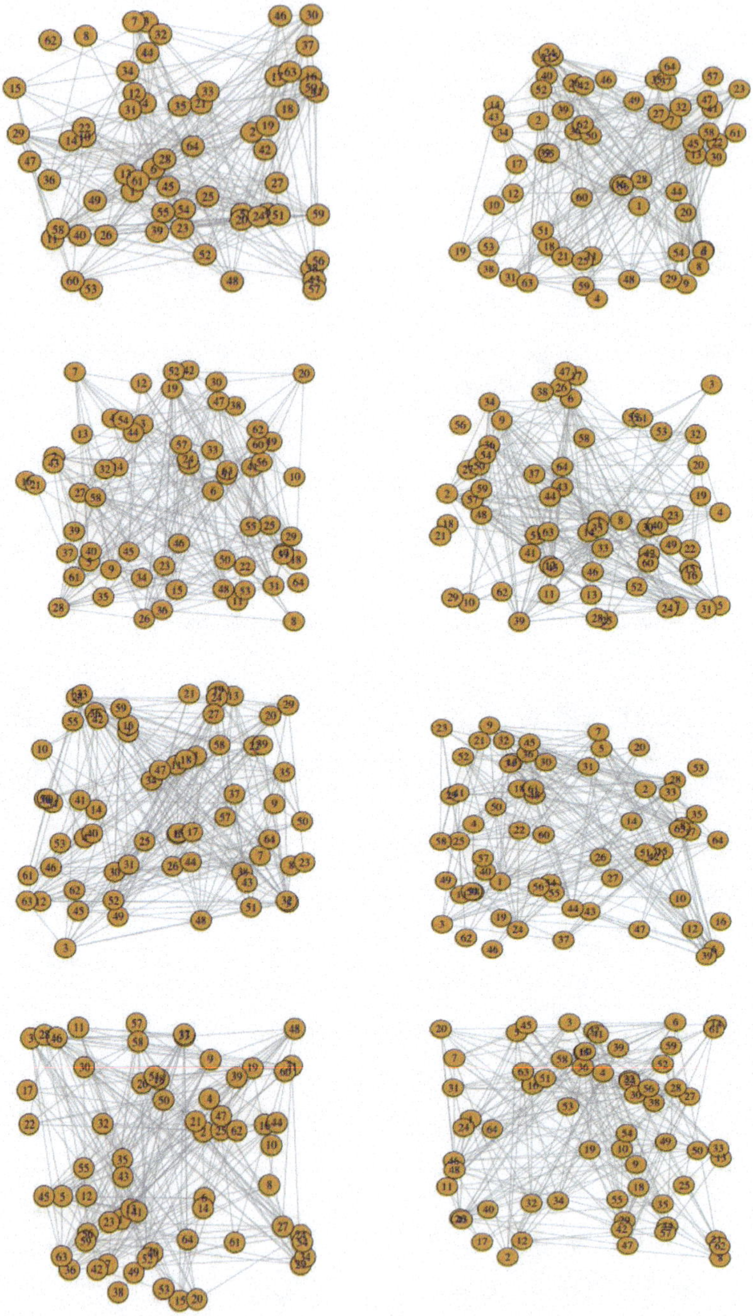

Fig. 42.1 (continued)

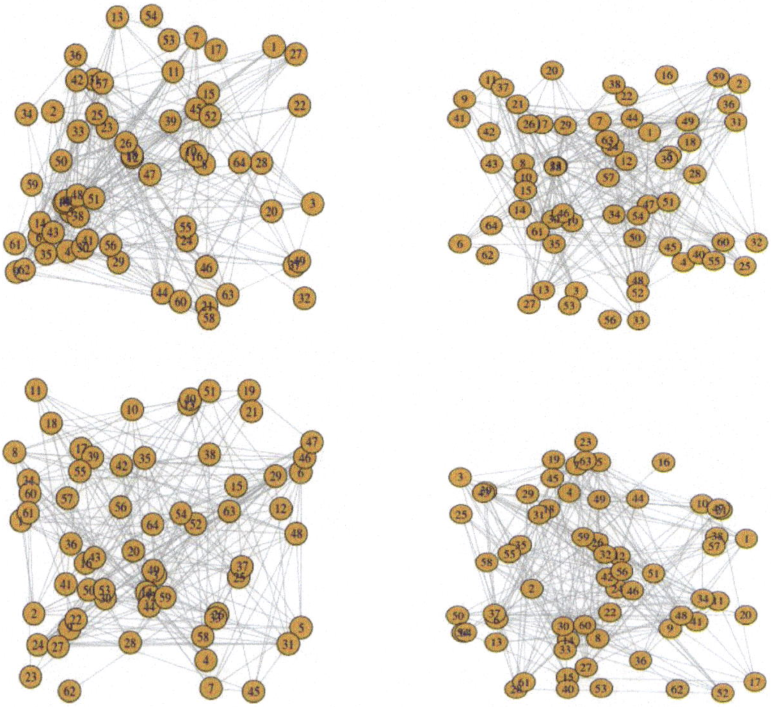

Fig. 42.1 (continued)

In this study, the modularity of Non-Alcoholic network is more than that of the Alcoholic network which means that Non-Alcoholic network has more robustness and resilience.

Transitivity—If the transitivity increases in a network, then local clustering also increases in a network which leads to the formation of small clusters in a network. In this study, the transitivity of Alcoholic network is more than that of the Non-Alcoholic network which means there is more local clustering in Alcoholic network.

Efficiency—If efficiency is more in a network then there is more in information flow in the network which means nodes are well connected in a network. Nodes are connected by the shortest path and that shortest path is less in that network which has more efficiency. In this study, efficiency of Alcoholic network is more than that of Non-Alcoholic network which there is an easy and more information flow in the Alcoholic network.

Assortativity—If an assortativity in a network is more, then that network is homogeneous and there is less assortativity in a network then that network is heterogeneous. Homogeneous means the same degree of nodes are connected and heterogeneous means different degree of nodes are connected. In this study, an assortativity of Alcoholic network is more than that of Non-Alcoholic network which means that an Alcoholic network is homogeneous.

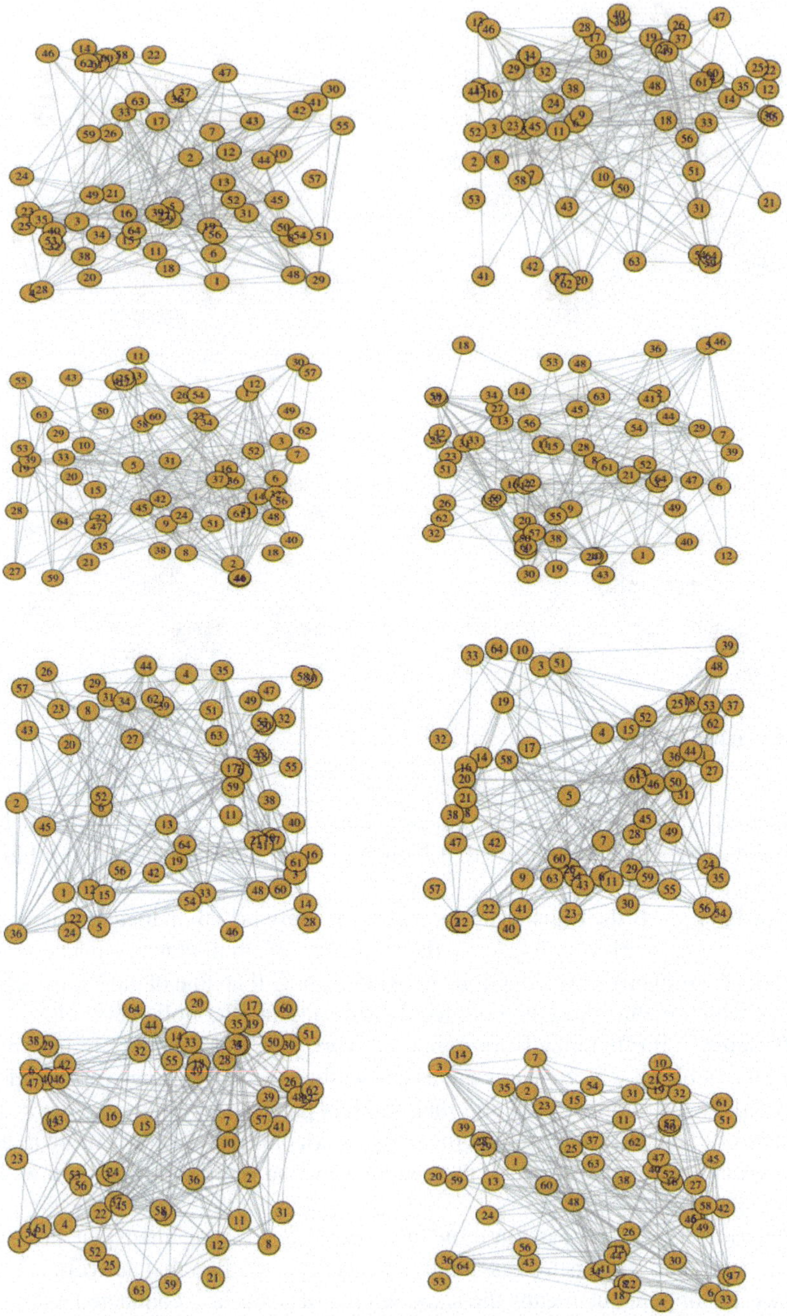

Fig. 42.2 *Alcoholic Graphs*: The following graphs are generated by EEG Alcoholic data. Further, these graphs are analyzed by using graph measures in Igraph R

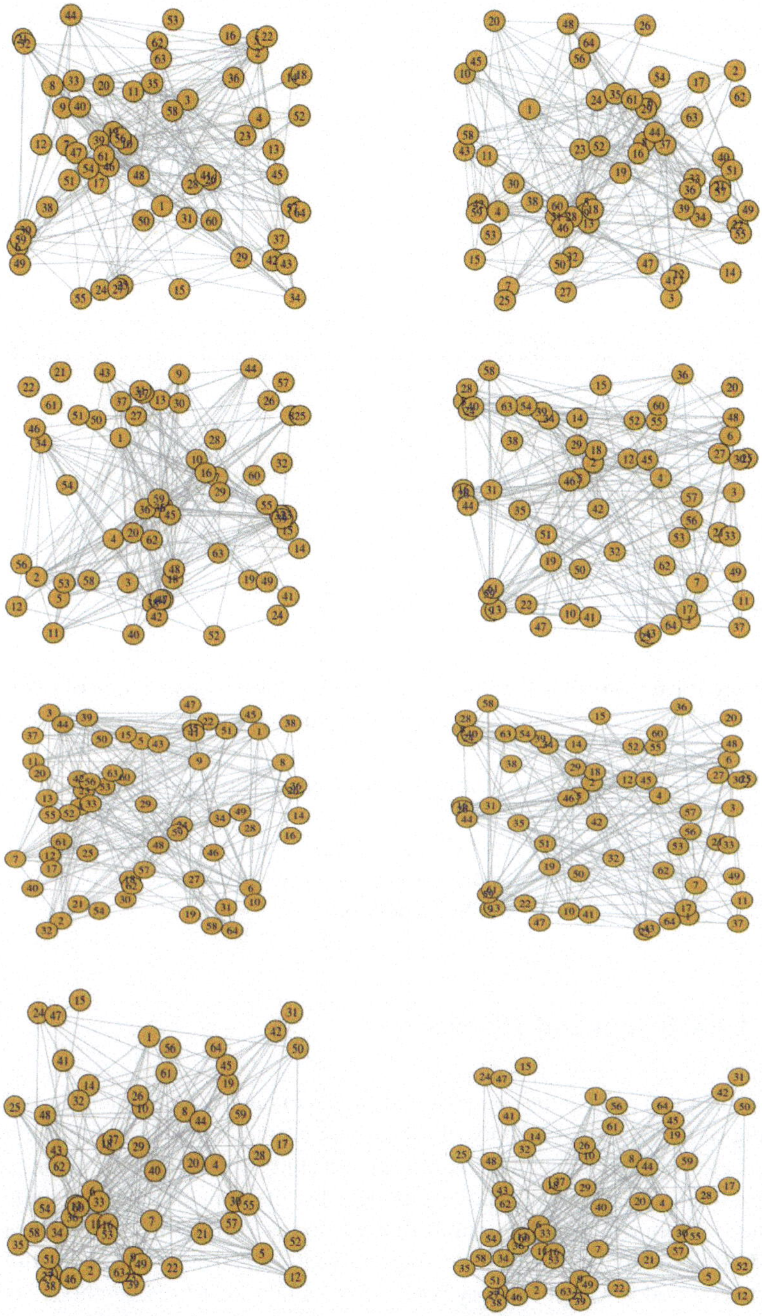

Fig. 42.2 (continued)

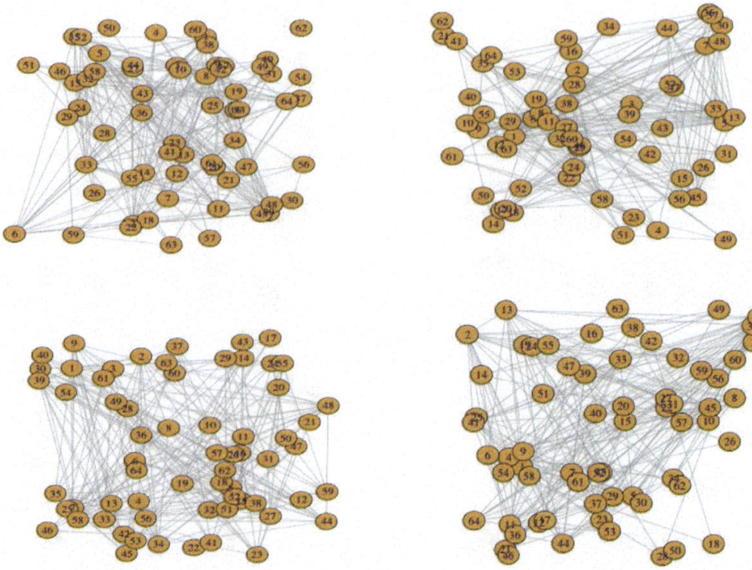

Fig. 42.2 (continued)

Average Path Length—If an average path length is more in a network then there is more information flow from one node to the other node in a network. In this study, average path length of the Alcoholic network is more than that of the Non-Alcoholic network which means that there is more information flow and connected nodes in the Non-lcoholic network.

Network Diameter—If network diameter is more in a network then there is more linear size of a network. In this study, the network diameter of Non-Alcoholic network is more than that of Alcoholic network.

42.5 Conclusion and Discussion

The human brain is one of the complex and largest organs of the Human Body. For the recording of the electrical activity of the human brain, the EEG monitoring method is used. Lately, EEG signals were helpful in the study of the functional connectivity in human brain. The previous studies reported that the human brain was studied as a connected system where nodes or units represented as a different region of the human brain and links or connection between regions represented as a communication pathway between nodes. Graph Theory was used to study measures, methods and tools that can be useful to analyze, model and study EEG Network [19, 25]. In previous studies, many computer scientists and neuroscientists used Graph-Theoretic measures to analyze the EEG data. Utianski et al. reported in a study that graph theory was used to evaluate Parkinson's Disease through EEG data. The main objective of this study was to determine the differences exist in graph theory network measures derived from EEG data, between the Parkinson's Disease and healthy controls. The EEG recording was used to assess Parkinson's Disease and healthy controls via Graph Theory measures. To quantify changes in the network, the global efficiency and local integration graph theory measures were used. In addition, minimal spanning tree analysis was also used. To assess the difference between groups and relationship with cognitive performance, T-tests and correlation were also used. The results revealed that bands with different patterns, network measure type alternation, the breakdown for PD and cognitive decline in PD. The distinct patterns of bands and network measures revealed specific ways by which interaction between cortical areas becomes abnormal and contributes to PD symptoms. In addition, the graph theory analysis by using EEG data advocates that the network alteration and breakdown are robust attributes of PD cortical dysfunction [24]. A study done by Mijalkov et al. reported to develop a graph theory software, a freeware MatLab-based software for brain network connectivity analysis derived from structural magnetic resonance imaging (MRI), functional MRI (fMRI), positron emission tomography (PET) and electroencephalography (EEG) data. BGRAPH software helps in calculating global and local network measures, performing non-parametric permutation for group comparisons, building connectivity matrices, identifying modules in the network and comparison of results with random networks. The two different studies were performed for the analysis of structural and functional graph theory of the network. In the first study, they used MRI data to identify the differences in global and local network topology in healthy controls and patients with Alzheimer's disease. In the second study, they used resting-state fMRI data to compare healthy controls and Parkinson's patients with mild cognitive impairment [17]. Cao et al. reported in a study, a graph-theoretic analysis was done of distributed connectivity of EEG functional networks in alcoholism. The main aim of this study was to identify the differences in the global properties of functional brain networks of alcoholics and healthy individuals under memory task. In this study, they have constructed EEG functional network of 28 alcoholics and 28 healthy controls by applying scalp electrodes and synchronization likelihood (SL). The results suggested that by using graph theory analysis, the brain networks of alcoholics had smaller clustering coefficients in Beta 1 bands, shorter characteristic path length, increased global efficiency, but similar small-world properties. The abnormal topological structure of the alcoholics may be related to the local-function brain damage. The study concluded the new perspective for Alcohol-Related Brain Damage [8]. In the previous studies Graph theory measures were used to identify and analyze the brain-related diseases through EEG, fMRI, PET and MRI data. In addition to this, several studies were conducted using graph theory softwares were also used to identify and analyze the brain-related diseases. In one of the studies, researchers used graph theory measures to evaluate alcoholic and non-alcoholic EEG data. In this work, graph-theoretic analysis is done on functional EEG Network in Igraph R of Alcoholic and Non-Alcoholic subject. It was found that the functioning connectivity of the brain networks of alcoholics is less different from that of a non-alcoholic group of 'UCI, Machine Learning Repository, Centre for Machine Learning and Intelligent System'. EEG Network follows small-world property as the mean degree of Alcoholic and Non-Alcoholic network which we computed in Igraph R is bigger than 2*log(64), i.e. 3.612. The measurements we calculated from Igraph R of Alcoholic and Non-Alcoholic network states that the measurement of Mean Degree, Modularity and Network Diameter of Non-Alcoholic is more than measurement of Alcoholic network and the measurement of Degree Centrality, Transitivity, Efficiency, Assortativity, Mean Path Length of Alcoholic network is less than measurement of Non-Alcoholic network. Thus we can conclude that Non-Alcoholic network of S1 (only one stimulus) leads to more nodes connection which causes more functionality, more robustness and more resilience, more linear network. The Alcoholic network of S1 matched condition (only one stimulus) leads to the formation of a small cluster, proper information flow, homogeneous network and more information flow from one node to another. We can say that the Alcoholic network lacks functionality, robustness, resilience and linear network. The present study reports that the Alcoholic network is unable to recover from some adverse conditions like mutation, disease or failure. In addition, the Alcoholic network has unsubstantial depletion in the functionality of communication between several nodes in the network or less communication. Moreover, the Alcoholic network has less linearity in its network. Thus alcohol causes much minute damage in the human brain due to which the human brain lacks normal functioning.

Acknowledgments The authors are thankful to DST-Serb (ECR/2017/003480), UPOE-II (id 257) and DST purse for providing research facility. The authors also express their deep gratitude to anonymous reviewers, editors for their valuable suggestions and comments.

References

1. Achard S, Bullmore E (2007) Efficiency and cost of economical brain functional networks. PLoS Comput Biol 3(2):e17. https://doi.org/10.1371/journal.pcbi.0030017
2. Achard S, Salvador R, Whitcher B, Suckling J, Bullmore E (2006) Aresilient, low frequency, small world human brain functional network with highly connected association cortical hubs. J Neurosci 26(1):63–72. https://doi.org/10.1523/jneurosci.3874-05.2006
3. Assortativity I graph. http://igraph.org/r/doc/assortativity.html. Accessed 22 July 2019
4. Average Path Length Wikipedia. https://en.wikipedia.org/wiki/Average_path_length. Accessed 22 July 2019
5. Bullmore E, Sporns O (2009) Complex brain networks: graph theoretical analysis of structural and functional systems. Nat Rev Neurosci 10(3):186–198. https://doi.org/10.1038/nrn2575
6. Bullmore E, Sporns O (2012) The economy of brain network organization. Nat Rev Neurosci 13(5):336–349. https://doi.org/10.1038/nrn3214
7. Bullmore ET, Bassett DS (2011) Braingraphs: graphical models of the human brain connectome. Annu Rev Clin Psychol 7:113–140. https://doi.org/10.1146/annurev-clinpsy-040510-143934
8. Cao R, Wu Z, Li H, Xiang J, Chen J (2014) Disturbed connectivity of EEG functional networks in alcoholism: a graph-theoretic analysis. Bio-Med Mater Eng 24:2927–2936. https://doi.org/10.3233/bme-141112

9. Centrality Wikipedia, https://en.wikipedia.org/wiki/Centrality. Accessed 22 July 2019
10. Deo N (2012) Graph theory with applications to engineering and computer science. PHI Learning Pvt. Ltd India
11. Diameter Wikipedia. https://en.wikipedia.org/wiki/Network_science#Diameter_of_a_network. Accessed 22 July 2019
12. Efficiency Wikipedia. https://en.wikipedia.org/wiki/Efficiency. Accessed 22 July 2019
13. Eubank S et al (2004) Modeling disease outbreaks in realistic urban social networks. Nature 429(6988):180–184. https://doi.org/10.1038/nature02541
14. Groppe DM, Bickel S, Keller CJ et al (2013) Dominant frequencies of resting human brain activity as measured by the electrocorticogram. Neuroimage 79:223–233. https://doi.org/10.1016/j.neuroimage.2013.04.044
15. Igraph. https://en.wikipedia.org/wiki/Igraph. Accessed 22 July 2019
16. Mean Degree Wikipedia. https://en.wikipedia.org/wiki/Degree_(graph_theory). Accessed 22 July 2019
17. Mijalkov M, Kakaei E, Pereira JB, Westman E, Volpe G (2017) BRAPH: a graph theory software for the analysis of brain connectivity. PLoS ONE 12(8):e0178798. https://doi.org/10.1371/journal.pone.0178798
18. Modularity Wikipedia. https://en.wikipedia.org/wiki/Modularity. Accessed 22 July 2019
19. Nantia D (2017) Graph theory at the service of electroencephalograms. Brain Connect 7(3). https://doi.org/10.1089/brain.2016.0426
20. R programming language. https://en.wikipedia.org/wiki/R_(programming_language). Accessed 22 July 2019
21. Stam CJ, Reijneveld JC (2007) Graph theoretical analysis of complex networks in the brain. Nonlinear Biomed Phys 1:3. https://doi.org/10.1186/1753-4631-1-3
22. Transitivity I graph. http://igraph.org/r/doc/transitivity.html. Accessed 22 July 2019
23. UCI, Machine learning repository, Centre for Machine Learning and Intelligent System. https://archive.ics.uci.edu/ml/datasets/EEG+Database. Accessed 22 July 2019
24. Utianski RL et al (2016) Graph theory network function in Parkinson's disease assessed with electroencephalography. Clin Neurophysiol 127(5):2228–2236. https://doi.org/10.1016/j.clinph.2016.02.017
25. Zhang J, Wang J, Wu Q et al (2011) Disrupted brain connectivity networks in drug-naive, first-episode major depressive disorder. Biol Psychiat 70(4):334–342. https://doi.org/10.1016/j.biopsych.2011.05.018

The manufacturer's authorised representative in the EU is Springer Nature Customer Service Centre GmbH, Europaplatz 3, 69115 Heidelberg, Germany. If you have any concerns regarding our products, please contact ProductSafety@springernature.com

Printed and bound by CPI Group (UK) Ltd, Croydon, CR0 4YY
06/10/2026
02225252-0016